BIOACTIVES AND PHARMACOLOGY OF LAMIACEAE

AAP Focus on Medicinal Plants

BIOACTIVES AND PHARMACOLOGY OF LAMIACEAE

Edited by

T. Pullaiah, PhD

APPLE ACADEMIC PRESS

First edition published 2024

Apple Academic Press Inc.
1265 Goldenrod Circle, NE,
Palm Bay, FL 32905 USA

760 Laurentian Drive, Unit 19,
Burlington, ON L7N 0A4, CANADA

CRC Press
2385 NW Executive Center Drive,
Suite 320, Boca Raton FL 33431

4 Park Square, Milton Park,
Abingdon, Oxon, OX14 4RN UK

Apple Academic Press exclusively co-publishes with CRC Press, an imprint of Taylor & Francis Group, LLC

Library and Archives Canada Cataloguing in Publication

Title: Bioactives and pharmacology of lamiaceae / edited by T. Pullaiah, PhD.
Names: Pullaiah, T., editor.
Series: AAP focus on medicinal plants.
Description: First edition. | Series statement: AAP focus on medicinal plants | Includes bibliographical references and index.
Identifiers: Canadiana (print) 20230445837 | Canadiana (ebook) 20230445896 | ISBN 9781774912935 (hardcover) | ISBN 9781774912942 (softcover) | ISBN 9781003346142 (ebook)
Subjects: LCSH: Plant bioactive compounds. | LCSH: Lamiaceae—Therapeutic use. | LCSH: Pharmacology.
Classification: LCC QK898.B54 B58 2024 | DDC 583/.96—dc23

Library of Congress Cataloging-in-Publication Data

CIP data on file with US Library of Congress

ISBN: 978-1-77491-293-5 (hbk)
ISBN: 978-1-77491-294-2 (pbk)
ISBN: 978-1-00334-614-2 (ebk)

AAP FOCUS ON MEDICINAL PLANTS

ABOUT THE SERIES

This new book series, edited by T. Pullaiah, focuses on bioactives and pharmacology of medicinal plants.

For over a millennia, medicinal plants have been a valuable source of therapeutic agents, and still many of today's drugs are based on plant-derived natural products or their derivatives. Bioactive compounds typically occur in small amounts, and they have more subtle effects than nutrients. They influence cellular activities that modify the risk of disease and help alleviate disease symptoms. Bioactive compounds have potentially important health benefits, and these compounds can act as antioxidants, enzyme inhibitors and inducers, inhibitors of receptor activities, and inducers and inhibitors of gene expression among other actions. A wide array of biological activities and potential health benefits of medicinal plants have been reported, which include antiviral, antibacterial, antifungal, antioxidant, anticancer, anti-inflammatory, antidiabetic, hepatoprotective, cardioprotective, nephroprotective properties as well as other protective effects on the liver, kidney, heart, and nervous system.

The volumes in this series aim to be comprehensive desk references on bioactives and pharmacology of a wide range of medicinal plants. They will also be important sourcebooks for the development of new drugs.

Book Series Editor

Prof. T. Pullaiah
Department of Botany
Sri Krishnadevaraya University,
Anantapur 515003, A.P., India
Email: pullaiah.thammineni@gmail.com

Book in the Series

- Bioactives and Pharmacology of Lamiaceae
- Bioactives and Pharmacology of Legumes
- Bioactives and Pharmacology of Medicinal Plants, 2-volume set
- Biomolecules and Pharmacology of Medicinal Plants, 2-volume set

- Frankincense–Gum Olibanum: Botany, Oleoresin, Chemistry, Extraction, Utilization, Propagation, Biotechnology, and Conservation
- Phytochemical Composition and Pharmacy of Medicinal Plants, 2-volume set
- Phytochemistry and Pharmacology of Medicinal Plants, 2-volume set

OTHER BOOKS FROM AAP BY DR. T. PULLAIAH

Ethnobotany of India, 5-volume set:
Editors: T. Pullaiah, PhD, K. V. Krishnamurthy, PhD, and Bir Bahadur, PhD

- Volume 1: Eastern Ghats and Deccan
- Volume 2: Western Ghats and West Coast of Peninsular India
- Volume 3: North-East India and the Andaman and Nicobar Islands
- Volume 4: Western and Central Himalaya
- Volume 5: Indo-Gangetic Region and Central India

Global Biodiversity, 4-volume set:
Editor: T. Pullaiah, PhD

- Volume 1: Selected Countries in Asia
- Volume 2: Selected Countries in Europe
- Volume 3: Selected Countries in Africa
- Volume 4: Selected Countries in the Americas and Australia

Handbook of Research on Herbal Liver Protection: Hepatoprotective Plants
T. Pullaiah, PhD and Maddi Ramaiah, PhD

Bio-Inspired Technologies for the Modern World: Emerging Trends and Applications
R. Ramakrishna Reddy, PhD and T. Pullaiah, PhD

Biodiversity of Hotspots–36 volumes
Editor: T. Pullaiah, PhD

ABOUT THE EDITOR

T. Pullaiah, PhD

Dr. T. Pullaiah is a former Professor at the Department of Botany at Sri Krishnadevaraya University in Andhra Pradesh, India, where he has taught for more than 35 years. He has held several positions at the university, including Dean, Faculty of Biosciences, Head of the Department of Botany, Head of the Department of Biotechnology, and a member of the Academic Senate. He was the President of the Indian Botanical Society (2014) and the Indian Association for Angiosperm Taxonomy (2013), and a Fellow of the Andhra Pradesh Akademi of Sciences. He was awarded Panchanan Maheshwari Gold Medal, Prof. P.C. Trivedi Medal, Dr. G. Panigrahi Memorial Lecture award of the Indian Botanical Society, and Prof. Y.D. Tyagi Gold Medal of the Indian Association for Angiosperm Taxonomy, and the Best Teacher Award from the Government of Andhra Pradesh. Under his guidance, 54 students obtained their doctoral degrees. He has authored 52 books, edited 23 books, and published over 330 research papers. His books include *Advances in Cell and Molecular Diagnostics* (published by Elsevier), *Camptothecin and Camptothecin producing Plants* (Elsevier) *Ethnobotany of India* (5 volumes published by Apple Academic Press), *Global Biodiversity* (4 volumes, Apple Academic Press), *Red Sanders: Silviculture and Conservation* (Springer), *Genetically Modified Crops* (2 volumes, Springer), *Monograph on Brachystelma and Ceropegia in India* (CRC Press), *Flora of Andhra Pradesh* (5 volumes), *Flora of Eastern Ghats* (4 volumes), *Flora of Telangana* (3 volumes), *Encyclopaedia of World Medicinal Plants* (7 volumes, 2nd edition), and *Encyclopaedia of Herbal Antioxidants* (3 volumes). He was also a member of the Species Survival Commission of the International Union for Conservation of Nature (IUCN). Professor Pullaiah received his PhD from Andhra University, India, and attended Moscow State University, Russia, and worked as a postdoctoral fellow from 1976 to 1978.

CONTENTS

CONTRIBUTORS

Jasha Momo H. Anal
Natural Product Chemistry Division, CSIR-Indian Institute of Integrative Medicine, Jammu, Jammu and Kashmir, India

Gadwal Shaik Nishat Anjum
VIT School of Agricultural Innovations and Advanced Learning (VAIAL), Vellore Institute of Technology, Tamil Nadu, India

Sharmila Arunagiri
Department of Biotechnology, School of Biosciences and Technology, Vellore Institute of Technology, Tamil Nadu, India

Rajeshwar K. K. Arya
Department of Pharmaceutical Sciences, Sir J.C. Bose Technical Campus, Kumaun University, Bhimtal, Uttarakhand, India

Dharam Chand Attri
G. B. Pant National Institute of Himalayan Environment, Kosi-Katarmal, Uttarakhand, India;

G. Balakrishna
Defence Institute of Bio-Energy Research (DRDO) Field Station, Pithoragarh, Uttarakhand, India

Emiliano Barreto
Laboratory of Cell Biology, Federal University of Alagoas, Alagoas, Brazil

Urooj Hassan Bhatt
Department of Botany, Faculty of Life Sciences, Aligarh Muslim University, Aligarh, Uttar Pradesh, India

Mudasir Nazir Bhat
Plant Sciences (Biodiversity and Applied Botany Division), CSIR–Indian Institute of Integrative Medicine, Jammu, Jammu and Kashmir, India
Academy of Scientific and Innovative Research (AcSIR), Ghaziabad, Uttar Pradesh, India

Pradeep Bhat
ICMR–National Institute of Traditional Medicine, Belagavi, Karnataka, India

Rafael Vrijdags Calado
Laboratory of Cell Biology, Federal University of Alagoas, Alagoas, Brazil

Sobia Aman Chishti
Department of Botany, Faculty of Life Sciences, Aligarh Muslim University, Aligarh, Uttar Pradesh, India

Vishal P. Deshmukh
Department of Botany, Jagadamba Mahavidyalaya, Achalpur city, Amravati, Maharashtra, India

B. Dinesh
Centre for Applied Genetics, Department of Studies in Zoology, Bangalore University, Jnana Bharathi Campus, Bengaluru, Karnataka, India

Jamylle Nunes de Souza Ferro
Laboratory of Cell Biology, Federal University of Alagoas, Alagoas, Brazil

Reema Gabrani
Department of Biotechnology, Jaypee Institute of Information Technology, Noida, Uttar Pradesh, India

Varughese George
Amity Institute for Herbal and Biotech Products Development, Thiruvananthapuram, Kerala, India

Savaliram G. Ghane
Department of Botany, Shivaji University, Vidyanagar, Kolhapur, Maharashtra, India
Plant Physiology Laboratory, Department of Botany, Shivaji University, Kolhapur, Maharashtra, India

Samrat Goswami
Taxonomy and Biosystematics Laboratory, Post Graduate Department of Botany,
Ramakrishna Mission Vivekananda Centenary College, Kolkata, West Bengal, India

Anchala Guglani
Defence Institute of Bio-Energy Research (DRDO) Field Station, Uttarakhand, India
Department of Pharmaceutical Sciences, Sir J.C. Bose Technical Campus, Kumaun University, Bhimtal, Uttarakhand, India

Vrushali Manoj Hadkar
Department of Biotechnology, School of Biosciences and Technology, Vellore Institute of Technology, Tamil Nadu, India

Harsha V. Hegde
ICMR – National Institute of Traditional Medicine, Belagavi, Karnataka, India

Thadiyan Parambil Ijinu
Amity Institute for Herbal and Biotech Products Development, Thiruvananthapuram, Kerala, India
Naturæ Scientific, Kerala University Business Innovation and Incubation Centre, Thiruvananthapuram, Kerala, India

Maria Ishaque
Department of Biotechnology, Jaypee Institute of Information Technology, Noida, Uttar Pradesh, India

Jenny Jacob
School of Biosciences, Mar Athanasios College for Advanced Studies Tiruvalla, Pathanamthitta, Kerala, India

Nasreen Jahan
Department of Ilmul Advia (Pharmacology), National Institute of Unani Medicine, Bengaluru, Karnataka, India

Kanishka Jain
Department of Biotechnology, Jaypee Institute of Information Technology, Noida, Uttar Pradesh, India

Santoshkumar Jayagoudar
Department of Botany, G. S. S. College & Rani Channamma University P. G. Centre, Belagavi, Karnataka, India

M. Masroor A. Khan
Department of Botany, Faculty of Life Sciences, Aligarh Muslim University, Aligarh, India

Satish Kumar
Botanic Garden Division, CSIR-National Botanical Research Institute (NBRI), Lucknow, Uttar Pradesh, India

Sunil Kumar
Department of Chemistry, Ma. Kanshiram Government Degree College, Farrukhabad, Uttar Pradesh, India
Plant Sciences (Biodiversity and Applied Botany Division), CSIR–Indian Institute of Integrative Medicine, Jammu, Jammu and Kashmir, India
Academy of Scientific and Innovative Research (AcSIR), Ghaziabad, Uttar Pradesh, India

Irin Kurikesu
Inter University Centre for Evolutionary and Integrative Biology, University of Kerala, Thiruvananthapuram, Kerala, India

C. Lalmuansangi
Department of Zoology, Mizoram University (A Central University), Aizawl, Mizoram, India

Marina Lalremruati
Department of Zoology, Mizoram University (A Central University), Aizawl, Mizoram, India

P. Mahesh
Centre for Applied Genetics, Department of Studies in Zoology, Bangalore University, Jnana Bharathi Campus, Bengaluru, Karnataka, India

Shaikh Ajij Ahmed Makbul
Department of Ilmul Advia (Pharmacology), National Institute of Unani Medicine, Kottigepalya, Bengaluru, Karnataka, India

Y. S. Meenakshi
Department of Studies in Biochemistry, Maharani Lakshmi Ammanni College for Women, Malleswaram, Bengaluru, Karnataka, India

Moinuddin
Botany section, Women's College, Aligarh Muslim University, Aligarh, Uttar Pradesh, India

M. Muniraju
Department of Studies in Botany, Bangalore University, Jnana Bharathi Campus, Bengaluru, Karnataka, India

Devika R. Nath
School of Agricultural Innovations and Advanced Learning, Vellore Institute of Technology, Vellore, Tamil Nadu, India

F. Nghakliana
Department of Zoology, Mizoram University (A Central University), Aizawl, Mizoram, India

Sankaralingam Nivitha
School of Agricultural Innovations and Advanced Learning, Vellore Institute of Technology, Vellore, Tamil Nadu, India

Shreedhar S. Otari
Plant Physiology Laboratory, Department of Botany, Shivaji University, Kolhapur, Maharashtra, India

H. K. Pandey
Defence Institute of Bio-Energy Research (DRDO) Field Station, Pithoragarh, Uttarakhand, India

Sudarsan Parvathy
Department of Agricultural Biotechnology and Molecular Biology, Dr. Rajendra Prasad Central Agricultural University, Pusa, Bihar, India

Suraj B. Patel
Plant Physiology Laboratory, Department of Botany, Shivaji University, Kolhapur, Maharashtra, India

K. Peddanna
Department of Biochemistry, Sri Venkateswara University, Tirupati, Andhra Pradesh, India

Felipe Lima Porto
Laboratory of Cell Biology, Federal University of Alagoas, Alagoas, Brazil

M. G. Prasad
Botanic Garden Division, CSIR–National Botanical Research Institute (NBRI), Lucknow, Uttar Pradesh, India

Maheswari Priya Rani
Phytochemistry and Phytopharmacology Division, Jawaharlal Nehru Tropical Botanic Garden and Research Institute, Thiruvananthapuram, Kerala, India

Palpu Pushpangadan
Amity Institute for Herbal and Biotech Products Development, Thiruvananthapuram, Kerala, India

Ragesh Raveendran Nair
Department of Botany, NSS College Nilamel, Kollam, Kerala, India

Sayeedur Rahman
Department of Ilmul Advia (Pharmacology), National Institute of Unani Medicine, Kottigepalya, Bangalore, Karnataka, India

A. Rajasekhar
Department of Biochemistry, Sri Venkateswara University, Tirupati, Andhra Pradesh, India

S. Rajashekara
Department of Studies in Zoology, Bangalore University, Jnana Bharathi Campus, Bengaluru, Karnataka, India

Nutan Rajput
Department of Botany, Government College, Manawar, Madhya Pradesh, India

C. Appa Rao
Department of Biochemistry, Sri Venkateswara University, Tirupati, Andhra Pradesh, India

Kallipudi Charishma Reddy
VIT School of Agricultural Innovations and Advanced Learning (VAIAL), Vellore Institute of Technology, Vellore, Tamil Nadu, India

Maria Danielma dos Santos Reis
Laboratory of Cell Biology, Federal University of Alagoas, Alagoas, Brazil

Biswajit Roy
Taxonomy and Biosystematics Laboratory, Post Graduate Department of Botany, Ramakrishna Mission Vivekananda Centenary College, Kolkata, West Bengal, India

Oishika Sah
VIT School of Agricultural Innovations and Advanced Learning (VAIAL), Vellore Institute of Technology, Vellore, Tamil Nadu, India

Parasurama Deepa Sankar
VIT School of Agricultural Innovations and Advanced Learning, Vellore Institute of Technology, Vellore, Tamil Nadu, India

Parameswaran Sasikumar
Drug Testing Laboratory, Department of Rasasastra and Bhaishajyakalpana, Government Ayurveda College, Thiruvananthapuram, Kerala, India

Veena Satya
Department of Botany, Shaheed Bheema Nayak Government Post Graduate College, Madhya Pradesh, India

Chinnadurai Immanuel Selvaraj
VIT School of Agricultural Innovations and Advanced Learning, Vellore Institute of Technology, Vellore, Tamil Nadu, India

Balan Kallullathil Shajima
Food Corporation of India, Food Storage Depot-Thikkodi, Kozhikode, Kerala, India

M. Shanthala
Centre for Applied Genetics, Department of Studies in Zoology, Bangalore University, Jnana Bharathi Campus, Bengaluru, Karnataka, India

Mayank Shukla
Indian Pharmacopoeia Commission, Ministry of Health and Family Welfare, Ghaziabad, Uttar Pradesh, India

Bikarma Singh
Botanic Garden Division, CSIR–National Botanical Research Institute (NBRI), Lucknow, Uttar Pradesh, India

Sangram Singh
Department of Botany, Faculty of Life Sciences, Aligarh Muslim University, Aligarh, India

Sarika Singh
Department of Botany, Faculty of Life Sciences, Aligarh Muslim University, Aligarh, India

Nayan Kumar Sishu
Department of Biotechnology, School of Biosciences and Technology, Vellore Institute of Technology, Vellore, Tamil Nadu, India

Sneha
Botanic Garden Division, CSIR–National Botanical Research Institute (NBRI), Lucknow, Uttar Pradesh, India

Tayhana Priscila Medeiros Souza
Laboratory of Cell Biology, Federal University of Alagoas, Alagoas, Brazil

Sasidharan Lathikakumari Sreejith
Central Analytical Facility, Sree Chitra Tirunal Institute for Medical Sciences and Technology, Thiruvananthapuram, Kerala, India

Sharad Srivastava
Pharmacognosy Division, CSIR–National Botanical Research Institute, Council of Scientific and Industrial Research, Lucknow, Uttar Pradesh, India

Babu Subramanian
School of Agricultural Innovations and Advanced Learning, Vellore Institute of Technology, Vellore, Tamil Nadu, India

P. Suneetha
Department of Studies in Biochemistry, Maharani Lakshmi Ammanni College for Women, Malleswaram, Bengaluru, Karnataka, India

Opender Surmal
Plant Sciences (Biodiversity and Applied Botany Division), CSIR—Indian Institute of Integrative Medicine, Jammu, Jammu and Kashmir, India
Academy of Scientific and Innovative Research (AcSIR), Ghaziabad, Uttar Pradesh, India

Shri Krishna Tewari
Botanic Garden Division, CSIR–National Botanical Research Institute (NBRI), Lucknow, Uttar Pradesh, India

Parthasarathi Theivasigamani
School of Agricultural Innovations and Advanced Learning, Vellore Institute of Technology, Vellore, Tamil Nadu, India

Deepika Tripathi
G. B. Pant National Institute of Himalayan Environment, Kosi-Katarmal, Almora, Uttarakhand, India

Treesa Varghese
School of Biosciences, Mar Athanasios College for Advanced Studies Tiruvalla, Pathanamthitta, Kerala, India

Rahul Zanan
Department of Botany, Elphinstone College, Dr. Homi Bhabha State University, Madam Cama Road, Mumbai, Maharashtra, India

Mary Zosangzuali
Department of Zoology, Mizoram University (A Central University), Aizawl, Mizoram, India

Zothansiama
Department of Zoology, Mizoram University (A Central University), Aizawl, Mizoram, India

ABBREVIATIONS

AA	arachidonic acid
AD	Alzheimer's disease
AGA	against androgenic alopecia
ALP	alkaline phosphatase
ALT	alanine aminotransferase
AMPK	activated protein kinase
AR	androgen receptors
ASAT	aspartate aminotransaminase
AST	aspartate aminotransferase
BBM	brush border membrane
BHT	butylated hydroxytoluene
CA	chromosome aberration
CA1	cornu ammonis 1
CA3	cornu ammonis 3
CBMN	cytochalasin B-blocked micronucleus assay
CFA	complete Freund's adjuvant
CMS	chronic mild stress
CNS	central nervous system
CUMS	chronic unpredictable mild stress
Cyp	cyclophosphamide
DAL	Dalton's ascetic lymphoma
DCM	dichloromethane
DG	dentate gyrus
DHA	dihydrotestosterone
DML	dimethyl lithospermate
DTH	delayed-type hypersensitivity
EAC	Ehrlich's ascetic carcinoma
EO	essential oil
ERK	extracellular-regulated kinase
ERα	estrogen receptor alpha
ESBL	extended-spectrum beta-lactamase
FFA	free fatty acids
FRAP	ferric reducing ability of plasma
FST	forced swimming test

GABA	γ-aminobutyric acid
GC-MS	gas chromatography-mass spectrometry
HA	humoral antibody
HPBLs	human peripheral blood lymphocytes
HUVECs	human umbilical vein endothelial cells
IL	interleukin
iNOS	inducible nitric oxide synthase
KOR	kappa-opioid receptor
LHAE	leaf hydroalcoholic extract
LP	lipid peroxidation
LPO	levels of lipid peroxide
LPS	lipopolysaccharide
LT	leukotriene
MAO	monoamine oxidase
MBC/MFC	minimum bactericidal/fungicidal
MCEs	methanolic crude extracts
MDR	multidrug resistant
ME	methanol extract
MI	myocardial infarction
MIC	minimum inhibitory concentration
MMP-2	matrix metalloproteinase-2
MN	micronucleus
MTT	3-[4,5-dimethylthiazol-2-yl]-2,5 diphenyl tetrazolium bromide
NAFLD	non-alcoholic fatty liver disease
NDI	nuclear division index
NOS	nitric oxide synthase
NRU	neutral red uptake
PHA	phytohemagglutinin
PI	proliferation index
PRRS	porcine reproductive and respiratory syndrome
PTZ	pentylenetetrazole
RLE	*Salvia rosmarinus* leaf extract
ROS	reactive oxygen species
RSC	radical-scavenging capability
SCE	sister chromatid exchange
SEM	scanning electron microscopy
SF	*Salvia fruticosa*
SFD	*S. fruticosa*'s dichloromethane extract

SFM	*Salvia fruticosa* methanol extract
SFP	*S. fruticosa* petroleum ether extract
STZ	streptozotocin
TC	total cholesterol
TCM	traditional Chinese medicine
TE	Trolox equivalents
TG	triglycerides
THP-1	human monocytic cells
TPA	tetradecanoyl phorbol-13 acetate
TPCs	total phenolic contents
UHPLC-ESI-MS	ultra-high performance liquid chromatography electrospray ionization mass spectrometry
VSMC	vascular smooth muscle cells
XO	xanthine oxidase

PREFACE

Lamiaceae is a family of flowering plants that produces essential oils. These oils are produced in quite a good number of species. This book is a continuation of AAP book series on medicinal plants. In this book, the bioactives and pharmacology of some important medicinal plants of Lamiaceae are given.

For each species, a brief introduction about its characters, distribution, and traditional uses are detailed. Under each species, bioactive compounds are listed and their chemical structures are explained. This is followed by their pharmacological activities. All the published literature on the pharmacological activities on that species is then reviewed. I hope that this will be a sourcebook for the development of new drugs. I request the readers to give their suggestions for the improvement of the forthcoming volumes and next edition.

I wish to express my thanks to all the authors who contributed to the reviewed chapters.

CHAPTER 1

BIOLOGICAL AND CHEMICAL ASPECTS OF THE GENUS *AJUGA* L.

BIKARMA SINGH[1,5], SUNIL KUMAR[2], JASHA MOMO H. ANAL[3], OPENDER SURMAL[4,5], MUDASIR NAZIR BHAT[4,5], M. G. PRASAD[1], SNEHA[1], SATISH KUMAR[1], and SHRI KRISHNA TEWARI[1]

[1]*Botanic Garden Division, CSIR-National Botanical Research Institute (NBRI), Lucknow, Uttar Pradesh, India*

[2]*Department of Chemistry, Ma. Kanshiram Government Degree College, Ninowa, Farrukhabad, Uttar Pradesh, India*

[3]*Natural Product Chemistry Division, CSIR-Indian Institute of Integrative Medicine, Jammu, Jammu and Kashmir, India*

[4]*Plant Sciences (Biodiversity and Applied Botany Division), CSIR-Indian Institute of Integrative Medicine, Jammu, Jammu and Kashmir, India*

[5]*Academy of Scientific and Innovative Research (AcSIR), Ghaziabad, Uttar Pradesh, India*

ABSTRACT

The genus *Ajuga* L. belongs to the family Lamiaceae and is one of the highly medicinal group of plants represented by 64 species distributed worldwide. Several species of this genus have traditional importance and are used for the treatment of different types of fevers, dysentery, malaria, hypertension, diabetes and gastrointestinal disorders. In this chapter, we have tried to summarize and provide traditional applications, biology and phytochemistry

Bioactives and Pharmacology of Lamiaceae. T. Pullaiah, PhD (Ed.)

of common medicinal *Ajuga's* growing in Asia and adjoining areas. Fifty six neo-clerodanediterpenoids have been reported from different plant parts of *Ajuga* species, and therefore, neo-clerodanediterpenoids considered as a major class of compounds found in most of the *Ajuga* species. The analgesic, antiarthritic, antimicrobial, antiestrogenic, anti-inflammatory, antihypertensive, antileukemic, antimalarial, antioxidant, antipyretic, cardiovascular, cytotoxic, antidiabetic, and vasorelaxation activities have been reported from different extracts of the plant parts of *Ajuga* based on its biological compounds. These data will provide baseline information for future research on Ajuga species for value addition and product development.

1.1 INTRODUCTION

The plant species member of the genus *Ajuga* L. belongs to the mint family, Lamiaceae, comprised of rhizomatous annual and perennial evergreen herbaceous flowering plants (Li and Hedge, 1994; Singh, 2020; Singh and Sharma, 2021). Most of the species under this genus are reported as the plants native to subtropical and temperate regions of Africa, Asia, Australia, Europe, and North America (Coll and Tendron, 2008; Israili and Lyoussi, 2009), and as time passes it spreads throughout the world as introduced or invasive plant species. The genus comprises 64 species worldwide (POWO, 2021). The member species of the genus can be separated from other Lamiaceae members by having a combination of reticulately sculptured mericarps, which get separated late in plant development producing a prominent areole at the point of attachment, presence of a persistent corolla base that often leaves a sheath around the mature mericarps (Harley et al., 2004; Kim et al., 2013). The other distinguished characters include the habit that varies from 5 to 50 cm height having attractive opposite leaves and two lipped tubular flowers of different colors, which mostly are blue, purple, or yellow in appearance (Li and Hedge, 1994; Hsieh and Huang, 1995). Some species of *Ajuga* plants are used in horticulture as groundcover or border and some species have a large number of varieties (cultivars) used in gardens as ornamental plant because of their varied blooms of different colors (Israili and Lyoussi, 2009). Most of the species of this genus usually bear flowers between January and March.

Many plant species of the *Ajuga* genus are reported to be used in the folklore/traditional system of medicines employed in the treatment of rheumatic fevers, dysentery, malaria, hypertension, diabetes, and gastrointestinal disorders (Qing et al., 2017; Singh et al., 2021). Table 1.1 is presented with traditional utilization of *Ajuga* plants across the globe. In China, the herb *A. macrosperma* is used in folk medicine to alleviate fever and remove phlegm and it has also been reported to be used medicinally for nephritis (Yunnan Institute of Botany, 1977). In East Africa, *Ajuga* plants are employed in the treatment of dysentery, fever, toothache, and blood pressure (Githinji and Kokwaro, 1993; Israili and Lyoussi, 2009), whereas the North African tribes utilized the *Ajuga* plants in the treatment of sugar disease diabetes and hypertension (Jouad et al., 2001; Bnouham et al., 2002; El-Hilaly et al., 2003; Tahraoui et al., 2007; Eddouks et al., 2007; Israili and Lyoussi, 2009) and gastrointestinal disorders (Bellakhdar et al., 1991). In Chinese pharmacopoeia, *Ajuga* is reported to produce a diuretic effect (Aliotta and Pollio, 1994) when consumed. The Himalayan regions are home to many endemic plant species due to varying climatic conditions, difficult terrains, and forest types (Chopra et al. 1986; Singh et al., 2016; Singh and Bedi, 2017). The plant species, *A. integrifolia* (=synonym *A. bracteosa*) and *A. macrosperma*, are the only two most common species found in high-altitude regions of Indian Himalaya, whereas the *A. brachystemon* and *A. parviflora* occur as less known from eastern Himalaya only (Haq et al., 2021). *A. integrifolia* is a well-studied plant from phytochemical perspectives; however, *A. macrosperma* is commonly called large-seed bugleweed that is a less studied plant from the chemistry point of view. However, both these species have a wide distribution in India, Bangladesh, China, Laos, Myanmar, Nepal, Taiwan, Thailand, and Vietnam. In India, *Ajuga* is reported from Kumaon, Kashmir, and Northeast India between the altitudes of 400 and 2600 m a.m.s.l. Traditionally, the Indian *Ajuga* plants are used in the treatment of nephritis, whereas the juice extracted from the roots is employed to cure boils and pimples (Singh, 2020). According to Hossan et al. (2009), the roots of *A. macrosperma* are dried, powdered, and made into tablets to size of bean seeds, and two tablets are taken twice daily for a month to get ride off respiratory problems such as asthma, breathing difficulty, and pneumonia. Khanavi et al. (2014) reported that *Ajuga* is used for the treatment of joint pain, gout, and jaundice in traditional Iranian medicine. Many of these plants have been used in traditional medicine as a remedy for fever, toothache, dysentery, malaria, high blood pressure, diabetes, gastrointestinal disorders, as anthelmintic, diuretic and antifungal, anti-inflammatory, and antimycobacterial agents. They are also used as insect growth inhibitors (Israili and Lyoussi, 2009; Singh et al., 2020).

TABLE 1.1 Ethnobotanical Uses of *Ajuga* Plants Across the World.

Plant species	Part used	Disease	Country	Citation
A. australis R.Br.	Leaves	Boils, sores, wounds	Australia, Tasmania, Hawaii	Brown (1995), Johnson (1999)
A. cham-aepitys (L.) Schreb.	Whole plant	Diuretic, stimulant, menstrual flow stimulant, gout, sclerosis, jaundice, cancer, rheu-matism, dropsy, antiarthritic, wound healing	Europe (France), Africa, Asia, Hawaii, Israel, Turkey, Spain, Jordan,	Grieve (1984), Brown (1995), Johnson (1999), Khalil et al. (2007), Marc et al. (2008)
A. ciliata Bunge	Root, leaves	Hemolysis, sore throat, tonsillitis,	China, Taiwan, Japan	Johnson (1999)
A. decumbens Thunb.	Whole plant	Anti-inflammatory, expectorant, gall bladder ailments, depurative, joint pain, diarrhea, eye troubles, febrifuge, hemostatic, sore throat, stomachache	East Asia, China, Japan, Taiwan, Korea	Duke and Ayensu (1985), Johnson (1999), Konoshima et al. (2000), Ono et al. (2008)
A. forrestii Diels	Root, leaves	Dysentery and ascariasis	China, Korea, Nepal	Johnson (1999)
A. genevensis L.	Aerial parts	Anticancer	Europe, England, Belgium, France, Germany, Italy, Sweden, China	Graham et al. (2000)
A. integrifolia Buch-Ham. ex D.Don (Syn.: *A. bracteosa* Wall. ex Benth.)	Whole plant	Diarrhea, stomach disorders, malaria, swollen legs, hypertension, wound healing, veterinary use, fever, common cold, fatigue, hemorrhoids, earache, ascariasis, rabies	Ethiopia, Kenya, N Africa, Africa (Congo, Kenya, Ethiopia), Nepal, India, Uganda	Cantrell et al. (1999), Asres et al. (2001), Gedif and Hahn (2003), Hamill et al. (2003), Matu and van Staden (2003), Debella et al. (2005), Muregi et al. (2007), Yineger et al. (2008), Singh (2020)

TABLE 1.1 *(Continued)*

Plant species	Part used	Disease	Country	Citation
A. iva (L.) Schreb.	Whole plant	Asthma, bronchitis, earache, fever, phthisis, diabetes, hypertension, antiulcer, antimalarial, mental disorders, skin disease, gastrointestinal disorders, wound healing	Europe, Spain, Israel, North Africa, Malta, Malaysia, Hawaii	Boulos (1983), Bellakhdar et al. (1991), Brown (1995), Ziyyat et al. (1977), Johnson (1999), Tahraoui et al. (2007), González-Tejero et al. (2008), Marc et al. (2008)
A. macrosperma Wall. ex Benth.	Whole plant	Fever, phlegm nephritis	China, Bhutan, India, Laos, Myanmar, Nepal, Thailand, Vietnam, Taiwan	Yunnan Institute of Botany (1977), Johnson (1999)
A. multiflora Bunge	Root	Diuretic	Russia, Taiwan, China, Korea, Siberia, Mongolia	Johnson (1999)
A. nipponensis Makino	Root	Traumatic injury, inflammation	Asia, China, Japan, Korea, Taiwan	Johnson (1999)
A. ophrydris Burch. ex Benth.	Juice/ leaves	Dysmenorrhea, sterility	Africa, Natal	Johnson (1999)
A. orientalis L.	Aerial parts	Skin disorders (wound)	Europe, Hawaii, Israel, Greece, Palestine, Iran	Boulos (1983), Ali-Shtayeh et al. (2000)
A. pantantha Hand.- Mazz.	Leaves/ juice	Fever, phlegm	China Sino-Himalaya	Yunnan (1977), Johnson (1999)
A. parviflora Benth.	Whole plant	Fever, stomach disorder, ulcer, colic, jaundice, and throat infections	Afghanistan, Pakistan, India, Nepal	Khan et al. (1999)

TABLE 1.1 *(Continued)*

Plant species	Part used	Disease	Country	Citation
A. reptans L.	Whole plant	Hemorrhage, cough, spitting of blood, jaundice, ulcers, heart tonic, hallucinogenic, gout, wound, skin disease, cardio-vascular disease, carminative, fever, vulnerary, cancer, healing	Asia, Africa, Iran, Europe, Turkey, America, Hawaii Mediterranean	Chung et al. (1980), Stuart (1987), Breschi et al. (1992), Diggs et al. (1999), Johnson (1999), Kuria et al. (2002)
A. taiwanensis Nakai ex Murata	Aerial parts	Hepatitis and hepatoma	Taiwan, Philippines, Ryukyus	Hou (1996)
A. turkestanica (Regel) Briq.	Aerial parts	Heart disease, muscle aches, stomach problems	Turkey, Uzbekistan	Mamatkhanov et al. (1998), Abdukadirov et al. (2004)

1.2 BIOACTIVES

Phytochemical studies on genus *Ajuga* lead to the isolation and characterization of a large number of compounds that include phytoecdysteroids, triterpines, diterpinoids, and neo-clerodane-diterpines class of compounds. These results are supported by several investigations carried out across the different parts of the world (Camps et al., 1979, 1981, 1982, 1984, 1987, 1990, 1991; Shimomura et al., 1987; Bhakuni et al., 1991; Shen et al., 1993; Beauchamp et al., 1996; Chen et al., 1995, 1996, 1997; Amano et al. 1997; Bremner et al., 1998; Malakov and Papanov, 1998; Cantrell et al., 1999; Ben-Jannet et al., 1999, 2000; Chaari et al., 2000; Kökdil et al., 2002; Castro et al., 2008; Coll and Tandrón, 2008; Israili and Lyoussi, 2009). The presence of other active constituents in the plants of *Ajuga* species was supported by other pioneer workers such as the presence of sterols by Cantrell et al. (1999), Fujimoto et al. (2000), Akbay et al. (2003) and Chan (2005); presence of anthocyanidin-glucosides and iridoid glycosides by Shimomura et al. (1987), Takasaki et al. (1998), Konoshima et al. (2000), Terahara et al. (2001), Chaari et al. (2002); presence of quinols by Khan et al. (1999); presence of withanolides by Khan et al. (1999) and Nawaz et al. (1999).

Neo-clerodane diterpenoids are a special class of bioactive secondary metabolites that have been isolated from the genus *Ajuga* (Ben-Jannet et al., 2000; Coll and Tandrón, 2008). Biosynthetically, they are structurally formed by 20 carbons with a decalin core in basic skeleton *via* different fashions (Fig. 1.1).

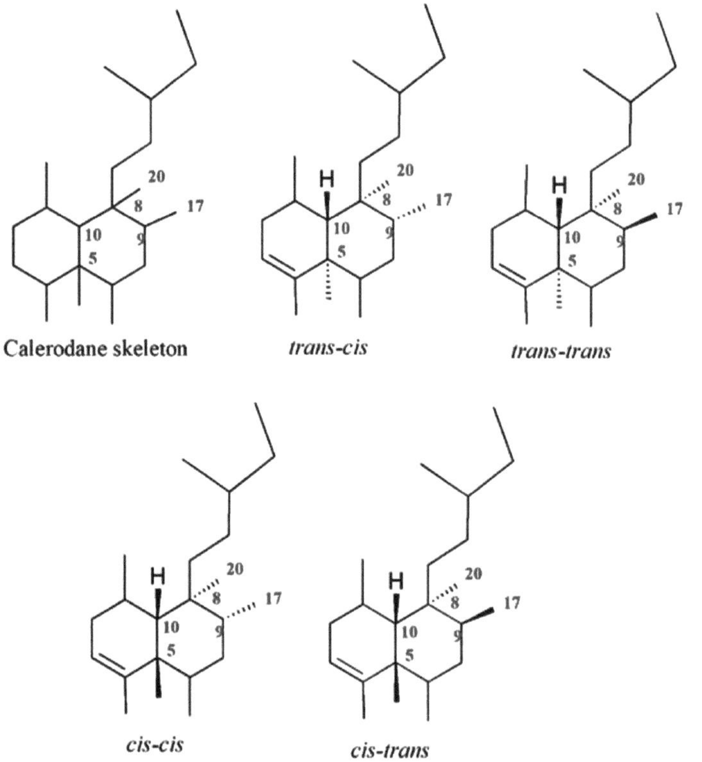

FIGURE 1.1 Biosynthetically synthesized stereoisomers of clerodane diterpenoids skeleton.

Shen et al. (1993) have studied the phytochemistry of *A. macrosperma* for the first time and isolated ajugamacrins A and B in a methanolic extract from the whole plant of *A. macrosperma*. Similarly, five *neo*-clerodane diterpenoids have been isolated and characterized (Xiaoyu et al., 1993). Six *neo*-clerodane diterpenoids (ajugaflorins A–F) have been characterized by Castro et al. (2015) along with six known compounds (ajugarin I, ajugalides B and C, ajugamarin F4, ajugamacrin E, and ajugatakasin B) reported from *Ajuga*. Three new compounds (ajugacetalsterones C, D, and breviflorasterone) were isolated and characterized from the roots of *A. macrosperma* along with five known compounds (20 hydroxyecdysone, cyasterone, makisterone A, 20-hydroxyecdysone 3-acetate, and 20-hydroxyecdysone 2-acetate) that are also isolated (Castro et al., 2008). *A. integrifolia* is yet another important plant species and bracteonin-A 6 f-acetoxy, 15-methoxy, 18-(4'- hydroxy, 3'- g -methyl, 3'- f -acetoxy, butyryloxy) neoclerodane, 14, 15- dihydroajugapitin, 14-hydro-15-hydroxy ajugapitin, g-sitosterol, and stigmasterol have been characterized. The other compounds identified were 14, 15-dihydroajugapitin, 14-hydro-15-hydroxy ajugapitin, g -sitosterol, and stigmasterol. The structural representation of some well-known compounds characterized from genus *Ajuga* is given in Figures 1.2–1.6.

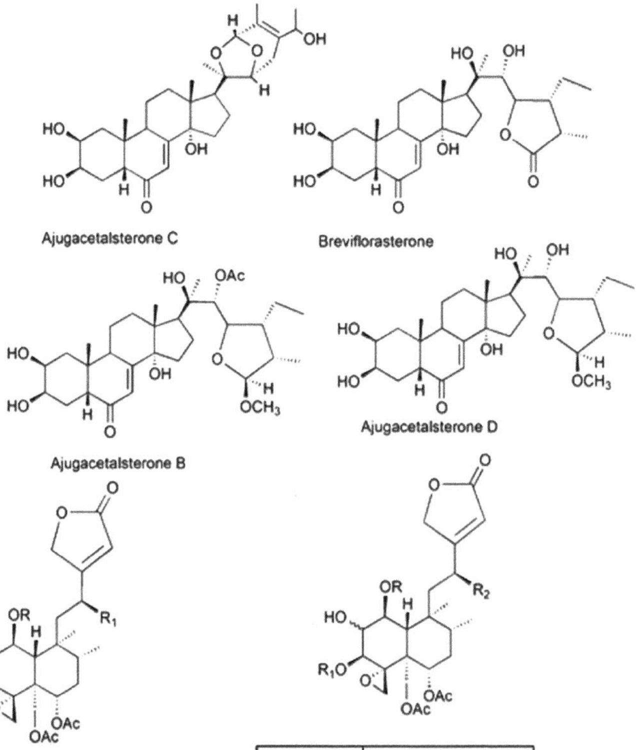

Compound	R	R_1
Ajugamacrin A	Ac	'Bu
Ajugamacrin B	Ac	'MB
Ajugamacrin C	iBu	iBu
Ajugamacrin D	iBu	MB
Ajugamacrin E	MB	iBu

Compound	R	R_1	R_3
Ajugaflorin A	Ac	MeBu	OTig
Ajugaflorin B	Ac	'Bu	Ac
Ajugaflorin C	MeBu	H	'Bu
Ajugaflorin D	MeBu	H	Ac
Ajugaflorin E	Ac	MeBu	Tig
Ajugaflorin F	Ac	MeBu	H

FIGURE 1.2 *Neo*-clerodane diterpenoids from *Ajuga macrosperma.*

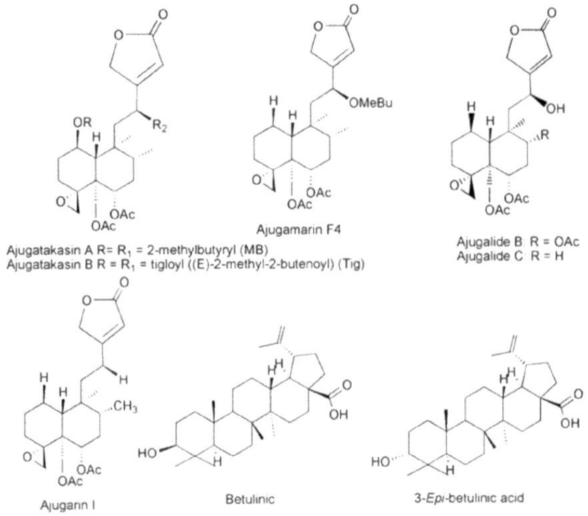

FIGURE 1.3 *Neo*-clerodane diterpenoids from *Ajuga* spp.

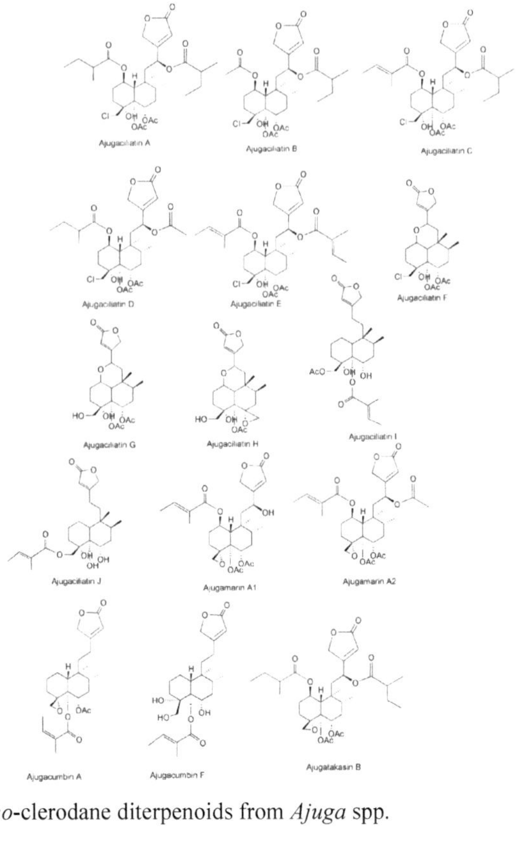

FIGURE 1.4 *Neo*-clerodane diterpenoids from *Ajuga* spp.

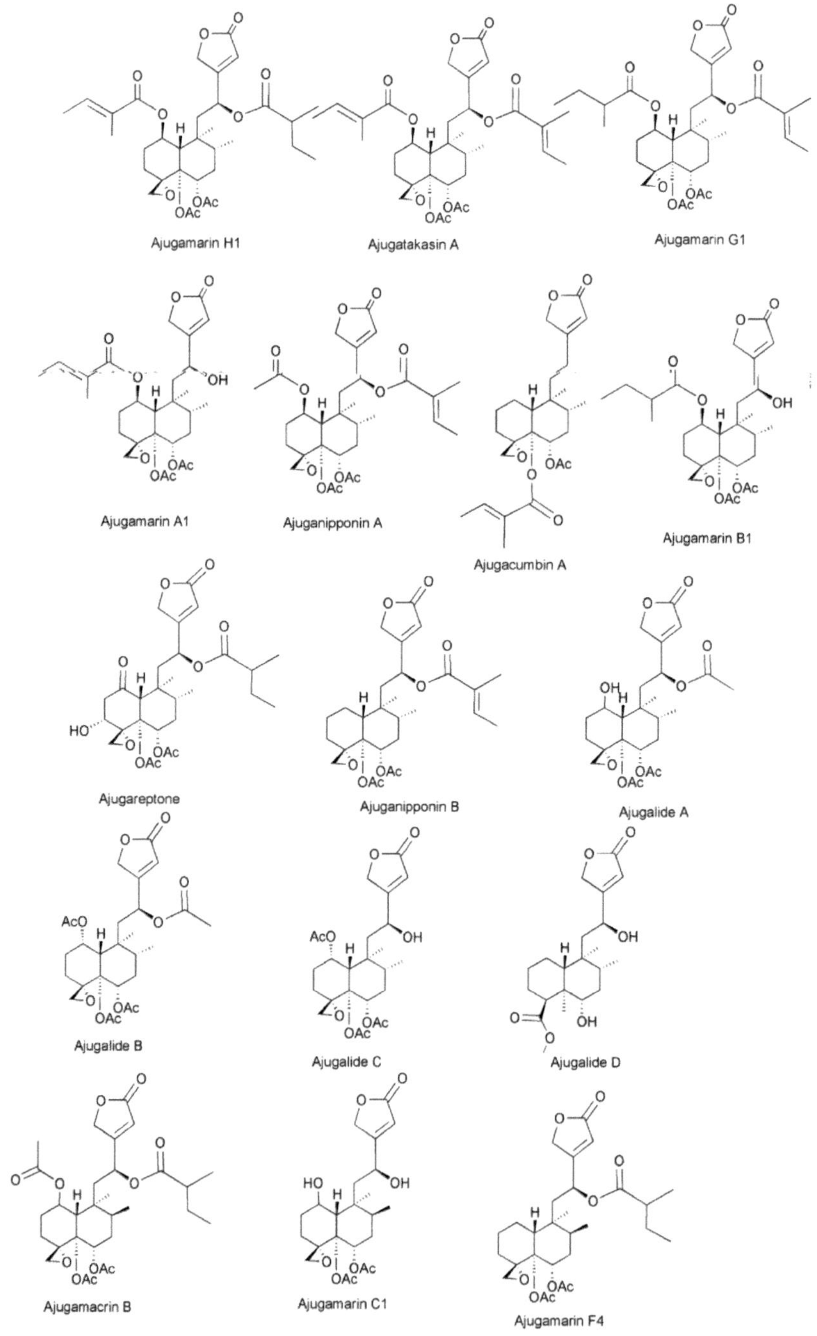

FIGURE 1.5 *Neo*-clerodane diterpenoids from *Ajuga* spp.

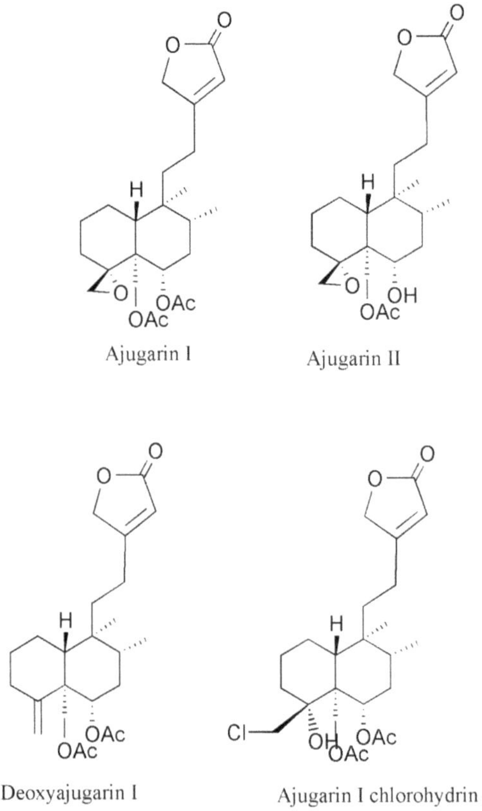

FIGURE 1.6 *Neo*-clerodane diterpenoids from *Ajuga* spp.

Essential oils were also reported from the genus *Ajuga*. The major constituents are limonene, α-humulene, β-myrcene, elemol, camphene, β-caryophellene, α-phellendrene, etc. that were examined by gas chromatography and gas chromatography mass spectrometry (GC/MS) (Baser et al., 2001). Besides, different neo-clerodane diterpenoids were isolated from different species of *Ajuga*.

1.3 PHARMACOLOGY

Several investigations of the genus *Ajuga* have been carried out with the plants' crude extracts/or molecule studied showing different pharmacological properties. Israili and Lyoussi (2009), while reviewing the ethnopharmacological aspects of the genus *Ajuga*, reported a wide variety of biological functions that include anabolic, analgesic, antiarthritic, antibacterial, antiestrogenic, antifungal, anti-inflammatory, antihypertensive, antileukemic, antimalarial antimycobacterial, antioxidant, antipyretic, cardiotonic, cytotoxic, hypoglycemic, and vasorelaxing activity, as well as antifeedant and insect growth-inhibitory properties. Other researchers' studies supporting the various aspects of pharmacological properties include antibacterial (Chen et al., 1997), antifungal (Kariba, 2001), antihypertensive (Odek-Ogunde and Rajab, 1993), anti-inflammatory (Marc et al., 2008), antimalarial (Kareru et al., 2007), antioxidant (Chenni et al., 2007), antipyretic (Baytop, 1984), antitumor (Rodriguez-Hahn et al., 1994), larvae and insect antifeedant (Camps and Coll, 1993; Ben-Jannet et al., 2000; Fekete et al., 2004), and insect growth inhibitor (Camps and Coll, 1993) activity from the different species of *Ajuga*. The most active isolated compounds from the *Ajuga* plants such as phytoecdysteroids, neo-clerodane-diterpenes and diterpenoids, triterpenes, sterols, anthocyanidin-glucosides, and iridoid glycosides, withanolides, flavonoids, triglycerides, and essential oils are reported to possess significant broad-spectrum pharmacological properties such as anabolic, analgesic, antibacterial, antiestrogenic, antifungal, anti-inflammatory, antihypertensive, antileukemic, antimalarial, antimycobacterial, antioxidant, antipyretic, cardiotonic, cytotoxic, hypoglycemic, and vasorelaxant activity, as well as antifeedant and insect growth-inhibitory properties (Manguro et al., 2006; Vanyolos et al., 2009; Castro et al., 2015). Some well-known pharmacological functions of different species of *Ajuga* are given in Table 1.2.

TABLE 1.2 Pharmacological Functions of Different *Ajuga* Species.

Pharmacological properties	Species employed	Plant extract/ compounds	Citations
Analgesic activity	*A. integrifolia* Buch.-Ham. ex D.Don (=syn. *A. remota* Benth.; *A. bracteosa* Wall. ex Benth.)	Aqueous and ethyl alcohol extract (mice)	Debella et al. (2003)
Anti-arthritic activity	*A. decumbens* Thunb.	Aqueous extract (rats)	Ono et al. (2008)

TABLE 1.2 *(Continued)*

Pharmacological properties	Species employed	Plant extract/ compounds	Citations
Antibacterial activity	*A. lupulina* Maxim. *A. iva* (L.) Schreb. *A. integrifolia* Buch.-Ham. ex D.Don *A. turkestanica* (Regel) Briq.	Neo-clerodane diterpenoids (lupulins A, B, and D), and phytoecdysteroid (in vitro); Steroids, cinnamic acid derivatives; phytoecdysteroids (in vitro); plant products (in vitro) 8-O-acetylharpagide (in vitro); phytoecdysteroids (in vitro) 8-O-acetylharpagide (in vitro); phytoecdysteroids (in vitro)	Chen et al. (1996, 1997); Chaari et al. (2000); Ben-Jannet et al. (2006); Bennaghmouch et al. (2001); Shafi et al. (2004); Ulukanli et al. (2005)
Antifungal activity	*A. integrifolia* Buch.-Ham. ex D.Don	Aqueous extract (in vitro); 8-O-acetylharpagide (in vitro)	Kariba (2001); Shafi et al. (2004)
Antihypertensive activity	*A. integrifolia* Buch.-Ham. ex D.Don	Aqueous extract (rat); clerodane diterpene ajugarin I (rat)	Odek-Ogunde and Rajab (1994)
Anti-inflammatory effect	*A. reptans* L. *A. integrifolia* Buch.-Ham. ex D.Don *A. decumbens* Thunberg	Phenylpropanoid glycosides (verbascoside, teupolioside) (in vitro); aqueous-, methyl alcohol and hexane-extract (in vitro), aqueous extract (rats)	Debella et al. (2003); Matu and vanStaden (2003); Ono et al. (2008)
Antileukemic activity	*A. salicifolia* Schreber	Sterol glycosides	Akbay et al. (2002)
Antimalarial/antiplasmodial activity	*A. integrifolia* Buch.-Ham. ex D.Don	Ajugarin-1 and ergosterol-5,8- endoperoxide; aqueous, methyl and ethyl alcohol-plant extract (mice; in vitro)	Kuria et al. (2002); Muregi et al. (2007); Fennell et al. (2004); Sharma et al. (2004)
Antimicrobial	*A. alba* (Gtrke) Robyns	Aqueous extract	Hamill et al. (2003)

TABLE 1.2 *(Continued)*

Pharmacological properties	Species employed	Plant extract/ compounds	Citations
Antimycobacterial activity	*A. integrifolia* Buch.-Ham. ex D.Don	Ergosterol 5,8-endoperoxide	Cantrell et al. (1991)
Anti-osteoporosis activity	*A. decumbens* Thunberg	Aqueous extract (mice)	Ono et al. (2008)
Antioxidant /oxygen scavenging activity	*A. iva* (L.) Schreb. *A. reptans* L. *A. integrifolia* Buch.-Ham. ex D.Don *A. decumbens* Thunb.	Aqueous extract (rats); aqueous extract in diet; anthocyanins; ecdysteroids; verbascoside and teupolioside in diet (piglets); aqueous, methyl alcohol and hexane-extract (in vitro)	El-Hilaly et al. (2006); Chenni et al. (2007); Bouderbala et al. (2008); Li et al. (1999); Terahara et al. (2001); Corino et al. (2007); Matu and vanStaden (2003);
Antipyretic activity	*A. integrifolia* Buch.-Ham. ex D.Don	Aqueous extract; 8-O-acetylharpagide (mouse)	Debella et al. (2005); Shafi et al. (2004)
Antispasmodic properties	*A. integrifolia* Buch.-Ham. ex D.Don	8-O-acetylharpagide	Shafi et al. (2004)
Antiviral properties	*A. decumbens* Thunb. *A. integrifolia* Ham.- Buch.	Aqueous-plant extract (mouse); aqueous- and methyl alcohol extract (in vitro); flavones (in vitro)	Ma et al. (2002); Asres et al. (2001)
Antitumor-promotor and cancer- chemo-preventive effect	*A. decumbens* Thunb. *A. integrifolia* Buch.-Ham. ex D.Don	Cyasterone, 8-O-acetylharpagide; polypodine B, decumbesterone A, and loliolide (in vitro/mouse); 8-O-acetylharpagide (mouse)	Takasaki et al. (1998, 1999); Konoshima et al. (2000); Kuria et al. (2002)
Cholinesterase inhibition	*A. integrifolia* Buch.-Ham. ex D.Don *A. parviflora* Benth.	Withanolides (bractin A and B, bracteosin A, B, and C) (in vitro); diterpenoids (in vitro) ajugin E, ajugin F (in vitro)	Riaz et al. (2004, 2007); Choudhary et al. (2005); Nawaz et al. (1999)
Xanthine oxidase inhibition	*A. australis* R. Br.	Aqueous plant extract (in vitro)	Sweeney et al. (2001)

TABLE 1.2 *(Continued)*

Pharmacological properties	Species employed	Plant extract/ compounds	Citations
Cytochrome P450 activity	*A. reptans* L.	Ecdysone (in vitro)	Alekseeva (2004)
Growth promotion	*A. reptans* L.	Verbascoside, teupolioside (piglets)	Corino et al. (2007)
Hypoglycemic activity	*A. turkestanica* (Regel) Brig. *A. iva* (L.) Schreber	Ecdysteroids (rat) Aqueous extract (diabetic rat)	Kutepova et al. (2001); El-Hilaly and Lyoussi (2002); El-Hilaly et al. (2006); Chenni et al. (2007)
Hypolipidemic action	*A. iva* (L.) Schreb. *A. turkestanica* (Regel) Brig.	Aqueous extract; aqueous extract in diet Ecdysteroids	El-Hilaly et al. (2006); Chenni et al. (2007); Syrov et al. (1983)
Lipoxygenase inhibition	*A. integrifolia* Buch.-Ham. ex D.Don	Withanolides (bractin A and B, bractic acid) (in vitro)	Riaz et al. (2007)
Stimulation of aquaporins- human skin hydration	*A. turkestanica* (Regel.) Brig.	Aqueous- and EtOH-extract; phytoecdysteroids (in vitro)	Dumas et al. (2002, 2007)
Vasoconstrictor effect	*A. reptans* L.	Aqueous extract, methyl alcohol-extract; acetylharpagide 8-O-acetylharpagide (in vitro)	Breschi et al. (1992)
Vasodilator effect	*A. iva* (L.) Schreb.	Aqueous extract (in vitro, rats)	El-Hilaly et al. (2004)
Toxicology	*A. iva* L. Schreb. *A. nipponensis* Makino	Aqueous extract (rats/mice) Oral consumption of plant (human)	El-Hilaly et al. (2004, 2006); Liao et al. (2005); Coll et al. (2007)

TABLE 1.2 *(Continued)*

Pharmacological properties	Species employed	Plant extract/ compounds	Citations
DNA damaging effect	*A. postii* Briq.	Methyl alcohol extract, iridoid glucoside reptoside (in vitro)	Gören et al. (2005)
Toxicity against cancer cells (KB HeLa) and Jurkat T-cells	*A. salcifolia* Schreber	Ajugasalicigenin, ajugasaliciosides F,G &H (in vitro)	Akbay et al. (2003)
Stimulation of erythropoesis (experimental anemia)	*A. turkestanica* (Regel) Briq.	Silenoside A, turkesterone (in rats)	Syrov et al. (1997)
Renoprotection (in experimental uremia)	*A. turkestanica* (Regel) Briq.	Phytoecdysterone (ecdysterone and turkesterone) (in rats)	Syrov and Khushbaktova (2001)
Anabolic activity (growth promotion)	*A. turkestanica* (Regel) Briq.	Turkesterone; 2-deoxy-alpha- ecdysone (rat)	Syrov et al. (2001)
Stimulation of protein synthesis in the liver	*A. turkestanica* (Regel) Briq.	Phytoecdysteroids (cyasterone, ecdysterone, turkesterone) (in mouse); phytosteroids (in vitro and in vivo in rats); turkesterone (mouse)	Syrov et al. (1978); Gorelick-Feldman et al. (2008); Syrov et al. (1978)
Hepatoprotection (against liver toxins)	*A. turkestanica* (Regel) Briq.	Ecdysterone, turkesterone, cyasterone, iridoid glycosides (rats)	Syrov et al. (1986,1992a, 1998); Syrov and Khushbaktova (2001)
Increase in blood flow	*A. decumbens* Thunb. *A. reptan* L. *A. turkestanica* (Regel) Briq.	Ecdysteroids (rat)	Tsuji et al. (1999)

TABLE 1.2 *(Continued)*

Pharmacological properties	Species employed	Plant extract/ compounds	Citations
Cell energetics and lipid metabolism in the liver (diabetes)	*A. turkestanica* (Regel) Briq.	Phytoecdysteroides, ecdysterone and turkesterone (rats)	Tashmukhamedova et al. (1985); Syrov et al. (1992b)
Reversal of mito-chondrial dysfunction (hepatitis)	*A. turkestanica* (Regel) Briq.	Phytoecdysteroids (ecdysterone, turkesterone and cyasterone) (rats)	Tashmukhamedova et al. (1986a,b); Syrov et al. (1992b)
Increase in insulin sensitivity	*A. turkestanica* (Regel) Briq.	Ecdysterone, ecdysteroid (rats)	Kosovskii et al. (1989)
Cardiac stimulant	*A. integrifolia* Buch.-Ham. ex D.Don	Plant extract (in vivo); 8-O-acetylharpagide	Shafi et al. (2004)
Biliary secretagogue	*A. turkestanica* (Regel) Brig.	Phytoecdysteroids (ecdysterone, turkesterone and cyasterone) (rat)	Syrov et al. (1986)
Wound healing	*A. reptans* L. *A. chia* Schreb. *A. orientalis* L. *A. iva* (L.) Schreb. *A. turkestanica*	Phenylpropanoid glycosides (verbas-coside, teupolioside) (in vitro/mouse); aqueous extract (mouse); plant products (in vitro/ mouse); lipids (mice)	Khalil et al. (2007); Boulos (1983); Ali-Shtayeh et al. (2000); Syrov et al. (1994)
Larvae/insect antifeedant	*A. reptans* ev Catlins Giant; *A. iva* (L.) Schreb.	*neo*-clerodane diterpenoids; clero-dane diterpenoids (14,15-dihydroajuga-pitin, lupulin A) (in vitro); diglycerides (in vitro); ajugarin/ clerodane diterpe-noids (in vitro)	Ben-Jannet et al. (2000); Bondì et al. (2000); Caballero et al. (2001)
Inhibition of insect ecdysis (molting)	*A. decumbens* Thunb. *A. reptans* L.	Ajugalactone (in vivo)	Kubo et al. (1981)
Insecticidal effect	*A. iva* (L.) Schreb.	Methyl alcohol-plant extract	Jbilou et al. (2006, 2008)

The root extract of *A. macrosperma* possessed significant antihelmintic activity against *Ascaridia galli*, a poultry roundworm. It showed an IC_{50} value at 1.73 ± 0.02. Albendazole used as a standard was found to possess the lowest IC_{50} value 0.74± 0.05 (Agarwal et al., 2010). The iridoid glycosides isolated from *A. macrosperma* have been investigated by many researchers for their protective effects in various in vitro cancer models. It was reported that the iridoid glycoside and its derivatives isolated and identified from the flowering whole plant of *Ajuga* demonstrate a remarkable inhibitory effect on the two-stage carcinogenesis test of mouse skin tumors (Takasaki et al., 1999; Konoshima et al. 2000). The antiviral activity of methanol extracts of *Ajuga* against Hepatitis C Virus (HCV)-infected HepG2 cells has been reported by Yousaf et al. (2018). The results showed that the extracts have significant abilities to reduce the growth of the HCV virus after 24 h of treatment.

KEYWORDS

- **indigenous uses**
- **chemical constituents**
- **biological activity**
- **value addition**

REFERENCES

Abdukadirov, I. T.; Khodzhaeva, M. A.; Turakhozhaev, M. T.; Mamatkhanov, A. T. Carbohydrates from *Ajuga turkestanica*. *Chem. Nat. Compd.* **2004,** *40*, 85–86.

Agarwal, G.; Pant, A. K.; Hore, S. K. In Vitro Evaluation of Antihelmantic Efficacy of *Trichilia* and *Ajuga* Species on *Ascarida galli*. *Hygeia. J. D. Med.* **2010,** *2* (2), 43–53.

Akbay, P.; Çalis; I., Heilmann, J.; Sticher, O. New Stigmastane Sterols from *Ajuga salicifolia*. *J. Nat. Prod.* **2003,** *66*, 461–465.

Akbay, P.; Gertsch, J.; Çalis, I.; Heilmann, J.; Zerbe, O.; Sticher, O. Novel Antileukemic Sterol Glycosides from *Ajuga salicifolia*. *Helv. Chim. Acta.* **2002,** *85*, 1930–1942.

Alekseeva, L. I. Ecdysone 20-Monooxygenase Activity of Cytochrome P_{450} in *Ajuga reptans* L. Plants and Cell Culture. *Appl. Biochem. Microbiol.* **2004,** *40*, 135–139.

Aliotta, G.; Pollio, A. Useful Plants in Renal Therapy According to Pliny the Elder. *Am. J. Nephrol.* **1994,** *14*, 399–411.

Ali-Shtayeh, M. S.; Yaniv, Z.; Mahajna, J. Ethnobotanical Survey in the Palestinian Area: A Classification of the Healing Potential of Medicinal Plants. *J. Ethnopharmacol.* **2000,** *73*, 221–232.

Amano, T.; Nishida, R.; Kuwahara, Y. Ajugatakasins A and B, New Diterpenoids from *Ajuga decumbens*, and Feeding Stimulative Activity of Related *neo*-Clerodane Analogs Toward the Turnip Sawfly. *Biosci. Biotechnol. Biochem.* **1997,** *61*, 1518–1522.

Anonymous. *Ajuga macrosperma* Wall. ex Benth. (accessed on 13th Dec 2021).

Asres, K.; Bucar, F.; Kartnig, T.; Witvrouw, M.; Pannecouque, C.; DeClercq, E. Antiviral Activity Against Human Immunodeficiency Virus Type 1 (HIV-1) and Type 2 (HIV-2) of Ethnobotanically Selected Ethiopian Medicinal Plants. *Phytother. Res.***2001,** *15*, 62–69.

Baser, K. H. C.; Kurkcuoglu, M.; Erdemgil, F. Z. The Essential Oil of *Ajuga bombycina* from Turkey. *Chem. Nat. Compd.* **2001,** *37*, 242–244.

Baytop, T. *Therapy with Medicinal Plants (Past and Present)*; Istanbul University Publications: Istanbul, 1984.

Beauchamp, P. S.; Bottini, A. T.; Caselles, M. C.; Dev, V.; Hope, H.; Larter, M.; Lee, G.; Mathela, C. S.; Melkani, A. B.; Millar, P. D.; Miyatake, P. D.; Pant, A. K.; Raffel, R. J.; Sharma, V. K.; Wyatt, D. *Neo*-Clerodane Diterpinoids from *Ajuga parviflora. Phytochemistry* **1996,** *43*, 827–834.

Bellakhdar, J.; Claisse, R.; Fleurentin, J.; Younos, C. Repertory of Standard Herbal Drugs in the Moroccan Pharmacopoeia. *J. Ethnopharmacol.* **1991,** *35*, 123–143.

Bellés, X.; Camps, F.; Coll, J.; Piulachs, M. D. Insect Antifeedant Activity of Clerodane Diterpenoids Against Larvae of *Spodoptera littoralis* (Boisd.). *J. Chem. Ecol.* **1985,** *11*, 1573.

Ben-Jannet H.; Chaari, A.; Mighri, Z.; Martin, M. T.; Loukaci, A. Neo-Clerodane Diterpenoids from *Ajuga pseudoiva* Leaves. *Phytochemistry* **1999,** *52*, 1541–1545.

Ben-Jannet, H.; Chaari, A.; Bakhrouf, A.; Mighri, Z. Structure-Antibacterial Activity Relationship of Secondary Metabolites from *Ajuga pseudoiva* Rob. Leaves. *Nat. Prod. Res.* **2006,** *20*, 299–304.

Ben-Jannet, H.; Harzallah-Skhiri, F.; Mighri, Z.; Simmonds, M. S. J.; Blaney, W. M. Responses of *Spodoptera littoralis* Larvae to Tunisian Plant Extracts and to *Neo*-Clerodane Diterpenoids Isolated from *Ajuga pseudoiva* Leaves. *Fitoterapia* **2000,** *71*, 105–112.

Bennaghmouch, L.; Hajjaji, N.; Zellou, A.; Cherrah, Y. Pharmacological Study of *Ajuga iva. Ann. Pharmaceut. Francais.* **2001,** *59*, 284.

Bhakuni, R. S.; Shukla, Y. N.; Thakur, R. S. Chemical Constituents of *Ajuga bracteosa. Indian J. Pharm. Sci.* **1991,** *49*, 225–226.

Bnouham, M.; Mekhfi, H.; Legssyer, A.; Ziyyat, A. Medicinal Plants Used in the Treatment of Diabetes in Morocco. *Int. J. Diab.Metab.* **2002,** *10*, 33–50.

Bondì, M. L.; Al-Hillo, M. R. Y.; Lamara, K.; Ladjel, S.; Bruno, M.; Piozzi, F.; Simmonds, M. S. J. Occurrence of the Antifeedant 14,15-Dihydroajugapitin in the Aerial Parts of *Ajuga iva* from Algeria. *Biochem. Syst. Ecol.* **2000,** *2*, 1023–1025.

Bouderbala, S.; Lamri-Senhadji, M.; Prost, J.; Lacaille-Dubois, M. A.; Bouchenak, M. Changes in Antioxidant Defense Status in Hypercholesterolemic Rats Treated with *Ajuga iva. Phytomedicine* **2008,** *15*, 453–461.

Boulos, L. *Medicinal Plants of North Africa*; Reference Publications Inc.: Algonac, Michigan, 1983.

Bremner, P. D.; Simmonds, M. S. J.; Blaney, W.M; Veitch, N. C. *Neo*-Clerodane Diterpenoid Insect Antifeedants from *Ajuga reptans* cv Catlins Giant. *Phytochemistry* **1998,** *47*, 1227–1232.

Breschi, M.; Martinotti, E.; Catalano, S.; Flamini, G.; Morelli, I.; Pagni, A. Vasoconstrictor Activity of 8-O-Acetylharpagide from *A. reptans. J. Natl. Prod.* **1992,** *55*, 1145–1148.

Brown, D. *Encyclopaedia of Herbs and Their Uses*; Dorling Kindersley: London, 1995.

Caballero, C.; Castañera, P.; Ortego, F.; Fontana, G.; Pierro, P.; Savona, G.; Rodríguez, B. Effects of Ajugarins and Related Neoclerodane Diterpenoids on Feeding Behaviour of *Leptinotarsa decemlineata* and *Spodoptera exigua* Larvae. *Phytochemistry* **2001,** *58*, 249–256.

Camps, F.; Coll, J. Insect Allelochemicals from *Ajuga* Plants. *Phytochemistry* **1993,** *32*, 1361–1370.

Camps, F.; Coll, J.; Cortel, A. A. Allelochemicals on Insects Isolated from *Ajuga* Plants (Labiatae). *Rev. Latinoam. Quím* **1991,** *12*, 81–87.

Camps, F.; Coll, J.; Cortel, A. A. New Clerodane Diterpenoids from *Ajuga iva* (Labiatae). *Chem. Lett.* **1982,** *7*, 1053–1056.

Camps, F.; Coll, J.; Cortel, A. A.; Messeguer, A. Ajugareptansin, a New Diterpenoid from *Ajuga reptans. Tetrahedron Lett.* **1979,** *20*, 1709–1712.

Camps, F.; Coll, J.; Cortel, A. Allelochemicals on Insects Isolated from *Ajuga* Plants (Labiatae). *Rev. Latinoam. Quím.* **1981,** *12*, 81–87, 90.

Camps, F.; Coll, J.; Dargallo, O. *Neo*-Clerodane Diterpenoids from *Ajuga pseudoiva. Phytochemistry* **1984,** *23*, 387–389.

Camps, F.; Coll, J.; Dargallo, O.; Rius, J.; Miravitlles, C. Clerodane Diterpenoids from *Teucrium* and *Ajuga* Plants. *Phytochemistry* **1987,** *26*, 1475–1479.

Camps, F.; Coll., J.; Marco, M. P.; Tomas, J. Efficient Determination of Phytoecdysteroids from *Ajuga* Species and *Polypodium vulgare* by High-Performance Liquid Chromatography. *J. Chromatogr. A.* **1990,** *514*, 199–207.

Cantrell, C. L.; Rajab, M. S.; Franzblau, S. G.; Fronczek, F. R.; Fisher, N. H. Antimycobacterial Ergosterol-5,8- Endoperoxide from *Ajuga remota. Planta. Med.* **1991,** *65*, 732–734.

Castro, A.; Coll, J.; Pant, A. K.; Pakrash, O. neo-Clerodane Diterpenoids from *Ajuga macrosperma* var. *breviflora. Natl. Prod. Commun.* **2015,** *10* (6), 1934578X1501000614.

Castro, A.; Coll, J.; Tandrón, Y. A.; Pant, A. K.; Mathela, C. S. Phytoecdysteroids from *Ajuga macrosperma* var. *breviflora* Roots. *J. Natl. Prod.* **2008,** *71* (7), 1294–1296.

Chaari, A.; Jannet, H. B.; Mighri, M.; Lallemand, M.-C.; Kunesch, N. (2002) 7-*O*-6'-*O*-Malonylcachinesidic Acid, a New Macrocyclic Iridoid Ester of Malonic Acid from the Tunisian Plant *Ajuga pseudoiva. J. Nat. Prod.* **2002,** *65*, 618–620.

Chaari, A.; Jannet, H. B.; Mighri, Z. Isolation and Structural Determination of a Novel Anti-Bacterial Phytoecdysteroid and of Two Cinnamic Acid Derivatives Based on *Ajuga pseudoiva* Leaves. *J. Soc. Chim. Tunisie* **2000,** *4*, 789–799.

Chan, Y. Y. Neoclerodane Diterpenoids from *Ajuga taiwanensis. Chem. Pharm. Bull.* **2005,** *53*, 164–167.

Chen, H. M.; Min, Z. D.; Iinuma, M.; Tanaka, T. Clerodane Diterpenoids from *Ajuga decumbens. Chem. Pharm. Bull.* **1995,** *43*, 2253–2255.

Chen, H.; Tan, R. X.; Liu, Z. L.; Zhang, Y.; Yang, L. Antibacterial Neoclerodane Diterpenoids from *Ajuga lupulina. J. Nat. Prod.* **1996,** *59*, 668–670.

Chen, H.; Tan, R. X.; Liu, Z. L.; Zhao, C. Y.; Sun, J. A Clerodane Diterpene with Antibacterial Activity from *Ajuga lupulina. Acta Cryst. C.* **1997,** *53*, 814–816.

Chenni, A.; Yahia, D. A.; Boukortt, F. O.; Prost, J.; Lacaille- Dubois, M. A.; Bouchenak, M. Effect of Aqueous Extract of *Ajuga iva* Supplementation on Plasma Lipid Profile and Tissue Antioxidant Status in Rats Fed a High-Cholesterol Diet. *J. Ethnopharmacol.* **2007,** *109*, 207–213.

Chopra, R. N.; Nayar, S. L.; Chopra, I. C. *Glossary of Indian Medicinal Plants (Including the Supplement).* Council of Scientific and Industrial Research, New Delhi, India, 1986.

Choudhary, M. I.; Nawaz, S. A.; Haq, Z. U.; Lodhi, M. A.; Ghayur, M. N.; Jalil, S.; Riaz, N.; Yousuf, S.; Malik, A.; Gilani, A. H.; Rahman, A. U. Withanolides, a New Class of Natural Cholinesterase Inhibitors with Calcium Antagonistic Properties. *Biochem. Biophys. Res. Commun.* **2005,** *334*, 276–287.

Chung, B. S.; Lee, H. K.; Kim, J. W. Iridoid Glycoside (I): Studies of the Iridoid Glycoside of *Ajuga spectabilis* Nakai. *Kor. J. Pharmacog.* **1980,** *11*, 15–23.

Coll, J.; Tandrón, Y. A. *Neo*-clerodane Diterpenoids from *Ajuga*: Structural Elucidation and Biological Activity. *Phytochem. Rev.* **2008,** *7*, 25–49.

Coll, J.; Tandrón, Y. A.; Zeng, X. N. New Phytoecdysteroids from Cultured Plants of *Ajuga nipponensis* Makino. *Steroids* **2007,** *72*, 270–277.

Corino, C.; Rossi, R.; Musella, M.; Cannata, S.; Pastorelli, G. Growth Performance and Oxidation Status in Piglets Supplemented with *Verbascoside* and *teupolioside. Ital. J. Anim. Sci.* **2007,** *6*, 292–294.

Debella, A.; Makonnen, E.; Abebe, D.; Teka, F.; Kidanemariam, A. T. Pain Management in Mice Using the Aqueous and Ethanol Extracts of Four Medicinal Plants. *East Afr. Med. J.* **2003,** *80*, 435–439.

Debella, A.; Makonnen, E.; Zerihun, L.; Abebe, D.; Teka, F. In Vivo Antipyretic Studies of the Aqueous and Ethanol Extracts of the Leaves of *Ajuga remota* and *Lippia adoensis. Ethiopian Med. J.* **2005,** *43*, 111–118.

Diggs, G. M.; Lipscomb, B. L.; O'Kennon, R. J. *Illustrated Flora of North Central Texas*; Botanical Research Institute: Texas, 1999; p 1626.

Duke, J. A.; Ayensu, E. S. *Medicinal Plants of China*; Algonac, Reference Publications Inc.: Michigan, 1985.

Dumas, M,; Sadick, N. S.; Noblesse, E.; Juan, M.; Lachmann-Weber, N.; Boury-Jamot, M.; Sougrat, R.; Verbavatz, J. M.; Schnebert, S.; Bonte, F. Hydrating Skin by Stimulating Biosynthesis of Aquaporins. *J. Drugs Dermatol.* **2007,** *6*, s20–s24.

Dumas, M.; Gondran, C.; Barre, P.; Sougrat, R.; Verbavatz, J. M.; Heusele, C.; Schnebert, S.; Bonte, F. Effect of an *Ajuga turkestanica* Extract on Aquaporin 3 Expression, Water Flux, Differentiation and Barrier Parameters of the Human Epidermis. *Eur. J. Dermatol.* **2002,** *12*, 25–26.

Eddouks, M.; Ouahidi, M. L.; Farid, O.; Moufid, A.; Khalidi, A.; Lemhadri, A. The Use of Medicinal Plants in the Treatment of Diabetes in Morocco. *Phytotherapie.* **2007,** *5*, 194–203.

El-Hilaly, J.; Lyoussi, B. Hypoglycaemic Effect of the Lyophilised Aqueous Extract of *Ajuga iva* in Normal and Streptozotocin Induced Diabetic Rats. *J. Ethnopharmacol.* **2002,** *80*, 109–113.

El-Hilaly, J.; Hmamouchi, M.; Lyoussi, B. Ethnobotanical Studies and Economic Evaluation of Medicinal Plants in Taounate Province (Northern Morocco). *J. Ethnopharmacol.* **2003,** *86*, 149–158.

El-Hilaly, J.; Israili, Z. H.; Lyoussi, B. Acute and Chronic Toxicological Studies of *Ajuga iva* in Experimental Animals. *J. Ethnopharmacol.* **2004,** *91*, 43–50.

El-Hilaly, J.; Tahraoui, A.; Israili, Z. H.; Lyoussi, B. Hypolipidemic Effects of Acute and Sub-Chronic Administration of an Aqueous Extract of *Ajuga iva* L. Whole Plant in Normal and Diabetic Rats. *J. Ethnopharmacol.* **2006,** *105*, 441–448.

Fekete, G.; Polgár, L.; Báthori, M.; Coll, J.; Darvas, B. Peros Efficacy of *Ajuga* Extracts Against Sucking Insects. *Pest Manag. Sci.* **2004,** *60*, 1099–1104.
Fennell, C. W.; Lindsey, K. L.; McGaw, L. J.; Sparg, S. G.; Stafford, G. I.; Elgorashi, E. E.; Grace, O. M.; vanStaden, J. Assessing African Medicinal Plants for Efficacy and Safety: Pharmacological Screening and Toxicology. *J. Ethnopharmacol.* **2004,** *94*, 205–217.
Fujimoto, Y.; Ohyama, K.; Nomura, K.; Hyodo, R.; Takahashi, K.; Yamada, J.; Morisaki M. Biosynthesis of Sterols and Ecdysteroids in *Ajuga* Hairy Roots. *Lipids*, **2000,** *35*, 279–288.
Gedif, T.; Hahn, H. J. The Use of Medicinal Plants in Self-Care in Rural Central Ethiopia. *J. Ethnopharmacol.* **2003,** *87* (2–3), 155–161. DOI: 10.1016/s0378–8741(03)00109–0
Githinji, C. W.; Kokwaro, J. O. Ethnomedicinal Study of Major Species in the Family Labiatae from Kenya. *J. Ethnopharmacol.* **1993,** *39*, 197–203.
González-Tejero, M. R.; Casares-Porcel, M.; Sánchez-Rojas, C. P.; Ramiro-Gutiérrez, J. M.; Molero-Mesa, J.; Pieroni, A.; Censorii, E.; dePasquale, C.; Della, A.; Paraskeva-Hadijchambi, D.; Hadjichambis, A.; Houmani, Z.; El-Demerdash, M.; El-Zayat, M.; Hmamouchi, M.; El-Johrig, S. Medicinal Plants in the Mediterranean Area: Synthesis of the Results of the Project Rubia. *J. Ethnopharmacol.* **2008,** *116*, 341–357.
Gorelick-Feldman, J.; Maclean, D.; Ilic, N.; Poulev, A.; Lila, M. A.; Cheng, D.; Raskin, I. Phytoecdysteroids Increase Protein Synthesis in Skeletal Muscle Cells. *J. Agric. Food Chem.* **2008,** *56*, 3532–3537.
Gören, A. C.; Zhou, B. N.; Topçu, G.; Kökdil, G.; Kingston, D. G. I. DNA Damaging, Activitics of Mcthanol Extract of *Ajuga postii* and Iridoid Glucoside Reptoside. *Nat. Prod. Res.* **2005,** *19*, 457–460.
Graham, J. G.; Quinn, M. L.; Fabricant, D. S.; Farnsworth, N. R. Plants Used Against Cancer—An Extension of the Work of Jonathan Hartwell. *J. Ethnopharmacol.* **2000,** *73*, 347–377.
Grieve, M. A. *Modern Herbal*; Penguin Books Ltd.: Middlesex, UK, 1984.
Hamill, F. A.; Apio, S.; Mubiru, N. K.; Bukenya-Ziraba, R.; Mosango, M.; Maganyi, O. W.; Soejarto, D. D. Traditional Herbal Drugs of Southern Uganda, II: Literature Analysis and Antimicrobial Assays. *J. Ethnopharmacol.* **2003,** *84*, 57–78.
Haq, S. M.; Hamid, M.; Lone, F. A.; Singh, B. Himalayan Hotspot with Alien Weeds: A Case Study of Biological Spectrum, Phenology and Diversity of Weedy Plants of High Altitude Mountains in Remote District Kupwara of J&K Himalaya, India. *Proc. Natl. Acad. Sci. India, Sect. B. Biol. Sci.* **2021,** *91*, 139–152. https://doi.org/10.1007/s40011–020–01219–6.
Harley, R. M.; Atkins, S.; Budantsev, A.; Cantino, P. D.; Conn, B.; Grayer, R. J.; Harley, M. M.; Kok, R. D.; Krestovskaja, T.; Morales, A.; Paton, A. J.; Ryding, O.; Upson, T. *Labiatae*; Springer: Berlin, 2004.
Hossan, M. S.; Hanif, A.; Khan, M.; Bari, S.; Jahan, R.; Rahmatullah, M. Ethnobotanical Survey of the Tripura Tribe of Bangladesh. *Am. Eurasian J. Sustain. Agric.* **2009,** *3* (2), 253–261.
Hou, D. *Flora of Taiwan*, Vol. 4; Editorial Committee of the Flora of Taiwan: Taipei, Taiwan, 1996; p 442.
Hsieh, T. H.; Huang, T. C. Notes on the Flora of Taiwan (22)-the Genus Ajuga L. (Lamiaceae). *Taiwania* **1995,** *40* (2), 157–172.
Israili, Z. H.; Lyoussi, B. Ethnopharmacology of the Plants of Genus *Ajuga. Pak. J. Pharm. Sci.* **2009,** *22* (4), 425–462.
Jbilou, R.; Amri, H.; Bouayad, N.; Ghailani, N.; Ennabili, S.; Sayah, F. Insecticidal Effects of Extracts of Seven Plant Species on Larval Development, α-Amylase Activity and Offspring

Production of *Tribolium castaneum* (Herbst) (Insecta: Coleoptera: Tenebrionidae). *Bioresour. Technol.* **2008,** *99*, 959–964.

Jbilou, R.; Ennabili, A.; Sayah, F. Insecticidal Activity of Four Medicinal Plant Extracts Against *Tribolium castaneum* (Herbst) (Coleoptera: Tenebrionidae). *Afr. J. Biotechnol.* **2006,** *5*, 936–940.

Johnson, T. *CRC Ethnobotany Desk Reference*; CRC Press: Boca Raton, FL, 1999.

Jouad, H.; Haloui, M.; Rhiouani, H.; El-Hilaly, J.; Eddouks, M. Ethnobotanical Survey of Medicinal Plants Used for the Treatment of Diabetes, Cardiac and Renal Diseases in the North Centre Region of Morocco (Fez- Boulemane). *J. Ethnopharmacol.* **2001,** *77*, 175–182.

Kareru, P. G.; Kenji, G. M.; Gachanja, A. N.; Keriko, J. M.; Mungai, G. Traditional Medicines Among the Embu and Mbeere Peoples of Kenya. *Afr. J. Trad. Complem. Altern. Med.* **2007,** *4*, 75–86.

Kariba, R. M. Antifungal Activity of *Ajuga remota*. *Fitoterapia* **2001,** *72*, 177–178.

Khalil, E. A.; Afifi, F. U.; Al-Hussaini, M. Evaluation of the Wound Healing Effect of Some Jordanian Traditional Medicinal Plants Formulated in Pluronic F127 Using Mice (*Mus musculus*). *J. Ethnopharmacol.* **2007,** *109*, 104–112.

Khan, P. M.; Ahmad, S.; Nawaz, H. R.; Ullah, N.; Malik, A. New Acetylated Quinols from *Ajuga parviflora*. *Fitoterapia* **1999,** *70*, 229–232.

Khanavi, M.; Davoodipoor, A. M.; Sadati, S. N.; Ardekani, M. R.S; Sharifzadeh, M. Antinociceptive Effect of Some Extracts from *Ajuga chamaecistus* Ging. ssp. *tomentella* (Boiss.) Rech.f. Aerial Parts. *Daru J. Pharm. Sci.* **2014,** *22*, 56. https://doi.org/10.1186/2008–2231–22–56.

Kim, S. Y.; Moon, S. H.; Kim, J. S.; Kim, J. H.; Lee, B. Y. First Record of *Ajuga nipponensis* Makino (Lamiaceae) from Korea. *Korean J. Pl. Taxon.* **2013,** *43* (3), 165–167.

Kökdil, G.; Topçu, G.; Gören, A. C.; Voelter, W. Steroids and Terpenoids from *Ajuga relicta*. *Z. Nat. Forsch., B. J. Chem. Sci.* **2002,** *57*, 957–960.

Konoshima, T.; M. Takasaki, H.; Tokuda, H. Nishino. Cancer Chemopreventive Activity of an Iridoid Glycoside, 8-Acetylharpagide, from *Ajuga decumbens*. *Cancer Lett.* **2000,** *157*, 87–92.

Kosovskii, M. I.; Syrov, V. N.; Mirakhmedov, M. M.; Katkova, S. P.; Khushbaktova, Z. A. The Effect of Nerobol and Ecdysterone on Insulin-Dependent Processes Linked Normally and in Insulin Resistance. *Probl. Endokrinol.* **1989,** *35*, 77–81.

Kubo, I.; Klocke, J. A.; Asano, S. Insect Ecdysis Inhibitors from the East African Medicinal Plant *Ajuga remota* (Labiatae). *Agric. Biol. Chem.* **1981,** *45*, 1925–1927.

Kuria, K. A. M.; Chepkwony, H.; Govaerts, C.; Roets, E.; Busson, R.; deWitte, P.; Zupko, I.; Hoornaert, G.; Quirynen, L.; Maes, L.; Janssens, L.; Hoogmartens, J.; Laekeman, G. The Antiplasmodial Activity of Isolates from *Ajuga remota*. *J. Nat. Prod.* **2002,** *65*, 789–793.

Kutepova, T. A.; Syrov, V. N.; Khushbaktova, Z. A.; Saatov, Z. Hypoglycemic Activity of the Total Ecdysteroid Extract from *Ajuga turkestanica*. *Pharm. Chem. J.* **2001,** *35*, 608–609.

Li, J. X.; Xin, D.; Li, H.; Lu, J. F.; Tong, C. W.; Gao, J. N.; Chan, K. M. Effect of Verbascoside Decreasing Concentration of Oxygen Free Radicals and Lipid Peroxidation in Skeletal Muscle. *Acta Pharmacol. Sin.* **1999,** *20*, 126–130.

Li, X. W.; Hedge, I. C. *Ajuga* L. In *Flora of China*; Wu, Z. Y., Raven, P. H., Hong, D. Y., Eds., Vol. 17; Science Press and Missouri Botanical Garden Press: Beijing & St. Louis, 1994; pp 63–69.

Liao, S. C.; Chiu, T. F.; Chen, J. C.; Lin, C. C. *Ajuga nipponensis* Makino Poisoning. *Clin. Toxicol.*, **2005,** *43*, 583–585.

Ma, S. C.; Du, J.; But, P.P-H.; Deng, X-L.; Zhang, Y-W.; Ooi, V.E-C.; Xu, H-X.; Lee, S.H-S.; Lee, S. F. Antiviral Chinese Medicinal Herbs Against Respiratoty Syncytial Virus. *J. Ethnopharmacol.* **2002,** *79*, 205–211.

Malakov, P. Y.; Papanov, G. Y. Areptins A and B Two New *Neo*-Clerodane Diterpenoids from *Ajuga reptans*. *Phytochemistry* **1998,** *49*, 2443–2447.

Mamatkhanov, A. U.; Yakubova, M. R.; Syrov, V. N. Isolation of Turkesterone from the Epigeal Part of *A. turkestanica* and Its Anabolic Activity. *Chem. Nat. Compd. (Khim Prir Soedin)*, **1998,** *34*, 150–154.

Manguro, L. O.; Wagai, S. O.; Lemmen, P.. Flavonol and Iridoid Glycosides of *Ajuga remota* Aerial Parts. *Phytochemistry* **2006,** *67*, 830–837.

Marc, E. B.; Nelly, A.; Annick, D. D.; Frederic, D. Plants Used as Remedies Antirheumatic and Antineuralgic in the Traditional Medicine of Lebanon. *J. Ethnopharmacol.* **2008,** *120*, 315–334.

Matu, E. N.; vanStaden, J. Antibacterial and Anti- Inflammatory Activities of Some Plants Used for Medicinal Purposes in Kenya. *J. Ethnopharmacol.* **2003,** *87*, 35–41.

Muregi, F. W.; Ishih, A.; Miyase, T.; Suzuki, T.; Kino, H.; Amano, T.; Mkoji, G. M.; Terada, M. Antimalarial Activity of Methanolic Extracts from Plants Used in Kenyan Ethnomedicine and Their Interactions with Chloroquine (CQ) Against a CQ-Tolerant Rodent Parasite, in Micc. *J. Ethnopharmacol.* **2007,** *111*, 190–195.

Nawaz, H. R.; Malik, A.; Khan, P. M.; Ahmed, S. Ajugin E and F: Two Withanolides from *Ajuga parviflora*. *Phytochemistry* **1999,** *52*, 1357–1360.

Odek-Ogunde, M.; Rajab, M. S. Antihypertensive effect of the Clerodane Diterpene Ajugarin I on Experimentally Hypertensive Rats. *East Afr. Med. J.* **1994,** *71*, 587–590.

Odek-Ogunde, M.; Rajab, M. S.; Migwi, G. J.; Ndegwa, J. M. Blood Pressure Responses to an Extract of *Ajuga remota* in Experimentally Hypertensive Rats. [Letter]. *Planta Med.* **1993,** *59*, 573–574.

Ono, Y.; Fukaya, Y.; Imai, S.; Yamakuni, T. Beneficial Effects of *Ajuga decumbens* on Osteoporosis and Arthritis. *Biol. Pharm. Bull.* **2008,** *31*, 1199–1204.

POWO. *Plants of the World Online.* Facilitated by the Royal Botanic Gardens, Kew, 2021. http://www.plantsoftheworldonline.org/.

Qing, X.; Yan, H. N.; Ni, Z. Y.; Vavricka, C. J.; Zhang, M. L.; Shi, Q. W.; Gu, Y. C.; Kiyota, H. Chemical and Pharmacological Research on the Plants from Genus *Ajuga* Chemical and Pharmacological Research on the Plants from Genus *Ajuga*. *Heterocycl. Commun.* **2017,** *23* (4), 245–268.

Riaz, N.; Malik, A.; Aziz-ur-Rehma;, Nawaz, S. A.; Muhammad, P.; Choudhary, M. I. Cholinesterase-Inhibiting Withanolides from *Ajuga bracteosa*. *Chem. Biodivers.* **2004,** *1*, 1289–1295.

Riaz, N.; Nawaz, S. A.; Mukhtar, N.; Malik, A.; Afza, N.; Ali, S; Ullah, S.; Muhammad, P.; Choudhary, M. I. Isolation and Enzyme-Inhibition Studies of the Chemical Constituents from *Ajuga bracteosa*. *Chem. Biodivers.* **2007,** *4*, 72–83.

Rodriguez-Hahn, L.; Esquivel, B.; Cárdenas, J. Clerodane Diterpenoids in Labiatae. *Prog. Chem. Nat. Prod.* **1994,** *63*, 107–196.

Shafi, N.; Khan, G. A.; Arfan, M.; Ahmad, K. U.; Gilani, N. D. Isolation and Pharmacological Screening of 8- O-Acetyl Harpagide from *Ajuga bracteosa* Wall. *Pak. J. Sci. Industr. Res.* **2004,** *47*, 176–179.

Sharma, P.; Mohan, L.; Srivastava, C. N. Larval Susceptibility of *Ajuga remota* Against Anopheline and Culicine Mosquitos. *Southeast Asian J. Trop. Med. Public Health.* **2004,** *35*, 608–610.

Shen, X.; Isogai, A.; Furihata, K.; Sun, H.; Suzuki, A. Two *Neo*-Clerodane Diterpenoids from *Ajuga macrosperma. Phytochemistry* **1993,** *33*, 887–889.

Shimomura, H.; Sashida, Y.; Ogawa, K. Iridoid Glucosides and Phenylpropanoid Glycosides in *Ajuga* Species of Japan. *Phytochemistry* **1987,** *26*, 1981–1983.

Shimomura, H.; Sashida, Y.; Ogawa, K.; Iitaka, Y. Ajugamarin, a New Bitter Diterpene from *Ajuga nipponensis* Makino. *Tetrahedron Lett.* **1981,** *22*, 1367–1369.

Singh, B. *Botanical Leads for Drug Discovery*; Springer Nature Pvt. Ltd: Singapore, 2020.

Singh, B.; Bedi Y. S. Eating from Raw Wild Plants in Himalaya: Traditional Knowledge Documentary on Sheena Tribe Along LoC Border in Kashmir. *Indian J. Nat. Products Resour.* **2017,** *8* (3), 269–275.

Singh, B.; Sharma, Y. P. *Plants for Novel Drug Molecule-Ethnobotany to Ethnopharmacology*; New India Publishing Agency (NIPA): New Delhi, India, 2021.

Singh, B.; Singh, B.; Kishor, A.; Singh, S.; Bhat, M. N.; Surmal, O.; Musarella, C. M. Exploring Plant-Based Ethnomedicine and Quantitative Ethnopharmacology in Protected Area: Ethnobotanical Study of Medicinal Plants Utilized by Population of Jasrota Hill in Western Himalaya, India. *Sustainability* **2020,** *12* (18), 7526. https://doi.org/10.3390/su12187526

Singh, B.; Singh, S.; Kishor, A.; Singh, B. Traditional Usage of Medicinal Plants in Humans and Animals Health Care and Their Chemical Constituents from Hills and Valleys of Jammu province, Western Himalaya. *Indian J. Nat. Prod. Resour.* **2021,** *22* (1), 84–100.

Singh, B.; Sultan, P.; Hassan, Q. P.; Gairola, S.; Bedi, Y. S. Ethnobotany, Traditional Knowledge and Diversity of Wild Edible Plants and Fungi: A Case Study in the Bandipora District of Kashmir Himalaya, India. *J. Herbs Spices Med. Pl.* **2016,** *22* (3), 247–248. https://doi.org/10.1080/10496475.2016.1193833.

Stuart, M. (ed.) *The Encyclopedia of Herbs and Herbalism*; Orbis Publishing: London, 1987; p 304.

Sweeney, A. P.; Wyllie, S. G.; Shalliker, R. A.; Markham, J. L. Xanthine Oxidase Inhibitory Activity of Selected Australian Native Plants. *J. Ethnopharmacol.* **2001,** *75*, 273–277.

Syrov, V. N.; Khushbaktova, Z. A. Experimental Study of Pharmacotherapeutic Effect of Phytoecdysteroids and Nerobol in Toxic Liver Damage. [Russian]. *Eksp. Klin. Farmakol.* **2001,** *64*, 56–58.

Syrov, V. N.; Khushbaktova, Z. A.; Abzalova, M. Kh.; Sultanov, M. B. Hypolipidemic and Antiatherosclerotic Properties of Phytoecdysteroids. *Doklady Akademii Nauk UzSSR.* **1983,** *9*, 44–45.

Syrov, V. N.; Khushbaktova, Z. A.; Nabiev, A. N. An Experimental Study of the Hepatoprotective Properties of Phytoecdysteroids and Nerobol in Carbon Tetrachloride-Induced Liver Lesion. [Russian]. *Eksp Klin Farmakol.*, **1992a,** *55*, 61–65.

Syrov, V. N.; Tashmukhamedova, M. A.; Khushbaktova, Z. A.; Mirtalipov, D. T.; Mamatkhanov, A. U. Effect of Phytoecdysteroids and Nerobol on Parameters of Carbohydrate and Lipid Metabolism and Phospholipid Spectrum of Liver Mitochondrial Membrane in Experimental Diabetes Mellitus of Rats. *Ukr. Biokhim. Zh.* **1992b,** *64*, 61–67.

Syrov, V. N.; Khushbaktova, Z. A.; Tolibaev, I.; Eletskaya, N. V.; Mamatkhanov, A. D. Effect of a Lipid Component from the Above Ground Portion of *Ajuga turkestanica* on the

Metabolic Process and Dynamics of Healing Skin Wounds Experimentally. *Pharm. Chem. J.* **1994,** *28*, 837–840.

Syrov, V. N.; Kurmukov, A. G.; Sakhibov, A. D. Effect of Turkesterone and Nerobol on the Activity of the Protein Synthesizing System of Mouse Liver [Russian]. *Vopr. Med. Khim.* **1978,** *24*, 456–460.

Syrov, V. N.; Nabiev, A. N.; Khushbaktova, Z. A.; Zakhidov, U. V.; Maksudov, M. S.; Saatov, Z. Hepatoprotector Activity of Iridoid Glycosides with Respect to Heliotrine-Induced Acute Toxic Liver Damage in Mice. *Pharm. Chem. J.* **1998,** *33* (8), 410–412.

Syrov, V. N.; Nabiev, A. N.; Sultanov, M. B. Action of Phytoecdysteroids on the bile-Secretory Function of the Normal Liver and in Experimental Hepatitis. [Russian]. *Farmakol. Toksikol.* **1986,** *49*, 100–103.

Syrov, V. N.; Nasyrova, S. S.; Khushbaktova, Z. A. The Results of Experimental Study of Phytoecdysteroids as Erythropoiesis Stimulators in Laboratory Animals. [Russian] *Eksp. Klin. Farmakol.* **1997,** *60*, 41–44.

Syrov, V. N.; Saatov, Z.; Sagdullaev, Sh.Sh.; Mamatkhanov, A. U. Study of the Structure—Anabolic Activity Relationship for Phytoecdysteroids Extracted from Some Plants of Central Asia. *Pharmaceut. Chem. J.* **2001,** *35*, 667–671.

Tahraoui, A.; El-Hilaly, J.; Israili, Z. H.; Lyoussi, B. Ethnopharmacological Survey of Plants Used in the Traditional Treatment of Hypertension and Diabetes in South-Eastern Morocco (Errachidia Province). *J. Ethnopharmacol.* **2007,** *110*, 105–117.

Takasaki, M.; Tokuda, H.; Nishino, H.; Konoshima, T. Cancer Chemopreventive Agents (Antitumor- Promoters) from *Ajuga decumbens. J. Nat. Prod.* **1999,** *62*, 972–975.

Takasaki, M.; Yamauchi, I.; Haruna, M.; Konoshima, T. New Glycosides from *Ajuga decumbens. J. Nat. Prod.***1998,** *61*, 1105–1109.

Tashmukhamedova, M. A.; Almatov, K. T.; Khushbaktova, Z. A.; Syrov, V. N. Effect of Phytoecdysteroids and Steranobols on the Activity and Stability of Membrane- Bound Enzymes of Liver Mitochondria in Experimental Hepatitis. *Vopr. Med. Khim.***1986a,** *32*, 81–84.

Tashmukhamedova, M. A.; Almatov, K. T.; Syrov, V. N.; Sultanov, M. B.; Abidov, A. A. Effect of Phytoecdisteroids and Anabolic Steroids on Liver Mitochondrial Respiration and Oxidative Phosphorylation in Alloxan Diabetic Rats. *Nauchnye. Doki. Vyss. Shkoly. Biol. Nauki.* **1985,** *9*, 37–39.

Tashmukhamedova, M. A.; Almatov, K. T.; Syrov, V. N.; Sultanov, M. B.; Abidov, A. A. Comparative Study of the Effect of Ecdysterone, Turkesterone and Nerobol on the Function of Rat Liver Mitochondria in Experimental Diabetes. *Vopr. Med. Khim.*, **1986b,** *32*, 24–28.

Terahara, N.; Callebaut, A.; Ohba, R.; Nagata, T.; Ohnishi- Kameyama, M.; Suzuki, M. Acetylated Anthocyanidin 3-sophoroside-5-Glucosides from *Ajuga reptans* Flowers and the Corresponding Cell Cultures. *Photochemistry* **2001,** *58*, 493–500.

Tsuji, K.; Shibata, J.; Okada, M.; Inaoka, Y. Blood Flow Amount-Improving Agent Comprising Steroid Derivative and Cosmetic Using Same. US Patent # 5976515 to Pola Chemical Industries, Inc., 1999.

Ulukanli, Z.; Ulukanli, S.; Ozbay, H.; Ilcim, A.; Tuzcu, M. Antimicrobial Activities of Some Plants from the Eastern Anatolia Region of Turkey. *Pharm. Biol.* **2005,** *43*, 334–339.

Vanyolos, A.; Simon, A.; Toth, G.; Polgar, L.; Kele, Z.; Ilku, A.; Matyus, P.; Bathori, M. C-29 Ecdysteroids from *Ajuga reptans* var. *reptans*. *J. Nat. Prod.* **2009,** *72*, 929–932.

Xiaoyu, S.; Isogai, A.; Furihata, K.; Handong, S.; Suzuki, A. Neo-Clerodane Diterpenoids from *Ajuga macrosperma* and *Ajuga pantantha*. *Phytochemistry* **1993,** *34* (4), 1091–1094.

Yineger, H.; Kelbessa, E.; Bekele, T.; Lulekal, E. Plants Used in Traditional Management of Human Ailments at Bale Mountains National Park, Southeastern Ethiopia. *J. Med. Plants Res.* **2008,** *2*, 132–153.

Yousaf, T.; Rafique, S.; Wahid, F.; Rehman, S.; Nazir, A.; Rafique, J.; Aslam, K.; Shabir, G.; Shah, S. M. Phytochemical Profiling and Antiviral Activity of *Ajuga bracteosa*, *Ajuga parviflora*, *Berberis lycium* and *Citrus lemon* Against Hepatitis C Virus. *Microb. Pathog.* **2018,** *118*, 154–158.

Yunnan Institute of Botany, Ed. *Flora Yunnan Ica*; Thomas Wall Science Press: Peking, 1977.

Ziyyat, A.; Legssyer, A.; Mekhfi, H.; Dassouli, A.; Serhrouchni, M.; Benjelloun, W. Phytotherapy of Hypertension and Diabetes in Oriental Morocco. *J. Ethnopharmacol.* **1977,** *58*, 45–54.

CHAPTER 2

A SYSTEMATIC REVIEW ON ETHNOPHARMACOLOGY, PHYTOCHEMICAL, AND PHARMACOLOGICAL STUDIES ON *AJUGA INTEGRIFOLIA* BUCH-HAM. (SYN.: *AJUGA BRACTEOSA* WALL. EX BENTH.)

ANCHALA GUGLANI[1,2], H. K. PANDEY[1], RAJESHWAR K.K. ARYA[2], and G. BALAKRISHNA[1]

[1]*Defence Institute of Bio-Energy Research (DRDO) Field Station, Pithoragarh, Uttarakhand, India*

[2]*Department of Pharmaceutical Sciences, Sir J.C. Bose Technical Campus, Bhimtal, Kumaun University, Uttarakhand, India*

ABSTRACT

The Himalayan herb, *Ajuga integrifolia*, is extensively used for traditional and Ayurvedic medicine due to its tremendous medicinal potential towards various diseases for centuries. *A. integrifolia* is an important source of medicine with antidiabetic, anti-inflammatory, antioxidant, antiarthritic, antimicrobial, anti-spasmodic, immunomodulatory, antitumor, and antimalarial properties. In the present chapter traditional uses, phytochemistry and pharmacological properties of *Ajuga integrifolia* are reviewed.

2.1 INTRODUCTION

Ajuga integrifolia Buch.-Ham. (Syn.: *Ajuga bracteosa* Wall. ex Benth.) (Ratpatiya) belongs to the family Lamiaceae. The genus *Ajuga* is known for its medicinal

Bioactives and Pharmacology of Lamiaceae. T. Pullaiah, PhD (Ed.)

and economical values and is mostly used for the treatment of fever, rheumatism, analgesia, hypertension, amenorrhea, inflammation, sclerosis, etc. (Luan et al., 2019). *Ajuga integrifolia* is commonly known as Neelkanthi, Ratpatiya, Bungle, and Kori-butti. It is native to subtropical and temperate regions of Asia, Europe, North America, Africa, and Australia. In India, the plant is distributed in the upper Gangetic plans at 1300–2000 m altitude and abundant in the temperate regions of Kashmir, western Himalayas and the plains of Punjab (Hussain et al., 2016; Khare, 2007; Chandel and Bagai, 2010; Chauhan, 1999; Upadhyay et al., 2012). *A. integrifolia* is indexed as endangered species based on its conservation status (Hussain et al., 2016; Ahmad et al., 2012). It is a small herb with erect or ascending stem covered with soft hair, growing up to 1–6 cm tall. Leaves are oblanceolate, opposite, or sub-spatulate, 3.0–12 cm long. The papery leaf blade is having dentate to incised margin. The hermaphrodite flowers are white or purplish tinged and axillary whorls in spike. It flowers from March to June and grows in sandy loamy soil (Luan et al., 2019; Hussain et al., 2016; Deka and Deka, 2015).

The Himalayan herb is extensively used for traditional and Ayurvedic medicine due to its tremendous medicinal potential toward various diseases for centuries. The ethnomedicinal importance of the plant promotes its excessive use, which posed consequential threat regarding its extinction. The investigations on the ethnobotanical, phytochemical, and ethnopharmacological aspects of the plant are gaining interest in recent years due to their medicinal properties. According to the extensive phytochemical and pharmacological studies, *A. integrifolia* is an important source of medicine with antidiabetic, anti-inflammatory, antioxidant, antiarthritic, antimicrobial, and cytotoxic properties and their effective application in food and pharmaceutical industries (Hussain et al., 2016; Pala et al., 2011).

2.2 ETHNOPHARMACOLOGY AND TRADITIONAL USES

Traditionally, the genus *Ajuga* has been utilized as an herbal medicine for the treatment of fever, dysentery, diabetes, toothache, high blood pressure, anthelmintic, malaria, diuretic, and gastrointestinal disorders (Israili and Lyoussi, 2009). The aromatic leaves of *A. integrifolia* have traditionally been used as an astringent, febrifuge, diuretic, stimulant, and aperient (Kumar et al., 2010). It also has various ethnopharmacological effects as its leaves have been used to cure malaria, diabetes, and digestion-related problems (Barkatullah et al., 2015; Ahmad et al., 2012). Traditionally, the leaves' extracts have been used to detoxify the blood and for acne, ear infections, hypertension, constipation, headache, measles, jaundice, stomach hyperacidity, and sore throats (Ibrar et al., 2007; Barkatullah and Hussain, 2009; Qureshi et al., 2009). Moreover, the root extracts have also been used for digestive problems (Ziyyat et al., 1997). The plant is also used traditionally for the cure of asthma, fever, jaundice (Singh et al., 2006), wounds, arthritis, cancer (Khalil et al., 2007), eye irritation, stomach pain, and liver problems (El-Beyrouthy et al., 2008). In other countries, the whole plant has been used to cure various inflammatory disorders like, pneumonia, hepatitis, and bone diseases (Hsieh and Huang, 1995). The root juice is used in diarrhea and dysentery conditions. The leaves are used as a substitute for quinine for the treatment of fever. The leaves' decoction in combination with honey and ginger juice is used to cure high fever and respiratory congestion (Hsieh et al., 2011). In Ayurveda, the plant is reported for the treatment of gout, rheumatism, amenorrhea, and palsy (Kaithwas et al., 2012). In India, the leaves, barks, and flowers' decoction are reported to cure cancer, diabetes, malaria, and inflammation (Chandel and Bagai, 2011; Nisar et al., 2014; Pal et al., 2014). *A. integrifolia* is utilized for various applications as medicine since ancient times. The plant is reported in ethno-medicine as astringent, antibacterial, anti-inflammatory, hypoglycemic, anthelmintic, and antifungal (Israili and Lyoussi, 2009). The use of plant is also reported in the traditional Chinese medicine for the detoxification of the body, relieving rheumatic pain, alleviating sore throat, reducing blood pressure, improving eyesight, treating tuberculosis, phlegm, dyspepsia, and fever (China Pharmacopoeia Commission, 2015).

2.3 PHYTOCHEMISTRY

The genus *Ajuga* possesses various phytoconstituents. Various active constituents were isolated and characterized by the researchers in this plant and then they evaluated them for a variety of biological activities in a systematic manner. The genus *Ajuga* showed the presence of majority of the phytoecdysteroids, flavonoids, neo-clerodane diterpenes, and glycosides-related compounds, which have been extracted and characterized from the whole plants or aerial parts or leaves (Luan et al., 2019). Many studies on the chemical composition of *A. integrifolia* revealed the presence of alkaloids, tannins, glycosides, flavonoids, triterpenoids, steroids, saponins, phenolic compounds, α-sitosterol, β-sitosterol, ceryl alcohol, cerotic acid, oleic acid, palmitic acid, and linoleic acid (Chauhan, 1999; Chopra et al., 1998; Rastogi and Mehrotra, 1993). Several therapeutically important compounds have been isolated from the plant and these include 20-hydroxyecdysone (20-HE) (Kayani et al., 2014), Ajuganane, 7-hydroxy-3,6,3′,4′-tetramethoxyflavone, 3,4′-dihydroxy-3,6,7 trimethoxyflavone, ursolic acid (Hussain et al., 2012), ajubractins A-E,clerodin, ajugapitin, 3-epi-caryoptin,3-epi-14,15-dihydro-caryoptin, ivain II,14,15-dihydroclerodin, 14,15-dihydro ajugapitin (Castro et al., 2011), tetracosanoic acid, and hexacosanol (Bhakuni et al., 1987). The dichloromethane extract fraction showed the presence of two diterpenes compound, i.e., clerodin and dihydroclerodin (Hussain et al., 2012). Two new sphingolipids (Bractin A and B) were also isolated from the plant along with the bractic acid and other diterpenoids (Riaz et al., 2007). Three new withanolides bracteosin A-C, Dihydroclerodin-1, dihydroajugapitin, lupulin A, and clerodinin A were also isolated from the whole plant (Riaz et al., 2004). The withanolides, a group of 28 steroidal lactones, is rearranged as an ergostane framework. These compounds are responsible for various biological activities like anti-inflammatory, hepatoprotective, antimicrobial, immunomodulatory, antitumor, and cytotoxic (Upadhyay et al., 2012). The hexane extract yields 1, 2-benzenedicarboxylic acid bis (2S-methyl heptyl) ester, a new phthalic acid ester; furthermore, the chloroform and methanolic extracts yielded two iridoid glycosides, 8-O-acetyl harpagide and reptoside, and neo-clerodanediterpene ajugarin-I (Singh et al., 2006). The other compounds isolated were bracteonin-A, linalylacetate (perfumery compound), 14-hydro-15-hydroxy ajugapitin, 14,15-dihydroajugapitin, and stigmasterol (Verma et al., 2002; Deka and Deka, 2015). A new compound triacontanyl docosanoate is isolated and characterized from the aerial parts of the plant. The chemical composition of the essential oil isolated from the aerial parts of *A. integrifolia* yields borneol, hexadecanoic acid, 9,12-octadecadienoic acid, germacrene-D, β-cubeben, α-pinene, β-caryophyllene, camphene, γ-terpinene, limonene, linalool, carvone, carvacrol, eugenol, and α-humulene (Ramzi et al., 2012). Some other bioactive compounds characterized from different parts of the plant are ecdysone, 20- hydroxy-ecdysterone, cyasterone, 3-Epicyasterone, 3-epi-22-acetylcyasterone, poly-podine, ajugalactone, 29-norcyasterone, ponosterone, 29-norsengosterone, ajugasterone C, etc. (Vohra and Kaur, 2011). The structures of important isolated compounds from *A. integrifolia* are given below.

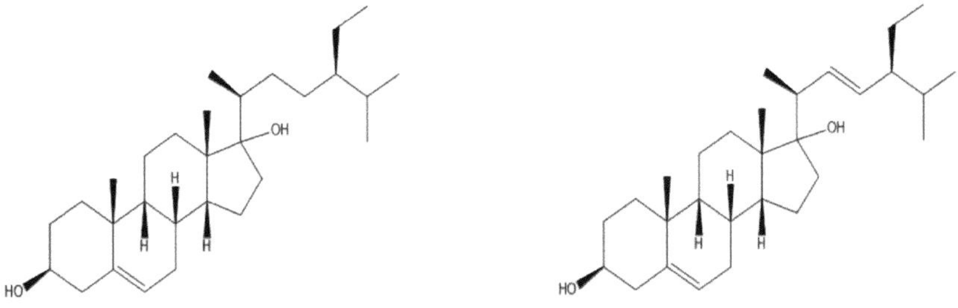

β-sitosterol Stigmasterol

FIGURE 2.1 Phytosterols isolated from *A. integrifolia*.

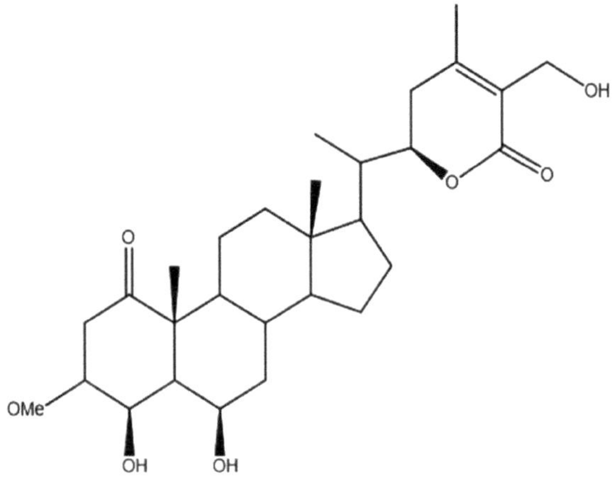

FIGURE 2.2 Bracteosin C.

2.4 PHARMACOLOGY

A. integrifolia possesses various pharmacological activities that were known traditionally and have been validated and documented by various researchers through the modern scientific procedures. Some of the important pharmacological activities of the *A. integrifolia* are asserted as follows.

2.4.1 ANTI-INFLAMMATORY ACTIVITY

The ethanolic extract of *A. integrifolia* showed a significant anti-inflammatory activity using 12-O-tetradecanoyl phorbol-13 acetate (TPA)-induced ear edema at 25 and 50 μg/mL concentrations. The effect is produced by the inhibition of COX-1 and COX-2. The study also revealed that the isolated compounds withaferin A, lupulin A, Aajugarin I, 6-deoxyharpagide, and reptoside exhibit anti-inflammatory activity. The study showed that the ethanolic extract possesses anti-inflammatory activity in a dose-dependent manner and validates the traditional use of A. *integrifolia* in the treatment of inflammatory diseases (Gautam et al., 2011).

The methanolic extract of *A. integrifolia* exhibited anti-inflammatory activity at different doses against carrageenan-induced paw edema and egg albumin-induced inflammation in rats and *in vitro* membrane stabilization method. The extracts showed a potent anti-inflammatory action and therefore supported its traditional use (Singh et al., 2012).

2.4.2 ANALGESIC ACTIVITY

The chloroform and water extracts of the aerial part of *A. integrifolia* showed a significant analgesic activity in a dose-dependent manner using the acetic acid-induced writhing test and the tail immersion test. The results indicated that the *A. integrifolia* extracts showed an opioid receptors mediated analgesic effect. The study confirms the traditional use in the treatment of painful conditions (Pal and Pawar, 2011).

2.4.3 ANTIARTHRITIC ACTIVITY

The ethanolic extract of *A. integrifolia* showed a significant dose-dependent inhibitory effect against complete Freund's adjuvant (CFA)-induced chronic immunological arthritis and turpentine oil-formaldehyde-induced acute nonimmunological arthritis in rats. The study supported the traditional claim of *A. integrifolia* for rheumatism and other inflammatory diseases (Kaithwas et al., 2012).

2.4.4 ANTIDIABETIC ACTIVITY

The hydroalcoholic leaves' extract of *A. integrifolia* showed a significant hypoglycemic activity using streptozotocin-induced diabetic rats. The study showed that the hydroalcoholic leaves' extract significantly reduces the blood glucose levels. Therefore, the findings suggested that the hydroalcoholic extract is a potent hypoglycemic agent (Singh et al., 2016).

The hydroalcoholic leaves' extract of *A. integrifolia* showed a significant antidiabetic potential using alloxan-induced diabetes in chicks. The study revealed that the leaves' extract showed significant reduction in blood glucose levels in diabetic chicks (Kumar and Bisht, 2017).

2.4.5 ANTIOXIDANT ACTIVITY

The *A. integrifolia* exhibited the antioxidant potential. The ethanol extract showed a significant free radical scavenging activity with an IC_{50} value of 14.40 μg/mL (Zahra et al., 2017). In addition, the methanolic roots' extract of *A. integrifolia* also possesses a significant free radical scavenging activity by the DPPH assay with an IC_{50} value of 36.90 μg/mL (Kayani et al., 2016).

2.4.6 ANTIMICROBIAL ACTIVITY

The ethanolic leaves' extract of *A. integrifolia* showed chemosuppression in a dose-dependent manner at dose levels of 250, 500, and 750 mg/kg/day during early stage, besides significant repository activity. The ethanolic extract showed maximum chemosuppression (77.7%) and (68.8%) during early and in established infections at 750 mg/kg/day dose, respectively. The ethanolic extract also showed a significant repository activity with an ED_{50} of 300 mg/kg body weight (Chandel and Bagai, 2010).

The different extracts of *A. integrifolia* roots, leaves, and bark showed significant antibacterial activity against various bacterial strains. The active constituents, withaferin A, and its compound from the plant are reported for their antibacterial activity against the gram-positive bacteria. The finding revealed that the active compound, 20-deoxywithanolide-D, has possessed the maximum antibacterial activity among other compounds (Pala et al., 2011).

2.4.7 IMMUNOMODULATORY ACTIVITY

The active constituent of *A. integrifolia*, that is withanolides, has been reported to possess immunomodulatory activity. The withaferin-A inhibited the graft versus host reaction in chicks and adjuvant arthritis in rats, which revealed the immunosuppressive property of the active compound. The proliferation of murine spleen cell culture is also inhibited by the withaferin A. The glucowithanolides, sitoindoside have been reported for their adaptogenic and immunostimulating properties (Pala et al., 2011).

2.4.8 ANTISPASMODIC ACTIVITY

Spasms are an episodic or colic pain in a particular organ due to spasms of smooth muscle. The active constituent of *A. integrifolia*, that is 8-O-acetylharpagide, has been reported for their significant antispasmodic activity (Pala et al., 2011).

2.4.9 CARDIO TONIC ACTIVITY

The alkaloidal fraction of *A. integrifolia* is reported for the cardiostimulant effects on the frog heart and rat ventricle. Dichloroisoprenaline antagonized the effects, which may result from the release of catecholamine stores in the heart and did not occur in reserpine-treated animals (Patel et al., 1962).

2.4.10 ANTITUMOR ACTIVITY

The methanolic, aqueous, and petroleum ether extracts of *A. integrifolia* have been evaluated for the antitumor activity against HepG2 and MCF-7 tumor cell lines by the thiazolyl blue test (MTT). The study revealed that the methanolic extracts exhibited a significant cytotoxic activity against HepG2 and MCF-7 tumor cell lines. The IC_{50} values for ethanolic, aqueous, and petroleum ether extracts were 5, 15, 30 μg/mL and 10, 70, 65 μg/mL on Hep-2 and MCF-7 cells at 24 h, respectively (Pal et al., 2014).

The aqueous extract of *A. integrifolia* significantly inhibited the growth of nasopharyngeal (Hone-1) and pharyngeal (Detroit 562) cancer cell lines at doses of 0, 1.56, 3.13, and 6.25 mg/mL and is also reported to induce cancer cell death (Lin et al., 2017).

2.4.11 ANTIMALARIAL ACTIVITY

A study by Chandel and Bagai (2011) demonstrated the antiplasmodial efficacy of the traditionally important plant *A. integrifolia.* The extract possesses a significant in-vitro antiplasmodial activity with an IC_{50} value of 10.0 μg/mL. The study was conducted to meet out the requirement of the newer antiplasmodial drug due to high risk of resistance for classical antimalarial drugs.

KEYWORDS

- ***Ajuga integrifolia***
- **ethnopharmacology**
- **pharmacology**
- **phytochemistry**
- **ethnobotaany**

REFERENCES

Ahmad, K. S.; Kayani, W. K.; Hameed, M.; Ahmad, F.; Nawaz, T. Floristic Diversity and Ethnobotany of Senhsa, District Kotli, Azad Jammu & Kashmir (Pakistan). *Pak. J. Bot.* **2012,** *44*, 195–201.

Ali, T.; Shah, Z.; Bashir, R. A Review on Phytochemical and Ethnopharmacological Studies of *Ajuga bracteosa* Wall. ex Benth. *J. Drug Deliv. Therap.* **2019,** *9* (2), 489–492.

Barkatullah, M.; Hussain, F. Ethnobotanical Studies of Plants of Charkotli Hills, Batkhela District, Malakand, Pakistan. *Front. Biol.* **2009,** *4*, 539–548.

Barkatullah, M., Ibra, A.; Rauf, T.; Ben Hadda, M. S.; Patel, S. Quantitative Ethnobotanical Survey of Medicinal flora Thriving in Malakand Pass Hills, Khyber Pakhtunkhwa, Pakistan. *J. Ethnopharmacol.* **2015,** *169*, 335–346.

Bhakuni, R. S.; Shukla, Y. N.; Thakur, R. S. Chemical Constituents of *Ajuga bracteosa. Indian J. Pharm. Sci.* **1987,** *49* (6), 225–226.

Castro, A.; Coll, J.; Arfan, M. Neo-Clerodanediterpenoids from *Ajuga bracteosa. J. Nat. Prod.* **2011,** *74* (5), 1036–1041.

Chandel, S.; Bagai, U. Screening of Antiplasmodial Efficacy of *Ajuga bracteosa* Wall ex. Benth. *Parasitol Res.* **2011,** *108* (4), 801–805.

Chandel, S.; Bagai, U. Antiplasmodial Activity of *Ajuga bracteosa* Against *Plasmodium berghei* Infected BALB/c Mice. *Indian J. Med Res.* **2010,** *131*, 440–444.

Chauhan, N. S. *Medicinal and Aromatic Plants of Himachal Pradesh*; Indus Publishing Company: New Delhi, 1999.

China Pharmacopoeia Commission. *Chinese Pharmacopeia;* China Medical Science Press: Beijing, China, 2015; p 346.

Chopra, R. N.; Chopra, I. C.; Verma, B. S. *Supplement to Glossary of Indian Medicinal Plants*, Reprinted 2nd March; NISC, CSIR: New Delhi, 1998; p 4.

Deka, A.; Deka, P. A Review on *Ajuga bracteosa*, a Traditionally Important Medicinal Plant. *Int. J. Emerg. Res. Manag. Tech.* **2015,** *4* (12), 203–206.

El-Beyrouthy, M.; Arnold, N.; Delelis-Dusollier, A.; Dopont, F. Plants Used as Remedies Antirheumatic and Antineuralgic in the Traditional Medicine of Lebanon. *J. Ethnopharmacol.* **2008,** *120*, 315–334.

Gautam, R.; Jachak, S. M.; Saklani, A. Anti-Inflammatory Effect of *Ajuga bracteosa* Wall ex Benth. Mediated Through Cyclooxygenase (COX) Inhibition. *J. Ethnopharmacol.* **2011,** *133*, 928–30.

Hsieh, T. H.; Huang, T. C. Notes on the flora of Taiwan (22)-The Genus *Ajuga* L. (Lamiaceae). *Taiwania* **1995,** *40*, 157–172.

Hsieh, W. T.; Liu, Y. T.; Lin, W. C. Anti-Inflammatory Properties of *Ajuga bracteosa* In Vivo and In Vitro Study and Their Effects on Mouse Model of Liver Fibrosis. *J. Ethnopharmacol.* **2011,** *135* (1), 116–125.

Hussain, J.; Begum, N.; Hussain, H.; Khan, F. U.; Rehman, N. U.; Al-Harrasi, A.; Ali, L. Ajuganane: A New Phenolic Compound from *Ajuga bracteosa. Nat. Prod. Commun.* **2012,** *7* (5), 615–616.

Hussain, M.; Bibi, Y.; Raja, N. I.; Iqbal, M.; Aslam, S.; Tahir, N.; Imran, M.; Iftikhar, A. A. Review of Therapeutic Potential of *Ajuga bracteosa:* A Critically Endangered Plant from Himalayas. *J. Coast. Life Med.* **2016,** *4* (11), 918–924.

Ibrar, M.; Hussain, F.; Sultan, A. Ethnobotanical Studies on Plant Resources of Ranyal Hills, District Shangla, Pakistan. *Pak. J. Bot.* **2007,** *39*, 329–337.

Israili, Z. H.; Lyoussi, B. Ethnopharmacology of the Plants of Genus *Ajuga. Pak. J. Pharm. Sci.* **2009,** *22*, 425–462.

Kaithwas, G.; Gautam, R.; Jachak, S. M.; Saklani, A. Antiarthritic Effects of *Ajuga bracteosa* Wall ex Benth. in Acute and Chronic Models of Arthritis in Albino Rats. *Asian Pac. J. Trop. Biomed.* **2012,** *2* (3), 185–188.

Kayani, W. K.; Rani, R.; Ihsanul, H.; Mirza, B. Seasonal and Geographical Impact on the Morphology and 20- Hydroxyecdysone Content in Different Tissue Types of Wild *Ajuga bracteosa* Wall. ex Benth. *Steroids*. **2014,** *87*, 12–20.

Kayani, W. K.; Dilshad, E.; Ahmed, T.; Ismail, H.; Mirza, B. Evaluation of *Ajuga bracteosa* for Antioxidant, Anti-Inflammatory, Analgesic, Antidepressant and Anticoagulant Activities. *BMC Complement. Altern. Med.* **2016,** *16*, 375.

Khalil, E. A.; Afifi, F. U.; Al-Hussaini, M. Evaluation of the Wound Healing Effect of Some Jordanian Traditional Medicinal Plants Formulated in Pluronic F127 Using Mice (*Mus musculus*). *J. Ethnopharmacol.* **2007,** *109*, 104–112.

Khare, C. P. *Indian Medicinal Plants—An Illustrated Dictionary*, 1st Indian Reprint; Springer (India): New Delhi, India, 2007; p 28.

Kumar, D.; Bisht, R. Antidiabetic Potency of Leaves of *Ajuga bracteosa* on Hypoglycemic Activity in Alloxan Induced Diabetic Chicks. *Int. J. Adv. Res.* **2017,** *5* (12), 836–841.

Kumar, G.; Kanaujia, N.; Bafana, A. Functional and Phylogenetic Diversity of Root-Associated Bacteria of *Ajuga bracteosa* in Kangra Valley. *Microbiol. Res.* **2010,** *167*, 220–225.

Lin, Y. T.; Lin, C. C.; Wang, H. C.; Hsu, Y. C. Induction of Mitotic Delay in Pharyngeal and Nasopharyngeal Carcinoma Cells Using an Aqueous Extract of *Ajuga bracteosa. Int. J. Med. Sci.* **2017,** *14*, 462–469.

Luan, F.; Han, K.; Li, M.; Zhang, T.; Liu, D.; Yu, L.; Lv, H. Ethnomedicinal Uses, Phytochemistry, Pharmacology, and Toxicology of Species from the Genus *Ajuga* L.: A Systematic Review. *Am. J. Chin. Med.* **2019,** *47* (5), 959–1003.

Nisar, A.; Akhtar, N.; Hassan, A.; Banday, T.; Wani, B.; Zargar, M. A. Effect of *Ajuga bracteosa* on Systemic T-Cell Immunity in Balb/C Mice: Dual Th1/Th2 Immunostimulatory Effects. *Am. J. Chin. Med.* **2014,** *42* (2), 375–392.

Pal, A.; Pawar, R. S. A Study on *Ajuga bracteosa* Wall ex. Benth for Analgesic Activity. *Int. J. Cur. Bio. Med. Sci.* **2011,** *1* (2), 12–14.

Pal, A.; Toppo, F. A.; Chaurasiya, P. K.; Singour, P. K.; Pawar, R. S. In Vitro Cytotoxicity Study of Methanolic Fraction from *Ajuga bracteosa* Wall ex. Benth on MCF-7 Breast Adenocarcinoma and Hep-2 Larynx Carcinoma Cell Lines. *Pharmacognosy Res.* **2014,** *6* (1), 87–91.

Pala, A.; Jadona, M.; Katarea, Y. K.; Singoura, P. K.; Rajakb, H.; Chaurasiya, P. K.; Patila, U. K.; Pawar, R. S. *Ajuga bracteosa* Wall: A Review on Its Ethnopharmacological and Phytochemical Studies. *Der Pharm. Sinica.* **2011,** *2* (2), 1–10.

Patel, D. G.; Gulati, O. D.; Gokhale, S. D. Positive Inotropic Action of an Alkaloidal Fraction from *Ajuga bracteosa* Wall ex Benth. *Indian J. Physiol. Pharmacol.* **1962,** *6*, 224–30.

Qureshi, R.; Waheed, A.; Arshad, M.; Umbreen, T. Medico-Ethnobotanical Inventory of Tehsil Chakwal, Pakistan. *Pak. J. Bot.* **2009,** *41*, 529–538.

Ramzi, A. M.; Mansour, S.; Hasoon, S. S.; Mohamed, A. A. Antimicrobial and Antioxidant Activities and Gas Chromatography Mass Spectrometry (GC/MS) Analysis of the Essential Oils of *Ajuga bracteosa* Wall. ex Benth. and *Lavandula dentata* L. Growing Wild in Yemen. *J. Med. Plants Res.* **2012,** *6* (15), 3066–3071.

Rastogi, R. P.; Mehrotra, B. N. *Compendium of Indian Medicinal Plants*; CDRL: Lucknow and NISC: New Delhi, 1993.

Riaz, N.; Malik, A.; Nawaz, S. A.; Muhammad, P.; Choudhary, M. I. Cholinesterase-Inhibiting with Anolides from *Ajuga bracteosa. Chem. Biodivers.* **2004,** *1* (9), 1289–1295.

Riaz, N.; Nawaz, S. A.; Mukhtar, N.; Malik, A.; Afza, N.; Ali, S.; Ullah, S.; Muhammad, P.; Choudhary, M. I. Isolation and Enzyme-Inhibition Studies of the Chemical Constituents from *Ajuga bracteosa. Chem. Biodivers.* **2007,** *4* (1), 72–83.

Singh, K. P.; Meena, H.; Pandey, H. K.; Mohd. Nasim. *Ajuga bracteosa*: A Central Himalayan Plant Shows Potent Hypoglycemic Effect in Rats. *Indian J. Pharm. Pharmacol.* **2016,** *3* (4), 210–213.

Singh, N.; Mahmood, U.; Kaul, V. K.; Jirovetz, L. A New Phthalic Acid Ester from *Ajuga bracteosa. Nat. Prod. Res.* **2006,** *20*, 593–597.

Singh, R.; Patil, S. M.; Pal, G.; Ahmad, M. Evaluation of In Vivo and In Vitro Anti-Inflammatory Activity of *Ajuga bracteosa* Wall ex Benth. *Asian Pac. J. Trop. Dis.* **2012,** S404–S407.

Upadhyay, S. U.; Patel, V. B.; Patel, A. A.; Upadhyay, U. L.; Patel, N. M. *Ajuga bracteosa*- A Promising Herb. *Int. J. Pharm. Sci. Res.* **2012,** 2080–2088.

Verma, V. H.; Mahmood, U.; Singh, B. Clerodanediterpenoids from *Ajuga bracteosa* Wall. *Nat. Prod. Lett.* **2002,** *16* (4), 255–259.

Vohra, A.; Kaur, H. Chemical Investigation of Medicinal Plant *Ajuga bracteosa. J. Nat. Prod. Plant Resour.* **2011,** *1* (1), 37–45.

Zahra, S. S.; Ahmed, M.; Qasim, B.; Gul, M.; Zia, B.; Haq, M.; Haq, I. U. Polarity Based Characterization of Biologically Active Extracts of *Ajuga bracteosa* Wall. ex Benth. and RP-HPLC Analysis. *BMC Complement. Altern. Med.* **2017,** *17*, 443.

Ziyyat, A.; Legssyer, A.; Mekhfi, H.; Dassouli, A.; Serhrouchni, M.; Benjelloun, W. Phytotherapy of Hypertension and Diabetes in Oriental Morocco. *J. Ethnopharmacol.* **1997,** *58*, 45–54.

CHAPTER 3

A REVIEW ON PHARMACOLOGICAL ACTIVITIES OF *ANISOMELES MALABARICA* (L.) R.BR. EX SIMS

C. APPA RAO, K. PEDDANNA, and A. RAJASEKHAR

Department of Biochemistry, Sri Venkateswara University, Tirupati, Andhra Pradesh, India

ABSTRACT

Anisomeles malabarica R.Br. ex Sims belongs to Lamiaceae family. It is aromatic, perennial plant and distributed in Southeast Asia and Tropical Australia. This plant is commonly used in Ayurveda, Unani, Traditional Chinese and Korean medicine. Several investigators reported the presence of bioactive compounds, which include terpenoids, alkaloids, saponins, carbohydrates, tannins, flavonoids, and amino acids. These compounds have antidiabetic, antimicrobial, anti-inflammatory, antioxidant, antiepileptic, antilarvicidal and anticancer activities in vitro and in vivo models. In this review, we have discussed the phytochemicals in *A.malabarica* and their biological activities. Bioactive compounds of *A. malabarica* can be potential therapeutic agents for the treatment of diabetes, inflammation, microbial infections and cancers.

3.1 INTRODUCTION

Anisomeles malabarica R.Br. ex Sims (family Lamiaceae) is an aromatic, pubescent, perennial herb with a height of around 1.2–2.0 m. This plant's leaves are simple, opposite, very thick, aromatic, oblong-lanceolate, acute,

Bioactives and Pharmacology of Lamiaceae. T. Pullaiah, PhD (Ed.)

pale above, white below, crenate-serrate, and woolly; its flowers are purple, in dense whorls of more or less interrupted spikes; fruits' nutlets are bearing ellipsoid and compressed seeds (Joshi, 2000). It is distributed in S. E. Asia, Malaysia, and Tropical Australia. It is commonly distributed in major parts of India, especially in South India. It is a traditional medicinal plant commonly known as Peymarutti (Tamil), Gouzaban (Hindi), Chodhara (Marathi), Karithumbi (Kannada), and Malabar catmint (English). The synonyms of *A. malabarica* are *Nepeta malabarica* L., *Stachys mauritiana* Pers., *Craniotome mauritianum* (Pers.) Bojer, *Nepeta pallida* Salisb., and *Ajuga fruticosa* Roxb.

This plant is commonly used as a medicinal plant in traditional communities for the alleviation of intermittent fever, colic dyspepsia, cold, cough, liver disorders, stomach ailments, and wound healing (Vijayalakshmi and Ranganathan, 2012). *A. malabarica* plant's parts are most commonly used in traditional medicine such as Ayurveda, Unani, traditional Chinese, and Korean medicine to cure various diseases like convulsions, dyspepsia in intermittent fevers, colic, boils and tetanus fever, skin diseases, abdominal pain, psoriasis, etc. (Apparananthem et al., 1982, Tripathi et al., 1996; Reddy et al., 1991). Several evidences show that the pharmacological activities of *A. malabarica,* include antidiabetic, antimicrobial, anti-inflammatory, antioxidant, and anticancer activities. In this review, we discuss the traditional uses, phytochemistry and pharmacological activities of *A. malabarica.*

3.2 BIOACTIVES

Aniosmeles malabarica has phytochemical constituents such as terpenoids, alkaloids, saponins, carbohydrates, tannins, flavonoids, and amino acids (Packialakshmi and Nilofer Nisha, 2014). Earlier studies have reported the presence of phytochemical compounds such as anisomeric acid, anisomelolide, 2-acetoxymalabaric acid, anisomelyl acetate anisomelin, betulinic acid, ß-sitosterol, citral, gerainic acid, malabaric acid, ovatodiolide, and triterpene betulinic acid in *A. malabarica* (Annapoorani, 2019; Choudhary et al., 2011). In our study also LC-SRM/MS analysis revealed the presence of antidiabetic compounds in *A. malabarica* such as quinic acid, caffeic acid, luetolin, and kaempferol (Kotha et al., 2017). In this study, we also identified essential and unsaturated fatty acids by GC-MS analysis which includes eicosanoic acid, 2-methyl-1-hexadecanol, methyl 13-methylpentadecanoate, palmitic anhydride/hexadecanoic anhydride, oleic acid, methyl 11-(3-pentyl-2-oxiranyl) undecanoate, 6-octadecenoic acid, hexadecanoic acid, 2-hydroxy-1-(hydroxymethyl) ethyl ester, 9-hexadecanoic acid, 10,13-octadecadienoic acid methyl ester., 4-hexyl-1-(7-methoxycarbonylheptyl)[4,4,0]deca-2,5,7-triene, and 38,8,8-trimethoxy-3-piperidyl-2,2-binapthalene-1,1,4,4-tetrone. In addition to that, by GC-MS analysis antimicrobial compounds such as Phenol,2,4-bis(1,1-dimethylethyl)- 1H-Imidazo(4,55-c)pyridine,2-(3,4-dimethoxyphenyl)-, 3,7-Dimethyl-6-nonen-1-ol acetate, Hexadecanoic acid methyl ester, Pentadecanoic acid methyl ester, Oleic acid, Octadecanoic acid, Di-n-octyl phthalate, Phenol,2,6-bis(1,1-dimethylethyl)-4-[(4-hydroxy-3,5-dimethylphenyl)methyl]-, 2H-Naphtalen-1-one and 3,4-dihydro-6-methoxy-2-(4-methoxybenzylideno)-Coumarine,3-[2-(1-methyl-2-imidazolylthio)-1-oxoethyl]were identified (Krishna et al., 2019). Moreover, α-thujone (17.6%), terpenyl acetate (16.45%), and δ-cadinene (11.5%) have been identified in the essential oil of aerial parts of *A. malabarica* (Yogesh and Krishnakant, 2011).

3.3 PHARMACOLOGY

3.3.1 ANTIBACTERIAL ACTIVITY

Several pieces of evidence suggested that *A. malabarica* has a potential antibacterial activity. One study proved that the leaves and boiled leaves of *A. malabarica* significantly inhibited the growth of pathogenic bacteria such as *Staphylococcus aureus* and *Pseudomonas aeruginosa* (Packialakshmi and Nilofer Nisha, 2014). Furthermore, the ethanolic extract of the leaves of *A. malabarica* has inhibitory effects on *Staphylococcus aureus*, *Bacillus subtilis,* and *Proteus vulgaris* (Kavitha et al., 2012; Krishna et al., 2019; Yogesh and Krishnakant, 2011). The aqueous extract of the leaves of *A. malabarica* at 50 μL concentration significantly attenuated the growth of soil-borne bacteria such as *B. subtilis, P. fluorescens*, and *S. aureus* compared with synthetic antibiotics Gentamicin and Tetracycline (Kiran et al., 2012). The methanolic extract also showed an inhibitory effect on Gram-positive and negative bacteria (Boobalan and Jeganathan, 2010). Furthermore, the aerial parts of *A. malabarica* were used to isolate the hydro distilled essential oils and evaluate the antibacterial activity by using the paper disc method. In this study, compared to gentamicin, the essential oil has strongly inhibited *Staphylococcus aureus, Escherichia coli, Pseudomonas aeruginosa,* and *Bacillus pumilus* (Yogesh and Krishnakant, 2011). Nisha Nilofer and

Packialakshmi (2014) have investigated the antibacterial activity of aqueous extracts of leaf, flower, and stem of *A. malabarica* by disc diffusion methods and the results confirmed that the extracts have an inhibitory effect on five pathogenic organisms such as *Staphylococcus aureus*, *Staphylococcus epidermis*, *E. coli*, *Pseudomonas aeruginosa*, and *Bacillus subtilis* except for *Proteus vulgaris*.

3.3.2 CYTOTOXIC ACTIVITY

The previous in vitro studies demonstrated the cytotoxic activity of *A. malabarica*. In an earlier study, the sequential extracts of *A. malabarica* were used to evaluate the antiproliferative activity in HPV16-positive human cervical cancer cells by using the MTT assay. This study found that n-hexane and chloroform extracts inhibited proliferation and induced death in HPV16-positive cervical cancer cells (Preethy et al., 2012). Also, the same group isolated the Anisomelic acid from the leaves of *A. malabarica* and tested for cytotoxic activity in breast and cervical cancer cells such as MDA-MB-231, MCF-7, SiHa, and ME180 (Preethy et al., 2013). Anisomelic acid significantly inhibited the growth of breast and cervical cancer cells. Also, Acridine Orange and Ethidium Bromide (AO & EB) and Hoechst 33258 staining showed that anisomelic acid mediates the condensation of chromatin and apoptosis in cancer cell lines. These results suggest that Anisomelic acid has antiproliferative activity and induces apoptosis in cancer cell lines.

3.3.3 ANTI-INFLAMMATORY ACTIVITY

Inflammation is a complex process in various chronic diseases and infections, which involves the increase of the number of leukocytes and mediators (cytokines, prostaglandins, and ROS) causing tissue or cellular damages (Abdulkhaleq et al., 2018). However, there are several glucocorticoids, steroidal, and nonsteroidal anti-inflammatory drugs that have been used in the treatment of inflammation in various diseases (Nunes et al., 2020). It has been reported that *A. malabarica* showed anti-inflammatory activity in inflammatory animal models. The in vitro study demonstrated that the methanolic extract of *A. malabarica* leaves has an anti-inflammatory effect by protecting human red blood cell membrane stabilization against hypotonicity-induced membrane lysis (Lavanya et al., 2010). This study also showed that the methanolic extract of *A. malabarica* leaves possesses antiplatelet and antiarthritic activities. Similarly, the in vitro study also supported the previous study, which has shown that the methanolic extracts of flowers and leaves induced the anti-inflammatory effect (Sheikh et al., 2016). Srinivasan et al. (2010) reported anti-inflammatory and antipyritic activity of *A. malabarica* in carrageenan-induced hind paw edema and cotton pellet granuloma in rats (Srinivasan et al., 2010). They showed that oral administration of ethanol and aqueous extracts at doses of 50,100, and 200 mg/kg had significant anti-inflammatory activity in a dose-dependent manner. Also, the methanolic extracts of aerial parts, leaves, and roots of *A. malabarica* showed inhibition of tumor necrosis factor-α (TNF-α) production in lipopolysaccharide (LPS)-mediated signaling in macrophage and mouse connective tissue cell cultures (Shareef et al., 2013). Methanolic extract of aerial parts at a dose of 200 μg/mL has a strong anti-inflammatory effect by 38.75% inhibition of TNF-α production. Furthermore, the bioassay-guided fractionation of chloroform extract of *A. malabarica* leaves also has an effective anti-inflammatory activity by the inhibition of lipoxygenase (LOX). In this study the two LOX inhibitors such as 3, 4 dihydroxy benzoic acid and 4′, 5, 7-trihydroxyflavone were isolated (Sudha and Srinivasan, 2014).

3.3.4 ANTIOXIDANT ACTIVITY

Antioxidant activity of *A. malabarica* has been reported by many in vitro and in vivo studies. Methanolic extract of *A. malabarica* leaves has a strong inhibitory effect on 2,2-diphenyl-1-picryl hydrazyl (DPPH)-induced free radical release when compared with quercetin (Dixit and Sharma, 2014; Lavanya et al., 2010; Vinod et al., 2014), as well as by other in vitro methods such as superoxide radical (O^{2-}) and 2,22′-azinobis- (3-ethyl-enzothiazoline-6-sulfonic acid) (ABTS) radical scavenging assays, phosphomolybdenum reduction, and Fe reduction assays. It is confirmed that the ethanolic extract of the leaves of *A. malabarica* has antioxidant activity (Boobalan and Jeganathan, 2010; Krishna et al., 2019). The methanolic extract of the whole plant of *A. malabarica* has free radical scavenging and antioxidant activity by the reduction of free radicals, superoxides and increased the iron-chelating and total antioxidant activity in a dose-dependent manner (Vijayalakshmi and Ranganathan, 2011b). Also, ethyl acetate extract of leaves of *A. malabarica* potentially scavenges hydroxyl, superoxide anion radicals, DPPH, and ABTS radicals (Vijayalakshmi and Ranganathan, 2011a).

Further, we also reported the antioxidant activity of *A. malabarica* leaves in streptozotocin-induced diabetic rats (Kotha et al., 2019). In our study, we evaluated the antioxidant activity of the antihyperglycemic fraction of ethyl acetate extract of leaves of *A. malabarica* in hepatic and renal tissues of streptozotocin-induced diabetic rats. In this study, we found that *A. malabarica* significantly increased the antioxidant status in diabetic treated rats by reducing lipid peroxides/thiobarbituric acid reactive substances (TBARS) and enhanced the activities of antioxidant enzymes such as Glutathione peroxidase (GPx), glutathione-S-transferase (GST), catalase, and superoxide dismutase (SOD).

3.3.5 ANTIDIABETIC ACTIVITY

Recently, we have reported the antidiabetic activity of *A. malabarica* leaves in STZ-induced diabetic rats (Kotha et al., 2017). In this study, the active (antihyperglycemic) fraction was isolated by the bioassay-guided fractionation of ethyl acetate extract of *A. malabarica* leaves using silica gel column chromatography and it has shown a significant antihyperglycemic activity at a dose of 50 mg/kg b.w. in diabetic rats. In chronic treatment also *A. malabarica* significantly improved the FBG (Fasting blood glucose), HbA1c, plasma insulin levels, and insulin sensitivity/resistance indicators such as glucose tolerance, HOMA-IR, HOMA-β, and QUICKI. Furthermore, the isolated active fraction has flavonoids and fatty acids identified as antidiabetic active compounds.

We also elucidated the mechanism of antidiabetic action of the active fraction of *A. malabarica* (AMAF) in diabetic rats (Kotha et al., 2019). The AMAF treatment resulted in an increase in body weights, hepatic and renal protein, and tissue glycogen levels in diabetic treated rats compared to diabetic rats. Also, the treatment improved activities of carbohydrate metabolic enzymes, antioxidant enzymes, and liver and renal function markers in the AMAF-treated diabetic rats. Gene expressions of key carbohydrate metabolic enzymes/factors glucokinase, glucose transporter protein (GLUT-2), and phosphoenolpyruvate carboxykinase (PEPCK) were also normalized upon AMAF treatment in diabetic rats.

3.3.6 ANTIEPILEPTIC ACTIVITY

Epilepsy is a common neurological disorder caused by a genetic defect or an acquired brain injury, such as a trauma or stroke. However, epilepsy can be treated with medication and in some cases by surgery, devices, or dietary changes. Ethnobotanically, *A. malabarica* has been documented in traditional medicines against convulsions. In previous studies also the antiepileptic activity of the leaves of *A. malabarica* in Maximal electroshock seizure (MES) and Pentylenetetrazole (PTZ)-induced convulsions in Wistar rats was reported (Choudhary et al., 2011; Singh et al., 2010). In these studies, Singh et al. (2010) reported that higher doses of (400 mg/kg, p.o.) chloroform and ethyl acetate extracts significantly decreased the extent of MES- and PTZ-induced convulsions. On the other hand, ethyl acetate extract at lowest and medium selected doses (i.e., 100 mg/kg, p.o. and 200 mg/kg, p.o., respectively) for 7 days had also significantly attenuated PTZ-induced convulsions (Singh et al., 2010). After this study, Choudhary et al. (2011) demonstrated the antiepileptic potential of flavonoids fractions from the leaves of *A. malabarica* in MEZ and PTZ Wistar rats (Choudhary et al., 2011). In this study, *A. malabarica* flavonoid fraction (AMFF) and tannins fraction (AMTF) was used to evaluate the antiepileptic activity against the PTZ- and MES model in Wistar rats. Pretreatment with AMFF at lower doses (6.25 and 12.5 mg/kg, i.p.) for 1 week has a significant effect on both MES and PTZ-induced convulsions without causing neurotoxic effects.

3.3.7 ANTI-LARVICIDAL ACTIVITY

Previous studies determined the anti-larvicidal activity of *A. malabarica* against malarial vector *Anopheles stephensi* (Jayakumar et al., 2014; Unpaprom et al., 2015). The in vitro larvicidal assay was performed against the mosquito species at various concentrations ranging from 20 to 100 ppm. Probit analysis was used to determine the LC50 and LC90 values of the methanolic extract of leaf and inflorescence of *A. malabarica*. The results concluded that the *A. malabarica* leaf and inflorescence crude extracts have the efficacy to be used as an ideal eco-friendly approach for the control of mosquito vectors (Unpaprom et al., 2015). A combination of *A. malabarica* and *Phyllanthus emblica* methanolic extract has a synergic effect against malarial vector, *Anopheles stephensi*, and exhibited the highest larval mortality (Jayakumar et al., 2014). Moreover, the methanolic extract of *A. malabarica* has larvicidal activity against the third instar larvae of the dengue vector mosquito, *Aedes aegypti* (Kamatchi et al., 2016).

KEY WORDS

- ***Anisomeles malabarica***
- **bioactives**
- **lamiaceae**
- **pharmacology**
- **phytochemicals**

REFERENCES

Abdulkhaleq, L.; Assi, M.; Abdullah, R.; Zamri-Saad, M.; Taufiq-Yap, Y.; Hezmee, M. The Crucial Roles of Inflammatory Mediators in Inflammation: A Review. *Vet. World* **2018,** *11*, 627–635.

Annapoorani, S. A Review on *Anisomeles malabarica* and Their Usage. *Int. J. Life Sci.* **2019,** *7*, 140–142.

Apparananthem, T.; Chelladurai, V.; Subramanian, V. Some Tribal Folk Medicines of Point Calimere (Kodikkarai) in Tamil Nadu. *Bull. Med. Ethnobot. Res.* **1982,** *3*, 173–177.

Boobalan, R.; Jeganathan, M. Evaluation of In Vitro Antioxidant and Antimicrobial Potential of Methanolic Leaf Extract of *Anisomeles malabarica* (Linn.). *J. Pharm. Res.* **2010,** *3*, 1471–1474.

Choudhary, N.; Bijjem, K. R. V.; Kalia, A. N. Antiepileptic Potential of Flavonoids Fraction from the Leaves of *Anisomeles malabarica. J. Ethnopharmacol.* **2011,** *135*, 238–242.

Dixit, P.; Sharma, A. Screening of Antibacterial and Antioxidant Activity of *Anisomeles indica. Int. J. Sci. Res.* **2014,** *3*, 2194–2196.

Jayakumar, J.; Elumalai, K.; Dalmin, G.; Krishnamurthy, R. Larvicidal and Synergistic Activity of *Anisomeles malabarica* and *Phyllanthus emblica* Against the Larvae of Common Malarial Vector, *Anopheles stephensi Liston.* Diptera: Culicidae. *Discovery* **2014,** *20*, 36–41.

Joshi, S. G. *Medicinal Plants*; Oxford and IBH Publishing Co Pvt.Ltd.: NewDelhi, 2000.

Kamatchi, P.; Subramanian, A.; Rajan, M. Evaluation of Larvicidal Efficacy of Plant Extracts of *Justicia adhatoda* Linn., *Murraya koenigii* (Linn.) Spreng and *Anisomeles malabarica* (Linn.) Against Dengue Vector Mosquito, *Aedes aegypti* (Linn.)(Diptera: Culicidae). *J. Appl. Zool. Res.* **2016,** *27*, 117–122.

Kavitha, T.; Nelson, R.; Thenmozhi, R.; Priya, E. Antimicrobial Activity and Phytochemical Analysis of *Anisomeles malabarica* (L) R. Br. *J. Microbiol. Biotech. Res.* **2012,** *2*, 1–5.

Kiran, B., Lalitha, V., Raveesha, K. In Vitro Evaluation of Antibacterial Activity of *Anisomeles malabarica* (L.) Sims Leaf Against Important Soil Borne Bacteria. *World J. Pharm. Res* **2012,** *1*, 1077–1083.

Kotha, P.; Badri, K. R.; Nagalapuram, R.; Allagadda, R.; Chippada, A. R. Anti-Diabetic Potential of the Leaves of *Anisomeles malabarica* in Streptozotocin Induced Diabetic Rats. *Cell Physiol. Biochem.* **2017**, *43*, 1689–1702.

Kotha, P.; Marella, S.; Allagadda, R.; Badri, K. R.; Chippada, A. R. Evaluation of Biochemical Mechanisms of Anti-Diabetic Functions of *Anisomeles malabarica. Biomed. Pharmacother.* **2019,** *112*, 108598.

Krishna, S.; Chandrasekaran, S.; Dhanasekar, D.; Perumal, A. GCMS Analysis, Antioxidant and Antibacterial Activities of Ethanol Extract of *Anisomeles malabarica* (L.) R. Br. ex. Sims Leaves. *Asian J. Pharm. Pharmacol.* **2019,** *5*, 180–187.

Lavanya, R.; Maheshwari, S.; Harish, G.; Raj, J.; Kamali, S.; Hemamalani, D.; Varma, J.; Reddy, C. In Vitro Antioxidant Activity of Methanolic Extract in Leaves of *Anisomeles malabarica* Linn. *Res. J. Pharm. Biol. Chem. Sci.* **2010,** *1*, 737–744.

Nisha Nilofer, H. M.; Packialakshmi, N. Analysis of Anti-Bacterial and Phytochemical Screening by Using Different *Anisomeles malabarica* Samples. *Int. J. Pharm. Res.* **2014,** *4*, 22–24.

Nunes, C.d.R.; Barreto Arantes, M.; Menezes de Faria Pereira, S.; Leandro da Cruz, L.; de Souza Passos, M.; Pereira de Moraes, L.; Vieira, I. J. C.; Barros de Oliveira, D. Plants as Sources of Anti-Inflammatory Agents. *Molecules* **2020,** *25*, 3726.

Packialakshmi, N.; Nilofer Nisha, H. M. Bioautography Screening of *Anisomeles malabarica* Leaves and Boiled Leaves. *Pharma Innov.* **2014,** *3*, 77–80.

Preethy, C. P.; Alshatwi, A. A.; Gunasekaran, M.; Akbarsha, M. A. Analysis of the Cytotoxic Potential of Anisomelic Acid Isolated from *Anisomeles malabarica. Scientia Pharma.* **2013,** *81*, 559–566.

Preethy, C. P.; Padmapriya, R.; Periasamy, V. S.; Riyasdeen, A.; Srinag, S.; Krishnamurthy, H.; Alshatwi, A. A.; Akbarsha, M. A. Antiproliferative Property of N-Hexane and Chloroform Extracts of *Anisomeles malabarica* (L). R. Br. in HPV16-Positive Human Cervical Cancer Cells. *J. Pharmacol. Pharmacother.* **2012,** *3*, 26–34.

Reddy, M. B.; Reddy, K. R.; Reddy, M. N. Ethnobotany of Cuddapah District, Andhra Pradesh, India. *Indian J. Pharmacog.* **1991,** *29*, 273–280.

Shareef, M. I.; Leelavathi, S.; Shariff, A. A. Inhibition of In Vitro TNF-α Production by *Anisomeles malabarica* R. Br. Reinforces Its Anti-Rheumatic and Immunomodulatory Properties. *Proc. Natl. Acad. Sci. India Section B: Biol. Sci.* **2013,** *83*, 187–191.

Sheikh, M. V.; Devadiga, N.; Hate, M. An In Vitro Anti-Inflammatory and Anti-Oxidant Activity of *Anisomeles malabarica* R. Br. ex Sims. *J. Chem. Pharm. Res.* **2016,** *8*, 1062–1067.

Singh, I.; Singh, M.; Singh, T.; Bijjem, K. R. V.; Kalia, A. N. Anticonvulsant Potential of *Anisomeles malabarica* Leaves Against Experimentally Induced Convulsions in Rats. *Int. J. Green Pharm.* **2010,** *4*, 199–205.

Srinivasan, P.; Bharathajothi, P.; Manik, R.; Arulvasu, C. Effects of Anti-Inflammatory and Anti-Pyretic Activity of *Anisomeles malabarica* R. Br. *J. Pharm. Res.* **2010,** *7*, 1598–1601.

Sudha, A.; Srinivasan, P. Bioassay-Guided Isolation, Identification and Molecular Ligand-Target Insight of Lipoxygenase Inhibitors from Leaves of *Anisomeles malabarica* R. Br. *Pharmacogn. Mag.* **2014,** *10*, S596–S605.

Tripathi, Y. C.; Prabhu, V. V.; Pal, R. S.; Mishra, R. N. Medicinal Plants of Rajasthan in Indian System of Medicine. *Anc. Sci. Life* **1996,** *15*, 190–212.

Unpaprom, Y.; Ramaraj, R.; Kadarkarai, M. Mosquito Larvicidal Properties of *Anisomeles malabarica* (L.) Extracts Against the Malarial Vector, *Anopheles stephensi* (Liston). *Chiang Mai J. Sci.* **2015,** *42*, 148–155.

Vijayalakshmi, R.; Ranganathan, R. Evaluation of In Vitro Antioxidant Potential of Ethyl Acetate Leaf Extract of *Anisomeles malabarica* (Linn.). *Int. J. Chem. Anal. Sci.* **2011a,** *2*, 93–96.

Vijayalakshmi, R.; Ranganathan, R. In Vitro Antioxidant and Free Radical Scavenging Activity of Various Extracts from Whole Plant of *Anisomeles malabarica* (Linn.) R. Br. *J. Pharm. Res.* **2011b,** *4*, 3535–3537.

Vijayalakshmi, R.; Ranganathan, R. Chemopreventive Effect of *Anisomeles malabarica* Whole Plant Extracts During DMBA Induced Hamster Buccal Pouch Carcinogenesis. *Asian J. Pharm. Clin. Res.* **2012,** *5*, 185–188.

Vinod, G.; Ramesh, B.; Suvarna, M.; Hanumanthappa, K. In Vitro Antioxidant Potential of Solvent Extracts from *Anisomeles malabarica. J. Pharmacogn. Phytochem.* **2014,** *3*, 99–103.

Yogesh, U.; Krishnakant, P. Chemical Composition and Antibacterial Activity of Essential Oil from *Anisomeles* Species Grown in India. *Pharmacognosy J.* **2011,** *2*, 55–59.

CHAPTER 4

AN UPDATED REVIEW ON *GMELINA ARBOREA* ROXB.

P. SUNEETHA[1], Y. S. MEENAKSHI[1], and S. RAJASHEKARA[2]

[1]*Department of Studies in Biochemistry, Maharani Lakshmi Ammanni College for Women, Malleswaram, Bengaluru, Karnataka, India*

[2]*Centre for Applied Genetics, Department of Studies in Zoology, Bangalore University, Jnana Bharathi Campus, Bengaluru, Karnataka, India*

ABSTRACT

Gmelina arborea is one of those medicinal plants belonging to the family Lamiaceae found in the soils with high fertility preferably near abundant water source. The medicinal properties are contributed by various parts of the plants such as the extracts of fruits, roots, and flowers. This plant consists of several properties such as anti-inflammatory, antioxidant and aphrodisiac, antidiabetic, antidiuretic, cardioprotective, hepatoprotective, anti-helminthic, anticonvulsant, antifungal, antibacterial, immunomodulatory, anti-hyperlipidaemic, anti-pyretic, anti-ulcer, analgesic, and antitumor due to the presence of several bioactive phytochemicals.

4.1 INTRODUCTION

Over the centuries, the ancient ayurvedic system has explored medicinal utilizations of numerous native plants and also confirmed to consume a great impact on betterment of human health, and was found to be advantageous over conventional way of treatment to cure diseases which were affecting the human society. *Gmelina arborea* (family Lamiaceae) is found to be

Bioactives and Pharmacology of Lamiaceae. T. Pullaiah, PhD (Ed.)

flourishing in the soils with high fertility preferably near abundant water source like valleys (Pathala et al., 2015). This deciduous tree accomplishes the development from modest to longer height up to 15–20 m. *G. arborea* is known as "Gambhari" plant in India. The pharmacological constituents of this tree are profoundly used in Ayurveda, the native medicine of India, as rejuvenate, memory enhancer, and aphrodisiac. Other than the uses mentioned, it also finds its applications as anthelminthic, antimicrobial, antidiabetic, diuretic, anticonvulsant, hepatoprotective, and antiepileptic agent (Nayak et al., 2012). The medicinal properties are contributed by various parts of the plants; fruit extracts exhibited antioxidant and aphrodisiac properties, the roots are anthelminthic and laxative, and the flowers are used to treat skin diseases and leprosy (Warrier et al., 2020). In the field of Ayurveda, the medicine Dashamoolarishtam, which uses *Gmelina* in its preparation, shows many medicinal properties.

4.2 BIOACTIVES

Phytochemical screening performed by Florence and Jeeva (2015) using various spectroscopic methods exhibited many phytochemicals.

A wide range of chemical constituents is reported to be present in *G. arborea* (Table 4.1). It is described that various groups of phytoconstituents are dispersed crosswise the entire plant. The bark of the plant shows the presence of flavonoids, saponins, terpenoids, and cardioglycosides. Sitosterol, quercetagetin, kaempferol, and apigenin are the flavones; coumarin glycosides and sterols are also reported from roots of the plant. Luteolin, a flavonoid isolated from the leaves is known to exhibit proinflammatory response. The fruit oil of *Gmelina* contains aliphatic alcohols like methyl arachidate, methyl palmitate, methyl oleate, methyl stearate, methyl gondoate, and methyl behenate which are found to be analgesic (Arora and Tamrakar, 2017; Muritala et al., 2018).

Apigenin, kaempferol, luteolin, quercetin, hentriacontanol, and β-sitosterol have been reported from the leaves of *G. arborea* (Rao and Rao, 1970). The chemical structure of the fruit oil of *G. arborea* was inspected by GC and GC/MS. One hundred parts were reported for 92.7% of the oil together with 24 trace substances. The universal terpenes are in smaller quantity. The fruit oil consists of plenty of (Z)-3-hexenol (17.9%), 1-octen-3-ol (8.4%), and hexanol (6.1%); heptacosane (5.6%), pentacosane (3.8%), and 1-pentacosene (3.2%); and nonanal (8.7%) and (E)-2-decenal (3.0%) (Moronkola et al., 2009).

Chellappan and Pemiah (2014) performed tests for the physicochemical coefficients; initial phytochemical and in vitro antioxidant test by utilizing the hydroalcoholic extract of *G. arborea* revealed the presence of tannins, saponins, glycosides, alkaloids, phytosterols, flavonoids, and terpenoids. The extract when screened for antioxidant potential exposed to the hydroalcoholic extract exhibited noteworthy gastroprotective usefulness at the dose of 250 mg/kg against ethanol-persuaded gastric ulcer model in experimental animals. Therefore, the hydroalcoholic extract of *G. arborea* upper portion performs the probable pharmacological, antioxidant, and antiulcer action.

Quantitative analysis of hydroacetonic crude extract of *G. arborea* leaves using the ultraperformance liquid chromatography tandem mass spectrometry by N'gaman et al. (2014) revealed the presence of flavonoids (75.50%) and hydroxycinnamic acid derivatives (24.49%).

TABLE 4.1 Phytochemicals Present in Various Extracts of *G. arborea*.

Solvents for extraction	Compounds present
Ethanol	Alkaloids, Cardiac glycosides, Anthraquinone glycosides, Gums, Mucilages, Tannins, Phenolic compounds, Steroids and Sterols, and Flavonoids
Ethyl acetate	Cardiac glycosides, Proteins, Amino acids, Gums, Mucilages, Steroids and Sterols
n- butanol	Alkaloids, Anthraquinone glycosides, Gums, Mucilages, Proteins, Amino acids, Phenolic compounds, Steroids and Sterols, Triterpenoids, Saponins, and Flavonoids
Petroleum ether	Alkaloids, Carbohydrates, Cardiac glycosides, Anthraquinone glycosides, Gums, Mucilages, Proteins, Amino acids, Steroids and Sterols, Triterpenoids, and Saponins

The presence of the flavonoids can be detected in the plant extract by the qualitative method (Chothani and Patel, 2018). Kaempferol, a polyphenol antioxidant, was reported from the plant. Several studies have defined the useful properties of nutritional kaempferol in treating chronic ailments, including cancer. Epidemiological studies have demonstrated that the intake of kaempferol source tends to reduce the cancer enhancement. Luteolin 7-galactoside is a flavoglycoside. Most of these flavonoids contribute for the antioxidant properties of the plant extract. The saponins were perceived by Pardeshi et al. (2018) and the triterpenoid saponin such as hupehensis was found in the plant extracts. Alkaloids were extracted using petroleum ether, ethanol, and n-butanol (Chothani and Patel, 2018) that revealed the presence of Premnazole.

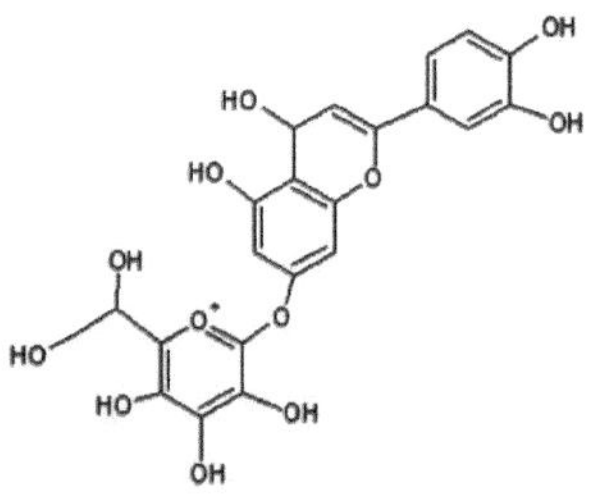

Kaempferol

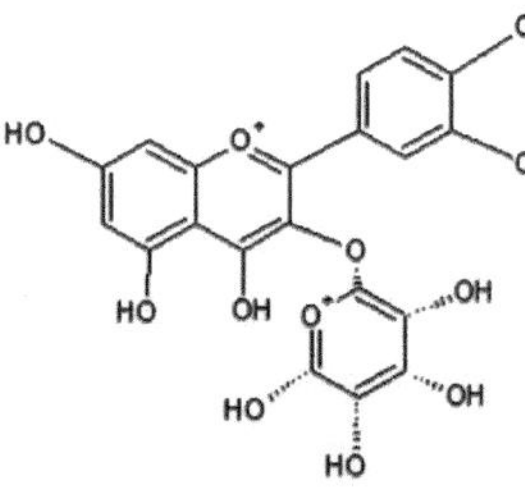

Luteolin 7-galactoside

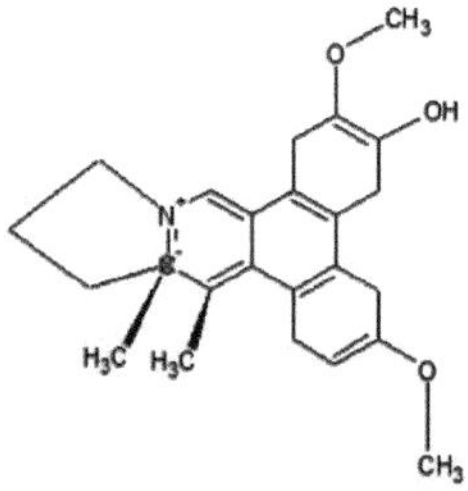

Premnazole

4.3 PHARMACOLOGY

4.3.1 ANTI-INFLAMMATORY ACTIVITY

Premnazole, an alkaloid from the methanolic extract of *G. arborea,* which is a type of isoazole is known to have anti-inflammatory response at acute and subacute levels. The fruit and bark extracts of *G. arborea* displayed anti-inflammatory action against the induced inflammation by egg albumin. The activity of fruit extracts of *G. arborea* was evaluated at 200 mg/kg concentration and exhibited the inhibition of about 30–40%, and at 400 mg/kg concentration proven to increase its anti-inflammatory activity (Muritala et al., 2018).

In carrageenan-induced edema, at low (750 mg/kg) and high (1000 mg/kg) concentrations tested, lower dosage showed a mild anti-inflammatory consequence, although advanced concentration verified to be active against irritation and edema. This is an indication that the second phase of inflammation was suppressed by the inhibition of released prostaglandins (Gandigawad et al., 2018).

4.3.2 ANTIOXIDANT ACTIVITY

5,7,3′,4′-tetrahydroxyflavone (luteolin), luteolin-4′-O-β-D-C-galactopyranoside, 3,5,7,4′-tetrahydroxy-flavone (kaempferol), quercetin-

3-O-β-D-4C-glucopyranoside (isoquercetin), quercetin-3-O-α-L-1C rhamnopyranosyl-(1‴ 6″)-O-β-D-4C glucopyranoside (rutin), luteolin-7-O-β-D-4C-galactoside, and 3-O-α-L-1C-rhamnopyranosyl-(1‴ 6)-β-D-4C-galactopyranoside quercetin 3-O-β-D-robinobioside were isolated from leaves of *G. arborea* which are known to exhibit free radical scavenging potency in action (Ghareeb et al., 2018).

Hexane extract of *G. arborea* showed antioxidant activity and restored HDL and LDL levels. It also showed increased levels of superoxide dismutase and catalase. This antioxidant activity is partially attributed to the flavonoids. A concentration of 150 mg/kg of hexane extract of *G. arborea* was found to be effective as an antioxidant by regulating plasma protein stages, lipid outline, and nitric oxide stages (Wansi et al., 2012).

Leaves of *G. arborea* are recognized to comprise polyphenols which are identified to display antioxidant actions, which is evaluated by DPPH radical scavenging activity. A concentration of 14.10 mg/ml in butanol extract of plant model was found to be maximum active in contrast with other solvent extracts (Shoeb et al., 2014). Methanolic extract of *G. arborea* is known to have an effective scavenging activity against free radicals, nitric oxide radical, peroxide radical, and hydroxyl radical (Patil et al., 2009).

The effect of bark and fruit aqueous extracts of *G. arborea* on paraquat- and H_2O_2-induced oxidative stress using liver slice culture was studied by Sinha et al. (2006). Paraquat and H_2O_2 were found to be cytotoxic as assessed by issue of lactate dehydrogenase from liver slice culture. Accumulation of bark and fruit extracts laterally with these cytotoxic agents directed to a decline in lactate dehydrogenase release. Actions of SOD, CAT, and GR occurred to incline on dealing with these prooxidants. Accumulation of the plant extracts along with the prooxidants repressed the enzyme actions. The extracts also showed antioxidant action in in vitro radical rummaging analyses (Sinha et al., 2006).

Attanayake et al. (2015) examined the antioxidant actions of the aqueous bark extract. The in vitro total antioxidant actions of the warm water bark extract of *G. arborea* were assessed by DPPH, ferric falling antioxidant potential (FRAP), and NO prevention analyses. The in vivo antioxidant action was estimated by the actions of liver enzymes, antioxidant enzymes, and level of LPO in the liver of STZ-diabetic rats. The in vitro antioxidant analyses (DPPH, FRAP, and NO) evidently confirmed the antioxidant latent of *G. arborea* extract. The *G. arborea* extract reduced LPO by 27%; actions of ALT, AST, and ALP declined by 29, 23, and 29% correspondingly. The liver reduced GSH, activities of GR, GPx, and GST of plant extract on experimented diabetic rats improved to 606.47 μg/g liver tissue, 7.92, 8.56, and 7.44 nmol/min/mg protein correspondingly. The extract was more active than glibenclamide in reestablishing the liver antioxidant enzymes in STZ-diabetic rats.

4.3.3 ANTIDIABETIC ACTIVITY

Aqueous extract of *G. arborea* bark of concentration 250 and 500 mg/kg was effective against diabetes-induced rats. The antidiabetic properties of *G. arborea* extract was compared with reference to glibenclamide. The bark of *G. arborea* showed the presence of tyrosol, balanophonin, gmelinol, 6 dimethoxy-p-benzoquinone, and 3,4,5-trimethoxyphenol (Kulkarni and Veeranjaneyulu, 2009, 2013).

Flavonoids present in *G. arborea* are accountable for controlling oxidative stress and reducing LDL levels. It is deduced that higher concentration of *G. arborea* is known to regulate carbohydrate metabolism (Kumaresan et al., 2014).

The benzene, chloroform ethanolic, and water extracts of the bark of *G. arborea* for antidiabetic action in Streptozotocin (STZ)-tempted diabetic rats were studied by Pattanayak et al. (2011) and highlighted the mechanism of its antidiabetic stuff. The preliminary blood glucose stages of the diabetic rats designated were extended from 240 to 280 mg/100 ml. At the end of the test, blood glucose level was 123.66 in the diabetic mice experimented with 420 mg/kg body mass of the EtOH extracts of the stem bark. The experiential noteworthy antidiabetic action of the ethanolic extract might be accredited to the improved blood GSH levels emphasizing the role of GSH as free radical rummager and in the restoration of free radical triggered biological injury.

When the aqueous extracts were orally administered, the plasma glucose level reduced in the diabetic rats. The continuous running of the water extract meaningfully condensed plasma glucose stages of the diabetic rats. Moreover, the water extract also decreased the loss of body mass and reduced diet and water consumption in the diabetic animals. In this manner, the water extract of *G. arborea* bark had antihyperglycemic action against streptozotocin (STZ)-encouraged diabetes in rats (Kulkarni and Veeranjaneyulu, 2011).

The antidiabetic actions of *G. arborea* bark extracts using EtOH, ethyl acetate, n-butanol, and petroleum ether were assessed by Nayak et al. (2015a). The examinations for cardiac glycosides and steroids were optimistic for entirely the extracts. The ethanol and n-butanol extracts comprised maximum phytochemicals, while ethyl acetate extracts were encompassing minimum number of phytochemicals. Altogether, the extracts decreased the sugar level in blood. The ethanol extract has better antidiabetic action in contrast to other extracts.

4.3.4 ANTIDIURETIC ACTIVITY

An enlarged quantity of sodium and potassium was detected in the urine samples collected from the experimental animals after the administration of *G. arborea* fruit extracts of ethanol, ethyl acetate, n-butanol, and petroleum ether. The n-butanol extract is known to exhibit antidiuretic property in comparison to other extracts (Nayak et al., 2013). The effects of *G. arborea* methanolic extract had exposed important diuretic action on albino rats. The examination extracts were managed with the dosage of 250 and 500 mg/kg body mass. Na^+, K^+ and Cl^- production in urine distinctly improved as related to regular saline. The *G. arborea* extract employed its diuretic action. There was a rise in the proportion of dosage of excreted sodium and potassium ions after ethanol extract of *G. arborea* treatment (Sravani et al., 2011).

4.3.5 CARDIOPROTECTIVE ACTIVITY

A common chemotherapeutic drug, doxorubicin is known to induce cardiotoxicity. This condition can be treated by administering ethanolic extract of *G. arborea* (250–500 mg/kg) which is known to increase the levels of glutathione responsible for antioxidant activity; it checks on the lipid profile and has hypocholesterolemic activity due to the presence of palmitic and linoleic acid such as n-hexadecanoic acid, 9,12-octadecatrienoic acid (Z,Z,Z)-, and oleic acid. The cellular damage induced by anthracyclines was minimized by *G. arborea* extracts. This property is achieved due to the presence of flavonols, flavanones, and alkaloids (Vijay et al., 2011).

4.3.6 HEPATOPROTECTIVE ACTIVITY

The liver diseases such as liver cirrhosis, oxidative stress-induced liver damage, and hepatitis B and C induced jaundice are the commonly found liver disorders. This creates a need for an antioxidant suitable to treat the oxidative stress in liver. The ethanol and chloroform extract of aerial parts of *G. arborea* exhibited the hepatoprotective properties (Merlin and Parthasarathy, 2011).

Iridoid glycosides, 1-O-trans-cinnamoyl-a-L-rhamnopyranose, 3-O-trans-p-methoxycinnamoyl-a-L-rhamnopyranose, and verbose were isolated from *G. arborea* flowers. The compounds exhibited cytoprotective activities against carbon tetrachloride-induced cytotoxicity (Gu et al., 2013).

4.3.7 ANTHELMINTHIC ACTIVITY

The bark of *G. arborea* was extracted using variety of solvents. The extract of all the solvents was effective as anthelminthic when compared to the standard drugs albendazole and piperazine citrate. A concentration of 25 and 10 mg/ml was used, the earlier was found to be maximum active than the later. In comparison of the effectiveness of all the extracts, n-butanolic extract was found to be more potent as the anthelminthic property of this extract was evaluated at 10 mg/ml, which is much less than lethal dose (Nayak et al., 2015b).

Panda et al. (2015a) made the phytochemical investigation and evaluation for the anthelminthic action of ethanol, ethyl acetate, n-butanol, and petroleum ether extracts of root of *G. arborea* utilizing Indian earthworms *Pheretima posthuma* as an experimental animal. Many dosages were tested. Albendazole (60 mg/ml) was employed as a typical control. Amongst the extract, ethanol and petroleum ether displayed dosage reliant on and noteworthy anthelminthic action. Thus, petroleum ether exhibited improved action as compared to control drug albendazole (Panda et al., 2015b).

The anthelminthic activity of *G. arborea* bark extracts employing ethanol, ethyl acetate, n-butanol, and pet ether as solvents was determined by Nayak et al. (2015a). Entirely extracts were having capacities to display anthelminthic action at 10 mg/ml dose and were analogous with the typical drugs.

Ambujakshi et al. (2009) demonstrated the effect of anthelminthic activity of EtOH and water extracts of leaves of *G. arborea*. These plant extracts exhibited the anthelminthic activity in concentration reliant on way providing through time of immobilization and death associated to piperazine citrate particularly with 100 mg/ml dose for *P. posthuma* and *Ascardia galii* worms by increasing chloride ion conduction of worm muscle membrane manufactures hyperpolarization and decreases nervousness that leads to muscle easing and flaccid paralysis.

4.3.8 ANTICONVULSANT ACTIVITY

The bark extract of *G. arborea* in methanol was found to be effective as anticonvulsant. An effective concentration of 500 mg/kg body weight was administered prior to inducing seizure using pentylenetetrazole (PTZ) by delaying the period of onset of seizures by 3.494 min, similarly for a lower concentration of 250 mg/kg body weight a delay onset of seizures by 2.408 min. The strychnine (STR)-induced model exhibited even better delay of onset periods for the same concentrations of extract. A delay of 11.81 min was recorded with 250 mg/kg methanol bark extract being administered. In case of 500 mg/kg, 80% protection against seizures was seen. This action was because of the existence of flavonoids and phenolics in appropriate amounts (Acharya et al., 2015).

4.3.9 ANTIFUNGAL ACTIVITY

The ethyl acetate extract of heartwood of *G. arborea* was found to be effective as antifungal agent against *Trametes versicolor* and *Fomitopsis palustris*. The presence of certain bioactive compounds in the heartwood extract such as 7-O-ethyl arboreol, paulownin, gmelinol, epieudesmin, and β-sitosterol in ethyl acetate extract contributed to their antifungal activity (Kawamura et al., 2004).

4.3.10 ANTIBACTERIAL ACTIVITY

The crude extract from leaves of *G. arborea* with various solvents was tested for antibacterial activity against *Escherichia coli, Klebsiella pneumoniae, Proteus mirabilis, Shigella dysenteriae*, and *Salmonella typhi* enterobacteria samples. *Salmonella typhi* occurred to be the maximum delicate bacteria for the leaf extract of *G. arborea*. The antimicrobial activity of *G. arborea* was deduced because of the presence of carbohydrates, alkaloids, phenols, tannins, saponins, anthracenes, and cardiac glycosides. However, anthraquinones were found to be present in the bark of the plant, similar results can be observed when bark extracts are employed for a similar study (El-Mahmood et al., 2010).

4.3.11 IMMUNOMODULATORY PROPERTY

The immunomodulatory activity of roots of *G. arborea* was investigated by Shukla et al. (2010) who assessed the pharmacological action employing the methanolic extract of *G. arborea* and its ethyl acetate portion. The adjusting impact was assessed on humoral and cell-mediated immune response employing animal models such as cyclophosphamide-induced myelosuppression, delayed-type hypersensitivity (DTH) response, and humoral antibody (HA) titer. Together test extracts showed noteworthy rise in HA titer, DTH response, and stages of total white blood cell count. Thus, drug was found to be a potential immunostimulant.

4.3.12 ANTIHYPERLIPIDEMIC ACTIVITY

Punitha et al. (2012) assessed the antihyperlipidemic properties of alcoholic leaf extract of *G. arborea* in male Wistar albino rats. The ethanolic extract of *G. arborea* at a concentration of 150 mg/kg of body mass and typical drug glibenclamide at the concentration of 100 μg/kg were specifically given to the animal replicas. The extract displayed noteworthy hypoglycemic action in animal replicas when compared with a normal antidiabetic drug glibenclamide. The hypoglycemia showed by the extract might be because of enlarged uptake of glucose at tissue level and or rise in pancreatic β-cell working or because of prevention of intestinal glucose absorption. The lipid outline such as TC, TG, and LDL levels were meaningfully inclined in diabetic regulator animals, while HDL levels were reduced when compared to the controller rats. The noteworthy changes were also found in the rats treated with ethanolic *G. arborea* extract. Hence, the ethanolic extract of *G. arborea* exhibited important antihyperglycemic action in streptozotocin (STZ)-encouraged diabetic rat which is analogous to that of the glibenclamide (Punitha et al., 2012).

4.3.13 ANTIPYRETIC ACTIVITY

Parhi et al. (2011) inspected the analgesic and antipyretic action of benzene, chloroform, ethanol, and water extracts of the heart wood bark of *G. arborea*. Vigorous albino mice considering about 30–40 g were utilized for estimation of the analgesic action. Wistar albino rats with yeast-induced pyrexia were managed with benzene, chloroform, alcoholic, and water bark extracts of the plant *G. arborea* (420 mg/kg). The water and alcoholic extracts of bark of the plant *G. arborea* displayed noteworthy antipyretic action as equivalent to paracetamol (50 mg/kg bw). But, the analgesic action of trial mixtures was found to be more important on acetic acid-induced test than tail flick test and therefore it seems that the trial substances prevent principally the peripheral pain mechanism. In this way, the barks of the plant *G. arborea* have important analgesic and antipyretic action as equivalent to the typical drug.

4.3.14 ANTIULCER ACTIVITY

Giri et al. (2009) reported the antiulcer action of *G. arborea* leaves in Wistar rats. The consequence of hydroethanolic extracts of leaves of *G. arborea* on gastric ulcers was assessed by means of various experimental replicas such as aspirin-encouraged ulcers, pylorus ligation-encouraged ulcers, alcohol-induced ulcers, and cold restrain stress-induced ulcers. The hydroalcoholic leaf extracts of *G. arborea* (286 and 667 mg/kg) displayed a noteworthy consequence on ulcer encouraged by various ulcer induced models. The leaf extract of *G. arborea* was active in ulcer therapeutic. In this way, the leaves of *G. arborea* raised the therapeutic of gastric ulcers and also avoided the growth of gastric ulcers in rats.

4.3.15 ANALGESIC ACTIVITY

Kulkarni et al. (2013) examined the toxicity and analgesic action of the alcoholic extract (AlcE) of *G. arborea* stem bark and its different portions. The acute toxicity of AlcE and its portions was examined in female Swiss albino mice by employing acetic acid-induced wriggling technique. In the acute toxicity trial, mice were experimented with AlcE and its portions (2000 mg/kg), orally. Animals were detected sporadically during the first 24 h after injection and regular then for 14 days. The AlcE (250–500 mg/kg) and its portions (200 mg/kg) were injected to animals earlier 45 min of running of acetic acid. AlcE and its fractions did not produce mortality, changes in behavior or any other physiological actions in mice at designated concentration. The AlcE at 250 and 500 mg/kg displayed a noteworthy decline in writhes as 13.4 and 12.2, correspondingly when associated with control (19.2). The n-butanol fraction presented higher prevention of wriggles

(57.37%) as associated to other portions. Consequently, AlcE and its portions are harmless after oral injection and have noteworthy analgesic action.

Gangwar et al. (2013) studied the peripheral analgesic activity of *G. arborea* (100, 300, and 1000 mg/kg) in mice using acetic acid-induced wriggling by Randall–Selitto test. The essential pain relieving of alcoholic extract of *G. arborea* was examined utilizing hot plate technique and tail clip technique. Alcoholic extract of *G. arborea* suggestively reduced the wriggling actions in mice in acetic acid-encouraged wriggling trial. Alcoholic extract of *G. arborea* (1000 mg/kg) pointedly improved the pain edge capacity in rats in Randall–Selitto test and the response time in hot plate trial.

4.3.16 TOXICITY

Kulkarni and Veeranjaneyulu (2012) designed methods for the evaluation of acute and frequent concentration toxicity of the methanol extract (ME) of the stem bark. The injection of ME from *G. arborea* bark at 300–5000 mg/kg did not yield mortality or noteworthy variations in the scientific signs and deprived of adverse result level (NOAEL) of ME was 5000 mg/kg. There were no noteworthy variances in the universal condition, development, organ weights, hematological limitations, clinical chemistry values, or gross and minute presence of the organs from the dealing groups as associated to the control group. From this, the ME of *G. arborea* was found harmless in acute and frequent dosage toxicity studies when established in mice and rats.

4.3.17 ANTICANCER ACTIVITY

N'gaman et al. (2014) showed that the hydroacetonic crude extract had a good anticancer activity against C6 glioma cells.

KEYWORDS

- ***Gmelina arborea***
- **antibacterial activity**
- **anti-inflammatory activity**
- **analgesic activity**
- **anticancer activity**

REFERENCES

Acharya, N. S.; Acharya, S. R.; Kumar, V.; Barai, P. Anticonvulsant and Antioxidant Effects of Methanol Extract of Stems of *Gmelina arborea* Roxb. *J. Nat. Remedies* **2015,** *15*(1), 23–32.

Ambujakshi, H. R.; Thakkar, H. S. Anthelmintic Activity of *Gmelina arborea* Roxb. Leaves Extract. *Indian J. Pharma. Res. Dev.* **2009,** *1*, 1–5.

Arora, C.; Tamrakar, V. *Gmelina arborea:* Chemical Constituents, Pharmacological Activities and Applications. *Int. J. Phytomedicine* **2017,** *9*, 528–542.

Attanayake, A. P.; Jayatilaka, K. A. P. W.; Pathirana, C.; Mudduwa, L. K. B. Antioxidant Activity of *Gmelina arborea* Roxb. (Verbenaceae) Bark Extract: In Vivo and In Vitro Study. *J. Med. Nutr. Nutraceut.* **2015,** *4*, 32–38.

Chellappan, D. R.; Pemiah, B. Pharmacognostical, Phytochemical and *In Vivo* Gastroprotective Investigation of *Gmelina arborea. Int. J. Pharm. Pharm. Sci.* **2014,** *6*, 153–157.

Chothani, D. L.; Patel, N. M. Phytochemical Screening and Quantification of Phytoconstituents in *Gmelina arborea* Fruits Extracts. *J. Med. Plants Stud.* **2018,** *6*(4), 31–35.

El-Mahmood, A. M.; Doughari, J. H.; Kiman, H. S. *In Vitro* Antimicrobial Activity of Crude Leaf and Stem Bark Extracts of *Gmelina arborea* (Roxb) Against Some Pathogenic Species of Enterobacteriaceae. *Afr. J. Pharm. Pharmacol.* **2010,** *4*(6), 355–361.

Florence, A. R.; Jeeva, S. FTIR and GC-MS Spectral Analysis of *Gmelina Asiatica* L. Leaves. *Sci. Res. Reporter* **2015,** *5*(2), 125–136.

Gandigawad, P.; Poojar, B., Hodlur, N.; Sori, R. K. Evaluation of Anti-inflammatory Activity of Ethanolic Extract of *Gmelina arborea* in Experimental Acute and Sub-acute Inflammatory Models in Wistar Rats. *Int. J. Basic & Clin. Pharmacol.* **2018,** *8*(1), 128.

Gangwar, A. K.; Ghosh, A. K.; Hoque, M.; Saxena, V. Analgesic Activity of *Gmelina arborea* Roxb in Colony Bred Swiss Mice and Wister Rats. *Int. J. Phcog. Phytochem. Res.* **2013,** *5*(1), 66–67.

Ghareeb, M. A.; Shoeb, H. A.; Madkour, H. M. F.; Refahy, L. A.; Mohamed, M. A.; Saad, A. M. Antioxidant and Cytotoxic Activities of Flavonoidal Compounds from *Gmelina arborea* Roxb. *Global J. Pharmacol.* **2014,** *8*(1), 87–97.

Giri, M.; Divakar, K.; Goli, D.; Dighe, S. B. Anti-ulcer Activity of Leaves of *Gmelina arborea* Plant in Experimentally Induced Ulcer in Wistar Rats. *Pharmacol. Online* **2009,** *1*, 102–110.

Gu, W.; Hao, X.-J.; Liu, H.-X.; Wang, Y.-H.; Long, C.-L. Acylated Iridoid Glycosides and Acylated Rhamnopyranoses from *Gmelina arborea* Flowers. *Phytochem. Lett.* **2013,** *6*(4), 681–685.

Kawamura, F.; Ohara, S.; Nishida, A. Antifungal Activity of Constituents from the Heartwood of *Gmelina arborea*: Part 1. Sensitive Antifungal Assay Against Basidiomycetes. *Holzforschung* **2004,** *58*, 189–192.

Kulkarni, Y. A.; Chavan, D.; Shah, K. Toxicity Study and Analgesic Activity of *Gmelina arborea* Extract and Its Fractions. *J. Nat. Pharm.* **2013,** *4*, 71–74.

Kulkarni, Y. A.; Veeranjaneyulu, A. Antidiabetic Activity of Aqueous Extract of *Gmelina arborea* Bark in Rats. *Altern. Ther. Health Med.* **2009,** *15*, 183.

Kulkarni, Y. A.; Veeranjaneyulu, A. Effect of *Gmelina arborea* Extract in STZ Induced Type I Diabetic Rats. *FASEB J.* **2011,** *25*, 805.

Kulkarni, Y. A.; Veeranjaneyulu, A. Effects of *Gmelina arborea* Extract on Experimentally Induced Diabetes. *Asian Pacific J. Trop. Med.* **2013,** 602–608.

Kulkarni, Y. A.; Veeranjaneyulu, A. Toxicological Evaluation of the Methanol Extract of *Gmelina arborea* Roxb. Bark in Mice and Rats. *Toxicol Int.* **2012,** *19*, 125–134.

Kumaresan, P.; Jeyanthi, K. A.; Kalaivani, R. Biochemical Evaluation of Antidiabetic Activity of Aqueous Extract of *Gmelina arborea* in Alloxan Induced Albino Rats. *Int. J. Herbal Med.* **2014,** *2*(2), 90–94.

Merlin, N. J.; Parthasarathy V. Antioxidant and Hepatoprotective Activity of Chloroform and Ethanol Extracts of *Gmelina Asiatica* Aerial Parts. *J. Med. Plants Res.* **2011,** *5*(4), 533–538.

Moronkola, D. O.; Ogunwande, I. A.; Ba!er, K. H. C.; Ozek, T.; Ozek, G. Essential Oil Composition of *Gmelina arborea* Roxb., Verbenaceae, from Nigeria. *J. Essent. Oil Res.* **2009,** *21*, 264–266.

Muritala, A.; Adebay, Avoseh, O. N.; Openiyi, I. O.; Lawal, O. A.; Isiaka A. O. Anti-nociceptive and Anti-inflammatory Effects of *Gmelina arborea* Roxb. *Investigational Med. Chem. Pharmacol.* **2018,** *1*(2), 13.

Nayak, B. S.; Dinda, S. C.; Ellaiah, P. Evaluation of Diuretic Activity of *Gmelina arborea* Roxb. Fruit Extract. *Asian J. Pharm. Clin. Res.* **2013,** *6*(1), 111–113.

Nayak, B. S.; Ellaiah, P.; Dinda, S. C. Antibacterial, Antioxidant and Antidiabetic Activities of *Gmelina arborea* Roxb. Fruit Extract. *Int. J. Green Pharmacy* **2012,** *6*(3), 224–230.

Nayak, B. S.; Dash, M. R.; Sahu, A.; Sethy, S. Anthelmintic and Antidiabetic Activities of *Gmelina arborea* Roxb. Bark Extracts. *Innoriginal Int. J. Sci.* **2015a,** *2*(2), 13–17.

Nayak, B. S.; Rath, A. K.; Pattnaik, S. C. Traditional Use, Toxicity Study and Phytochemical Screening of *Gmelina arborea* Roxb. Extracts. *Int. J. Phytopharmacy Res.* **2015b,** *6*(2), 78–81.

N'gaman, K. C. C.; Kabran, G. R. M.; Kadja, B. A.; Mamyrbékova-Békro, J. A.; Pirat, J. L.; Lecouvey, M.; Sainte-Cathérine, O.; Sommerer, N.; Verbaere, A.; Meudec, E.; Békro, Y. A. ULPC-MS/MS Phenolic Quantification and In Vitro Anticancer Potential of *Gmelina arborea* Roxb. (Verbenaceae). *Der Chemica Sinica.* **2014,** *5*, 13–17.

Panda, S. K.; Das, D.; Tripthathy, N. K. Evaluation of Laxative and Antipyretic Activity of Various Root Extracts of *Gmelina arborea* Roxb. *World J. Pharm. Pharmaceut. Sci.* **2015b,** *4*, 1105–1112.

Panda, S. K.; Das, D.; Tripthathy, N. K. Phytochemical Investigation and Anthelmintic Activity of Various Root Extracts of *Gmelina arborea* Roxb. *Asian J. Plant Sci. Res.* **2015a,** *5*, 54–58.

Pardeshi, J.; Katariya, M. V.; Amolik, K.; Sane, B.; Mahapure, A. Physico-chemical and Preliminary Phytochemical Screening of Roots of Some Species of Dashmoola. *J. Pharmacognosy Phytochem.* **2018,** *7*(3), 3583-3586.

Parhi, K. P.; Pattanayak, P.; Misra, P.; Pani, M. K. An In-vivo Study on Analgesic and Antipyretic Activity of Bark Extracts of *Gmelina arborea. Int. J. Pharm. Sci. Rev. Res.* **2011,** *10*, 78.

Pathala, D.; Harini, A.; Hegde, P. L. A review on Gambhari (*Gmelina arborea* Roxb.). *J. Pharmacogn. Phytochem.* **2015,** *4*(2), 127–132.

Patil, S. M.; Kadam, V. J., Ghosh, R. *In Vitro* Antioxidant Activity of Methanolic Extract of Stem Bark of *Gmelina arborea* Roxb (Verbenaceae). *Int. J. Pharm. Tech. Res.* **2009,** *1*(4), 1480–1484.

Pattanayak, P.; Parhi, P. K.; Mishra, S. K.; Khandei, P. K. Screening of Anti-diabetic Activity of Bark Extracts of *Gmelina arborea* in Streptozotacin Induced Diabetic Rats. *Int. J. Pharm. Sci. Rev. Res.* **2011,** *8*, 130–132

Punitha, D.; Thandavamoorthy, A.; Arumugasamy, K.; Suresh, S. N.; Danya, U.; Udhayasankar, M. R. Anti-hyperlipidemic Effect of Ethanolic Leaf Extract of *Gmelina arborea* in Streptozotocin Induced Male Wistar Albino Rats. *Intern. J. Life Sci. Pharma Res.* **2012,** *2*, 46.

Rao, D. V.; Rao, E. V. Chemical Components of the Leaves of *Gmelina arborea* Linn. *Indian J. Pharm.* **1970,** *32*(5), 140.

Shoeb, H. A.; Madkour, H. M. F.; Refahy, L. A.; Mohamed, M. A.; Saad, A. M.; Ghareeb, M. A. Antioxidant and Cytotoxic Activities of *Gmelina arborea* Roxb. Leaves. *British J. Pharm. Res.* **2014,** *4*(1), 125–144.

Shukla, S. H.; Saluja, A. K.; Pandya, S. S. Modulating Effect of *Gmelina arborea* Linn. On Immunosuppressed Albino Rats. *Phcog. Res.* **2010,** *2*, 359.

Sinha, C.; Dixit, P.; Bhargava, S.; Devasagayam, T. P.; Ghaskadbi, S. Bark and Fruit Extracts of *Gmelina arborea* Protect Liver Cells from Oxidative Stress. *Pharm. Biol.* **2006,** *44*, 237.

Sravani, P.; Murali, C. M.; Syed, S.; Sadik, B. S.; Soubia, S. N. Evaluation of Diuretic Activity of *Gmelina arborea* Roxb. *Intern. J. Adv. Pharmaceut. Res.* **2011,** *2*(4), 157–161.

Vijay, T.; Rajan, M. S. D.; Sarumathy, K.; Palani, S.; Sakthivel, K. Cardioprotective, Antioxidant Activities and Phytochemical Analysis by GC-MS of *Gmelina arborea* (GA) in Doxorubicin-induced Myocardial Necrosis in Albino Rats. *J. Appl. Pharm. Sci.* **2011,** *1*(5), 198–204.

Wansi, S. L.; Nyadjeu, P.; Nguelefack, T. B.; Fodouop, S. F. K.; Donatien, A. A.; Kamanyi, A. *In Vivo* Antioxidant and Vasodilating Activities of *Gmelina arborea* (Verbenaceae) Leaves Hexane Extract. *J. Comple. Integrative Med.* **2012,** *9*(1), 26.

Warrier, R. R.; Priya, S. M.; Kalaiselvi, R. *Gmelina arborea* – An Indigenous Timber Species of India with High Medicinal Value: A Review on its Pharmacology, Pharmacognosy and Phytochemistry. *J. Ethnopharmacol.* **2021,** *267*, 113593.

CHAPTER 5

PHYTOCHEMISTRY AND PHARMACOLOGY OF *HYPTIS CRENATA* POHL EX BENTH.: A REVIEW

VISHAL P. DESHMUKH

Department of Botany, Jagadamba Mahavidyalaya, Achalpur, Amravati, Maharashtra, India

ABSTRACT

Hyptis crenata Pohl ex Bentham is an important annual herbaceous plant belonging to family Lamiaceae, reported widely from Brazil. It has common names like "Salva-do-marajó," "salva," "hortelã brava" (wild peppermint), "hortelã-do-campo," and "hortelãzinha". *H. crenata* plant parts are used to cure fever, rheumatism, gastrointestinal disorders, inflammation and as contraceptive. Plants parts are rich source of essential oil and more than 33 compounds are reported from it. Plants have antimicrobial, antioxidant, antiulcer, hepatoprotective, neuroprotective and anti-inflammatory potential. Present review focuses on phytochemical diversity and pharmacological potential of *Hyptis crenata*.

5.1 INTRODUCTION

Hyptis crenata Pohl ex Bentham is an important member of the family Lamiaceae (McNeil et al., 2011). It is an annual herbaceous plant with height less than a meter. It widely breeds near banks of rivers in sandy soil mostly in Marajó Island, State of Pará, Brazil (Rebelo et al., 2009). "Salva-do-marajó,"

Bioactives and Pharmacology of Lamiaceae. T. Pullaiah, PhD (Ed.)

"salva," "hortelã brava" (wild peppermint), "hortelã-do-campo," and "hortelãzinha" are the common trivial names used for *Hyptis crenata* in Northeast part of Brazil (Coelho-de-Souza et al., 2021; Scramin et al., 2000; Zoghbi et al., 2002). In Brazil, decoctions and infusions of *H. crenata* are used to cure fever, contraceptive, rheumatism, gastrointestinal disorders, and inflammation (Ribeiro et al., 2017; Elisabetsky and Posey, 1989; Di Stasi et al., 2002; de Jesus et al., 2009; Van den Berg and da Silva, 1988).

5.2 BIOACTIVES

Hydroalcoholic extraction of flowers and leaves of *Hyptis crenata* using ultrasound for a period of 45 min produces excellent amounts of phenolics, and flowers showed superior levels of phenolic compounds than leaves (Cardoso et al., 2020). The GC-MS analysis of essential oil extracted from leaves and branches of *H. crenata* documented total 13 compounds like, camphor, 1,8-cineole (Eucalyptol), α-pinene, β-caryophyllene, p-cymene, l-limonene, camphene, γ-terpinene, aromadendrene, ledene, viridiflorol, 10-epi-γ-eudesmol, and caryophyllene oxide (Fig. 5.1) (Coelho-de-Souza et al., 2021). Steam distilled essential oil extracted from aerial parts of *H. crenata* identified 14 compounds, namely, myrcene, α-terpinene, 4-terpineol, aromadendrene, α-pinene, viridiflorol, 1,8-cineole, ledene, γ-terpinene, camphene, p-cymene, camphor, β-caryophyllene, and caryophyllene oxide (Diniz et al., 2013). The GC-MS investigation of fresh and dried leaves and stem extract of *H. crenata* yielded total 23 volatile components such as γ-terpinene, α-thujene, 4-terpineol, α-longipinene, camphene, sabinene, β-chamigrene, myrcene, p-cymene, α-terpinene, α-pinene, 1,8-cineole, β-pinene, terpinolene, camphor, borneol, α-phellandrene, limonene, α-terpineol, β-caryophyllene, aromadendrene, β-himachalene, and himachalol. Among these α-pinene and 1,8-cineole shared major portions (Rebelo et al., 2009). Decoction, infusion, and cold extraction with water and alcohol of *H. crenata* collected from two regions of Brazil confirmed camphor and eucalyptol as dominant volatile compounds (Rocha et al., 2009). Essential oil of *H. crenata* flowering aerial parts collected from Brazilian Pantanal, were dominated by compounds like α-pinene, β-pinene, camphor, and β-caryophyllene, additionally camphene, myrcene, δ-3-carene, p-cymene, 1,8-cineole, limonene, γ-terpinene, terpinolene, isobomeol, terpinen-4-ol, and α-terpineol were also isolated (Scramin et al., 2000). Essential oil extracted from *H. crenata* aerial parts accounted 33 compounds on GC-MS investigation with predominance of monoterpenes like 1,8-cineole, α-pinene, γ-terpinene, borneol, and p-cymene (Violante et al., 2012a). Methanol extract of *H. crenata* yielded bis-seco-abietane diterpenoid called "hyptisolide A" which is extremely oxidized in nature (Yun et al., 2014). Total 95 compounds were isolated from essential oil of *H. crenata* leaves and branches collected from four different sites of Brazil. Sample "A" contains 47 compounds of which terpinolene occupies major portion (37.8%), sample "B" accounted 40 compounds with major portion of 1,8-cineole (23.9%), sample "C" comprises 21 compounds having α-pinene as prime compound (51.1%), and sample "D" with 44 compounds dominated by 1,8-cineole (36.7%) (Zoghbi et al., 2002).

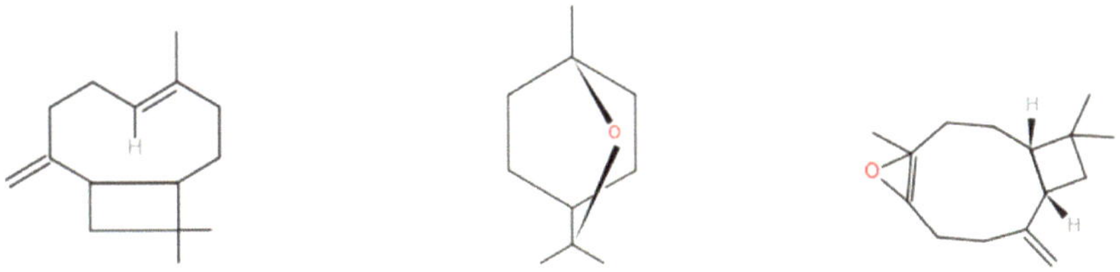

β-caryophyllene **1,8-cineole** **Caryophyllene oxide**

FIGURE 5.1 Structures of major compounds found in essential oil of *Hyptis crenata* (Bridi et al., 2021; Xu et al., 2013).

5.3 PHARMACOLOGY

5.3.1 ANTIBACTERIAL ACTIVITY

Aqueous extract of *H. crenata* leaves inhibited the growth of *E. coli* and *P. aeruginosa* with 12.50 and 16.00 mm zones of inhibition, respectively (Mota et al., 2015). Essential oil extracted from aerial parts of *H. crenata* revealed potent antibacterial activity against three out of the five bacteria explored, and strongest activity with 62.5 mg/mL MIC was recorded against *S. aureus* and *E. faecalis* (Violante et al., 2012a). Dichloromethane and ethyl acetate fractions of *H. crenata* revealed strongest antibacterial activity against *E. faecalis* with MIC of 62.5 and 31.3 mg/mL, respectively; four out of five fraction of *H. crenata* showed activity against *S. aureus* with MIC range of 62.5–250 mg/mL (Violante et al., 2012b).

5.3.2 ANTIFUNGAL ACTIVITY

Ethanol extract of *H. crenata* aerial parts showed antifungal potential against 75 colonies of *Sphaerotheca fuliginea* (Stadnik et al., 2003). Essential oil derived from *H. crenata* aerial parts showed antifungal activity against all tested fungal species with MIC 250 mg/mL and strongest activity was observed against *C. neoformans* with MIC 62.5 mg/mL (Violante et al., 2012a). Ethanol extract, dichloromethane, and ethyl acetate fraction of *H. crenata* showed substantial activity against *C. krusei* with MIC range of 125–250 mg/mL. Ethanol extract and hexane fraction of *H. crenata* also showed promising activity (MIC = 125 mg/mL) against *Cryptococcus neoformans* (Violante et al., 2012b).

5.3.3 ANTIOXIDANT EFFICACY

Hydroalcoholic extract of *H. crenata* flower and leaves prepared using ultrasound for 45 min showed best antioxidant potential in ABTS assay as compared to other tested time period of ultrasound extraction. Among them flower had upper edge over leaf (84.46 ± 1.93 μM trolox/g and 65.81 ± 2.06 μM trolox/g, respectively) (Cardoso et al., 2020). In DPPH assay, essential oil extracted from fresh and dry parts of *H. crenata* showed low inhibition potential ranging between 42.6–79.9% and 24.5–71.4% with EC_{50} values of 1951.5 ± 54.5 and 6882.7 ± 522.9 mg/mL respectively; on the contrary, methanol extract of *H. crenata* dry parts showed significant antioxidant potential ranges from 14.2 to 94% and EC_{50} = 16.7 ± 0.4 mg/mL (Rebelo et al., 2009).

5.3.4 ANTIEDEMATOGENIC EFFECT

In dextran-induced edema, essential oil derived from leaves and branches of *H. crenata* showed high (59.24 ± 2.13%) inhibition at 300 mg/kg dose, whereas at the same dose in carrageenan-induced edema the highest (82%) inhibition was observed. Additionally, histamine, serotonin, and bradykinin-induced edema was significantly inhibited by *H. crenata* essential oil (Coelho-de-Souza et al., 2021).

5.3.5 ANTIULCER ACTIVITY

In absolute ethanol-induced mice, oral administration of essential oil extracted from *H. crenata* aerial parts at 300 mg/kg dose significantly declined (1.71

± 0.54 mm^2) gastric lesion and reduced TBARS level (2.4 ± 0.1 μmols/mg tissue) compared to control. Against indomethacin-induced gastric lesions a significant gastroprotective activity at 300 mg/kg dose of essential oil was recorded. Animals treated with essential oil of same dose revealed a declined rate of gastric emptying as compared to control. On the contrary, there was no change in gastric juice pH, gastric volume, and mucous amount of gastric mucosa of animals at any dose of essential oil (Diniz et al., 2013).

5.3.6 HEPATOPROTECTIVE ACTIVITY

Oral administration of *H. crenata* essential oil declines the sepsis-induced elevated serum ALP, ALT, and total bilirubin compared to control. The liver of essential oil-treated septic rats revealed a significantly normal tissue pattern with declined hydropic degeneration and existence of glycogenosis (Lima et al., 2018).

5.3.7 NEUROPROTECTIVE EFFECT

In cecal ligation and puncture sepsis operated rats, *H. crenata* essential oil at 300 mg/kg dose hampered elevated IL-1β and TNF-α levels and epithelial loss preventing inflammatory cell infiltration in the hypothalamus. *H. crenata* essential oil significantly restrains enhanced hypothalamic lipid peroxidation due to sepsis (Lima et al., 2020).

5.3.8 TOXICOLOGICAL ASSAY

Aqueous extract of *H. crenata* leaves showed poor in vitro toxicity against microcrustacean *Artemia salina* Leach with LC_{50} = 1028.30 μg/mL proving it as nontoxic (Mota et al., 2015).

5.3.9 CYTOTOXIC ACTIVITY

In brine shrimp bioassay, essential oil extracted from fresh parts of *H. crenata* revealed LC_{50} = 6.7 ± 0.2 μg/mL which is two times greater than methanol extract of dry parts with LC_{50} = 13.0 ± 3.7 μg/m (Rebelo et al., 2009). Ethanol extract and its fractions failed to show any toxicity against

A. salina having $LD_{50} > 1000$ mg/mL in brine shrimp lethality bioassay (Violante et al., 2012b).

5.3.10 ANTINOCICEPTIVE EFFECT

Oral administration of *H. crenata* decoction to C57/BL6 mice showed significant antinociceptive effects in phasic pain and tonic pain model with zero mortality as compared to indomethacin (positive control) with 38% of mortality (Rocha et al., 2009).

5.3.11 RENAL PROTECTION

Prolong treatment of mice with *H. crenata* essential oil hardly fluctuates thiol levels and catalase activity in the kidney, however essential oil at 300 mg/kg dose found to reduce nitric production as compared to vehicle (Vasconcelos et al., 2016).

5.3.12 ANTI-INFLAMMATORY EFFECT

In carrageenan-induced paw edema model *H. crenata* essential oil reported significant anti-inflammatory activity in mice with reduced myeloperoxidase (59.48%) and nitrite levels (41.08%), respectively as compared to control (Soares et al., 2016).

5.3.13 ANTI-TICK POTENTIAL

In adult immersion test, *H. crenata* essential oil was tested against engorged females of cattle tick *Rhipicephalus (Boophilus) microplus*. Although essential oil failed to show hatchability inhibition at all tested concentrations, it significantly retarded (2.8 ± 8.4%) egg conversion at 2.5% concentration and showed higher product effectiveness (94.4 and 76.1%) at 2.5 and 1.0% concentration, respectively (Violante et al., 2012a).

5.3.14 BIOACTIVITY

"Hyptisolide A" is a bis-seco-abietane diterpenoid isolated from methanol extract of *H. crenata* found to possess CRE-dependent transcription induction in mouse Neuro2A cells with three times enhanced luciferase activity (Yun et al., 2014).

KEYWORDS

- **essential oil**
- ***Hyptis crenata***
- **neuroprotective**
- **phytochemical**
- **rheumatism**

REFERENCES

Bridi, H.; Meirelles, G. D. C.; Lino von Poser, G. Subtribe Hyptidinae (Lamiaceae): A Promising Source of Bioactive Metabolites. *J. Ethnopharmacol.* **2021,** *264*, 113225.

Cardoso, R. V.; Lopes Santos, S. D. J.; de Araújo, A. L.; de Araújo, J. A.; Gandra, E. A. Effect of the Ultrasound Assisted Extraction on the Content Phenolic Compounds and Antioxidant Activity from Leaf and Flower Extracts of Salvado-Marajó (*Hyptis crenata* Pohl ex Benth.). *Braz. J. Develop.* **2020,** *6*, 61533–61542.

Coelho-de-Souza, A. N.; Alves-Soares, R.; Oliveira, H. D.; Gomes-Vasconcelos, Y. A.; Souza, P. J. C.; Santos-Nascimento, T.; Oliveira, K. A.; Diniz, L. R. L.; Guimarães-Pereira, J.; Leal-ardoso, J. H. The Essential Oil of *Hyptis crenata* Pohl ex Benth. Presents an Antiedematogenic Effect in Mice. *Braz. J. Med. Biol. Res.* **2021,** *54*, e9422.

de Jesus, N. Z. T.; Lima, J. C. D. S.; Da Silva, R. M.; Espinosa, M. M.; Martins, D. T. D. O. Levantamento Etnobotanico de Plantas Popularmente Utilizadas Como Antiúlceras e Antiinflamatorias Pela Comunidade de Pirizal, Nossa Senhora do Livramento-MT. Brasil. *Brazilian J. Pharmacogn.* **2009,** *19*, 130–139.

Di Stasi, L. C.; Oliveira, G. P.; Carvalhaes, M. A.; Queiroz-Junior, M.; Tien, O. S.; Kakinami, S. H.; Reis, M. S. Medicinal Plants Popularly Used in the Brazilian Tropical Atlantic Forest. *Fitoterapia* **2002,** *73*, 69–91.

Diniz, L. R. L.; Vieira, C. F. X.; Santos, E. C. D.; Lima, G. C.; Aragao, K. K. V.; Vasconcelos, R. P.; Araújo, P. C. D. C.; Vasconcelos, Y. D. A. G.; de Oliveira, A. C.; de Oliveira, H. D.; Portella, V. G.; Coelho-de-Souza, A. N. Gastroprotective Effects of the Essential Oil of *Hyptis crenata* Pohl ex Benth on Gastric Ulcer Models. *J. Ethnopharmacol.* **2013,** *149*, 694–700.

Elisabetsky, E.; Posey, D. A. Use of Contraceptive and Related Plants by the Kayapo Indians (Brazil). *J. Ethnopharmacol.* **1989,** *26*, 299–316.

Lima, G. C.; Araújo, P. C. C.; Silva, R. O.; Soares, P. M. G. Portella, V. G.; de Souza, A. N. C.; Diniz, L. R. L. The Effect of Essential Oil of *Hyptis crenata* on Sepsis-induced Hypothalamus Injury in Rats. *AJMP* **2020,** *8*, 88–95.

Lima, G. C.; Vasconcelos, Y. A. G.; de Santana Souza, M. T.; Oliveira, A. S.; Bomfim, R. R.; de Albuquerque Júnior, R. L. C.; Camargo, E. A.; Portella, V. G.; Coelho-de-Souza, A. N.; Diniz, L. R. L. Hepatoprotective Effect of Essential Oils from *Hyptis crenata* in Sepsis-induced Liver Dysfunction. *J. Med. Food.* **2018,** *21*, 709–715

McNeil, M.; Facey, P.; Porter, R. Essential Oils from the *Hyptis* Genus – A Review (1909–2009). *Nat. Prod. Commun.* **2011,** *6*, 1775–1796.

Mota, L. J. T.; da Silva, E. D.; de Almeida, S. S. M. D. Toxicity Test and Antimicrobial Activity of Aqueous Extract of Leaves of *Hyptis crenata* Pohl ex. Benth. *Revista de Biologia E Ciencias da Terra* **2015,** *1*, 59–63.

Rebelo, M. M.; da Silva, J. K. R.; Andrade, E. H. A.; Maia, J. G. S. Antioxidant Capacity and Biological Activity of Essential Oil and Methanol Extract of *Hyptis crenata* Pohl ex Benth. *Braz J. Pharmacogn.* **2009,** *19*, 230–235.

Ribeiro, R. V.; Bieski, I. G. C.; Balogun, S. O.; Martins, D. T. de O. Ethnobotanical Study of Medicinal Plants Used by Ribeirinhos in the North Araguaia Microregion, Mato Grosso. *Brazil. J. Ethnopharmacol.* **2017,** *205*, 69–102.

Rocha, G.; Roughan, J. V.; Leach, M. C.; Flecknell, P. A.; Ingram, C. D.; Brandt, K. Traditional Use, Chemical Analysis and Antinociceptive Effects of *Hyptis crenata* Pohl. *Planta Med.* **2009,** *75*, PA40.

Scramin, S.; Saito, M. L.; Pott, A.; Marques, M. O. M. Volatile Constituents of *Hypfis crenata* Pohl (Labiatae) Native in Brazilian Pantanal. *J. Essent. Oil Res.* **2000,** *12*, 99–101.

Soares, R. A. S.; Campos, D. C. O.; Vasconcelos, Y. A. G.; Oliveria, H. D.; Coelho-De-Sousa, A. The Essential Oil of *Hyptis crenata* Decreases of Myeloperoxidase and Nitritelevels in Mice Carrageenan-induced Paw Edema [Abstract], XIII SBBQ northeast Regional Meeting – Fortaleza, VI International Symposium in Biochemistry of Macromolecules and Biotechnology, **2016,** J-059.

Stadnik, M. J.; Bettiol, W.; Saito, M. L. Bioprospecting for Plant and Fungus Extracts with Systemic Effect to Control the Cucumber Powdery Mildew. *J. Plant Dis. Prot.* **2003,** *110*, 383–393.

Van den Berg, M. E.; da Silva, M. H. Contribuiçao ao Conhecimento da flora Medicinal de Roraima. *Acta Amazonica* **1988,** *18*, 23–35.

Vasconcelos, Y. A. G.; Pontes, E. O. B.; Carvalho, N. R. P.; Dourado, K. F.; Oliveira, H. D.; Coelho-de-Souza, A. N. Prolonged Treatment with Essential Oil of *Hyptis crenata* Reduces Nitrite Levels in Kidney of Rats [Abstract], XIII SBBQ Northeast Regional Meeting – Fortaleza, VI International Symposium in Biochemistry of Macromolecules and Biotechnology, **2016,** T-006.

Violante, I. M. P.; Garcez, W. S.; Barbosa, C. D. S.; Garcez, F. R. Chemical Composition and Biological Activities of Essential Oil from *Hyptis crenata* Growing in the Brazilian Cerrado. *Nat. Prod. Commun.* **2012a,** *7*, 1387–1389.

Violante, I. M. P.; Hamerski, L.; Garcez, W. S.; Batista, A. L.; Chang, M. R.; Pott, V. J.; Garcez, F. R. Antimicrobial Activity of Some Medicinal Plants from the Cerrado of the Central-western Region of Brazil. *Braz. J. Microbiol.* **2012b,** *43*, 1302–1308.

Xu, D. H.; Huang, Y. S.; Jiang, D. Q.; Yuan, K. The Essential Oils Chemical Compositions and Antimicrobial, Antioxidant Activities and Toxicity of Three *Hyptis* Species. *Pharm. Biol.* **2013,** *51*, 1125–1130.

Yun, Y. S.; Fukaya, H.; Nakane, T.; Takano, A.; Takahashi, S.; Takahashi, Y.; Inoue, H. A New Bis-*seco*-abietane Diterpenoid from *Hyptis crenata* Pohl ex Benth. *Org. Lett.* **2014,** *16*, 6188–6191.

Zoghbi, M. G.; Andrade, E. H.; da Silva, M. H.; Maia, J. G.; Luz, A. I.; da Silva, J. D. Chemical Variation in the Essential Oils of *Hyptis crenata* Pohl ex Benth. *Flavour Fragr. J.* **2002,** *17*, 5–8.

CHAPTER 6

AN OVERVIEW ON *ISODON RUGOSUS* (WALL. EX BENTH.) CODD (FAMILY: LAMIACEAE)

M. MUNIRAJU[1], B. DINESH[2], and S. RAJASHEKARA[2]

[1]*Department of Studies in Botany, Bangalore University, Jnana Bharathi Campus, Off Mysuru Road, Bengaluru, Karnataka, India*

[2]*Centre for Applied Genetics, Department of Studies in Zoology, Bangalore University, Jnana Bharathi Campus, Bengaluru, Karnataka, India*

ABSTRACT

Isodon rugosus is distributed widely throughout the northern temperate regions of the Himalaya and grow up to 1–5 ft tall. Stems are erect with rather slim quadrangular verdant branches. The medicinal properties are contributed by various parts of the plants. This plant has several properties such as anti-pyretic, anthelmintic, anti-aging, anti-nociceptive, antibacterial, analgesic and antioxidant due to the presence of several bioactive phytochemicals such as heptenal, N-butylpropargylamine, octanoic acid, 2-decenal, 2,4-decadienal, decanedioic acid, hexadecanoic acid, ethyl ester, linoleic acid ethyl ester, 9-octadecenoic acid, ethyl ester, 1,2-benzenedicarboxylic acid and bis(2-ethylhexyl) ester.

6.1 INTRODUCTION

Isodon rugosus (Wall. ex Benth.) Codd. is commonly known as Wrinkled Leaf Isodon, distributed widely throughout the northern temperate regions of

Bioactives and Pharmacology of Lamiaceae. T. Pullaiah, PhD (Ed.)

the Himalaya at the raise of 1500–2500 m and is a fragrant, much-expanded shrub, 1–5 ft tall. Stems are erected with rather slim quadrangular verdant branches. Leaves are broad-ovate, inconstant in size, wrinkled. Flowers are borne in cymes on thin sidelong branches, lax, forming often-leafy racemes. Nutlets are pale brown to dark brown, rhombus, 1.5 × 1 mm. Flowering during March–October. *Isodon rugosus* is used in traditional treatment in Dir (KPK), Pakistan for various sorts of pains (Zeb et al., 2017; Sadiq et al., 2018).

6.2 BIOACTIVES

Phytochemical and toxicological examinations of crude methanolic extracts exposed that *I. rugosus* encompasses various distinctive auxiliary, for example, alkaloids, anthraquinones, flavonoids, saponins, oils, glycosides, anthraquinones, tannins, terpenoids, and confirmed negative for the existence of sterols (Zeb et al., 2014a).

Description of the fundamental oil conformation of *I. rugosus* revealed that 83.9–92.6% of the total oil compositions and 92 constituents were identified. Essential oil components were α-pinene (0.8–19.3%), trans-ferruginol (3.0–17.1%), α-phellandrene (2.6–10.5%), (E)-caryophyllene (4.0–9.3%), germacrene D (1.9–8.8%), abietatriene (1.9–5.3%), β-phellandrene (3.1–4.4%), δ-cadinene (2.9–3.6%), limonene (1.3–3.6%), myrcene (0.2–3.6%), and p-cymene (2.9–3.5%). Moreover, 52 constituents including trans-ferruginol, abietatriene, totarene, phyllocladene, 4-epi-abietal, caryophyllene acetate, cubenol, allo-hedycaryol, δ-amorphene, transcadina-1 (6), 4-diene, cis-cadina-1 (6), 4-diene, α-muurolol, amorpha-4,9-dien-2-ol, (Z)-nuciferol acetate, and viridiflorol were documented in the *I. rugosus* essential oil (Verma et al., 2015).

In gas chromatography–mass spectroscopy (GC–MS) investigation, 83 compounds were recognized among which eight substances including benzyl alcohol, sebacic acid, myristic acid, phytol, sugiol, tocopherol, a-Amyrin, and stigmasterol were figured out as formerly described analgesic substances. Zeb et al. (2017) exposed that pain relieving of *I. rugosus* can be accredited to the existence of pain-relieving compounds. In this way, narcotics receptors are associated with the pain-relieving instrument of *I. rugosus* because of the viable threat of naloxone. The designs of bioactive substances have been portrayed in Figure 6.1.

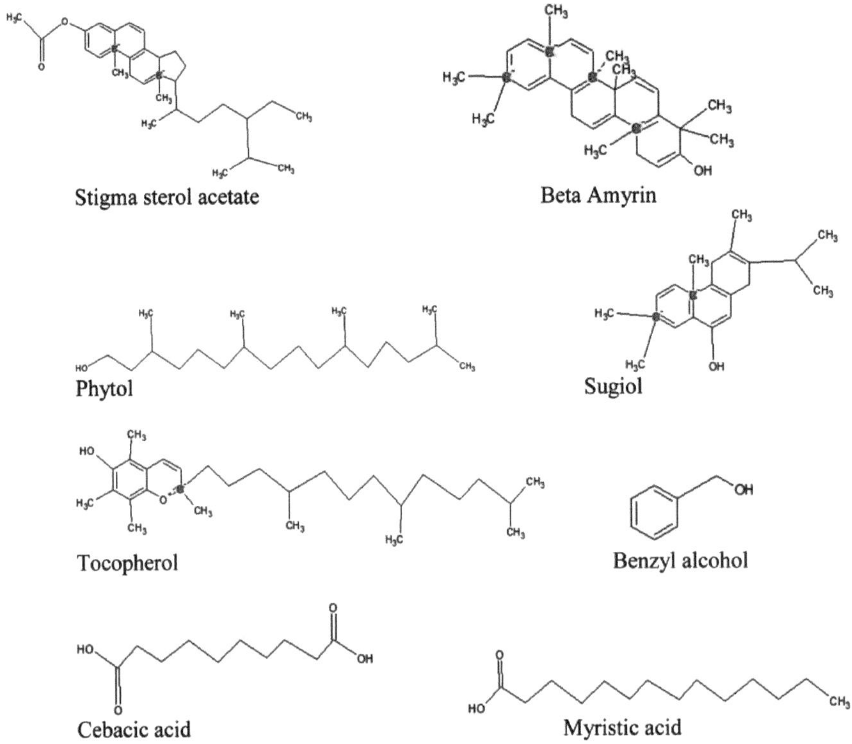

FIGURE 6.1 Bioactive of *I. rugosus.*

An aggregate of 19 ingredients was evaluated and related to various concentrations in fundamental oil, among which caryophyllene oxide was 28.99%. Limonene oxide cis was 26.52%, (Z, E)-Farnesol was 17.34%, trailed by 1-erpinen-4-ol 3.33%, o-cymene 2.04 whereas the concentrations of rest of ingredients were among 1.11–0.2% (Hussain et al., 2017).

The major compounds identified in *I. rugosus* chloroform fraction (Ir.Chf) are heptenal, N-butylpropargylamine, octanoic acid, ethyl ester, 2-decenal, 2,4-decadienal, decanedioic acid, hexadecanoic acid, ethyl ester, linoleic acid ethyl ester, 9-octadecenoic acid, ethyl ester, 1,2-benzenedicarboxylic acid, and bis (2-ethylhexyl) ester (Zeb et al., 2017).

Ullah et al. (2019) separated two diterpenes: Rugosodon 1 and oridonin 2 using the MeOH extract of *I. rugosus* roots. The new compound 1 was illuminated as 1-O-α-D-glucopyranosyl-7α,13β-dihydroxyabieta-8(9)-en-11-one and the recognized compound 2 as 7α,20-epoxy-1α,6β,7,14-tetrahydroxy-kaur-16-en-15-one (oridonin). Rugosodon showed noteworthy probable of

α-glucosidase inhibition with IC_{50} 0.453 mg/mL, as associated to the typical acarbose (IC_{50} 0.921 mg/mL).

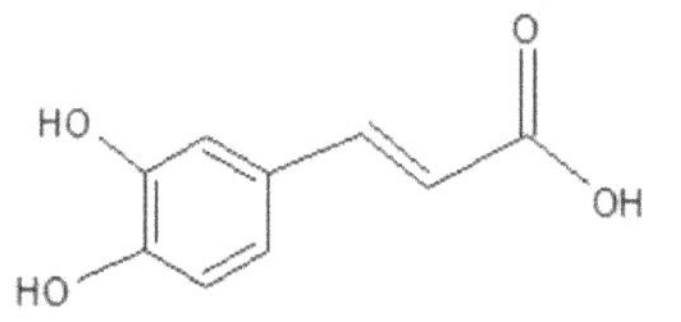

Caffeic acid

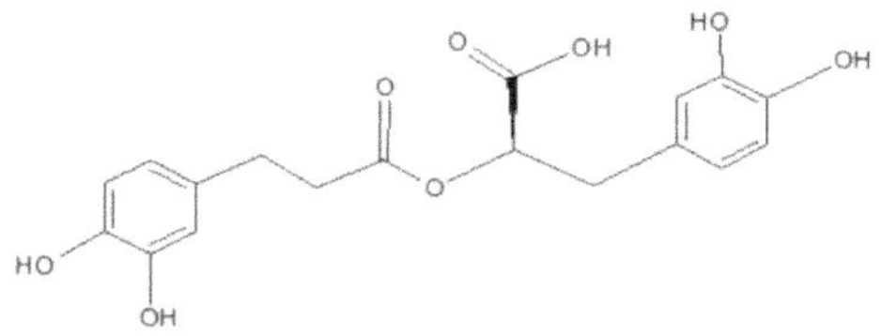

Rosmarinic acid

6.3 PHARMACOLOGY

6.3.1 ANTIPYRETIC ACTIVITY

The utilization of the raw methanol extract of the upper parts of *I. rugosus* to isolated rabbit jejunum preparations unveiled relaxant outcome (0.01–0.3 mg/mL). It also relaxed K^+-induced spastic reductions in isolated rabbit jejunum preparations and lifted the Ca^{+2} concentration–response curves toward the right (0.01–0.3 mg/mL). Additionally, the extract, when applied to the isolated rabbit tracheal preparations relaxed the carbachol (1 μM) along with K^+-induced contractions in a concentration array of 0.01–1.0 mg/mL. Additionally, it also relaxed the phenylephrine (1 μM)$^-$ and K^+ (80 mM)-induced reductions in an isolated rabbit aorta preparation. The Ir.Cr (80 mg/kg) confirmed antipyretic activity on pyrogen-induced pyrexia in rabbits as compared to aspirin as typical drug. The Ir.Cr also revealed antioxidant in addition to inhibitory consequence on acetyl- and butyrylcholinesterase and lipoxygenase (0.5 mg/mL). Along these lines, the noticed relaxant impact on isolated rabbit jejunum, trachea, and aorta arrangements brought about by Ir.Cr was facilitated through Ca^{+2} channel barrier and in this manner might give a logical premise to approve the folkloric employments of the plant in the administration of gastrointestinal, respiratory, and cardiovascular disorders. The experimental antioxidant activity along with the lipoxygenase inhibitory action authenticated the folklore practice in pain and swellings (Janbaz et al., 2014).

6.3.2 ANTHELMINTIC ACTIVITY

The raw saponins exhibited the highest anthelmintic action against *Pherethima posthuma* and *Ascaridia galli* with death times of 27.67 and 29.22

min, respectively, at concentrations of 40 mg/mL. Chloroform portion and saponins unveiled noteworthy outcomes against *Rhyzopertha dominica* (60 and 70%) and *Tribolium castaneum* (70 and 76%) at 200 mg/mL in insecticidal action. Subsequently, in anti-Pharaoh movement, the $CHCl_3$, ethyl acetate, and saponins fractions were maximum effective, each displaying LD50 of <0.1 mg/mL. In GC–MS examination, entire of 57 compounds were recognized. Some of the bioactive compounds recognized in GC–MS analysis are palmitic acid, hinokiol, α-amyrin, phytol, ethyl linolate, cyclohexanone, hinokione, methyl palmitate, ethyl palmitate, and stigmasterol acetate. Consequently, *I. rugosus* hold robust antibacterial, insecticidal, and anthelmintic abilities having crude saponins and ethyl acetate as the maximum vigorous portions. Ethyl acetate fraction is an appropriate mark for the isolation of varied collections of bioactive mixtures (Zeb et al., 2017).

6.3.3 ANTI-AGING ACTIVITY

I. rugosus contains a lot of phenolics and pentacyclic triterpenes. Abbasi et al. (2019) investigated the *in vitro* callus induction from stem and leaf explants of *I. rugosus* for the manufacture of antioxidant and anti-aging substances. Stem-derived callus culture exhibited the highest total phenolic content and antioxidant activity underneath finest hormonal combination. HPLC examination revealed the existence of plectranthoic acid (373.92 μg/g DW), oleanolic acid (287.58 μg/g DW), betulinic acid (90.51 μg/g DW), caffeic acid (91.71 μg/g DW), and rosmarinic acid (1732.61 μg/g DW). Whole antioxidant and anti-aging potential of extracts with very distinct phytochemical outlines were explored. Correlation investigates discovered rosmarinic acid as the chief supplier for antioxidant activity and anti-aging hyaluronidase, advance glycation end-products preventions, and SIRT1 activation, while pentacyclic triterpenoids were associated with elastase, collagenase, and tyrosinase preventions. In total, these consequences evidently showed the excessive valorization potential of *I. rugosus* calli for the manufacture of antioxidant and anti-aging bioactive extracts for beautifying applications (Abbasi et al., 2019).

6.3.4 ANTI-NOCICEPTIVE ACTIVITY

Sadiq et al., (2018) explored the antinociceptive, antioxidant, and anticholinesterase abilities of fundamental oils from *I. rugosus* (Ir.EO), and governed a conceivable mechanism of anti-nociception. The Ir.EO was exposed to GC–MS examination for governing its chemical compositions and analyzed for pain-relieving potential succeeding acetic acid-induced wriggling, formalin exam and hot plate techniques in animal replicas. The GC–MS investigation of Ir.EO exposed the documentation of 141 substances. Ir.EO established robust antinociceptive latent in all three *in-vivo* replicas. With the utilization of nalaxone, it was established that the fundamental oil was substituted on the vital trail of nociception. The Ir.EO also displayed robust free radicals rummaging potential, displaying IC_{50} values of 338 and 118 μg/mL for DPPH and ABTS free radicals correspondingly. In AChE and BChE repressing analyses, the detected IC_{50} values were 93.56 and 284.19 μg/mL, respectively. Therefore, inspiring antinociceptive, antioxidant, and anticholinesterase actions exposed that Ir.EO is a rich basis of bioactive compounds as noticeable from the GC–MS examination.

6.3.5 ANTIBACTERIAL ACTIVITY

Flavonoids and ethyl acetate segments displayed the maximum antibacterial actions among the verified samples of *I. rugosus* (Sadiq et al., 2018).

6.3.6 ANALGESIC ACTIVITY

Zeb et al. (2016) evaluated the analgesic action and confirmed a dosage relied on action. The average prevention of wriggles of positive regulator at the dosage of 10 mg/kg was 16.5 with 73.02% prevention. The *I. rugosus* chloroform portion (Ir.Chf) displayed average prevention of 28.33 with 53.67% prevention at the dosage of 100 mg/kg although at 10 mg/kg it displayed an average prevention of 39.83 with 34.87% prevention. The Ir.Chf exhibited a proportional response (53.67%) with the positive control (73.02%) at the dose of 100 mg/kg. In this way, pain relieving capability of *I. rugosus* can credit to the presence of pain-relieving substances.

6.3.7 ANTIOXIDANT ACTIVITY

Zeb et al. (2014b) assessed the crude methanolic extracts of *I. rugosus* in oxidative stress and neurodegenerative diseases. Among the different fractions, flavonoids (Ir.Flv) and chloroform (Ir.Cf) exhibited the highest inhibitory activity against Acetylecholinestrase (AChE) (87.44, 83.73%) and butyrylcholinesterase (BChE) (82.53, 88.55%) enzymes at 1 mg/mL with IC_{50} values of 45, 50 for AChE and 40, 70 μg/mL for BChE, respectively. The action of these portions was similar to galanthamine producing 96.00 and 88.61 prevention of AChE and BChE at 1 mg/mL dose with IC_{50} values of 20 and 47 μg/mL correspondingly. In antioxidant assays, Ir.Flv, Ir.Cf, and ethyl acetate (Ir.EtAc) confirmed the maximum radicals scavenging actions in DPPH and H_2O_2 assays which were analogous to ascorbic acid. Ir.Flv was seen as generally powerful with IC_{50} of 19 and 24 μg/mL against DPPH and H_2O_2 radicals correspondingly. While antioxidant triggers of plant samples against ABTS free radicals was reasonable. Ir.Cf, Ir.EtAc, and Ir.Cr showed the maximum phenolic and flavonoid constituents and dosages of these substances in various portions interconnected well to their antioxidant and anticholinestrase actions (Zeb et al., 2014b).

KEYWORDS

- **antioxidant activity**
- **antibacterial activity**
- **anti-pyretic activity**
- **analgesic activity**
- **anti-aging activity**

REFERENCES

Abbasi, B. H.; Siddiqua, A.; Tungmunnithum, D.; Bose, S.; Younas, M.; Garros, L.; Drouet, S.; Giglioli-Guivarc'h, N.; Hano, C. *Isodon rugosus* (Wall ex Benth.) Codd In Vitro Antioxidant and Antiaging Activities. *Int. J. Mol. Sci.* **2019,** *20*(2), 452.

Hussain, I.; Khan, A. U.; Ullah, R.; Alsaid, M. S.; Salman, S.; Iftikhar, S.; Marwat, G. A.; Sadique, M.; Jan, S.; Adnan, M.; Tariq, A. Chemical Composition, Antioxidant and Anti-bacterial Potential of Essential Oil of Medicinal Plant *Isodon rugosus*. *J. Essent. Oil Bear. Pl.* **2017,** *20*(6), 1607–1613.

Janbaz, K. H.; Arif, J.; Saqib, F.; Imran, I.; Ashraf, M.; Zia-Ul-Haq, M.; Jafar, H. Z. E.; De Feo, V. *In-vitro* and *In-vivo* Validation of Ethnopharmacological uses of Methanol Extract of *Isodon rugosus* Wall. ex Benth. (Lamiaceae). *BMC Compliment. Altern. Med.* **2014,** *14*, 71.

Sadiq, A.; Zeb, A.; Ullah, F.; Ahamad, S.; Ayaz, M.; Rashzid, U.; Muhammad, N. Chemical Characterization, Analgesic, Antioxidant, Anticolinesterase Potentials of Essential Oils from *Isodon rugosus* Wall. ex Benth. *Front. Pharmacol.* **2018,** *9*, 623.

Ullah, A.; Uddin, G.; Rashid, M. U.; Ismail, I.; Nasruddin, N.; Siddiqui, B. S.; Ullah, Z.; Alamzeb, M. A New Diterpene: Abietane Glycoside from the Roots of *Isodon rugosus* Wall ex Benth. *Rec. Nat. Prod.* **2019,** *13*(4), 287–295.

Verma, R. S., Pandey, V., Chauhan, A., Tiwari, R. Characterization of the Essential Oil Composition of *Isodon rugosus* (Wall. ex Benth.) Codd. from Garhwal Region of Western Himalaya. *Med. Aromat. Plants* **2015,** S3, 002.

Zeb, A.; Ahamad, S.; Ullah, F.; Ayaz, M.; Sadiq, A. Anti-nociceptive Activity of Ethnomedicinally Important Analgesic Plant *Isodon rugosus* Wall. ex Benth: Mechanistic Study and Identifications of Bioactive Compounds. *Front. Pharmacol.* **2016,** *7*, Article 200.

Zeb, A.; Sadiq, A.; Ullah, F.; Ahmad, S.; Ayaz, M. Phytochemical and Toxicological Investigations of Crude Methanolic Extracts, Subsequent Fractions and Crude Saponins of *Isodon rugosus. Biol. Res.* **2014a,** *47*, Article 57.

Zeb, A.; Sadiq, A.; Ullah, F.; Ahmad, S.; Ayaz M. Investigations of Anticholinesterase and Antioxidant Potentials of Methanolic Extract, Subsequent Fractions, Crude Saponins and Flavonoids Isolated from *Isodon rugosus. Biol. Res.* **2014b,** *47*, Article 76.

Zeb, A.; Ullah, F.; Ayaz, M.; Ahamad, S.; Sadiq, A. Demonstration of Biological Activities of Extracts from *Isodon rugosus* Wall. ex Benth.: Separation and Identification of Bioactive Phytoconstituents by GC-MS Analysis in the Ethyl acetate Extract. *BMC Compliment. Altern. Med.* **2017,** *17*, Article 284.

CHAPTER 7

PHYTOCONSTITUENTS AND PHARMACOLOGY OF *LEUCAS ASPERA* L.: AN IMPORTANT AROMATIC WEED

SHREEDHAR S. OTARI, SURAJ B. PATEL, and
SAVALIRAM G. GHANE

Plant Physiology Laboratory, Department of Botany, Shivaji University, Kolhapur, Maharashtra, India

ABSTRACT

Leucas aspera L. belongs to family Lamiaceae. It is a common aromatic herb found in India. It is branched, annual herb with a height of 15–60 cm with erect, stout and quadrangular stems. In several traditional medicines, it is used to treat cough, cold, painful swellings, wound, and skin disorders. It is also useful to treat the snake bite. Active compounds found in various plant parts showed significant medicinal potential. Variety of phytoconstituents such as triterpenoids, phenolics, flavonoids, glycosides, alkaloids, steroids, cardiac glycosides and diterpenes have been reported from the plant. Seed oil contains several fatty acids such as palmitic acid, stearic acid, oleic acid, linoleic acid, and linolenic acid. *L. aspera* possesses several potent biological activities like antioxidant, antibacterial, cytotoxic, larvicidal, antifungal, anticancer, anti-inflammatory etc.

7.1 INTRODUCTION

Leucas aspera L. belongs to family Lamiaceae. It is a common aromatic herb found in India, Africa, Asian temperate, and tropical countries. Plant

Bioactives and Pharmacology of Lamiaceae. T. Pullaiah, PhD (Ed.)

has been identified by several synonyms as *Leucas dimidiata* Benth., *Leucas dimidiata* (Roth) Spreng., *Leucas minahassae* Koord., *Leucas obliqua* Buch.-Ham. ex Dillwyn., *Leucas plukenetii* (Roth) Spreng., etc. (POWO, 2019). *Leucas aspera* is branched, annual herb with a height of 15–60 cm with erect, stout, and quadrangular stem. Stem showed light greenish-yellow color with rough surface, hairy nature along with four prominent furrows about 3–4 mm thick. Leaf yellowish-green in color, about 3–9 cm wide, and ovate in shape. Sessile inflorescence, white-colored flowers. Schizocarpic fruits with nutlets 3 mm. The nutlets showing near about 2.5 mm long, darker, elongated, smooth nature, found external face adjustment. Roots are cylindrical, smooth, along with numerous wire like appearance.

It is used as an analgesic, antipyretic, and anti-inflammatory. Plants paste prepared from the plant applied externally on inflamed area. The whole plant is used as an insecticide. In several traditional medicines, it is used to treat cough, cold, painful swellings, wound, and skin disorders. Plant is also useful to cure the snake bite (Chew et al., 2012; Kumar and Devanna, 2016, 2019). *Leucas aspera* possess several potent biological activities like antioxidant, antibacterial, cytotoxic, larvicidal, and antifungal (Rahman et al., 2013; Chew et al., 2012; Kayesh et al., 2013; Elumalai et al., 2015; Babu et al., 2015).

7.2 BIOACTIVES

Phytoconstituents such as triterpenoids, phenolics, flavonoids, glycosides, alkaloids, acidic compounds, steroids, cardiac glycosides, oleanolic acid, ursolic acid, β-sitosterol, nicotine, sterols, glucoside, and diterpenes have been reported from the aerial parts of plant (Kripa et al., 2016; Rahman et al., 2013; Gopi et al., 2014; Prajapati et al., 2010). Four new diterpenes such as leucasperone A, leucasperone B, leucasperol A, and leucasperol B have been isolated from whole plant (Sadhu et al., 2006). Three new isopimarane glycosides like leucasperoside A, -B, and -C along with the known compounds asperphenamate, maslinic acid, isololiolide, and linifolioside have been investigated (Sadhu et al., 2006). From the roots, triterpenoid named as leucolactone have been reported (Pradhan et al., 1990). TLC, HPTLC, and RP–HPLC analysis revealed the presence of nicotine alkaloid from the aerial parts of the plant (Mangathayaru et al., 2006). From the shoots of plant, two new aliphatic ketols such as 28 hydroxypentatriacontan-7-one and 7-hydroxydotriacontan-2-one has been isolated along with 5-acetoxytriacontane (Misra et al., 1992, 1993). Misra et al. (1992) reported compounds such as 1-hydroxytetratriacontan-4-one and 32 methyltetratriacontan-8-01 along with dotriacontanol from the shoots. Single compound belongs to sterol category (beta-sitosterol) have been reported from the seeds and aerial part (Misra et al., 1993; Jam et al., 1968; Gopi et al., 2014). From the leaves and flowers, several volatile constituents were reported using GC and GC/MS. Leaf showed presence of α-farnesene, α-thujene, and menthol; whereas amyl propionate and isoamyl propionate were found in the flowers (Kalachaveedu et al., 2006). Seed oil contains several fatty acids such as palmitic acid, stearic acid, oleic acid, linoleic acid, and linolenic acid (Jam et al., 1968). GC–MS analysis revealed major bioactive constituents such as 6-octadecenoic acid (32.47%), n-hexadecanoic acid (25.97%), and 17-octadecen-14-yn-1-ol (14.22%) from the plant (Gopi et al., 2014). In addition, GC–MS analysis of the methanolic extract of whole plant revealed various phytocompounds like 1-hexadecanol, 2-methyl, 3,7,11,15-tetramethyl-2-hexadecen-1-ol, catechin, 9,12-octadecadienoic acid(zz)-methyl ester, 9,12,15-octadecatrienoic acid, methyl ester, heptadecanoic acid, 9-methyl-, methyl ester, eicosanoic acid, methyl ester cholestan-3 ol, 2-methylene, (3a, 5a), aspidospermidne-17 ol, 1-acetyl, 19, 21-epoxy-15, 16-dimethoxy, tetradecane,2,6,10-trimethyl-oxiraneundecanoic acid, 3-pentyl, methyl ester, tetracosahexane, and 2, 6, 10, 15, 19, 23-hexamethyl (Elumalai et al., 2015).

7.3 PHARMACOLOGY

7.3.1 ANTIOXIDANT ACTIVITY

Antioxidant activity was assessed from the ethanol extract of whole plant. Different concentrations (25, 50, 100, 150, 200, 400, and 800 μg/mL) and ascorbic acid was used as positive control (1–100 μg/mL). The highest mean percentage free radical scavenging activity (86.62 ± 0.49%) was found at concentration of 800 μg/mL. Inhibitory concentration was assessed by plotting the graph of percentage scavenging activity against the log concentration of extract. IC_{50} value of plant extract was 99.58 ± 1.22 μg/mL whereas in case of ascorbic acid, it was 1.25 ± 0.95 μg/mL (Rahman et al., 2013).

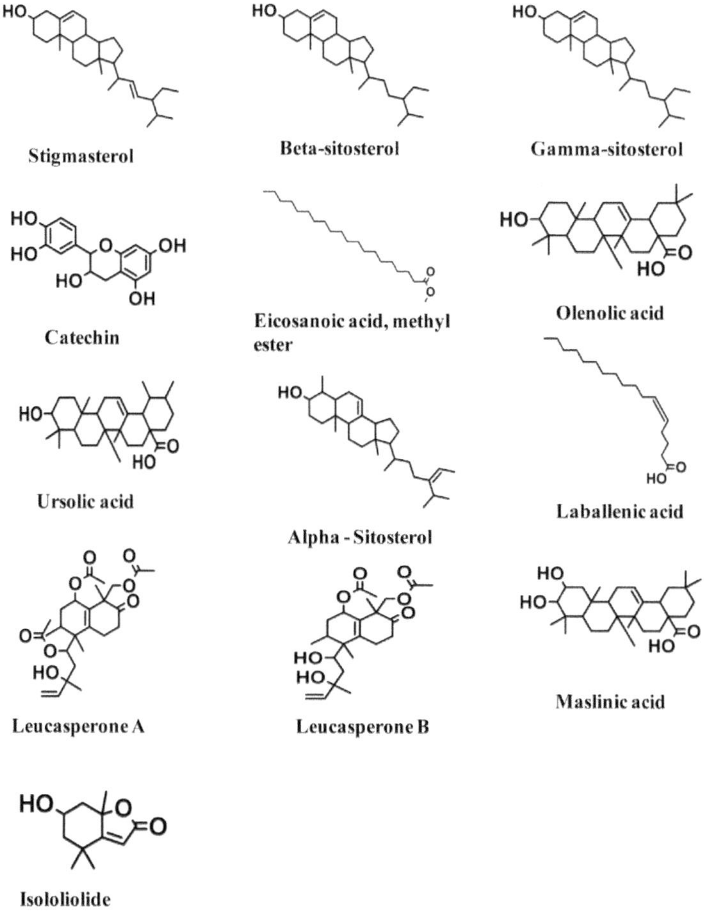

FIGURE 7.1 Bioactive compounds from *Leucas aspera.*

Similarly, different plants parts (stem, flower, leaf, and root) were extracted with 80% methanol and used to investigate antioxidant activity. Sources of positive control were BHT and Vitamin E. Results showed that the root extract possessed the highest activity as compared to remaining plant extracts. Root methanolic extract showed the highest DPPH radical scavenging activity (32.36 ± 1.19%). Similarly, stem, flower, and leaf extracts were noted with comparatively poor responses (26.39 ± 0.07, 13.42 ± 0.56, 17.04 ± 0.82%, respectively). At the same time positive controls, vitamin E, and BHT showed 41.67 ± 0.58 and 65.67 ± 0.58% inhibition of free radicals. Based on the results, authors suggested that the roots of *Leucas aspera* were good source of natural antioxidant as compared to other plant parts (Chew et al., 2012). Antioxidant activity from the leaves extracted with different solvents was studied. Methanolic crude extract of the leaves (40 ppm) revealed the highest (88.57 ± 0.31%) DPPH radical scavenging activity. At the same time, least activity was noted from acetone extract (9.47 ± 0.63%). Further, the superoxide radical scavenging activity found to be the best in methanolic crude extract of leaves (84.62 ± 0.08%) as compared to remaining extracts. The positive control BHA exhibited 88–89% DPPH radical scavenging activity. All the findings confirmed that the leaves of *Leucas aspera* exhibited the excellent antioxidant activity (Meghashri et al., 2010).

7.3.2 ANTIBACTERIAL ACTIVITY

Antibacterial activity of ethanol extract of whole plant was assessed using disc diffusion method. Several Gram positive and Gram negative bacterial strains were used. In tested Gram-positive bacterial strains, highest zone of inhibition was found in *Bacillus megaterium* (13.00 ± 1.50 mm) whereas least responses showed by *Staphylococcus aureus* (8.00 ± 0.50 mm). On other hand, in Gram-negative bacterial strain, maximum zone of inhibition was reported from *P. aeruginosa* (13.00 ± 1.00 mm) as compared to others. All the findings suggested that plant as a good antibacterial agent (Rahman et al., 2013). Chew et al. (2012) also studied the antibacterial activity of methanolic extract of different plant parts (root, flower, leaf, and stem) using the disc diffusion method. In this investigation, microorganism like *E. coli, S. aureus, S. choleraesuis, S. typhimurium, S. flexneri,* and *P. aeruginosa* were used. Results revealed that root extract exhibited the highest zone of inhibition studied microorganism. Obtained values highlighted that root extract showed significant activity against tested microorganism viz. *E. coli* (9.0 ± 0.5 mm), *S. aureus* (10.0 ± 0.6 mm), *S. choleraesuis* (11.0 ± 0.5 mm), *S. typhimurium* (11.0 ± 0.6 mm), *S. flexneri* (11.0 ± 0.5 mm), and *P. aeruginosa* (11.0 ± 0.6 mm) at a concentration of 100 mg/mL. Positive control chloramphenicol (30 μg/mL) showed 21 ± 0.4 to 25.0 ± 0.5 mm zone of inhibition against tested microorganism (Chew et al., 2012).

7.3.3 CYTOTOXIC ACTIVITY

Cytotoxic activity was carried out from ethanolic extract of whole plant using the method of brine shrimp lethality bioassay. In the same study, *Artemia salina* was used as a test organism. The ethanolic extract of whole plant showed LC_{50} value of 181.68 ± 2.15 µg/mL as compared to positive control vincristine sulfate (LC_{50} 0.76 ± 0.04 µg/mL). Cytotoxicity could be due to the secondary metabolites which present in the extract (Rahman et al., 2013).

Similarly, Ali et al. (2013) also investigated the cytotoxic activity from the crude methanolic extract of leaves using the brine shrimp lethality bioassay. In the same study, different concentrations of the extract were tested against organisms. Vincristine sulfate was used as a positive control. Results revealed that crude methanol extract of leaves found cytotoxic (LC_{50} 30.00 µg/mL) to *Artemia salina* which was comparable to that of the positive control vincristine sulfate (LC_{50} 10.44 µg/mL). All the findings revealed that the plant has promising cytotoxic potential (Ali et al., 2013). *In vitro* cytotoxic activity was also analyzed from the different extracts of leaves using brine shrimp lethality bioassay and percentage mortality was calculated. Results showed the highest LC_{50} value of 13.29 µg/mL from leaves of petroleum ether extract at 25 µg/mL concentration followed by methanol (LC_{50} 4.28 µg/mL) when compared to vincristine sulfate (LC_{50} 0.53 µg/mL). This data clearly disclosed the strongest cytotoxic property of the plant (Akter et al., 2012). Kayesh et al. (2013) also studied the cytotoxic activity from the methanolic extract of leaves (LC_{50} 0.796 µg/mL) and proved the cytotoxic potential.

7.3.4 LARVICIDAL ACTIVITY

Elumalai et al. (2015) assessed larvicidal activity from the different extracts of whole plant. Two organisms like *Anopheles stephensi* and *Culex quinquefasciatus* were used. At concentration of 500 ppm, it was observed that LC_{50} and LC_{90} values of ethanol extracts were 40.877, 34.359, and 72.903, 67.355 ppm against *Anopheles stephensi* after 24 and 48 h treatments, respectively. Similarly, in case of *Aedes aegypti* LC_{50} and LC_{90} values were 35.624 and 20.897 ppm after 24 h of treatment. After 48 h, LC_{50} and LC_{90} values were 64.260 and 60.096 ppm, respectively. Methanol extract was tested against *Culex quinquefasciatus* and showed appreciable LC_{50} and LC_{90} values (79.150 and 73.284 ppm, respectively) after 48 h of treatment. Maheshwaran et al. (2008) studied crude leaf extract of *Leucas apsera* tested for their larvicidal activity against the *C. quinquefasciatus* and *A. aegypti.* The hexane extract of *Leucas aspera* showed highest activity against the tested organisms. The LC_{50} values of *Leucas aspera* against first-, second-, third-, and fourth-stage instar larvae of *C. quinquefasciatus* were 122.50, 149.97, 193.43, and 230.71 ppm, respectively. Similarly, in case of *A. aegypti*, LC_{50}

values were 77.40, 144.00, 199.72, and 257.17 ppm, respectively. Ethanolic extract of whole plant also showed larvicidal and pupicidal activity against I, II, III, IV instar larvae of *Anopheles stephensi* with LC_{50} values of 9.695, 10.272, 10.823, 11.30, and 12.732%, respectively. These findings confirmed that the ethanolic extracts of whole could be an excellent source for controlling *A. stephensi* (Kovendan et al., 2012).

7.3.5 ANTIFUNGAL ACTIVITY

In vitro antifungal activity was assessed from the methanolic extract of leaves using agar well diffusion method and clotrimazele was used as positive control. In same study, several fungi like *Candida tropicalis*, *Candida albicans*, *Aspergillus*, and *Penicilium* sp. were used. Maximum zone of inhibition found in *Penicillium* sp. (7.5 mm) where least was observed in *Candida albicans* (3.0 mm). In control, zone of inhibition ranged from 5 to 10 mm. These finding revealed appreciable antifungal activity (Babu et al., 2015).

7.3.6 ANTHELMINTIC ACTIVITY

Anthelmintic activity from methanol extract of leaves (20 mg/mL) was investigated against adult earthworm (*Phertima posthuma*) (Kayesh et al., 2013). Albendazole was used as standard. Results indicated that standard albendazole showed paralysis and death of worm at 17.67 ± 0.54, 48 ± 0.47 min, respectively; whereas in case of plant extract comparatively more time required to paralyze and kill the tested organism (52 ± 0.50, 79.66 ± 0.30 min, respectively).

7.3.7 ANTIMICROBIAL ACTIVITY

Vasundha et al. (2019) investigated the antimicrobial activity from the different extract of root and leaves. Gram positive and Gram negative bacteria such as *Bacillus subtilis*, *Staphylococcus aureus*, *Escherichia coli*, and *Pseudomonas aeruginosa* were used in the same experiment. Streptomycin was used as standard. Methanolic leaf extract (100 µg/mL) showed the highest zone of inhibition against *S. aureus* (12.8 ± 0.31 mm) followed by hexane leaf extract against *S. subtilis* (8.1 ± 0.27 mm). Similarly, aqueous leaf extract was tested against *S. subtilis* and revealed 7.4 ± 0.29 mm zone of inhibition. Root methanolic and hexane extracts also showed the highest zone of inhibition against *S. aureus* (6.6 ± 0.32 and 6.0 ± 0.13 mm, respectively). However, minimum zone of inhibition was found from aqueous extract against *B. subtilis* (5.1 ± 0.29 mm).

7.3.8 ANTICANCER ACTIVITY

Anticancer activity was estimated from the aqueous and methanol extract of the leaves by using MTT assay. Results showed 43.4% death of HeLa cell lines (MCF-7) when treated with aqueous extract. Similarly, methanol extract also revealed 56.4% cell death of HeLa cell lines. Methanolic leaf extract showed comparatively stronger anticancer activity as compared to the aqueous extract (Suruthi et al., 2016).

7.3.9 ANTI-INFLAMMATORY ACTIVITY

Alcoholic extract of leaves of *Leucas aspera* was tested for anti-inflammatory activity using animal models. Different concentrations such as 100, 200, and 400 mg/kg were used. In the present investigation, carrageenan-induced paw edema and cotton pellet-induced granuloma rat models were used. Diclofenac was used as standard. All the 3 tested doses (100, 200, and 400 mg/kg), variable responses in anti-inflammatory activity at different time intervals (0–4 h) in each model were recorded. At 400 mg/kg dose of plant extract, maximum percentage inhibition 60.64% was observed in carrageenan model as compared standard diclofenac (60.70%). Similarly, in cotton pellet model 60.27% inhibition in inflammation was reported (Patil et al., 2014).

KEYWORDS

- ***Leucas aspera***
- **bioactive metabolites**
- **aromatic**
- **antioxidant**
- **antibacterial**
- **anticancer**

REFERENCES

Akter, M.; Khan, M.; Muhsin, M. A.; Hamid, K.; Ullah, M. O.; Bulbul, I. J.; Firoz, A.; Uddin, M. B.; Urmi, K. F. *In vitro* Studies on Antibacterial, Antifungal and Cytotoxic Properties of *Leucas aspera*. *Biol. Med.* **2012,** *4*(4), 183–187.

Ali, M. S.; Sayeed, M. A.; Nabi, M. M.; Rahman, M. A. A. In Vitro Antioxidant and Cytotoxic Activities of Methanol Extract of *Leucas aspera* Leaves. *J. Pharmacogn. Phytochem.* **2013,** *2*(1), 8–13.

Babu, A.; Mohamed, M. S. N.; Jaikumar, K.; Anand, D.; Saravanan, P. In-vitro Antifungal Activity of Leaf Extracts of *Leucas aspera* and *Leucas zeylanica*. *Int. J. Pharm. Sci. Res.* **2016,** *7*(2), 752–756.

Chew, A. L.; Jessica, J. J. A.; Sasidharan, S. Antioxidant and Antibacterial Activity of Different Parts of *Leucas aspera*. *Asian Pac. J. Trop. Biomed.* **2012,** *2*(3), 176–180.

Elumalai, D.; Hemalatha, P.; Kaleena, P. K. Larvicidal Activity and GC–MS Analysis of *Leucas Aspera* Against *Aedes Aegypti*, *Anopheles Stephensi* and *Culex Quinquefasciatus*. *J. Saudi Soc. Agric. Sci.* **2015,** *16*(4), 306–313. http://dx.doi.org/10.1016/j.jssas.2015.10.003.

Gopi, K.; Renu, K.; Jayaraman, G. Inhibition of *Naja naja* Venom Enzymes by the Methanolic Extract of *Leucas aspera* and Its Chemical Profile by GC–MS. *Toxicol. Rep.* **2014,** *1*, 667–673.

Jam, M. P.; Nath, H. B. Examination of the Component Fatty Acids of the Oil from the Seeds of *Leucas aspera*. *Lab. Dev.* **1968,** *6*, 34–36.

Kalachaveedu, M.; Ghosh, A.; Ranjan, R.; Vedam Venkat, K. Volatile Constituents of *Leucas aspera* (Willd.). *J. Essent. Oil Res.* **2006,** *18*, 104-105.

Kayesh, D. M. I.; Haque, M. M.; Alam, M. K.; Islam, M. N. Evaluation of Cytotoxic and Anthelmintic Activities of the Leaves of *Leucas aspera*. *Int. J. Adv. Pharm. Res.* **2013,** *4*(6), 1817–1822.

Kumar, G. V.; Devanna, N. An Update of *Leucas aspera* - A Medicinal Plant. *Int. J. Sci. Res.* **2016,** *5*(1), 485–503.

Kumar, E. V.; Nampalli, A.; Bakshi, V.; Gangarapu, K.; Boggula, N. A Review on *Leucas aspera* for Phytopharmacological Studies. *INNOSC Theranostics Pharmacol. Sci.* **2019,** *2*(1), 3–7.

Kovendan, K.; Murugan, K.; Vincent, S.; Barnard, D. R. Studies on Larvicidal and Pupicidal Activity of *Leucas aspera* Willd. (Lamiaceae) and Bacterial Insecticide, *Bacillus sphaericus*, Against Malarial Vector, *Anopheles stephensi* Liston. (Diptera: Culicidae). *Parasitol. Res.* **2012,** *110*, 195–203.

Kripa, K. G.; Sangeetha, R.; Chamundeeswari, D. Pharmacognostical and Physicochemical Evaluation of the Plant *Leucas aspera*. *Asian J. Pharm. Clin. Res.* **2016,** *9*, 263–268.

Maheswaran, R.; Sathish, S.; Ignacimuthu, S. Larvicidal Activity of *Leucas aspera* (Willd.) Against the Larvae of *Culex quinquefasciatus* Say. and *Aedes aegypti* L. *Int. J. Integr. Biol.* **2008,** *2*(3), 214–217.

Mangathayaru, K.; Thirumurugan, D.; Patel, P. S.; Pratap, D. V. V.; David, D. J.; Karthikeyan, J. Isolation And Identification Of Nicotine From *Leucas Aspera* (Willd) Link. *Indian J. Pharm. Sci.* **2006,** *68*(1), 88–90.

Meghashri, S.; Kumar, V. H.; Gopal, S. Antioxidant Properties of a Novel Flavonoid from Leaves of *Leucas aspera*. *Food Chem.* **2010,** *122*, 105–110.

Misra, T. N.; Singh, R. S.; Prasad, C.; Singh, S. Two Aliphatic Ketols from *Leucas aspera*. *Phytochemistry* **1993,** *32*(1), 199–201.

Misra, T. N.; Singh, R. S.; Pandey, H. S.; Singh, S. Long-chain Compounds from *Leucas aspera. Phytochemistry* **1992,** *31*(5), 1809–1810.

Patil, N.; Someshekar, H. S.; Reddy, S. K.; Nayak, V.; Narendranath, S.; Bairy, K. L. Evaluation of Anti-inflammatory Activity of Alcoholic Extract of Leaves of *Leucas aspera* in Albino Rats. *Int. J. Pharm. Pharm Sci.* **2014,** *6*, 715–719.

POWO. *Plants of the World Online*. Facilitated by the Royal Botanic Gardens, Kew. 2019, Published on the Internet; http://www.plantsoftheworldonline.org/ (accessed Mar 22, 2021).

Pradhan, B. P.; Chakraborty, D. K.; Subba, G. C. A Triterpenoid Lactone from *Leucas aspera. Phytochemistry* **1990,** *29*(5), 1693–1695.

Prajapati, M. S.; Patel, J. B.; Modi, K.; Shah, M. B. *Leucas aspera*: A Review. *Pharmacogn Rev*. **2010,** *4*(7), 85–87. DOI: 10.4103/0973-7847.65330.

Rahman, M. A.; Islam, M. S. Antioxidant, Antibacterial and Cytotoxic Effects of the Phytochemicals of Whole *Leucas aspera* Extract. *Asian Pac. J. Trop. Biomed.* **2013,** *3*(4), 273–279.

Sadhu, S. A.; Okuyama, E.; Fujimoto, H.; Ishibashi, M. Diterpenes from *Leucas aspera* Inhibiting Prostaglandin-induced Contractions. *J. Nat. Prod.* **2006,** *69*, 988–994.

Suruthi, M.; Sivabalakrishnan, S.; Yuvasri, G.; Ragunathan, R.; Johney, J. Antioxidant Anticancer Activity of *Leucas aspera* Plant Extract and Its DNA Damage Study on HeLa Cell Lines. *Res. J. Life Sci.* **2016,** *2*(2), 1–9.

Vasudha, K.; Archana, D.; Mutyalamma, B.; Kishori, B. Phytochemical Screening, Antimicrobial and Antioxidant Activities of Root and Leaf Extracts of *Leucas aspera. Asian J. Pharm. Clin. Res*. **2019,** *12*(3), 141–147.

CHAPTER 8

PHYTOCHEMICAL EVALUATION OF *LEUCAS URTICAEFOLIA* (VAHL) R.BR. EX SM. (FAMILY: LAMIACEAE)

VEENA SATYA[1] and NUTAN RAJPUT[2]

[1]*Department of Botany, Shaheed Bheema Nayak Government Post Graduate College, Barwani, Madhya Pradesh, India*

[2]*Department of Botany, Government College, Manawar, Dhar, Madhya Pradesh, India*

ABSTRACT

Leucas urticaefolia (Vahl) R.Br. ex Sm. is traditionally used for the treatment of various diseases. A few but remarkable compounds have been separated/discovered by several researchers from the plant like alkaloids, diterpene, fatty acids, flavonoids, proteins, steroids, tannins and triterpenes. These drugs hold a good promise for managing ailments like Alzheimer's disease; have various pharmacological effects like antihyperlipidemic activity, wound-healing activity, antimicrobial and antioxidant activity, and antinutritional property. This chapter reflects the information regarding different bioactives present in the plant and their pharmacological impact.

8.1 INTRODUCTION

Leucas urticaefolia (Vahl) R.Br. ex Sm. belongs to family Lamiaceae. Its vernacular name is "Kharkhuda or Kharkhumba." It is commonly found on wastelands near villages and old walls. It is an erect, tomentose, branched,

Bioactives and Pharmacology of Lamiaceae. T. Pullaiah, PhD (Ed.)

aromatic herb. Leaves are membranous, ovate, serrate, subacute, base cuneate, often rounded, and glabrescent. Flowers are white, dense, distant, and inflorescence with axillary globose whorls.

L. urticaefolia is used for treating arthritis, bronchial catarrh, cystitis, calculus, dizziness, diarrhea, dropsy, dysentery, fever, gravel, skin diseases, typhoid, uterine hemorrhages, and various types of mental disorders (Watt, 1890; Mhaskar et al., 1935).

Apical shoots with decoction of the leaves and "Gur" are used locally as an abortion-inducing drug, up to 3 months of pregnancy (Jafri et al., 1966). Apart from this, infusion of its flowers is given in cold and cough. Roasted leaves are bandaged on the swelling part (Katewa and Galaw, 2005).

8.2 BIOACTIVES

Until now, a few but noteworthy compounds have been separated/discovered from the plant. Alkaloids, flavonoids, proteins, and tannins are some of the important components of *L. urticaefolia*. High content (28% w/w) of allenic acid was found in *L. urticaefolia* seeds (Aitzetmuller et al., 2006). Flavonoidal glucosides and leufolin A and B were reported from the ethyl acetate fraction of methanolic extract (Noor et al., 2007). A new steroid leucisterol and a new peroxyacid, urticic acid along with methoxybenzyl benzoate, 4-hydroxybenzoic acid, β-sitosterol, and ursolic acid have been separated from the chloroform soluble fraction of the whole plant of *L. urticaefolia* (Fatima et al., 2008).

Habib et al. (2008) reported leucisterol, β-sitosterol and ursolic acid, urticic acid, methoxybenzyl benzoate, and 4-hydroxy benzoic acid. Presence of various classes of compounds has been emphasized during chemical studies, the main being diterpene, fatty acids, flavonoids, and triterpenes. Chouhan and Singh (2011) reviewed the genus *Leucas* and gave an account of variety of phytoconstituents, which include coumarins, fatty acids, flavonoids, lignans, steroids, terpenes, and aliphatic long-chain compounds.

Das et al. (2012) reviewed the ethnobotanical, phytochemical, and pharmacological attributes of various species of genus *Leucas*. Dixit et al. (2015) reported that total phenolic and flavonoid content were 0.71335% and 0.2594%, respectively. Values of secondary metabolites in the plant *L. urticaefolia* ranged from 0.302 for flavonoids, 0.577 for nitrates, 0.841 for oxalates, 0.862 for total phenols, 1.011 for saponins, and 4.239 for tannins (Ehsen et al., 2016).

According to Rajput and Satya (2019) alkaloids, phenolics and flavonoids in *L. urticaefolia* were 0.79, 1.75, and 0.93 mg/g, respectively. Leaf parts of the plant were found to have more concentration of phytochemicals analyzed. Most of the phytochemicals were found to be extractable in ethyl acetate solvent followed by methanol while aqueous extracts showed minimum number of phytochemicals.

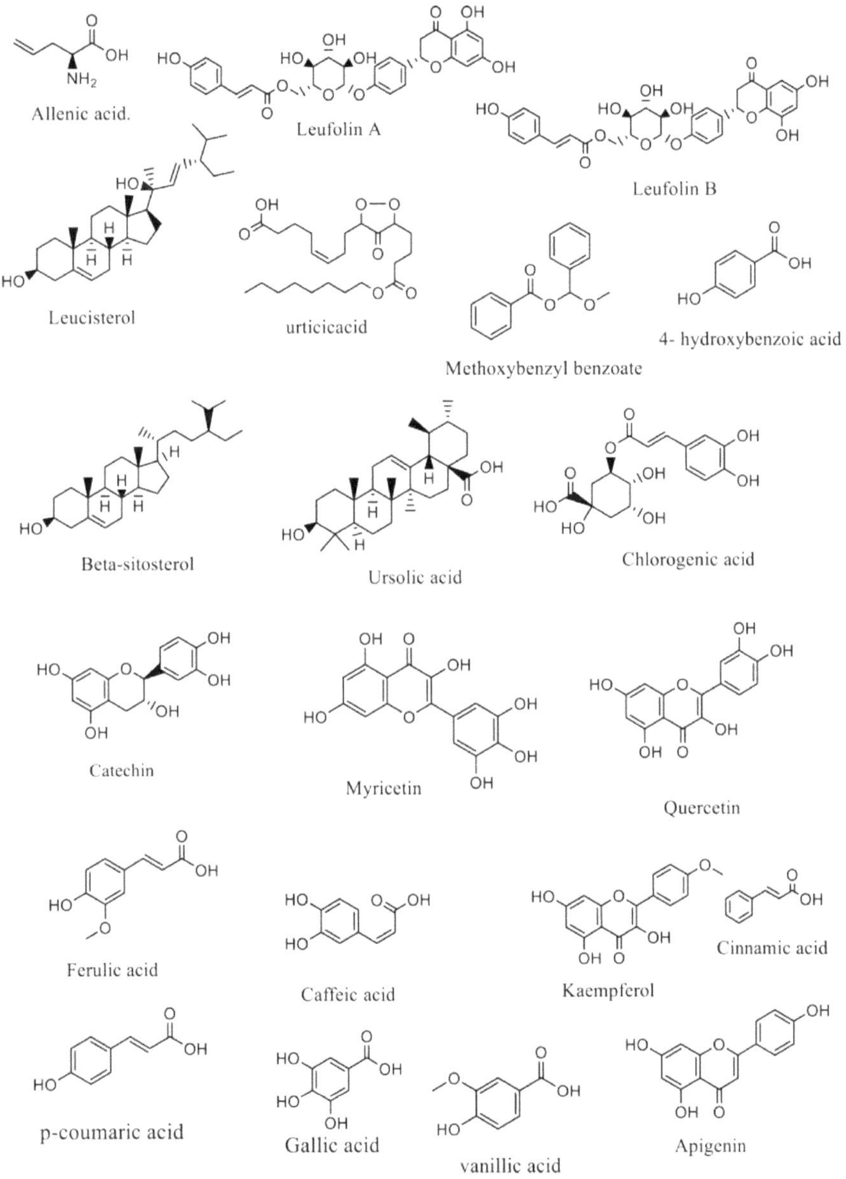

Chloroform:methanol (7.5:2.5) at 366 nm showed 12 distinct peaks. These compounds were chlorogenic acid, catechin, myricetin, quercetin, ferulic acid, caffeic acid, kaempferol, cinnamic acid, *p*-coumaric acid, gallic acid, vanillic acid, and apigenin (Rajput and Satya, 2019).

8.3 PHARMACOLOGY

8.3.1 ALZHEIMER'S DISEASE

Noor et al. (2007) reported that leufolins A and B are potent butyryl cholinesterases inhibiting flavonoid glucosides. Fatima et al. (2008) scrutinized their work on new butyrylcholinesterase (BChE)-repressive steroid and peroxy acid of *L. urticaefolia* from Karachi (Pakistan). Strong inhibitory activity was revealed by leucisterol against BChE enzyme. Phytochemistry and pharmacological property of *L. urticaefolia* scanned by Akhtar et al. (2010) from Pakistan found the similar results.

8.3.2 ANTIHYPERLIPIDEMIC ACTIVITY

The antihyperlipidemic activity of ethanolic *L. urticaefolia* extract in rats was reported by Suthar et al. (2012) from Jaipur (Rajasthan).

8.3.3 WOUND-HEALING ACTIVITY

The aqueous extract of *L. urticaefolia* showed wound-healing activity. Treated animals showed great wound-breaking strength, increased rate of wound contraction, and decline in period of epithelialization when matched with control group of animals. Acquired findings suggest that aqueous extract has significant wound-healing activity (Suthar et al., 2011). Similarly, ethanolic extract of *L. urticaefolia* has significant wound-healing activity (Suthar et al., 2012).

Phytochemical investigation exposed the presence of flavonoids in *L. urticaefolia.* Flavonoids are authenticated to possess powerful antioxidant and free radical scavenging effect, which is supposed to be one of the most important factors of wound healing (Suthar and Patel, 2012).

8.3.4 ANTIMICROBIAL AND ANTIOXIDANT ACTIVITY

Anti-inflammatory, analgesic, antidiarrheal, antimicrobial, and antioxidant activities have been reported in the extracts of different species of *Leucas* and its phytoconstituents by different workers and researchers (Chouhan and Singh, 2011).

Fifty-percent EtOH extract of this plant was tested against *Staphylococcus epidermidis*, *Salmonella typhi*, *Salmonella typhimurium*, *Candida krusei*, and *Aspergillus fumigatus* by disc-diffusion method. The extract revealed remarkable antimicrobial competence against *S. epidermidis*, *S. typhi*, and *C. krusei* in concentration-dependent mode.

A prominent correlation was witnessed between antioxidant and antimicrobial activity of 50% EtOH extract of *L. urticaefolia*. Important polyphenols appeared during HPLC analysis, which may be accountable for antimicrobial and antioxidant activity of the extract. The observation gives the impression that this plant might have potential source of crude bioactive molecules (Dixit et al., 2015).

8.3.5 ANTINUTRITIONAL FACTORS

Based on secondary metabolite accumulation, plants are divided into three antinutritive categories, that is, high (>10%), moderate (5–10%), and low (<5%). Most of the experimental species including *L. urticaefolia* lie in moderate category. Plants from low and moderate categories are within the tolerable limits for animal consumption.

Increased production of secondary metabolites is believed to not only defend plants from detrimental forces of environment to ensure their existence but also make them unsuitable as animal feed due to their harmful effects. Under critical environmental conditions, protective measures are mandatory for proper maintenance of resuscitative functions for the survival of plants like halophytes. Production of secondary metabolites may be a tactic of plants but their high quantity, when supplied over a long period, may disturb natural metabolism and obstruct animal growth (Ehsen et al., 2008).

KEY WORDS

- ***Leucas urticaefolia***
- **alkaloids**
- **diterpene**
- **fatty acids**
- **flavonoids**
- **wound-healing activity**
- **steroids**
- **tannins**
- **triterpenes**
- **pharmacological**

REFERENCES

Aitzetmuller, K.; Tsevegsuren, N.; Vosmann, K. A New Allenic Fatty Acid in *Phlomis* (Lamiaceae) Seed Oil. *Fet/Lipid.* **2006,** *99*, 74–78.

Akhtar, K.; Khan, S. B.; Malik, A. Phytochemistry and Pharmacological Aspects of *Leucas urticaefolia* (Vahl) Benth. *J. Pharmaceut. Res. Health Care.* **2010,** *2* (2), 197–203.

Chouhan, H. S.; Singh, S. K. A Review of Plants of Genus *Leucas. J. Pharmacogn. Phytother.* **2011,** *3* (3), 13–26.

Das, S. N.; Patro, V. J.; Dinda, S. C. Ethnobotanical Survey of Genus *Leucas. Pharmacogn. Rev.* **2012,** *6* (12), 100–106.

Dixit, V.; Siddiqi, S.; Agnihotri, P.; Paliwal, A. K.; Husain, T. Evaluation of Antioxidant and Antimicrobial Potential of *Leucas urticaefolia* (Lamiaceae). *J. App. Pharm. Sci.* **2015,** *5* (Suppl 1), 39–45.

Ehsen, S.; Qasim, M.; Abideen, Z.; Rizvi, R. F.; Gul, B.; Ansari, R.; Khan, M. A. Secondary Metabolites As Anti-Nutritional Factors in Locally Used Halophytic Forage/Fodder. *Pak. J. Bot.* **2016,** *48* (2), 629–636.

Fatima, I.; Ahmad, I.; Anis, I.; Malik, A.; Afza, N.; Iqbal, L. New Butyrylcholinesterase Inhibitory Steroid and Peroxy Acid from *Leucas urticaefolia. Arch. Pharm. Res.* **2008,** *31*, 999–1003.

Habib, R. S.; Jamshaid, M.; Tahir, M. N.; Khana, T. J.; Khan, I. U. (4R,5R,6S,7R,8S,9R,10S,13S)-7,8 *β*-Epoxymomilactone-A. *Acta Crystallographica.* **2008,** *64*, 0892. DOI: 10.1107/S1600536808010556

Harborne, J. B.; Baxter, H.; Moss, G. P. *Phytochemical Dictionary: A Handbook of Bioactive Compounds from Plants*; Tylor & Francis Ltd.: London, 1999.

Jafri, S. M. H. *Flora of Karachi*; The Book Corporation Karachi: Karachi, Pakistan, 1966.

Katewa, S. S.; Galav, P. K. Traditional Herbal Medicines from Shekhawati Region of Rajasthan. *Indian J. Tradit. Knowl.* **2005,** *4* (3), 237–245.

Mhaskar, K. S.; Blatter, E.; Caius, J. F. *Indian Medicinal Plants*, Vol. 9; Sri Satguru Publications: Delhi, India. 1935; p 2778.

Noor, A. T.; Fatima, I.; Ahmad, I.; Malik, A.; Afza, N.; Iqbal, L.; Latif, M.; Khan. Leufolins A and B, Potent Butyrylcholinesterase-Inhibiting Flavonoid Glucosides from *Leucas urticaefolia. Molecules* **2007,** *12*, 1447–1454.

Rajput N.; Satya V. Preliminary Pharmacognostic and Phytochemical Analysis of *Leucas urticaefolia* (Vahl) R. Br. ex Sm.: A Medicinal Herb. *J. Pharmacogn. Phytochem.* **2019,** *8* (2), 1752–1756.

Suthar S.; Patel R. Wound Healing Potency of Ethanolic Extract of *Leucas urticaefolia* in Experimental Animals. *Int. J. Pharm. Sci. Res.* **2012,** *3* (9), 3219–3224.

Suthar, S.; Patel, R.; Kumar, V.; Sharma, V. Wound Healing Activity of Aqueous Extract of *Leucas urticaefolia* Leaves in Experimental Animals. *Intern. J. Res. Pharmaceut. Sci.* **2011,** *1* (2), 141–150.

Suthar, S.; Patel, R.; Kumar, V.; Sharma, V. The Antihyperlipidaemic Activity of Ethanolic *Leucas urticaefolia* Extract in Triton wr-1339 Induced Hyperlipidaemic Rats. *Intern. J. Drug Res. Tech.* **2012,** *2* (4S), 371–377.

Watt, G. *Dictionary of the Economics Product of India*; Cosmo Publication, 1890.

CHAPTER 9

BIOACTIVES AND PHARMACOLOGY OF *LEUCAS ZEYLANICA* (L.) R.BR. (FAMILY: LAMIACEAE)

S. RAJASHEKARA[1], P. MAHESH[1], M. SHANTHALA[1], and M. MUNIRAJU[2]

[1]*Centre for Applied Genetics, Department of Studies in Zoology, Bangalore University, Jnana Bharathi Campus, Bengaluru, Karnataka, India*

[2]*Department of Studies in Botany, Bangalore University, Jnana Bharathi Campus, Bengaluru, Karnataka, India*

ABSTRACT

Leucas zeylanica is a tiny, terrestrial, herbaceous, upright or crested, hispid and fragrant plant of the family Lamiaceae. It is an unwanted plant of unused spaces and is native to Tropical Asia and Malaysia, Sri Lanka, India, spreading eastward through Malaya Peninsula up to China. The medicinal properties are contributed by various parts of the plants. This plant consists of several properties such as antioxidant, antifungal, anti-inflammatory, antidiabetic, antimicrobial, antianalgesic, anti-helminthic, anti-ulcer and genotoxic due to the existing of several bioactive phytochemicals.

9.1 INTRODUCTION

Leucas zeylanica (L.) B. Br., generally recognized as Ceylon slitwort, is a tiny, terrestrial, herbaceous, flowering once a year, upright or sometimes

Bioactives and Pharmacology of Lamiaceae. T. Pullaiah, PhD (Ed.)

crested, hispid, and fragrant plant of the family Lamiaceae. It is an unwanted plant of unused spaces and is native to Tropical Asia and Malaysia, Sri Lanka, India, and China. Synonyms of the plant include *Leucas bancana* Miq., *Phlomis zeylanica* L., and *Spermacoce denticulata* Walp.

A herb of about 50 cm tall. Leaves with a range of 0.3–1.0 cm elongated leafstalk; lamina 4–8 × 0.5–2.5 cm, roughly elliptic to elliptic-lanceolate, remotely serrate, acute, hirsute on abaxial surface and on the nerve's underneath. Flowering of terminal spirals, infrequently axillary. Nutlets ca. 0.2 × 0.1 cm, obovoid-oblong, angular on inward surface, rounded on outside, smooth, brownish black. Sap of entire plant is utilized in itches, skin diseases, in headache and cold (Chopra et al., 1996). The sap of the leaves is breathed as a therapy for snakebite (Caius, 1998).

9.2 BIOACTIVES

Triterpenoid glucoside, leuctriterpencoside, (–)-epiloliolide, and (E)-4-((1S,3R,4R)-1-hydroxy-4,5,5-trimethyl-7-oxabicyclo [4.1.0]heptan-1-yl) but-1-en-3-one were produced from *L. zeylanica.* They were assessed for their anti-inflammatory actions on LPS-induced NO manufacture in RAW264.7, and the new compound displayed modest repressive action (Zhang et al., 2016). Furthermore, Leuctriterpencoside (chemically known as 29-hydroxy-3-oxo-olean-12-en-28-*O*-a-L-rhamnopyranosyl ester) showed significant inhibitory activity against α-glucosidase (IC_{50} value of 0.85 ± 0.12 μM) (Chen et al., 2020).

Phytochemical screening of the ethanol extract of *L. zeylanica* indicated the presence of gums, carbohydrates, reducing sugar, phenols, saponins, cardiac glycosides, and tannins.

9.3 PHARMACOLOGY

9.3.1 *ANTIOXIDANT ACTIVITY*

Khan et al. (2010) investigated the antioxidant, neuropharmacological, analgesic, and antimicrobial actions of the ethanol extract of *L. zeylanica* plants. TLC-based antioxidant assay of DPPH-free radical revealed positive activity of the extract and in quantitative assay. IC_{50} value was 62.9 μg/ml.

Hossain et al. (2013) confirmed the existence of antioxidant phytochemicals, with polyphenols and flavonoids in *L. zeylanica* and henceforward deliberated safety against ethanol and H_2O_2-induced oxidative pressure on hepatic tissue. The antioxidative action of *L. zeylanica* was assessed through its capability to impede the hepatic levels of lipid peroxide (LPO), as indicator of oxidative pressure. The alanine aminotransferase (ALT) and aspartate aminotransferase (AST) actions and the levels of LPO of hepatic tissue were pointedly increased by oxidative pressure, though *L. zeylanica* pretreatment meaningfully suppressed the oxidative pressure on hepatic tissue, as designated by the declined actions of ALT and AST enzymes and levels of LPO. Studies of the phytochemicals exposed that the extract confined considerable quantities of polyphenols (74.32 ± 4.6 μg of pyrogallol equivalent/mg) and flavonoids (15.69 ± 2.2 μg quercetin equivalent/mg of extract).

The antioxidant action of leaves methanolic extract of the leaves of *L. zeylanica* was estimated utilizing isolated frog heart as a model. The frog heart model for orientation of oxidative pressure by H_2O_2 exhibited the negative inotropic and chronotropic properties and the cardiac seizure was formed at 20th min. In the existence of a methanolic extract of the leaves of *L. zeylanica*, the cardiac seizure was observed at 38th min, that is, heart was safeguarded longer period that designates antioxidant action which was associated with the normal ascorbic acid (Bindu, 2019).

9.3.2 ANTIFUNGAL ACTIVITY

Methanol leaf extract of *L. zeylanica* inhibited the development of fungal strains *Candida tropicalis*, *C. albicans*, *Trichophyton mentagrophytes*, *Aspergillus flavus*, and *Penicillium* sp. and hold latent antidermatophytic action (Babu et al., 2016).

9.3.3 ANTI-INFLAMMATORY ACTIVITY

A dichloromethane extract efficiently repressed 5-lipoxygenase (5-LO) action in inspired human neutrophils (IC_{50} = 5.5 μg/mL) and secluded human 5-LO and microsomal prostaglandin E_2 synthase (mPGES)-1 (IC_{50} = 2.2 and 0.4 μg/mL). Effective prevention of xanthine oxidase was detected by the same extract (IC_{50} = 47.5 μg/mL); only the *n*-hexane extract unveiled antibacterial action against *Staphylococcus aureus* and *Staphylococcus saprophyticus* with a MIC of 250 μg/mL, whereas the anti-*Candida* activity was modest. GC-MS investigation exposed the existence of phytosterols, fatty acids, sesquiterpenes, diterpenes, and numerous other kinds of secondary metabolites. Hence, the ethnopharmacological usages of *L. zeylanica* extracts as anti-inflammatory and antigout therapy are justified (Napagoda et al., 2018).

9.3.4 LARVICIDAL ACTIVITY

The fundamental oil from the upper portions of *L. zeylanica* was separated for mosquito larvicidal action against *Aedes aegypti* and *Aedes albopictus*. The major components of *L. zeylanica* essential oil from Vietnam are β-caryophyllene, α-humulene, germacrene D, β-selinene, caryophyllene oxide, 1-tetradecanol, and phytol. *L. zeylanica* essential oil showed good larvicidal activity against *Ae. albopictus* and *Ae. aegypti* with average lethal dosage—LC_{50} (24 h) values of 67 and 44 μg/mL, respectively. Hence, *L. zeylanica* fundamental oil characterizes a reasonable and ecologically gentle agent for regulating of the dengue virus vectors (Hung et al., 2019).

9.3.5 GENOTOXIC ACTIVITY

The consequence of water leaf extract of *L. zeylanica* was explored on the mitotic index and chromosomal abnormalities in the root tip cells of onion. The occurrence of cell division (mitotic index) and chromosomal abnormalities in various stages of mitotic cell division of each treatment exhibited that *L. zeylanica* leaf extracts are mitodepressive and persuading chromosomal anomalies like tenacity, sections, spans, slow pokes, micronuclei, upset metaphase/anaphase, and micronuclei. The chromosomal anomalies displayed a direct connection with the portions of the leaf extracts (Sinha and Kumar, 2014).

9.3.6 ANTIDIABETIC ACTIVITY

Dutta et al. (2016) evaluated antidiabetic activity of aqueous extract of *L. zeylanica* leaf by means of the alloxan-induced diabetic rats and associated the action with diabetic regulator and antidiabetic drug Glibenclamide. Leaf extract (50 mg/kg) of *L. zeylanica* and Glibenclamide were injected to standard and trial diabetic rat. *L. zeylanica* extract (50 mg/kg) knowingly depressed the fasting blood glucose level in the alloxan-induced diabetic rat model. Likewise, variations in body mass and liver glycogen levels evaluated in the extract treated to the diabetic rats were associated with diabetic control and normal rats. Remarkably, body mass was improved meaningfully in *L. zeylanica*–treated diabetic assembly.

9.3.7 ANTIMICROBIAL ACTIVITY

Antibacterial activities of *L. zeylanica* leaves' extract were detected utilizing disc dispersion test and 70 µg/µL of *L. zeylanica* extract was the best dosage to afford the effect of 11 mm restraint zone for *E. coli* and 15 mm restraint zone for *S. aureus*. Along these lines, *L. zeylanica* leaves might be the elements in the adhesive for "etak salai" preparation (Abdullah et al., 2019).

9.3.8 ANTI-ANALGESIC ACTIVITY

The methanolic crude extracts (MCEs) of *L. zeylanica* plants showed the maximum peripheral analgesic action among the trial samples with a prevention of pain impression of 40% (at 200 mg/kg body mass dosage, MCE_{200}) and 32% (at 100 mg/kg body mass dosage, MCE_{100}) in contrast to diclofenac sodium (66.67%). In central pain-relieving action exam, extension of glancing period was estimated at 30, 60, and 90 min of management of trials, wherever morphine was the positive regulator. Entirely extracts and fractionates exhibited noteworthy rise in tail-glancing period, though methanolic extract displayed the maximum action (at 60 min) amongst all test samples (15.44 ± 0.256). In antioxidant exams, butylated hydroxytoluene (BHT) was utilized as the positive regulator, wherever methanolic extracts displayed the maximum antioxidant capacities amongst the trails. IC_{50} (µg/ml) values were 19.61 (BHT), 28.46 (MCE), 65.61 (EASF), 84.75 (PESF), and 97.09 (chloroform-soluble fractions). Furthermore, whole plant extracts exhibited weak antimicrobial action. Subsequently, the methanolic extracts and various portions from *L. zeylanica* consume robust pain-relieving and antioxidant latent and can be a noteworthy basis of natural remedy (Islam et al., 2020).

9.3.9 ANTI-HELMINTHIC ACTIVITY

Radhika (2018) inspected the antihelminthic action of *L. zeylanica* leaves. The extracts were analyzed for antihelminthic action against Indian earthworms *Pheretima posthuma*; acetone and ethanolic extracts exhibited the dose-dependent action by paralysis trailed by the death of earthworms. The antihelminthic action of ethanol extract was stronger compared to the acetone extract. The earthworms were extra subtle to the extracts of ethanol at 20 mg/ml meditations as contrast to the drug albendazole (10 mg/ml).

9.3.10 ANTIULCER ACTIVITY

Pathakota et al. (2017) investigated the antiulcer action of the ethanolic and petroleum ether-dichloromethane extracts of the leaves of the folklore Indian medicinal plants *Raphanus sativus* and *L. zeylanica* at various dosages utilizing ranitidine as normal medicine. Alcohol persuaded ulcers and pyloric ligation techniques were modified to estimate the biological possible of the plant extracts utilizing male Wistar rat trial animal replicas. In both the techniques, the extracts were found to hold dosage-dependant antiulcer action. At the point when looked at, the petroleum ether dichloromethane extracts of both the plants at the dosage of 300 mg/kg displayed noteworthy action analogous to ranitidine medication.

KEYWORDS

- **antioxidant activity**
- **antibacterial activity**
- **antifungal activity**
- **antidiabetic activity**
- **antianalgesic activity**
- ***Leucas zeylanica***

REFERENCES

Abdullah, F.; Nasir, S. N. A. M. N.; Han, D. K.; Appalasamy, S.; Nor, M. M.; Rak, A.E. Potential of *Leucas zeylanica* Extract to Eliminate *E. coli* and *S. aureus* in *Corbicula fluminea* ("Etak") Tissue. *Malaysian J. Fundamental Appl. Sci.* **2019,** *15* (4), 597–599.

Babu, A.; Mohamed, M. S. N.; Jaikumar, K.; Anand, D.; Saravanan, P. In Vitro Antifungal Activity of *Leucas aspera* and *Leucas zeylanica*. *Intl. J. Pharm. Sci. Res.* **2016,** *7* (2), 752–756.

Bindu, Ch.S. Antioxidant Activity of Methanolic Extract of Leaves of *Leucas zeylanica* Linn. by Using Isolated from Heart. *Innovative J. Sci.* **2019,** *7* (1), 1–3.

Caius, J. F. *The Medicinal and Poisonous Plants of India*; Scientific Publishers, India, 1998.

Chen, G.-Y.; Zhang, B.; Zhao, T.; Nidhal, N.; Wang, J.-L.; Zhou, X.-M.; Dai, C.-Y. A New Triterpenoid Glucoside from *Leucas zeylanica*. *Nat. Prod. Res.*, **2020,** *34* (13), 1874–1878.

Chopra, R. N.; Nayar, S. L.; Chopra, I. C. *Glossary of Indian Medicinal Plants*; National Institute of Science Communication: New Delhi, India, 1996.

Dutta, M.; Waseque, M. M.; Uddin, M. Z.; Sultana, S.; Tazneen, N.; Barua, S. Antidiabetic Potentiality of Aqueous Leaf Extract of *Leucas zeylanica* in Alloxan-Induced Diabetic Rats. *J. Innov. Pharm. Biol. Sci.* **2016,** *3* (4), 121–124.

Hossain, S.; Rahman, M.; Fatima, N.; Haque, M.; Islam, J. *Leucas zeylanica* (L.) R. Br. Protects Ethanol and Hydrogen Peroxide-Induced Oxidative Stress on Hepatic Tissue of Rats. *Intl. Curr. Pharm. J.*, **2013,** *2* (9), 148–151.

Hung, N. H.; Chuong, N. T. H.; Satyal, P.; Hieu, H. V. et al. Mosquito Larvicidal Activities and Chemical Compositions of the Essential Oils of *Leucas zeylanica* Growing Wild in Vietnam. *Nat. Prod. Commum.* **2019,** *14* (6), 1–7.

Islam, M. M.; Parvin, E. S.; Zahir, R. S.; Mahmud, Z. A. Methanolic Extracts and Different Fractions of Whole Plants of *Leucas zeylanica* Show Promising Analgesic and Antioxidative Activities. *Egypt. Pharm. J. [NRC].* **2020,** *19* (3), 260–265.

Khan, M. A. A.; Karmakar, U. K.; Islam, M. I.; Biswas, N. N.; Ahmed, M. I.; Sadhu, S. K. Assessment of Phytochemical and Pharmacological Activities of *Leucas zeylanica* (L) R. Br. (Family: Lamiaceae). *Hamdard Med.* **2010,** *53*, 19–29.

Napagoda, M.; Gerstmeier, J.; Butschek, H.; Lorenz, S.; Kanatiwela, D.; Soyza, S. D.; Qader, M., Nagahawatte, A.; Wijayaratne, G. B.; Svatoš, A.; Jayasinghe, L.; Koeberle, A.; Werz, O. Lipophilic extracts of *Leucas zeylanica*, a Multi-Purpose Medicinal Plant in the Tropics, Inhibit Key Enzymes Involved In Inflammation and Gout. *J. Ethnopharmacol.* **2018,** *224*, 474–481.

Pathakota, R.; Anasuri, S.; Damu, U. S.; Kankanala, S.; Tekurala, M. Anti-Ulcer Activity of the Leaves of *Raphnus sativa* and *Leucas zeylanica*. *World J. Pharmaceut. Res.* **2017,** *6* (12), 955–963.

Radhika, B. Anti-Helminthic Activity of *Leucas zeylanica* Linn. *Int. J. Pharmaceut. Biol. Archive* **2018,** *9* (2), 70–73.

Sinha, V. S.; Kumar, N. Genotoxic Effects of *Ocimum basilicum* L and *Leucas zeylanica* L in Root Tip Cells of *Allium cepa* L. *J. Indian Bot. Soc.*, **2014,** *93* (1 and 2), 135–139.

Zhang, X., Gui, M.; Zhang, C.; Chen, C.; Yu, L.; Liu, J. Leucasinoside: A New Abietane Diterpenoid Glycoside from *Leucas zeylanica*. *Records Nat. Prod.* **2016,** *10* (5), 645–648.

CHAPTER 10

BIOMOLECULES AND THERAPEUTICS OF *MENTHA ROTUNDIFOLIA* (L.) HUDS.

VISHAL P. DESHMUKH

Department of Botany, Jagadamba Mahavidyalaya, Achalpur City, Dist. Amravati, Maharashtra, India

ABSTRACT

Mentha rotundifolia (L.) Huds. is a widely distributed, perennial, aromatic herb reported from major parts of world. Being important member of the family Lamiaceae, it has common names like "apple mint" or "green apple". *M. rotundifolia* plant parts are frequently used against skin pathogens, digestive, and respiratory disorders. *M. rotundifolia* is rich source of phenolics, flavonoids, carotenoids alkaloids, leucoanthocyanins, iridoids, saponins, anthocyanins, aldehydes and tannins. Essential oil is obtained from all plant parts with predominance of rotundifolone. *M. rotundifolia* revealed antimicrobial, antioxidant, insecticidal, anti-inflammatory, and acetylcholinesterase inhibitory potential.

10.1 INTRODUCTION

Mentha rotundifolia (L.) Huds. is a wild, perennial, aromatic herb and measures about 50–70 cm. It is native to Africa, Europe, and temperate Asia and is extensively scattered in America, Mediterranean basin, and in western Asia (Bouhabila et al., 2018; Khadraoui et al., 2014). It is an important

Bioactives and Pharmacology of Lamiaceae. T. Pullaiah, PhD (Ed.)

member of the family Lamiaceae and is evolved as a fertile hybrid between *Mentha longifolia* (L.) L. and *Mentha suaveolens* Ehrh. (Lorenzo et al., 2002). *M. rotundifolia* is popularly known as "apple mint" or "green apple" (Brahmi et al., 2016; Sbai et al., 2020), and throughout Algeria and northern Africa region, it is commonly known as "timarssad" or "timija" (Boussouf et al., 2017; Fatiha et al., 2015). Plant parts of *M. rotundifolia* are frequently used against skin pathogens, digestive, and respiratory disorders (Gonzalez-Tejero et al., 2008; El-Hilaly et al., 2003).

10.2 BIOACTIVES

Phytochemical screening of *M. rotundifolia* aerial parts' decoctions and infusions confirmed the presence of flavonoids, anthocyanins, total tannins, reducing sugars, alkaloids, coumarins, saponines, *O*-heterosides, *C*-heterosides, and sterols and triterpenes (Redhouane et al., 2020). Chemically *M. rotundifolia* is rich in phenolics, flavonoids, and tannins (Benabdallah et al., 2016; Abu-Qatouseh et al. 2019; Btissam et al., 2018). *M. rotundifolia* also revealed the existence of carotenoids and flavonoids (Brahmi et al., 2018). Chemical examination revealed the incidence of alkaloids, leucoanthocyanins, iridoids, saponins, anthocyanins, and aldehydes from *M. rotundifolia* (Btissam et al., 2018). Phenolic acids like chlorogenic acid, *p*-coumaric acid, caffeic acid, and rosmarinic acid along with flavonoids such as luteolin, diosmin, naringenin, apigenin, kaempferol, and diosmetin were acknowledged from ethanol extract of *M. rotundifolia* leaves using HPLC (Fatiha et al., 2015). LC–MS analysis of 70% ethanol extract of *M. rotundifolia* leaves identified polyphenolic compounds like caftaric acid, gentisic acid, caffeic acid, chlorogenic acid, *p*-coumaric acid, isoquercitrin, rutin, quercitrin, luteolin, and apigenin (Moldovan et al., 2014). UHPLC–HRESIMS/MS method quantified 17 polyphenoilc compounds from 10 populations of Tunisian *M. rotundifolia* including caffeic acid, luteolin-7-rutinoside, luteolin-7-glucoside, luteolin-7-glucuronide, isonaringin, hesperidin, apigenin-7-glucoside, salvianolic acid L, diosmin, rosmarinic acid, luteolin-3′-glucuronide, isosalvianolic acid B, isosalvianolic acid W, salvianolic acid B, luteolin, isosalvianolic acid A, and newly identified salvianolic acid W (Yahia et al., 2019c). Rotundifolone had been a major component of Japanese *M. rotundifolia* essential oil (Shimizu, 1957). Piperitone oxide and 1,2-epoxy menthyl acetate as prime chemical compounds were identified in the new chemical strain of *M. rotundifolia* (Shibata and Shimizu, 1973). Two new stereoisomers explained as (1*S*:2*S*:3*R*:4*R*)-(–)-1,2-epoxyisomenthyl acetate and (1*S*:2*S*:3*R*:4*R*)-(–)-1,2-poxyneoisomenthyl acetate were isolated from self-pollinated *M. rotundifolia* essential oil (Umemoto, 1998). Esculetin, an antioxidant, was extracted from *M. rotundifolia* leaves using pressurized fluid extraction and ultrasonic extraction methods (Dobiáš et al., 2010). *M. rotundifolia* essential oil had the highest percentage of rotundifolone (42.6%) (Yahia et al., 2019a). Maximum amount of rotundifolone is found in essential oil of *M. rotundifolia* gathered from high altitude upper semi-arid bioclimatic zone of Tunisia, whereas lower humid region samples are rich in rotundifolone or β-caryophyllene or germacrene D. Sample collected from lowest altitude revealed maximum quantity of *p*-menthane-1,2,3-triol (Yahia et al., 2019b). Aqueous and acetone extract of *M. rotundifolia* identified 15 and 16 phenolic compounds in LC–MS/MS analysis with predominance of rosmarinic acid in both cases (Siham et al., 2019). Tissue culture–raised plants of *M. rotundifolia* grown at elevated temperature accounted for higher essential oil content than wild plants (Benahmed et al., 2018). Essential oil of Algerian *M. rotundifolia* identified a total of 27 compounds, of which rotundifolone shared a major portion; additionally, γ-murolene and β-farnesene were also found in substantial amounts (Benabdallah et al., 2018). GC–MS analysis of essential oil extracted from leave stems and bunds of *M. rotundifolia* yielded 21, 13, and 21 compounds with major occupancy of germacrene, pulegone, and pulegone oxide in respective parts (Bouhabila et al., 2018). Out of 14 compounds, 3-cyclopenten-1-one and 2-hydroxy-3-(3-methyl-2-butenyl) were the major components detected in *M. rotundifolia* essential oil followed by β-terpinyl acetate, cinerolone, α-pinene, and β-pinene (Amina et al., 2018). GC–MS investigation of *M. rotundifolia* essential oil collected from two sites of Algeria accounted for 39 compounds each. Piperitenone oxide and cis-piperitone oxide are the main constituents of essential oil of respective localities (Brada et al., 2006). Using chromatographic and spectroscopic methods in combination, 59 compounds were isolated from *M. rotundifolia* essential oil having piperitenone oxide as the main constituent (Nagell and Hefendehl, 1974). GC–MS analysis of Cuban *M. rotundifolia* essential oil identified 50 compounds with a high proportion of 2,4(8),6-*p*-menthatrien-2,3-dio, germacrene D, β-caryophyllene, and α-cadinol (Pino et al., 1999). Essential oil from Uruguayan *M. rotundifolia* leaves recorded 23 compounds in GC–FID-MS investigation. Composition point-of-view piperitenone oxide was the major component of essential oil followed by (*Z*)-sabinene hydrate and 4-terpineol (Lorenzo et al., 2002). GC–FID and GC/MS study of essential oil of *M. rotundifolia* samples collected from Algeria distinctly discriminated into two chemotypes. The first group is dominated by piperitone oxide and piperitenone oxide, whereas other groups have major proportion of piperitenone and piperitenone oxide (Brada et al., 2007). GC–MS examination of *M. rotundifolia* essential oil confirms quantitatively higher portions of piperitol, rotundifolone, β-caryophyllene, and limonene (Pérez Raya et al., 1990). Essential oil obtained by hydrodistillation of *M. rotundifolia* leaves on GC–MS investigation recorded 66 compounds in which *trans*-piperitone epoxide, piperitone oxide, thymol, and terpinen-4-ol shared major portion (Brahmi et al., 2016). Essential oil of Moroccan *M. rotundifolia* leaves yielded 32 compounds in GC–MS investigation. Menthol found to be major component of essential oil, besides menthone, menthyl acetate, menthofuran, neomenthol, linalyl acetate, piperitone oxide, and piperitone were also observed in appreciable amount (Derwich et al., 2009, 2010a, 2010b). Essential oil of *M. rotundifolia* leaves on CG–FID–MS analysis documented that 19 components with predominance of piperitone oxide and compounds like 1,8-cineole, myrcene, limonene, and β pinene were also present in considerable amount (Gende et al., 2014). GC–MS analysis of field collected and somaclonal variants of *M. rotundifolia* yielded 18 and 20 components, respectively. Percentage point of view in both cases, carvone was a prime component (Jose and Thoppil, 2014). *M. rotundifolia* essential oil of aerial parts recorded 14 compounds on GC–MS investigation having 3-cyclopenten-1-one and 2-hydroxy-3-(3-methyl-2-butenyl) as predominant compound, also β-terpinyl acetate, α-pinene, cinerolone, and β-pinene located in trace amount (Leblalta et al., 2020). GC and GC–MS analyses detected 38 compounds in *M. rotundifolia* essential oil; among them, piperitenone oxide was the major component followed by piperitone oxide, terpinen-4-ol, pulegone, and limonene (Fatiha et al., 2016). *M. rotundifolia* essential oil on GC–MS analysis yielded 20 compounds having high proportion of linalool, ρ-menth-1-en-8-ol,(s)-(–)-, geranyl acetate, 1,8-cineole, and myrcene (Aziz and Abbass, 2010). Pulegone was the predominant compound, whereas methyl acetate, menthol, and menthone were observed in recognizable amounts on GC–MS investigation of *M. rotundifolia* essential oil (El Arch et al., 2003). GC–MS testing of *M. rotundifolia* (apple mint) essential oil revealed major occurrence of (+)-piperitenone oxide, germacrene D, and β-caryophyllene (Miyazawa et al., 1998). Isomers of 1,2-epoxymenthyl acetate and piperitone were the prime compounds identified from essential oil of *M. rotundifolia* foliage using GC–MS (Oka et al., 2000). Essential oil of Tunisian *M. rotundifolia* collected from "Beja" region scores 45 compounds with predominance of β-caryophyllene, germacrene D, carveol, and viridiflorol, whereas samples of "Bizerte" region yield 40 compounds in which pulegone, piperitenone oxide, 5-acetyl thiazole, and germacrene D shared a major portion of essential oil (Riahi et al., 2013). GC–MS analysis of essential oil derived from *M. rotundifolia* leaves showed high content of 2-isopropylidenecyclohexanone, eucarvone, γ-muurolene, 2-isopropyl-5-methyl-3-cyclohexen-1-one, and *p*-menthane-1,2,3-triol (Sbai et al., 2020). During analysis of Tunisian *Mentha* collection based on essential oil content, a "rotoundifolia" group is determined having five accessions and is characterized by the predominance of piperitone oxide subsequently menthol, limonene, and viridiflorol (Soilhi et al., 2019). Algerian *M. rotundifolia* essential oil on GC–MS investigation recorded 27 compounds, among them piperitenone oxide, caryophyllene oxide, cis-cinerolone, and isophorone were major components (Yakhlef et al., 2020). GC–MS analysis of *M. rotundifolia* essential oil identified 51 compounds belonging to oxygenated monoterpenes, monoterpene hydrocarbons, sesquiterpene hydrocarbons, and oxygenated sesquiterpenes groups (Yu et al., 2018).

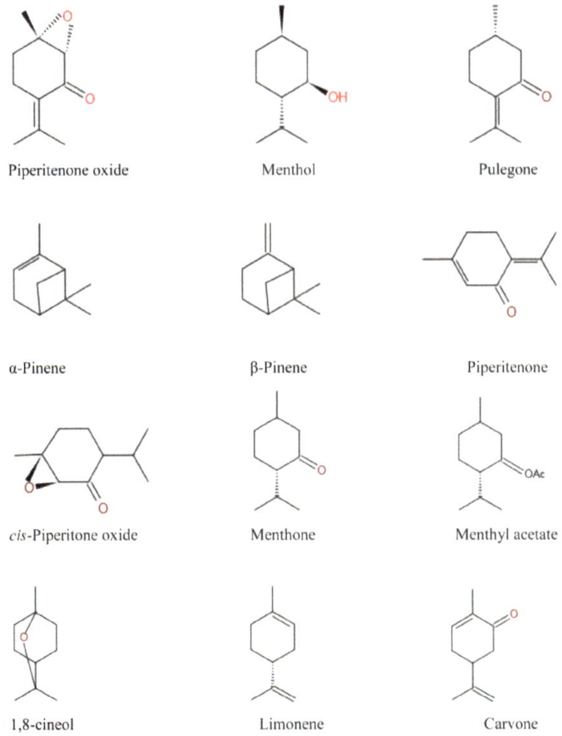

FIGURE 10.1 Structures of chemical compounds extracted from *M. rotundifolia* essential oil (de Sousa Barros et al., 2015; Mamadalieva et al., 2020; Kumar et al., 2011).

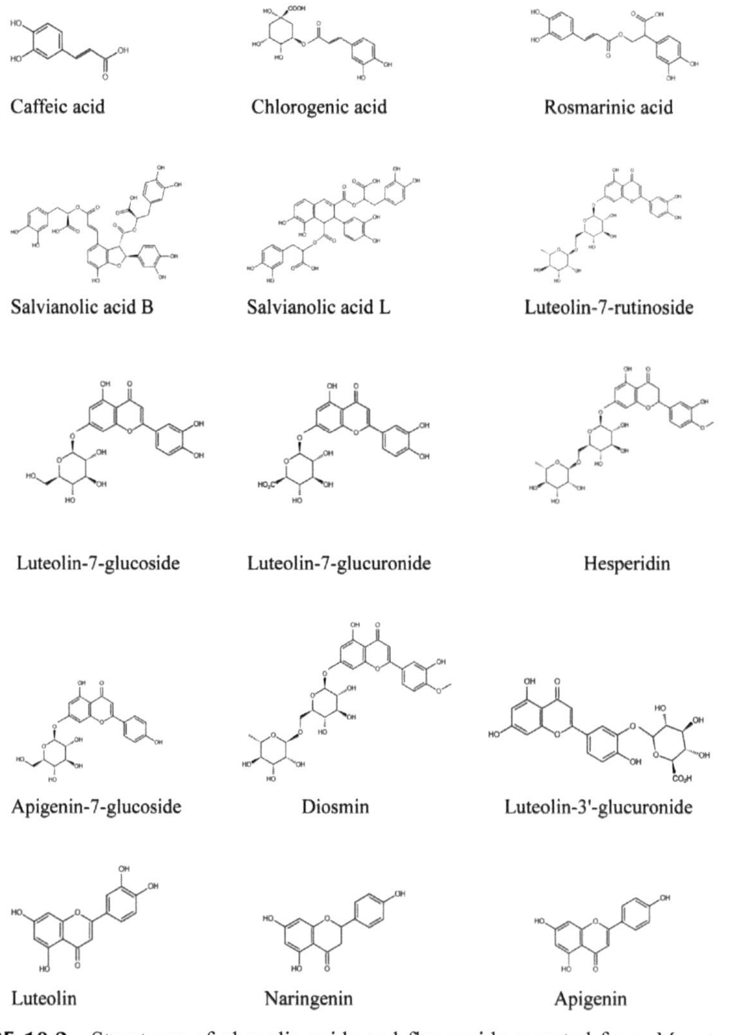

FIGURE 10.2 Structures of phenolic acids and flavonoids reported from *M. rotundifolia* (Fatiha et al., 2015; Hanafy et al., 2020; Yahia et al., 2019c).

10.3 PHARMACOLOGY

10.3.1 ANTIBACTERIAL ACTIVITY

Methanol phase of Jordanian *M. rotundifolia* leaves revealed moderate activity against *Propionibacterium acnes* (Abu-Qatouseh et al., 2019).

M. rotundifolia essential oil performs better in-spot method compared to disc-diffusion method and depicts significant activity against MRSA, *Staphylococcus aureus*, and *Bacillus subtilis* (Brahmi et al., 2016). Aqueous-ethanolic extract of *M. rotundifolia* showed bacteriostatic and bactericidal action against gram-negative and gram-positive bacterial strains (Btissam et al., 2018). *M. rotundifolia* leaves essential oil depicted strongest activity with 45, 34, and 31 mm zone of inhibition against *Escherichia coli*, *S. aureus*, and *S. intermedius*, respectively, whereas *Micrococcus luteus* and *Proteus mirabilis* were least affected (Derwich et al., 2010a). Essential oil of *M. rotundifolia* leaves had moderate activity against different strains of *Paenibacillus larvae* (white) with MIC range of 600–700 μg/mL; however, in bioautography method only, piperitone oxide showed inhibitory potential equal to chromatographic spot against most of strains (Gende et al., 2014). Essential oil from aerial parts of *M. rotundifolia* demonstrated best inhibitory action against *E. coli* and *P. mirabilis* with 24 and 22 mm zones of inhibition (Ladjel et al., 2011). *E. coli*, *B. subtilis*, and *S. aureus* were completely susceptible at 1/1000 v/v concentration of *M. rotundifolia* essential oil (El Arch et al., 2003). Seventy percent ethanol extract of *M. rotundifolia* leaves showed high, moderate, low, and no inhibitory activity against *L. monocytogenes*, *S. aureus*, *S. typhimurium*, and *E. coli*, respectively (Moldovan et al., 2014). Extract of *M. rotundifolia* collected from Bejaïa region of Algeria showed significant inhibitory effect against all tested bacterial strains than the same species collected from Jijel region (Redhouane et al., 2020). Essential oil of *M. rotundifolia* collected from both "Bizerte" and "Beja" sites of Tunisia showed strong antibacterial activity (30–34 mm of inhibition range) against gram-positive bacteria as compared to gram-negative bacteria (22–24 mm of inhibition range) (Riahi et al., 2013). Essential oil of *M. rotundifolia* revealed moderate antibacterial activity (Yahia et al., 2019a). *E. coli*, *K. pneumoniae*, *C. freundii*, *S. sushi*, *P. mirabilis*, *A. baumannii*, and *S. aureus* are found to be susceptible to *M. rotundifolia* leaves essential oil having inhibition diameter range of 8–56 mm (Sbai et al., 2020). Methanol as well as aqueous extracts of leaves and stem revealed meager antimicrobial activity against five bacterial strains with 6.1–10.4 mm range of inhibition zones (Seladji et al., 2014).

10.3.2 ANTIFUNGAL ACTIVITY

M. rotundifolia essential oil reported appreciable antifungal activity against *A. flavus* and *C. albicans* (Brahmi et al., 2016). Essential oil extracted from *M. rotundifolia* aerial parts showed significant antifungal activity at 0.1%

and above concentrations and completely inhibited *F. oxysporum* growth (Leblalta et al., 2020). *T. pini*, *P. parasiticus*, and *A. niger* were susceptible to *M. rotundifolia* essential oil at 1/1000 v/v concentration (El Arch et al., 2003). *M. rotundifolia* leaves' 70% ethanol extract showed moderate inhibitory activity against *C. albicans* with 20 ± 0.05 inhibition zone (Moldovan et al., 2014). Essential oil of Tunisian *M. rotundifolia* collected from "Bizerte" and "Beja" regions showed strong antifungal activity against *A. niger* and *C. albicans* (Riahi et al., 2013). *C. albicans* is the only fungi found susceptible to methanol and aqueous extract of *M. rotundifolia* leaves and stem, whereas *A. flavus* and A. *fumigatus* showed complete resistance (Seladji et al., 2014). Among tested *Mentha* species, *M. rotundifolia* revealed highest antifungal activity against tested fungi (Yahia et al., 2019a). *M. rotundifolia* essential oil revealed significant antifungal potential by inhibiting mycelium; however, inhibitory potential is more distinct in micro-atmosphere method (IC_{50} = 0.092 ± 0.006 μL/mL) than agar medium (IC_{50} = 0.254 ± 0.031 μL/mL) (Yakhlef et al., 2020).

10.3.3 ANTIOXIDANT EFFICACY

Hydromethanolic extract of Algerian *M. rotundifolia* revealed strong antioxidant activity in DPPH, β-carotene bleaching, and chelating ion ability assays with IC_{50} = 31.66 ± 2.16, 763.00 ± 0.70, and 1300.00 ± 0.60 μg/mL, respectively (Benabdallah et al., 2016). In DPPH, β-carotene bleaching, and chelating ability methods, *M. rotundifolia* essential oil observed high antioxidant activity with IC_{50} = 1.50 ± 0.05, 0.54 ± 0.03, and 4.03 ± 0.02 mg/mL, respectively (Benabdallah et al., 2018). Essential oil extracted from tissue culture–raised plants of *M. rotundifolia* showed wide range of antioxidant potential in DPPH (IC_{50} = 4.18–24.93 mg/mL), ABTS (IC_{50} = 0.51–1.56 mg/mL), and CUPRAC ($A_{0.50}$ = 0.34–2.71 mg/mL) assays (Benahmed et al., 2018). After Carrageenan injection, *M. rotundifolia* extract showed significant *in-vivo* antioxidant potential in the liver homogenate with enhanced catalase, superoxide dismutase, glutathione activities, and declined malondialdehyde level (Boussouf et al., 2017). Hydro-ethanolic extract of *M. rotundifolia* leaves revealed substantial antioxidant potential in KRL test, whereas in FRAP and DPPH assays, it showed the weakest activity as compared to other two *Mentha* species (Brahmi et al., 2018). *M. rotundifolia* essential oil showed reasonable antioxidant activity (IC_{50} = 138.2 ± 8.9 μg/mL) in ABTS assays; on the contrary in Kit Radicaux Libres test, it showed potent antioxidant activity

(58.8 ± 1.7 μg/mL) at merely 50 μg/mL concentration (Brahmi et al., 2016). In phosphomolybdate and DPPH assays, methanol extract of *M. rotundifolia* showed highest antioxidant potential as compared to other solvents (Brahmi et al., 2017a). Ethanol extract of *M. rotundifolia* leaves showed poor antioxidant potential in DPPH, ABTS, TEAC, and Quenching of electrogenerated O_2 assays (IC_{50} = 71.3 ± 2.6, 40.4 ± 0.7 μg/mL; 0.20 ± 0.0 mM and AI_{50} = 389 ± 52 μg/mL, respectively) as compared to other two *Mentha* species (Fatiha et al., 2015). Antioxidants extracted from *M. rotundifolia* leaves using pressurized fluid extraction method showed higher inhibition (90.3%) of DPPH-free radicals compared to ultrasonic extraction (81.7%) method (Dobiáš et al., 2010). In DPPH, ABTS, reducing power, and phosphomolybdenum assays, *M. rotundifolia* essential oil was found less efficient (IC_{50} = 2222.2 ± 25.2; 133.8 ± 4.8; 166.6 ± 1.9 and 45.2 ± 1.2 μg/mL respectively) compared to standard (Fatiha et al., 2016). Seventy percent ethanol extract of *M. rotundifolia* leaves exhibited utmost radical scavenging activity (IC_{50} = 104.74 ± 1.76 μg/mL) in DPPH assay (Moldovan et al., 2014). Ethanol extract of *M. rotundifolia* aerial parts showed moderate antioxidant potential in DPPH and ABTS assays with IC_{50} = 21.71 and 158.90 μg/mL, respectively (Nickavar et al., 2008). Pure methanol and 70% methanol extract of *M. rotundifolia* collected from Jijel and Bejaia regions of Algeria showed DPPH inhibition with IC_{50} = 1.41 and 1.27 mg/mL, respectively (Redhouane et al., 2020). In DPPH assay, essential oil of *M. rotundifolia* collected from "Bizerte" region of Tunisia revealed highest antioxidant activity (IC_{50} = 26.11 ± 1.04 μg/mL) than "Beja" site (IC_{50} = 29.52 ± 1.31 μg/mL). Regarding FRAP test, essential oil from "Bizerte" region revealed higher (390.91 ± 4.01 μmol Fe^{2+} E/g EO) ferric reducing antioxidant power than "Beja" site essential oil (330.46 ± 3.18 μmol Fe^{2+} E/g EO) (Riahi et al., 2013). Methanolic extract of *M. rotundifolia* leaves revealed the highest radical scavenging ability and high ferrous iron chelating activity with IC_{50} = 0.7 ± 0.028 and 0.6 mg/mL, respectively; In β-carotene-linoleic acid assay, methanolic and the aqueous extract of stem showed potent activity (IC_{50} = 0.62 ± 0.02 and 0.52± 0.02 mg/mL, respectively) as compared to leaves (Seladji et al., 2014). In DPPH assay, acetone extract of *M. rotundifolia* showed high radical-scavenging activity than aqueous extract (IC_{50} = 15.066 ± 0.449 and 97.862 ± 0.218 μg/mL, respectively); in ABTS assay, aqueous extract revealed substantial antioxidant potential than acetone extract (IC_{50} = 5.868 ± 0.487 and 22.026 ± 0.759 μg/mL, respectively); in case of $OH^{\cdot}$ scavenging activity assay, acetone extract demonstrates better $OH^{\cdot}$ radical scavenging potential compared to aqueous extract (IC_{50} = 204.355 ± 3.925 and 377. 232 ± 4.995 μg/mL, respectively); aqueous extract of *M. rotundifolia* showed strong chelating activity (IC_{50}= 128.155 ± 5.887 μg/mL) but acetone extract failed to reach 50% chelating activity; aqueous extract illustrated strong reducing power compared to acetone extract (EC_{50} = 18.809 ± 0.369 and 25.83 ± 0.208 μg/mL, respectively); in linoleic acid bleaching assay, acetone as well aqueous extract revealed nearly close percentage of inhibition ratio (79.778 ± 1.812% and 84.666 ± 0.963%, respectively) (Siham et al., 2019). Among 10 populations of *M. rotundifolia* collected from various climatic regions of Tunisia, population number 10 revealed the strongest antioxidant activity in DPPH (IC_{50}=3.66 ± 0.2 mg/ml) and FRAP (0.20 ± 0.0 mmol Fe^{2+}/g EO) assays; however, population eight showed highest (IC_{50} = 10.46 ± 0.0 mg/mL) β-carotene bleaching inhibition potential (Yahia et al., 2019b). Essential oil of *M. rotundifolia* showed premier antioxidant activity in FRAP and iron chelating assays (Yahia et al., 2019a). Out of 10 *M. rotundifolia* populations gathered from Tunisia, essential oil of MROT-1 population revealed the strongest antioxidant activity in DPPH, β-carotene (IC_{50} = 15.16 ± 0.16 and 106.66 ± 6.66 μg/mL, respectively), and FRAP assays (574.03 ± 0.98 μmol eq. Fe^{2+}/g) (Yahia et al., 2019c).

10.3.4 CYTOTOXIC EFFECTS

In MTT assay, aqueous ethanol extract of *M. rotundifolia* showed moderate activity with IC_{50} = 368.02 μg/mL. *M. rotundifolia* extract above 200 μg/mL concentration leads to significant elevation of PI positive cells (Brahmi et al., 2017b).

10.3.5 ANTIPROLIFERATIVE ACTIVITY

Hexane extract of *M. rotundifolia* leaves and flowers showed strong antiproliferative activity against KB and MCF-7 tumor cell lines with GI_{50} (MCF-7 = 13.09 mg/mL), TGI (KB = 37.76 mg/mL), and IL_{50} (KB = 291.07 mg/mL); additionally, same extract revealed similar activity with GI_{50} (183.65 mg/mL), TGI (280.54 mg/mL), and IL_{50} (384.59 mg/mL) against NIH 3T3 cells (Nedel et al., 2012).

10.3.6 CYTOPROTECTIVE ACTIVITY

M. rotundifolia leaves' extract showed dose-dependent cytoprotective effect in RAW 264.7 cells caused by 7-ketocholesterol (Brahmi et al., 2018).

10.3.7 INSECTICIDAL ACTIVITY

M. rotundifolia essential oil showed very strong contact and fumigant insecticidal activity against *Sitophilus granarius* with 100% mortality after 24 h of incubation at a dose of 4% and 1%, respectively (Amina et al., 2018). Essential oil obtained from *M. rotundifolia* leaves at 2 μL/mL concentration revealed the most excellent insecticidal activity against lesser grain borer *R. dominica* in fumigation bioassay (44.3% mortality) as compared to contact toxicity (30.7% mortality) and it also showed 47.54% of repellency after 96 h of treatment (Brahmi et al., 2016). Essential oil of leaves and stem of *M. rotundifolia* at 1% concentration drastically reduced the fecundity and total output of eggs (10.2% and 90.3% eggs/two females, respectively) of *C. maculatus*; it also lower down the incubation period and percent hatchability (5.31 days and 14.77%, respectively) of the eggs laid down by *C. maculatus*. Importantly oil also showed the highest mortality and longest larval and pupal durations (98.12%, 24.5 and 5.6 days, respectively) against *C. maculatus* (Aziz and Abbass, 2010). Dichloromethane extract of *M. rotundifolia* aerial parts at 5000 ppm dose had very high lethal impact (75% mortality) on *Tribolium castaneum* larvae and had sublethal impact on pupation time; on the contrary plant infusion at the same dose showed sublethal effect in both parameters (Clemente et al., 2003). Cereal insects like *R. dominica* and *S. oryzae* were completely wiped out at 6.5×10^{-2} mL/cm^3 concentration of *M. rotundifolia* essential oil after the first day of treatment (El Arch et al., 2003). *M. rotundifolia* essential oil revealed very high insecticidal activity against *S. granarius* and *T. confusum* after 3 h of exposure (LC_{50} = 0.195 and 0.561 μL/mL air, respectively, and LC_{99} = 1.072 and 1.530 μL/mL air, respectively). Essential oil at the lowest 0.1 μL/mL concentration showed mortality (LT_{50} = 2.93 and 4.13 h, respectively) against *S. granarius* and *T. confusum*; *M. rotundifolia* essential oil also showed significant repellent (63.75 ± 8.69% and 81.25 ± 4.84%) effect against *S. granarius* and *T. confusum* (Yakhlef et al., 2020).

10.3.8 ANTI-INFLAMMATORY EFFECTS

Methanol extract of *M. rotundifolia* leaves showed less inhibition of interleukin 1 alpha release in UVB-treated skin explants (Abu-Qatouseh et al., 2019). In the carrageenan-induced paw edema model, hydromethanolic extract of *M. rotundifolia* exhibits dose-dependent inhibition and highest (84.49%) inhibitory effect was documented with 600 mg/kg dose at 4 h (Boussouf et al., 2017). Hydro-ethanolic extract of *M. rotundifolia* leaves distinctly declined secretion of IL-6, MCP-1, and TNF-α in RAW 264.7 cells (Brahmi et al., 2018).

10.3.9 ACETYLCHOLINESTERASE INHIBITORY ABILITY

M. rotundifolia essential oil showed moderate (IC_{50} = 52.5d ± 1.44 μg/mL) acetylcholinesterase inhibitory potential as compared to other *Mentha* species (Benabdallah et al., 2018). Despite having a major portion of piperitenone oxide in *M. rotundifolia* essential oil, it demonstrated no inhibition of AChE activity (Miyazawa et al., 1998). MROT-3 and 5 populations of Tunisian *M. rotundifolia* revealed strongest (IC_{50} = 0.21 ± 0.01 mg/mL) acetylcholinesterase inhibitory activity (Yahia et al., 2019c).

10.3.10 PHYTOPHOTODERMATITIS EFFECT

Application of *M. rotundifolia* poultice for quite some time and exposure to sunlight showed phytophotodermatitis effect with shallow second-degree burn (Benhiba and Hassam, 2014).

10.3.11 ANALGESIC EFFECT

Against acetic acid-induced contraction in mice, hydromethanolic extract of *M. rotundifolia* follows dose-dependent analgesic effects and at 600 mg/kg bw dose highest inhibition (85.29%) was observed which was quite superior than Aspirin (Boussouf et al., 2017).

10.3.12 TYROSINASE INHIBITION ACTIVITY

Ethanol extract of *M. rotundifolia* leaves revealed poor (IC_{50} = 108 ± 20 μg/mL) antityrosinase activity as compared to kojic acid (IC_{50} = 23 ± 1 μg/mL), a potent tyrosinase inhibitor (Fatiha et al., 2015).

10.3.13 EFFECT ON APOPTOSIS AND AUTOPHAGY

Activation of apoptosis and autophagy by hydro-ethanolic extract of *M. rotundifolia* was confirmed through lack of cleaved caspase-3 and conversion of LC3-I to LC3-II (Brahmi et al., 2017b).

10.3.14 CNS ACTIVITY

Essential oil of *M. rotundifolia* was found to be less toxic with LD_{50} = 641.6 ± 2.4 mg/kg in acute toxicity assay; it also showed moderate percentage potentiation of barbiturate sleep and effects of curiosity in mice were found to be less lengthened with *M. rotundifolia*. Oral administration of *M. rotundifolia* essential oil revealed decreased body temperature at all applied doses; however, a major hypothermic effect was observed at 400 mg/kg dose after 3 h (Pérez Raya et al., 1990).

KEYWORDS

- **essential oil**
- **Insecticidal**
- ***Mentha rotundifolia***
- **phytochemical**
- **respiratory disorders**

REFERENCES

Abu-Qatouseh, L.; Mallah, E.; Mansour, K. Evaluation of Anti-*Propionibacterium acnes* and Anti-Inflammatory Effects of Polyphenolic Extracts of Medicinal Herbs in Jordan. *Biomed. Pharmacol. J.* **2019,** *12*, 211–217.

Amina, L., Daoud, H.; Ibrahim, D.; Cem, K, Ö.; Fatih, G. Chemical Composition and Insecticidal Activity of *Mentha rotundifolia* L. Essential Oil Against *Sitophilus granarius*. *Haya: Saudi J. Life Sci.* **2018,** *3*, 127–133.

Aziz, E. E.; Abbass, M. H. Chemical Composition and Efficiency of Five Essential Oil Against the Pulse Beetle *Callosobruchus maculatus* (F.) on *Vigna radiata* seeds. *Am.-Eurasian J. Agric. Environ. Sci.* **2010,** *8*, 411–419.

Benabdallah, A.; Boumendjel, M.; Aissi, O.; Rahmoune, C.; Boussaid, M.; Messaoud, C. Chemical Composition, Antioxidant Activity and Acetylcholinesterase Inhibitory of Wild *Mentha* Species from Northeastern Algeria. *S. Afr. J. Bot.* **2018,** *116*, 131–139.

Benabdallah, A.; Rahmoune, C.; Boumendjel, M.; Aissi, O.; Messaoud, C. Total Phenolic Content and Antioxidant Activity of Six Wild *Mentha* Species (Lamiaceae) from Northeast of Algeria. *Asian Pac. J. Trop. Biomed.* **2016,** *6*, 760–766.

Benahmed, A.; Harfi, B.; Benbelkacem, I.; Daas, A.; Laouer, H.; Belkhiri, A. In Vitro Propagated *Mentha rotundifolia* (L.) Huds and Antioxidant Activity of Its Essential Oil. *Curr. Issues Pharm. Med. Sci.* **2018,** *31*, 204–208.

Benhiba, H.; Hassam, B. Phytophotodermatitis Due to *Mentha rotundifolia*. *Pan Afr. Med. J.* **2014,** *18*, 192.

Bouhabila, A.; Zellagui, A.; Telci, I.; Maarfia, S.; Gherraf, N. Changeability of Essential Oils in Algerian *Mentha rotundifolia* L. (Lamiaceae) Growing in Sub Humid Area. *World J. Environ. Biosci.* **2018,** *7*, 45–47.

Boussouf, L.; Boutennoune, H.; Kebieche, M.; Adjeroud, N.; Al-Qaoud, K.; Madani, K. Anti-Inflammatory, Analgesic and Antioxidant Effects of Phenolic Compound from Algerian *Mentha rotundifolia* L. Leaves on Experimental Animals. *S. Afr. J. Bot.* **2017,** *113*, 77–83.

Brada, M.; Bezzina, M.; Marlier, M.; Carlier, A.; Lognay, G. Variabilité de la composition chimique des huiles essentielles de *Mentha rotundifolia* du Nord de l'Algérie. *Biotechnol. Agron. Soc. Environ.* **2007,** *11*, 3–7.

Brada, M.; Bezzina, M.; Marlier, M.; Lognay, G. C. Chemical Composition of the Leaf Oil of *Mentha rotundifolia* (L.) from Algeria. *J. Essent. Oil Res.* **2006,** *18*, 663–665.

Brahmi, F.; Abdenour, A.; Bruno, M.; Silvia, P.; Alessandra, P.; Danilo, F.; Drifa, Y.; Fahmi, E. M.; Khodir, M.; Mohamed, C. Chemical Composition and In Vitro Antimicrobial, Insecticidal and Antioxidant Activities of the Essential Oils of *Mentha pulegium* L. and *Mentha rotundifolia* (L.) Huds Growing in Algeria. *Ind. Crops Prod.* **2016,** *88*, 96–105.

Brahmi, F.; Dahmoune, F.; Kadri, N.; Chibane, M.; Dairi, S.; Remini, H.; Oukmanou-Bensidhoum, S.; Mouni, L.; Madani, C. Antioxidant Capacity and Phenolic Content of Two Algerian *Mentha* species *M. rotundifolia* (L.) Huds, *M. pulegium* L., Extracted with Different Solvents. *J. Complement. Integr. Med.* **2017a,** 20160064.

Brahmi, F.; Hadj-Ahmed, S.; Zarrouk, A.; Bezine, M.; Nury, T.; Madani, K.; Chibane, M.; Vejux, A.; Andreoletti, P.; Boulekbache-Makhlouf, L.; Lizard, G. Evidence of Biological Activity of *Mentha* Species Extracts on Apoptotic and Autophagic Targets on Murine RAW264.7 and Human U937 Monocytic Cells. *Pharm. Biol.* **2017b,** *55*, 286–293.

Brahmi, F.; Nury, T.; Debbabi, M.; Hadj-Ahmed, S.; Zarrouk, A.; Prost, M.; Madani, K.; Boulekbache-Makhlouf, L.; Lizard, G. Evaluation of Antioxidant, Anti-Inflammatory and Cytoprotective Properties of Ethanolic Mint Extracts from Algeria on 7-Ketocholesterol-Treated Murine RAW 264.7 Macrophages. *Antioxidants* **2018,** *7*, 184.

Btissam, R.; Fatima, E. M.; Kamal, E.; Hassane, G.; Mohamed, N. Composition and Antibacterial Activity of Hydro-Alcohol and Aqueous Extracts Obtained from the Lamiaceae Family. *Pharmacogn J.* **2018,** *10*, 81–91.

Clemente, S.; Mareggiani, G.; Broussalis, A.; Martino, V.; Ferraro, G. Insecticidal Effects of Lamiaceae Species Against Stored Product Insects. *Bol. San. Veg. Plagas.* **2003,** *29*, 421–426.

de Sousa Barros, A.; de Morais, S. M.; Ferreira, P. A. T.; Vieira, I. G. P.; Craveiro, A. A.; Fontenelle. R. O. D. S.; de Menezes, J. E. S. A.; da Silva, F. W. F.; de Sousa, H. A. Chemical Composition and Functional Properties of Essential Oils from *Mentha* species. *Ind. Crops Prod.* **2015,** *76*, 557–564.

Derwich, E.; Benziane, Z.; Boukir, A. Antibacterial Activity and Chemical Composition of the Leaf Essential Oil of *Mentha rotundifolia* from Morocco. *EJEAFChe.* **2010a,** *9*, 19–28.

Derwich, E.; Benziane, Z.; Boukir, A.; Benaabidate, L. GC–MS Analysis of the Leaf Essential Oil of *Mentha rotundifolia*, a Traditional Herbal Medicine in Morocco. *Chem. Bull. "POLITEHNICA" Univ. (Timisoara)* **2009,** *54*, 85–88.

Derwich, E.; Benziane, Z.; Taouil, R.; Senhaji, O.; Touzani, M. Comparative Essential Oil Composition of Leaves of *Mentha rotundifolia* and *Mentha pulegium* a Traditional Herbal Medicine in Morocco. *Am. Eurasian J. Sustain. Agric.* **2010b,** *4*, 47–54.

Dobiáš, P.; Pavlíková, P.; Adam, M.; Eisner, A.; Beňová, B.; Ventura, K. Comparison of Pressurised Fluid and Ultrasonic Extraction Methods for Analysis of Plant Antioxidants and Their Antioxidant Capacity. *Cent. Eur. J. Chem.* **2010,** *8*, 87–95.

El Arch, M.; Satrani, B.; Farah, A.; Bennani, L.; Boriky, D.; Fechtal, M.; Blaghen, M.; Talbi, M. Composition chimique et activités antimicrobienne et insecticide de l'huile essentielle de *Mentha rotundifolia* du Maroc. *Acta Bot. Gallica* **2003,** *150*, 267–274.

El-Hilaly, J.; Hmammouchi, M.; Lyoussi, B. Ethnobotanical Studies and Economic Evaluation of Medicinal Plants in Taounate Province (Northern Morocco). *J. Ethnopharmacol.* **2003,** *86*, 149–158.

Fatiha, B.; Didier, H.; Naima, G.; Khodir, M.; Martin, K.; Léocadie, K.; Caroline, S.; Mohamed, C.; Pierre, D. Phenolic Composition, In Vitro Antioxidant Effects and Tyrosinase Inhibitory Activity of Three Algerian *Mentha* Species: *M. spicata* (L.), M. *pulegium* (L.) and *M. rotundifolia* (L.) Huds (Lamiaceae). *Ind. Crop. Prod.* **2015,** *74*, 722–730.

Fatiha, B.; Khodir, M.; Nabila, D.; Sabrina, I.; Fatma, H.; Mohmed, C.; Moussa, B. Assessment of the Chemical Composition and In Vitro Antioxidant Activity of *Mentha rotundifolia* (L.) Huds Essential Oil from Algeria. *J. Essent. Oil Bear. Pl.* **2016,** *19*, 1251–1260.

Gende, L. B.; Mendiara, S.; Fernández, N. J.; Baren, C. V.; Lira, A. D. L.; Bandoni, A.; Fritz, R.; Floris, I.; Eguaras, M. Essentials oils of Some *Mentha* spp. and Their Relation with Antimicrobial Activity Against Paenibacillus Larvae, the Causative Agent of American Foulbrood in Honey Bees, by Using the Bioautography Technique. *Bull. Insectology* **2014,** *67*, 13–20.

Gonzalez-Tejero, M. R.; Casares-Porcel, M.; Sanchez-Rojas, C. P.; RamiroGutierrez, J. M.; Molero-Mesa, J.; Pieroni, A.; Giusti, M. E.; Censorii, E.; De Pasquale, C.; Della, A.; Paraskeva-Hadijchambi, D.; Hadjichambis, A.; Houmani, Z.; ElDemerdash, M.; El-Zayatf, M.; Hmamouchi, M.; El Johrig, S. Medicinal Plants in the Mediterranean Area: Synthesis of the Results of the Project Rubia. *J. Ethnopharmacol.* **2008,** *116*, 341–357.

Hanafy, D. M.; Burrows, G. E.; Prenzler, P. D.; Hill, R. A. Potential Role of Phenolic Extracts of *Mentha* in Managing Oxidative Stress and Alzheimer's Disease. *Antioxidants* **2020,** *9*, 631.

Jose, D. M.; Thoppil, J. E. Micropropagation and Phytochemical Analysis of *Mentha rotundifolia* (L.) Huds. *Indian J. Appl. Res.* **2014,** *4*, 41–44.

Khadraoui, A.; Khelifa, A.; Hamitouche, H.; Mehdaoui, R. Inhibitive Effect by Extract of *Mentha rotundifolia* Leaves on the Corrosion of Steel in 1 M HCl Solution. *Res Chem. Intermed.* **2014,** *40*, 961–972.

Kumar, P.; Mishra, S.; Malik, A.; Satya, S. Insecticidal Properties of *Mentha* Species: A Review. *Ind. Crops Prod.* **2011,** *34*, 802–817.

Ladjel, S.; Gherraf, N.; Hamada, D. Antimicrobial Effect of Essential Oils from the Algerian Medicinal Plant *Mentha rotundifolia* L. *J. Appl. Sci. Res.***2011,** *7*, 1665–1667.

Leblalta, A.; Daoud, H.; Semcheddine, C.; Demirtaş, I. Antifungal Activity of *Mentha rotundifolia* Essential Oil Against *Fusarium oxysporum*. *IJIAAR*, **2020,** *4*, 56–68.

Lorenzo, D.; Paz, D.; Dellacassa, E.; Davies, P.; Vila, R.; Cañigueral, S. Essential Oils of *Mentha pulegium* and *Mentha rotundifolia* from Uruguay. *Braz. Arch. Biol. Technol.* **2002,** *45*, 519–524.

Mamadalieva, N. Z.; Hussain, H.; Xiao, J. Recent Advances in Genus *Mentha*: Phytochemistry, Antimicrobial Effects, and Food Applications. *Food Front.* **2020,** *1*, 435–458.

Miyazawa, M.; Watanabe, H.; Umemoto, K.; Kameoka, H. Inhibition of Acetylcholinesterase Activity by Essential Oils of *Mentha* Species. *J. Agric. Food Chem.* **1998,** *46*, 3431–3434.

Moldovan, R. I.; Oprean, R.; Benedec, D.; Hanganu, D.; Duma, M.; Oniga, I.; Vlase, L. LC-MS Analysis, Antioxidant and Antimicrobial Activities for Five Species of *Mentha* Cultivated in Romania. *Dig. J. Nanomater. Biostruct.* **2014,** *9*, 559–566.

Nagell, V. A.; Hefendehl, F. W. Composition of the Essential Oil of *Mentha rotundifolia. Planta Medica* **1974,** *26*, 1–8.

Nedel, F.; Begnini, K.; de Azambuja Carvalho, P. H.; Lund, R. G.; Beira, F. T. A.; Del Pino, F. A. B. Antiproliferative Activity of Flower Hexane Extract Obtained from *Mentha spicata* Associated with *Mentha rotundifolia* Against the MCF7, KB, and NIH/3T3 Cell Lines. *J. Med. Food* **2012,** *15*, 955–958.

Nickavar, B.; Alinaghi, A.; Kamalinejad, M. Evaluation of the Antioxidant Properties of Five *Mentha* Species. *Iran. J. Pharmaceut. Res.* **2008,** *7*, 203–209.

Oka, Y.; Nacar, S.; Putievsky, E.; Ravid, U.; Yaniv, Z.; Spiegel, Y. Nematicidal Activity of Essential Oils and Their Components Against the Root-Knot Nematode. *Phytopathology* **2000,** *90*, 710–715.

Pérez Raya, M. D.; Utrilla, M. P.; Navarro, M. C.; Jiménez, J. CNS Activity of *Mentha rotundifolia* and *Mentha longifolia* Essential Oil in Mice and Rats. *Phytother Res* **1990,** *4*, 232–234.

Pino, J. A.; Rosado, A.; Fuentes, V. Chemical Composition of the Leaf Oil of *Mentha rotundifolia* (L.) Hudson from Cuba. *J. Essent. Oil Res.* **1999,** *11*, 241–242.

Redhouane, B.; Khaled, B.; Hadjira, B.; Imene, I.; Abderrezak, K.; Fethia, F.; Švarc-Gajić, J. Evaluation of Some Biological Activities of Phenolic Compounds Obtained from Two Algerian Medicinal Plants: *Mentha rotundifolia* and *Satureja calamintha. APTEFF* **2020,** *51*, 87–102.

Riahi, L.; Elferchichi, M.; Ghazghazi, H.; Jebali, J.; Ziadi, S.; Aouadhi, C.; Chograni, H.; Zaouali, Y.; Zoghlami, N.; Mliki, A. Phytochemistry, Antioxidant and Antimicrobial Activities of the Essential Oils of *Mentha rotundifolia* L. in Tunisia. *Ind Crops Prod* **2013,** *49*, 883–889.

Sbai, A.; Ouhssine, M.; Chaouch, A.; Oubihi, A.; El Hartiti, H.; Rhaim, N.; Talbi, F. Z. Chemical Composition and Antimicrobial Activities of the Oil Essential Oil Of *Mentha rotundifolia. Int. J. ERT*, **2020,** *9*, 955–958.

Seladji, M.; Belmekki, N.; Bekhechi, C.; Bendimerad, N. Antioxidant and Antimicrobial Activity of Aqueous and Methanolic Extracts of *Mentha rotundifolia* L. from Algeria. *Int. J. Pharm. Sci. Rev. Res.* **2014,** *26*, 228–234.

Shibata, H.; Shimizu, S. Absolute Configuration of 1,2-Epoxy Menthyl Acetate Isolated from New Chemical Strain of *M. rotundifolia. Agr. Biol. Chem.* **1973,** *37*, 2675–2676.

Shimizu, S. Studies on Essential Oil of *Mentha rotundifolia.* Part II. Structure of Rotundifolone, a New Terpenic Ketone. *Bull. Agr. Chem. Soc. Japan* **1957,** *21*, 107–114.

Siham, F.; Rachid, B.; Al-Zoubi, R. M. Chemical Composition and Antioxidant Effect of *Mentha rotundifolia* Extracts. *Pharmacog. J.* **2019,** *11*, 521–526.

Soilhi, Z.; Rhimi, A.; Heuskin, S.; Fauconnier, M. L.; Mekki, M. Essential Oil Chemical Diversity of Tunisian *Mentha* spp. Collection. *Ind. Crop. Prod.* **2019,** *131*, 330–340.

Umemoto, K. Two New Stereoisomers of 1,2-Epoxymenthyl Acetate from Self-Pollinated Plant Oils of *Mentha rotundifolia. Nat. Prod. Lett.* **1998,** *11*, 161–165.

Yahia, I. B. H.; Bouslimi, W.; Messaoud, C.; Jaouadi, R.; Boussaid, M.; Zaouali, Y. Comparative Evaluation of Tunisian *Mentha* L. Species Essential Oils: Selection of Potential Antioxidant and Antimicrobial Agents. *J. Essent. Oil Res.* **2019a,** *31*, 184–195.

Yahia, I. B. H.; Jaouadi, R.; Trimech, R.; Boussaid, M.; Zaouali, Y. Variation of Chemical Composition and Antioxidant Activity of Essential Oils of *Mentha x rotundifolia* (L.) Huds. (Lamiaceae) Collected from Different Bioclimatic Areas of Tunisia. *Biochem. Syst. Ecol.* **2019b,** *84*, 8–16.

Yahia, I. B. H.; Zaouali, Y.; Ciavatta, M. L.; Ligresti, A.; Jaouadi, R.; Boussaid, M.; Cutignano, A. Polyphenolic Profiling, Quantitative Assessment and Biological Activities of Tunisian Native *Mentha rotundifolia* (L.) Huds. *Molecules* **2019c,** *24*, 2351.

Yakhlef, G.; Hambaba, L.; Pinto, D. C. G. A.; Silva, A. M. S. Chemical Composition and Insecticidal, Repellent and Antifungal Activities of Essential Oil of *Mentha rotundifolia* (L.) from Algeria. *Ind. Crop. Prod.* **2020,** *158*, 112988.

Yu, X.; Liang, C.; Fang, H.; Qi, X.; Li, W.; Shang, Q. Variation of Trichome Morphology and Essential Oil Composition of Seven *Mentha* species. *Biochem. Syst. Ecol.* **2018,** *79*, 30–36.

CHAPTER 11

PHYTOCHEMICAL AND PHARMACOLOGICAL PROFILE OF *MENTHA* SPECIES

SHAIKH AJIJ AHMED MAKBUL, NASREEN JAHAN, and SAYEEDUR RAHMAN

Department of IlmulAdvia (Pharmacology), National Institute of Unani Medicine, Kottigepalya, Magadi Main Road, Bangalore, Karnataka, India

ABSTRACT

Mentha (also known as mint or pudina) is a common herb that has been used for therapeutic purposes since ancient times. *Mentha* is one of the most important Lamiaceae family member, with approximately 19 species and 13 natural hybrids. They are rapidly growing and invasive, and can endure a wide range of agro-climatic conditions. They are found in Europe, Africa, Asia, Australia, and North America. *M. piperita*, *M. spicata,* and *M. arvensis* are widely grown across the world for their essential oil. Mints are said to be excellent companion plants, repelling nuisance insects while attracting beneficial ones. The majority of *Mentha* species are perennial, contain essential oils, and are commonly farmed as industrial crops for the manufacture of essential oils. It also has an important role in Traditional medicine, has pharmacological action. *Mentha* species' leaves, blossoms, and stems have traditionally been used as herbal teas and spices in various meals to add aroma and flavour. Fresh and dried plant material, raw extracts and essential oils of mint plants are used as a part of confectionary, as taste enhancing agents in toothpastes, chewing gums and beverages, baking, cosmetics, as oral hygiene products, pharmaceuticals and pesticides. Even so, no such

Bioactives and Pharmacology of Lamiaceae. T. Pullaiah, PhD (Ed.)

comprehensive review on phytochemical, pharmacological and ethnomedicinal use of *Mentha* species is available. Therefore, this chapter has been compiled.

11.1 INTRODUCTION

The genus *Mentha* belongs to the family Lamiaceae and consists of 38 species. They are subcosmopolitan in distribution; they can survive a broad range of agroclimatic conditions (Kumar et al., 2011). *Mentha piperita* (mint) is a common plant that has been used for medical and aromatherapeutic purposes since ancient times (Kumar et al., 2011). Cool and wet places in partial shade are preferred by all mints, and they flourish there. Mints endure a broad range of environments and may even be grown in direct sunlight. They spread quickly, using a network of runners to expand their reach over surfaces. Due to their rapid growth, one plant of each chosen mint will give more than enough mint for household usage with a little maintenance (Abbaszadeh et al., 2009). The majority of *Mentha* species are commonly farmed as industrial crops for the manufacture of essential oils (EOs). *Mentha* species' leaves, blossoms, and stems have long been used as herbal teas and spices to add scent and taste to a variety of dishes (Salehi et al., 2018). *Mentha piperita*, *Mentha spicata*, and *Mentha arvensis* are widely grown across the world, mostly for their EO output (Bahadori et al., 2018). *Mentha longifolia* has a peppermint fragrance and is a varied perennial (Mikaili et al., 2013). *M. piperita*, found in Iran and other areas of the world, is famed for its flavoring, odor, and medicinal capabilities in meals and cosmetic industrial goods. *M. piperita* also has therapeutic characteristics in its leaves and flowers (Saharkhiz et al., 2012). The oil is found to possess astringent, antiseptic, antipyretic, stimulant, emmenagogue antispasmodic, antimicrobial, rubefacient, and antiaging properties (Zaidi and Dahiya, 2015). *M. spicata* L., known as spearmint, brown mint, garden mint, lady's mint, and sage of Bethehem (Kee et al., 2017), is a mint species found in North Africa, Egypt, and Morocco. It is an invasive species that was originally discovered in the Great Lakes area in 1843 (Sulieman et al., 2011). Spearmint is widely utilized in commercially manufactured product, cooking, and medicine for its aromatic and flavorsome qualities (Sulieman et al., 2011). It is an aromatic plant used widely in cosmetic, confectionary, chewing gum, food, toothpaste, and pharmaceutical industries and for volatile oil productions (Zaidi and Dahiya, 2015).

11.2 ETHNOPHARMACOLOGICAL USES

The herb *M. arvensis* L. is used to treat liver and spleen problems, as well as asthma and jaundice. Menthol is used in medicine to treat gastrointestinal problems and in headache ointments. Indigestion and rheumatic pain are treated with a leaf infusion (Akram et al., 2011). The leaves are commonly used in salads and vegetable dishes in China, as well as to cure fevers, flu, colds, nosebleeds, nose and throat ailments, snake and bug bites, and neurological problems in children. Peppermint oil is applied to burns to ease pain and destroys gallstones when consumed on a regular basis. Mint is used to tone the stomach, stimulate the mind and body, remove gas from the intestines, and treat muscle spasms in India. Mint is used by the Arabs for skin problems and as a general tonic (Akram et al., 2011). It gives minty flavor to various food products. It is also used in oral products.

Empirically mints in fresh, dried, and processed forms are used in herbal medicine for treatment of several health problems and discomforts like gastric acidities, aerophagia, nausea, flatulence, vomiting, ulcerative colitis, stomach aches, dyspepsia, diarrhea, jaundice, liver complaints, anorexia, cramps, bronchitis, throat infections, and viral hepatitis (Bahadori et al., 2018; Gulluce et al., 2007). The leaves are used in Turkish and Iraqi traditional medicine to treat mild sore throats and mouth or throat discomfort. It's also included in nasal decongestants and treatments for mild pains and sprains. *M. longifolia* is thought to aid in the development of the immune system. In addition, this herb is used to treat coughs, colds, and influenza. Wild mint is also used externally to heal wounds and inflamed glands. In South Africa, *M. longifolia* leaves are employed in food preparation to develop taste and appearance due to their color, scent, and flavor. The pungent fragrance of the leaves repels mosquitoes when applied on the body and bedding. It's also been used to keep rats away from grain in granaries (Okut et al., 2017).

M. piperita has a long history of safe usage as a flavoring ingredient and in medical products. On the smooth muscle of the gastrointestinal system, it exerts a spasmolytic action (Alves et al., 2012). Peppermint is not only a medicinal plant, but it is also a popular herb commonly employed throughout the world in teas, flavorings, and confections (Malekmohammad et al., 2019). *Mentha* × *piperita*, commonly identified as peppermint, has been widely used as a traditional medicine in Eastern and Western countries for the preparation of toothpaste and mouthwashes, aromatherapy, irritation, and inflammation. Traditionally, peppermint infusion, leaves, or EO can be used for respiratory-related diseases such as anticongestive and expectorants and as antispasmodic on the digestive and vascular systems (Mahendran and Rahman, 2020).

People on the Colombian Caribbean coast frequently use plants to relieve stress and sleeplessness. Leaves of *M. spicata* L. are used to prepare a popular tea locally known as "yerba buena" (Caro et al., 2018). *M. spicata* traditionally used as a remedy for treating of colds and flu, respiratory tract problems, bad breath, gastralgia, hemorrhoids, carminative, anti-spasmodic, diuretic and sedative agents, and stomach-ache (Kee et al., 2017; Mahboubi, 2017). It was used as a tea to manage a variety of digestive issues (Sulieman et al., 2011). It is used as a stimulant and carminative, both fresh and dried. The EO is produced by distillation from freshly picked mint leaves or dried leaves. The EO obtained having various pharmacological activities (Zaidi and Dahiya, 2015).

11.3 CHEMICAL CONSTITUENTS OF MENTHA SPECIES

Mentha species EO contains menthol (33–60%), menthone (15–32%), menthyl acetate (2–11%), isomenthone (2–8%), 1,8-cineol (5–13%), menthofuran (1–7%), limonene (1–7%), β-myrcene (0.1–1.7%), β-caryophyllene (2–4%), pulegone (0.5–1.6%), and carvone (1%) (Sevindik, 2018). EO of *Mentha aquatica* mainly consists of mono and sesquiterpenes. GC-MS of EO identified the main substances were menthofuran (51.26–58.59%), limonene (5.94–12.06%), trans-β-ocimene (5.59–6.10%), ledol (3.01–4.06%), β-caryophyllene (2.923–3.557%), linalyl acetate (26.1%), α-pinene (22.7%), linalol (13.755%), α-terpeneol (3.42%), and geranyl butyrate (3.39%) (Andro et al., 2013; Chaker et al., 2014). Numerous compounds such as1-*O*-linoleoyl-2-*O*-enadecanoyl-3-*O*-palmitoleoyl-*sn*-glycerol,1-*O*-linoleoyl-2-*O*palmitoleoyl-*sn*-glycerol, rosmarinic acid, oleanolic acid, corosolic acid, asiatic acid, choline, acetic acid, caffeic acid, formic acid, γ-aminobutyric acid (GABA), lactic acid, quinic acid, salicylic acid, succinic acid, fructose, glucose, sucrose, xylose, alanine, aspartic acid, glycine, isoleucine, leucine, phenylalanine, threonine, and valine have been reported (Venditti et al., 2017; Zebelo et al., 2011).

M. arvensis yields 70–90% of mentholthen menthone, isomenthone, limonene, neomenthol, methylacetate, β-caryophyllene, piperitone, α- and β-pinene, tannins, and flavonoids. The other components found are phellandrene, cineol, piperitone, and sesquiterpene (Akram et al., 2011). It also contains terpenes such as α-menthol, neomenthol, isomenthol, D-menthone, isomenthone, menthofuran, menthylacetate, carvomenthone, cineol, *p*-cymene, aromadendrene, limonine, α-phellandrene, pipertone, carvacrol, α-pinene, α-phellandrene, β-pinene, dipentene, cardinene, and -thujone. It also contains the flavonoids such as quercetin, menthoside, isorhoifolin, vitamin K, thymol, and eugenol (Thawkar et al., 2016). The leaves yield about 0.2–0.8% EO. Various constituents are cineole (3.5–14.0%), menthone (14.0–32.0%), isomenthone (1.5–10.0%), menthyl acetate (2.8–10.0%), isopulegol (max. 0.2%), menthol (30.0–55.0%), and carvone (max. 1.0%) (Saleem and Idris, 2016), piperitone, α- and β-pinene, tannins, and flavonoids. The components of EO are 4.5–10% esters-menthyl acetate and 15–20% of ketones. *M. arvensis* yields 70–90% of menthol. Akram et al. (2011) reported that the EO contains cineole, sesquiterpene, phellandrene, and piperitone.

GC-MS analysis of the oil of *M. longifolia* found that cis-piperitone epoxide, pulegone and piperitenone oxide piperitenon (43.9%), tripal (14.3%), oxathiane (9.3%), piperiton oxide (5.9%), and d-limonene (4.3%) are main components (Gulluce et al., 2007). Peppermint oil is composed primarily of menthol (37.4%), menthyl acetate (17.4%), and menthone (12.7%). Other constituents like pulegone, menthofuran, and limonene; flavonoid glycosides, namely, narirutin, hesperidin, luteolin-7-*o*-rutinoside, isorhoifolin, diosmin; 5,7-dihydroxycromone-7-orutinoside; polyphenols like eriocitrin, rosmarinic acid; cinnamic acid, caffeic acid, and naringenin-7-*o*-glucoside; and salvianolic acid, dedihydro-salvianolic acid, luteolin-glucouronide, luteolin-diglucoronide, luteolin-glucopyranosyl-rhamnopyranoside, and eriodictyolglucopyranosyl-rhamnopyranoside were also isolated from the aerial parts. Besides these, the herb also contains 1,8-cineole, α-amorphene, α-cadinene, α- carotene, α-copaene, α-gurjunene, α-pinene, α-terpinene, α-terpineol, α-thujone, α-tocopherol, amyl-valerate, anethole, azulene, β-betulenol, β-carotene, β-caryophyllene, β-copaene, β-ionone, β-pinene, β-thujone, β-ylangene, betaine, bicycloelemene, bisabolene, cadinene, camphene, carvacrol, carveol, carveol-acetate, carvone, caryophyllene-oxide, cedrene, cedrol, choline, cineole, cinerol, cis-piperitol, cis-roseoxide, cis-sabinol, citronellol, cryptone, flavonshymenoxin, isoamylphenylacetate, isomenthol, isomenthol-acetate, isomenthone, isomenthyl-acetate, isopulegol-acetate, isovaleraldehyde, isovaleric acid, isovaleric-acid-noctyl-ester, jasmone, lavandulol, ledol, limonene, linalool, luteolin, menthoside, menthyl-isovalerate, menthyl-valerate, myrcene, myrtenol, neoisomentholacetate, neomenthol, neomenthone, neomenthyl-acetate, nerolidol, nevadensin, octan-3-ol, *p*-cymene, *p*-cymol, pectin, pent-cis-2-en-1-ol, perillyl-alcohol, phellandrene, phenylethanols, phenyl-propyl-pyridines, pinene, piperitenone, piperitone, piperitone-oxide, pyridine sabinene, sabinene-acetate, sabinene-hydrate, salvigenin, sideritoflavone, terpinolene, thymol, trans-piperitol, trans-roseoxide, vanillin, viridiflorol, and xanthomicrol (Rita and Animesh, 2011). *M. spicata* contains volatile oils, resins, tannins, coumarins, flavonoids, steroids, and alkaloids. The principal constituent is carvone along with some other monoturpenic constituents like limonene, phellendrene, dihydrocarveol, dihydrocarveol, cetate, cineol, α-pinene, and linalool. The chemical components of peppermint leaves and oil vary with plant maturity, variety, geographical region, and processing conditions (Ullah et al., 2011). *M. spicata* EO contains limonene, 1,8-cineole, cisdihydrocarvone, dihydrocarveol, *trans*-carveol, cis-carveol, pulegone, carvone, iso-dihydrocarveol acetate, β-bourbonene, and ε-caryophyllene and shows mean percentages more than 1%, which represent 90% of the total oils (Ali-Shtayeh et al., 2019). The major minerals in dried peppermint leaves include K, Ca, Mg, and lower amounts of Na, along with smaller amounts of Fe, Mn, Zn, and Cu. Trace amounts of Cr, I, and Se are also present (McKay and Blumberg, 2006).

11.4 PHARMACOLOGICAL ACTIVITIES OF *MENTHA AQUATICA* L.

Mentha spp. possess various pharmacological properties due to active constituents (Adham, 2015). In a previous screening Southern African plants used for depression-like conditions for MAO-inhibitory activity, a 70% ethanol extract of *M. aquatica* had high activity (Olsen et al., 2008). *M. aquatica* has antimicrobial, antiviral, anti-inflammatory, mildly anesthetic, antispasmodic, antiulcer, cytoprotective, and hepatoprotective activities (Ferhat et al., 2017).

11.4.1 MONOAMINE OXIDASE (MAO) INHIBITORY ACTIVITY

Six *M. aquatica* extracts with varying polarity were tested for MAO bioassay. Ethanol extract (70%) had the highest inhibitory activity and (*S*)-Naringenin was also isolated from extract. The presence of naringenin could explain why it's used in traditional medicine to treat depression (Olsen et al., 2008).

11.4.2 ANTIBACTERIAL AND ANTIOXIDANT ACTIVITY

EO of *M. aquatica* growing in Ethiopia was studied for its antibacterial activity and showed the most susceptible for *Staphylococcus aureus*, *Sarcina luteus*, and *Bacillus pumilus*. The oil's minimum inhibitory concentration (MIC) against these microorganisms was reported to be less than 5 g/mL. Using 2,2-diphenyl-1-picrylhydrazyl (DPPH) and deoxyribose degradation tests, the oil was shown to have strong radical scavenging action, with IC50 values of 11.2 and 3.74 L/mL, respectively **(Getahun et al., 2008).**

M. aquatica was studied for its antioxidant, anticholinesterase, and antibacterial properties by Ferhat et al. (2017). Aerial part extracts, as well as roots extracts of *M. aquatica*, were tested for antioxidant activity using b-carotene-linoleic acid, DPPH and ABTS scavenging, CUPRAC, and metal chelating assays. AChE and BChE were used to test the anticholinesterase activity. MICs were determined against *Escherichia coli*, *S. aureus*, *P. aeruginosa*, *S. heidelberg*, *K. pneumoniae*, *E. aerogenes*, and *M. morganii* strains to determine antibacterial activity. AERMA has an IC50 of 4.21 0.28 lg/mL in the ABTS test. BChE inhibitory activity was also discovered, with an IC50 of 19.23 1.42 lg/mL. Antibacterial activity was found in the extracts that were tested (Ferhat et al., 2017).

11.5 PHARMACOLOGICAL ACTIVITIES *M. ARVENSIS* L.

11.5.1 ANTIBACTERIAL ACTIVITY

Ameen and Shafi (2016) showed that the ethanolic extract of *M. arvensis* inhibits *E. coli*, *S. aureus*, *P. aeruginosa*, *S. flexineri*, and *K. pneumoniae*. The EOs inhibited the growth of *Helicobacter pylori*, *Salmonella enteritidis*, *E. coli*, methicillin-resistant *S. aureus*, and methicillin-sensitive *S. aureus* in a dose-dependent manner (Thawkar et al., 2016).

11.5.2 ANTIOXIDANT ACTIVITIES

Cineole, an essential ingredient of mint extract, was shown to reduce ethanol-induced gastric mucosal damage in rats, and tests revealed that this action was owing to the test compound's antioxidant, lipoxygenase inhibitory, and ability to restore nonprotein sulfhydryl to normal levels. Mint contains antioxidants such as cineole, eugenol, quercetin, and thymol, which are said to reduce lipid peroxidation. Mint extract contains eugenol, terpenes, and flavonoids, which are excellent modulators of xenobiotic enzymes, particularly Phase-2 enzymes like GST (Thawkar et al., 2016).

11.5.3 ANTI-ALLERGIC AND ANTI-INFLAMMATORY ACTIVITY

Histamine-produced paw edema in mice and histamine release inhibition test were used to find out the anti-inflammatory and anti-allergic activities of alcoholic and water extracts (leaves, stem, and roots) of *M. arvensis*, respectively. *M. arvensis* (especially the leaves) are a rich source of secondary phytoconstituents, which have anti-allergic and anti-inflammatory properties. The results for anti-allergic action demonstrated that ethanolic extracts of leaf and root had significant inhibitory efficacy, with 57% and 53% inhibition, respectively. Compared to the percentage inhibitory potential of the standard medicine, diclofenac sodium, which causes 77.87% edema inhibition, ethanolic extracts of plant parts have an anti-inflammatory potential of leaf = 68.30 > root = 48.80 > stem = 10.70% (Thawkar et al., 2016).

11.5.4 INFLAMMATORY BOWEL SYNDROME

Oil of *M. arvensis* showed a protective role in the inflammatory bowel syndrome (Ameen and Shafi, 2016).

11.5.5 ANTIULCEROGENIC EFFECT

Petroleum ether, choloroform, and aqueous extract of *M. arvensis* were found to have a protective action against acid secretion and gastric ulcers in ibuprofen + pyloric ligation, 0.6 mol/L HCl and 90% ethanol-induced ulcers (Ameen and Shafi, 2016).

11.5.6 ANTIFUNGAL ACTIVITY

The fungicidal activity of EOs, menthol, and menthone against *Rhizoctonia solani* and *Fusarium moniliforme* was tested. When compared to EOs and menthone, menthol was the most effective. At all doses evaluated, they all showed less action than standard bavistin (Makkar et al., 2018).

11.5.7 CISPLATIN-INDUCED NEPHROTOXICITY

Hydroalcoholic extract of leaves of *M. arvensis* (200 and 400 mg/kg) was evaluated for the cisplatin-induced nephrotoxicity. Test extract found effective at both doses, but high dose (400 mg/kg) was found more effective (Singh et al., 2014).

11.5.8 HEPATOPROTECTIVE ACTIVITY

Methanol extract of *M. arvensis* tested for hepatoprotective activity against isoniazid and rifampicin-induced liver impairment in rats. Test drug-treated animals' liver toxicity was controlled significantly indicated by enhancement of the antioxidant state as well as the restoration of biochemical indicators to normal levels (Mir et al., 2020).

11.5.9 ANTIDIABETIC ACTIVITY

Methanol extract of *M. arvensis* root and aerial parts was found to possess significant hypoglycemic activity against alloxan-induced diabetes (Ameen and Shafi, 2016).

11.5.10 ANTI-CANDIDA ACTIVITY

A study on the EO and ethanolic extract of *M. arvensis* showed anti-*Candida* activity (Ameen and Shafi, 2016).

11.5.11 ANALGESIC ACTIVITY

The ethyl acetate, EtOH, and water extract of root and aboveground parts of *M. arvensis* showed analgesic activity (Ameen and Shafi, 2016). In another study, ethanolic extract of *M. arvensis* showed statistically significant analgesic effect in mice induced by acetic acid (Saleem and Idris, 2016).

11.5.12 INSECTICIDAL ACTIVITY

The EO of *M. arvensis* leaves was found to repel larvae and adults of *T. castaneum.* They strongly (97.66%) repel *T. castaneum* even at low concentration (Mohammed, 2013).

11.5.13 ANTI-ARTHRITIC ACTIVITY

The MeOH extracts of leaves of *M. arvensis* were reported to possess anti-arthritic activity (Ameen and Shafi, 2016).

11.5.14 IN-VITRO THROMBOLYTIC ACTIVITY

The ethanolic, methanolic, acetone, and chloroform extracts of *M. arvensis* are reported to possess significant blood clots lytic activity in vitro (Ameen and Shafi, 2016).

11.5.15 SEDATIVE AND HYPNOTIC

A study on methanolic and aqueous extracts of *M. arvensis* evaluated on mice showed potentiation of phenobarbitone-induced sleeping time (Verma et al., 2003; Ameen and Shafi, 2016).

11.5.16 RADIO-PROTECTIVE ACTIVITY

Pretreatment of mice with the aqueous and ethanolic extracts of *M. arvensis* resulted in a decrease in the severity of radiation sickness symptoms and death (Ameen and Shafi, 2016).

11.5.17 ACETYLCHOLINESTERASE INHIBITOR ACTIVITY

Linarin (acacetin-7-*O*-β-D-rutinoside) from the flower extract of *M. arvensis* showed selective dose-dependent inhibitory effect on acetylcholinesterase (Oinonen et al., 2006).

11.5.18 ANTI-INFLAMMATORY ACTIVITY

Pharmacological study was conducted on processed methanolic extract of the leaves. Albino rats were used to test anti-inflammatory activity. The activity was also compared to that of nimesulide, a common anti-inflammatory drug, and the percent inhibition of edema was calculated. When tested on mice, the sedative-hypnotic action demonstrated a potentiation of pentobarbitone-induced sleep time. The standard deviation from the mean was calculated using the average recovery time data (Verma et al., 2003).

11.6 PHARMACOLOGY OF *M. LONGIFOLIA* L.

11.6.1 ANTIMICROBIAL AND ANTIOXIDANT EFFECT

Gulluce et al. (2007) evaluated the antibacterial and antioxidant capabilities of the EO and methanol extract from *M. longifolia* ssp. *longifolia*. The EO exhibited antimicrobial activity against 30 microorganisms tested; however, the extract was almost inactive. The extract, on the other hand, inhibited the free radical DPPH (IC50 of 57.4 lg/mL) and b-carotene/linoleic acid (10700 lg/mL) systems better than the EO. These samples did not efficiently inhibit linoleic acid oxidation in the β-carotene/linoleic acid test, demonstrating only 24% and 36% inhibitions at 2 mg/mL, respectively; both were much below that of BHT (Gulluce et al., 2007).

In another research, the antibacterial activity of EO from *M. longifolia* against six bacterial strains as well as its free radical scavenging capability (RSC) were studied. EOs of the wild species *M. longifolia* from Bosnia and Herzegovina were examined. The scavenging activity of EOs on DPPH was used to determine RSC. EO showed high antimicrobial activity. Gram-negative microorganisms *E. coli*, *Pseudomonas aeruginosa*, and *Salmonella enterica* were used to test the EO's antibacterial activity. As an RSC, antioxidant activity was assessed. In a dose-dependent manner, the EO was able to convert DPPH radicals to the neutral DPPHH form (IC50 = 10.5 g/mL). The EO has a high RSC, which was linked to the presence of mostly monoterpene ketones and aldehydes (Nikšić et al., 2012).

11.6.2 ANTICANCER EFFECTS

M. longifolia crude extract (ME) was tested on two adrenocortical tumor cell types (H295R and SW13 cells). ME effects were detected in a brine shrimp fatality study at concentrations of >0.5 g/L. MTT, SRB, and Trypan blue tests were used to measure cell viability and vitality in H295R and SW13 cells. ME had a stronger antiproliferative effect in SW13 cells after 72 h (ME > 0.5 g/L). ME in combination with mitotane (an approved medicine for adrenocortical cancer) did not appear to increase the herb's efficacy. Human fibroblasts were treated with ME as a control and showed no influence on cell survival. The results of the clonogenic experiment were consistent with prior cell viability tests (ME > 0.5 g/L). Cell cycle analysis revealed an increase in the sub-G0/G1 phase, which is associated with cell death. Western blot analysis revealed that the MAPK and PI3k/Akt pathways were influenced (Patti et al., 1996).

11.6.3 HEPATOPROTECTIVE ACTIVITY

The effects of *M. longifolia* crude ethanolic extract were studied on CCl_4-induced liver injury in mice. Pretreatment with *M. longifolia* extracts significantly improved hepatic antioxidant status in mice, which was most pronounced in the reduction of CCl_4-mediated lipid peroxidation (Mimica-Dukić et al., 1999).

11.6.4 PHENOLIC COMPOSITION AND FUNCTIONAL PROPERTIES

An RP–HPLC–DAD system was used to assess the phenolic profile of *M. longifolia* infusion and ethanol extract. The extracts included 16 phenolic components (10 phenolic acids and 6 flavonoids), with sinapic acid (7132 g/g extract) and rosmarinic acid (6260 g/g extract) being the most prevalent. In the 1,1-diphenyl-2-picrylhydrazyl radical, 2,2-azino-bis, and other assays, strong antioxidant effects were reported. The extract also showed acetylcholinesterase inhibitory activity as well as substantial inhibitory capability for -amylase and -glucosidase (Bahadori et al., 2018).

11.7 PHARMACOLOGY OF *M. PIPERITA* L.

11.7.1 ANTIBACTERIAL AND DIPTERIAN ADULTICIDAL

Ethanolic and aqueous extract of *M. piperita* have been used for antibacterial and adulticidal study. Antimicrobial activity was determined by using agar well diffusion method, and "Sugar Bait" method for adulticidal potential. The aqueous (5% of 50 μL) and ethanolic (5% of 100 μL) showed antibacterial activity. Both the extracts showed resistance against *Salmonella typhimurium*. The aqueous extract was found with effective adulticidal potential (Patil et al., 2016).

11.7.2 ANTIFUNGAL AND ANTIBIOFILM ACTIVITIES

EO of *M. piperita* was tested for antifungal activity. An XTT reduction test was used to assess biofilm formation inhibition. The most abundant ingredient in the EO was menthol (53.28%), followed by menthyl acetate (15.1%) and menthofuran (13.08%) (11.18%). At doses 0.12–8.0 L/mL, the EO showed excellent antifungal activity against the tested fungi. In addition, at doses up to 2 L/mL, the EO prevented *Candida albicans* and *Candida dubliniensis* biofilm development (Saharkhiz et al., 2012).

11.7.3 TREATMENT OF INFANTILE COLIC

Alves et al. (2012) investigated the effect of *M. piperita* on infantile colic. Thirty infants were randomized in the treatment for 7 days with each medicine in a double-blind crossover study. The mean weeping time per day was 192 min (51.6) at baseline, and daily episodes of infantile colic were 3.9 (1.1). Daily episodes of colic dropped to 1.6 (0.6) and crying duration to 111 (28) min towards the conclusion of the research. All mothers reported a reduction in the frequency and duration of infantile colic episodes, and there were no variations in drug effects (Alves et al., 2012).

11.7.4 OXIDATIVE STRESS AND SURVIVAL IN LIPOPOLYSACCHARIDE-TREATED MACROPHAGES

Because of their antimicrobial properties and ability to link the innate and adaptive immune responses, macrophages are critical in bacterial infection management. In vitro, Arruda et al. (2017) looked at the effects of *M. piperita* leaf hydroalcoholic extract (LHAE) on cultured murine peritoneal macrophages activated or not with lipopolysaccharide (LPS). Cells that had been treated with a vehicle served as controls. Incubation with LHAE (1–30 g/mL) restored LPS-induced macrophage death. In the absence of LPS, higher quantities of the extract (100 g/mL) reduced macrophage viability by 49–57%. H_2O_2 (34.6–53.4%) production by these cells was reduced by LHAE (1–300 g/mL), but not nitric oxide production. The extract boosted the activity of superoxide dismutase (15.3–63.5-fold) and glutathione peroxidase (34.4–73.6-fold) in LPS-treated macrophages at comparable doses. When treated with LHAE (3–30 g/mL), only LPS-unstimulated macrophages showed increased phagocytosis (3.6–6.6-fold increase). Overall, *M. piperita*

LHAE inhibits inflammatory responses by activating the antioxidant system in these cells (Arruda et al., 2017).

11.7.5 TOXICOLOGICAL EFFECTS

Peppermint (*Mentha x piperita* L.) is also used as a tea infusion, herb, spice, or flavoring agent in other products, in addition to its therapeutic purposes. Peppermint and its principal constituents (pulegone, menthone, menthol, and menthofuran) are moderately toxic. Peppermint and its menthol isomers have no major innate mutagenic, genotoxic, or embryotoxic properties. In human liver microsomes, peppermint EO interacts with cytochrome P450 isoenzymes. It is not recommended for people who have bile-duct obstruction or gall bladder inflammation. In patients with gastrointestinal reflux or a hiatus hernia, it should be used with caution because it can increase the symptoms of gastrointestinal reflux (Malekmohammad et al., 2019; Shah and D'Mello, 2004).

11.7.6 ACTION ON LIPID PROFILE

Badal et al. (2011) investigated the effect of the aqueous extract *M. piperita* on serum lipid. On fructose-fed rats, extract (100 mg/kg and 250 mg/kg p.o. daily) was given for 3 weeks. The test extract had a substantial protective effect against fructose-induced hyperlipidemia and showed strong antioxidant activity, according to the findings. The aqueous extract reduced raised glucose, cholesterol, triglycerides, VLDL, LDL, and atherogenic index levels while simultaneously increasing high-density lipoprotein cholesterol levels and HDL ratio without changing serum insulin levels (Badal et al., 2011).

11.7.7 CUTANEOUS WOUND-HEALING POTENTIAL

Aqueous extract 3% ointment of *M. piperita* showed cutaneous wound-healing potential. A 3-cm piece of all skin thicknesses on days 10, 20, and 30 was prepared for biochemical and histological investigation of the cutaneous wound-healing trend. The application of extract ointment resulted in a significant decrease in the wound area, total cells, neutrophil, lymphocyte, and macrophage levels, as well as a notable increase in wound contracture, hydroxyl proline, hexosamine, hexuronic acid, fibrocyte, and the rate of fibrocyte to fibroblast conversion (Zangeneh et al., 2019).

11.7.8 HEPATOPROTECTIVE ACTIVITY

In-vitro antioxidant activity of *M. piperita* leaf essential oil (MpEO) against CCl_4-induced oxidative stress in rats was investigated. MpEO's in-vitro antioxidant activity was lower than silymarin's. In CCl_4-induced stress, pretreatment of rats with MpEO at a dose of 5 mg/kg had no effect on stress markers. At 15 and 40 mg/kg doses, EO significantly lowered stress markers. Compared to CCl_4-treated rats, MpEO (40 mg/kg) administration resulted in reduction in hepatic and kidney lipid peroxidation as well as an increase in antioxidant enzymes. Furthermore, pretreatment with MpEO at a dose of 40 mg/kg can significantly reduce the histological hepatic and renal lesions caused by CCl_4 (Bellassoued et al., 2018).

11.8 PHARMACOLOGICAL ACTIVITIES OF *M. SPICATA* L. (SPEARMINT)

11.8.1 ANTIMICROBIAL ACTIVITY

The antimicrobial potential of EOs from *M. spicata* and *M. piperita* was tested against selected clinical isolates using the agar well diffusion method. The TLC-bioautographic method was used to investigate the antibacterial impact. On 11 bacterial and 4 fungal clinical isolates, the antimicrobial activity of EOs from mint species was tested. Both EOs had the highest efficacy against *S. aureus*, with maximum inhibition zones of 210.09 mm in *M. spicata* and 19.20.07 mm in *M. piperita*. In *M. spicata* EO, thin layer chromatography and bioautography assays revealed well-defined inhibition zones against *Acinetobacter* spp. (Zaidi and Dahiya, 2015).

Ullah et al. (2011) used traditional literature methods to investigate the antibacterial activity of *M. spicata*. This activity is highest in crude extract samples, followed by ethyl acetate and lowest in aqueous extract. Antifungal activity was tested on the same results.

The growth of *E. coli*, *Bacillus subtilis*, *Aspergillus niger*, and *C. albicans* was inhibited by spearmint oil. The results showed that spearmint oil has potent antimicrobial activity against all organisms tested; the highest antibacterial effect was found against *E. coli*, while the highest inhibitory level was found against the mold *A. niger*, where the inhibition zone (19 mm) was found at a higher oil concentration (Sulieman et al., 2011).

11.8.2 USED IN STRESS AND INSOMNIA

Colombians use teas made from *M. spicata* to relieve stress and insomnia. Caro et al. (2018) tested anxiolytic and hypnotic activity in Wistar rats. The anxiolytic action of *M. spicata* extract (1000 mg/kg) was demonstrated by a considerable increase in exploration and time spent in the open arms. *M. spicata* extract (1000 mg/kg) significantly increased the sleeping period caused by pentobarbital, implying that the plants' extracts have a sedative and hypnotic action. The acute toxicological research revealed that the doses utilized had no influence on mortality or toxicity at the hepatic or renal levels. The activity appears to be due to a variety of compounds, the majority of which are phenolic compounds such flavonoids and tannins (Caro et al., 2018).

11.8.3 ANTIOXIDANT CAPACITY

The ORAC value for an aqueous solution *M. piperita* leaves was among the highest found in Zheng and Wang's (2001) analysis of popular medicinal herbs. *M. piperita* had an ORAC value of 15.84 (0.42) mol Trolox equivalents (TE)/g fresh weight, which was slightly lower than *M. aquatica* and *M. spicata* had ORAC values of 19.80 (0.43) mol TE/g and 8.10 (0.26) mol TE/g, respectively.

Dragland et al. (2003) discovered that the relative antioxidant value of dried *M. piperita* (78.5 mmol/g) was lower than *S. officinalis* (91.2 mmol/g) but greater than *T. vulgaris* (74.6 mmol/g) in a study employing the ferric reducing ability of plasma (FRAP) assay. High antioxidant concentrations are indicated by FRAP values >75 mmol/g (McKay and Blumberg, 2006). Seasonal fluctuations in antioxidant activity have been detected in *M. piperita*, as have seasonal variations in other substances (range 59.8–96.1 mmol/g) (Dragland et al., 2003).

The antioxidant activity of *M. piperita* oil was higher than that of *M. aquatica* or *M. longifolia*, according to Mimica-Dukic et al. (2003). *M. piperita* hindered the formation of the OH radical in the Fenton reaction by 24% and reduced the radical generator DPPH by 50% (IC50 = 2.53 g/mL) in their experiment. The antioxidant capacities of aqueous solutions of *M. piperita* (0.1%, 0.5%, and 1.0%) were among the highest of the tea infusions tested (Pazdzioch-Czochra and Widenska, 2002) in an assay based on the oxidation of homovanillic acid to its fluorescent biphenyl dimer in the presence of H_2O_2 and peroxidase. A 0.5% *M. piperita* infusion (the closest approach to a usual dose) inhibited fluorescence at a rate of 67%, which was lower than comparable amounts of black (78%) and green (81%) teas, but higher than other herb teas like hibiscus (56%) and rooibos (56%) (52%). These teas were rated similarly when the results were stated as Trolox equivalents. The capacity to assess solely the H_2O_2 scavenging potential of the tested herbs, rather than the scavenging of additional free radicals, is one limitation of this method.

Ethanol extracts of dried *M. piperita* were found to stabilize the autooxidation of kinetically pure triacylglycerols of sunflower oil (Yanishlieva and Marinova, 1995), as well as natural sunflower oil (Yanishlieva and Marinova, 1995; Marinova and Yanishlieva, 1997). Peppermint EO was found to have a stronger antioxidant impact than butylated hydroxytoluene against sunflower oil peroxidation (McKay and Blumberg, 2006).

11.8.4 ANTIOXIDANT AND ANTI-*VIBRIO* SPP. ACTIVITIES

The antioxidant and anti-*Vibrio* spp. properties of an EO extracted from *M. spicata* (spearmint) aerial parts were examined. Using the XTT assay, the effect of the EO on *Vibrio* spp. biofilm inhibition and eradication was investigated. Disc diffusion and microdilution experiments were used to assess antibacterial efficacy against 30 *Vibrio* spp. strains (16 species). All microbes were severely harmed, demonstrating that the oil has a significant antibiotic capability. Furthermore, as compared to BHT, the examined oil had high antioxidant potency, as determined by four distinct tests (Snoussi et al., 2015).

11.8.5 FUNGICIDAL AND ANTIMICROBIAL ACTIVITY

Pattnaik et al. (1996) found that peppermint oil has fungicidal activity against 11 of the 12 fungi tested, including *C. albicans*, *T. mentagrophytes*, *A. fumigatus*, and *C. neoformans*, with a MIC range of 0.25–10 L/mL. In other research, peppermint extracts were found to have a moderate effect against these and other pathologically relevant fungi. In food crops and foodstuffs, peppermint oil has also been proved to be an efficient antibacterial and pest control agent (McKay and Blumberg, 2006).

11.8.6 TREATMENT OF PAIN IN OSTEOARTHRITIS

The usage of EOs as a pain reliever has risen in recent years. The analgesic properties of spearmint EO are less well understood. To establish its analgesic properties, Mahboubi (2017) reviewed all relevant articles from a database. *M. spicata* EO and its compounds in reducing pain severity in osteoarthritis patients have been reported in several animal studies (Mahboubi, 2017).

11.8.7 ANTIMICROBIAL ACTIVITIES AND EFFICACY IN REPELLING MOSQUITO

P. aeruginosa and *B. subtilis* have been found to be resistant to the extract from *M. spicata* leaf. The growth of other two bacteria and fungi were inhibited by the extract. When the maximum concentration of *M. spicata* extract was added to the produced cream, it had the highest repellence period, lasting up to 4 h (Ojewumi et al., 2017).

11.8.8 EFFECTIVENESS IN FLATULENCE

In Iranian Traditional Medicine, flatulence is referred to as "Nafkh," and it is said to be caused by the amount of stomach temperature, food humidity, or aberrant humidity in the digestive tract, as well as their interactions. The EO of *M. spicata* is used to relieve flatulence. If *M. spicata* EO is used in suitable quantities, no harmful effects have been documented. Other digestive disorders involving discomfort and gas, such as IBS, also benefit from *M. spicata* EO (Mahboubi, 2018).

KEYWORDS

- ***Mentha* species**
- **mint**
- **Lamiaceae**
- ***M. arvensis***
- ***M. spicata***

REFERENCES

Abbaszadeh, B.; Valadabadi, S. A.; Farahani, H. A.; Darvishi, H. H. Studying of Essential Oil Variations in Leaves of *Mentha* Species. *Afr. J. Plant Sci.* **2009,** *3* (10), 217–221.

Adham, A. N. Synergistic Effects Between *Mentha piperita*, *Mentha longifolia* and *Ocimum basilicum* on Different Bacterial Strains. *Intern. J. Chem.* **2015,** *7* (2), 170–176. DOI: 10.5539/ijc.v7n2p170

Akram, M.; Uzair, M.; Malik, N. S.; Mahmood, A. M.; Sarwer, N.; Madni, A.; Asif, H. M. *Mentha arvensis* L.: A Review Article. *J. Med. Plants Res.* **2011,** *5* (18), 4499–4503.

Ali-Shtayeh, M. S.; Jamous, R. M.; Abu-Zaitoun, S. Y.; Khasati, A. I.; Kalbouneh, S. R. Biological Properties and Bioactive Components of *Mentha spicata* L. Essential Oil: Focus on Potential Benefits in the Treatment of Obesity, Alzheimer's Disease, Dermatophytosis, and Drug-Resistant Infections. *Evid. Based Compl. Altern. Med.* **2019,** Article ID 3834265, 11–1. https://doi.org/10.1155/2019/3834265.

Alves, J. G. B.; Moraes de Brito, R. C. C.; Cavalcanti, T. S. Effectiveness of *Mentha piperita* in the Treatment of Infantile Colic: A Crossover Study. *Evid. Based Compl. Altern. Med.* **2012,** Article ID 981352, 1–4. DOI: 10.1155/2012/981352.

Ameen, N.; Shafi, S. Pharmacognostical Review of *Mentha arvensis*. *Intern. J. Curr. Res.* **2016,** *8* (7), 34120–34123.

Andro, A.; Boz, I.; Zamfirache, M.; Burzo, L. Chemical Composition of Essential Oils from *Mentha aquatica* L. at Different Moments of the Ontogenetic Cycle. *J. Med. Plants Res.* **2013,** *7* (9), 470–473.

Arruda, M. O.; Mendes, S. J. F.; Teixeira, S. A.; Silva de Mesquita, L. S.; Ribeiro, M. N. S.; Galvão, S. S. L.; Muscará, M. N.; Fernandes, E. S.; Monteiro-Neto, V. The Hydroalcoholic Extract Obtained from *Mentha piperita* L. Leaves Attenuates Oxidative Stress and Improves Survival in Lipopolysaccharide-Treated Macrophages. *J. Immunol. Res.* **2017,** Article ID 2078794, 1–9. https://doi.org/10.1155/2017/2078794

Badal,R. M.; Badal, D.; Badal, P. B.; Khare, A.; Shrivastava, J.; Kumar. V. Pharmacological Action of *Mentha piperita* on Lipid Profile in Fructose-Fed Rats. *Iran. J. Pharma. Res.*.**2011,** *10* (4), 843–848.

Bahadori, M. B.; Zengin, G.; Bahadori, S.; Dinparast, L.; Movahhedin, N. Phenolic Composition and Functional Properties of Wild Mint (*Mentha longifolia* var. *calliantha* (Stapf) Briq.). *Intern. J. Food Prop.* **2018,** *21* (1), 183–193. DOI: 10.1080/10942912.2018.1440238.

Bellassoued, K.; Hsouna, A. B.; Athmouni, K.; Pelt, J. V.; Ayadi, F. M.; Rebai, T.; Elfeki, A. Protective effects of *Mentha piperita* L. Leaf Essential Oil Against CCl_4 Induced Hepaticoxidative Damage and Renal Failure in Rats. *Lipids Health Dis.* **2018,** *17* (9), 1–14. DOI: 10.1186/s12944–017–0645–9.

Caro, D. C.; Rivera, D. E.; Ocampo, Y.; Franco, L. A.; Salas, R. D. Pharmacological Evaluation of *Mentha spicata* L. and *Plantago major* L., Medicinal Plants Used to Treat Anxiety and Insomnia in Colombian Caribbean Coast. *Evid. Based Compl. Altern. Med.* **2018,** Article ID 5921514, 1–7. https://doi.org/10.1155/2018/5921514

Chaker, A. N.; Boukhebti, H.; Sahraoui, R.; Ramdhani, M. Essential Oils and Morphological Study of *Mentha aquatica*. *Pharmacogn. Commun.* **2014,** *4* (2) DOI: 10.5530/pc.2014.2.7.

Dai, D. N.; Thang, T. D.; Emmanuel, E. E.; Abdulkabir, O. O.; Ogunwande, I. A. Study on Essential Oil of *Mentha aquatica* L. from Vietnam. *Am. J. Essent. Oils Nat. Prod.* **2015,** *2* (4), 12–16.

Dragland, S.; Senoo, H.; Wake, K.; Holte, K.; Blomhoff, R. Several Culinary and Medicinal Herbs Are Important Sources of Dietary Antioxidants. *J. Nutr.* **2003,** *133*, 128–190.

Ferhat, M.; Erol, E.; Beladjila, K. A.; Çetintaş, Y.; Duru, M. E.; Öztürk, M.; Kabouche, A.; Kabouche, Z. Antioxidant, Anticholinesterase and Antibacterial Activities of *Stachys guyoniana*and *Mentha aquatica. Pharma. Biol.* **2017,** *55* (1), 324–329. DOI: 10.1080/13880209.2016.1238488.

Getahun, Z.; Asres, K.; Mazumder, A.; Bucar, F. Essential Oil Composition, Antibacterial and Antioxidant Activities of *Mentha aquatica* Growing in Ethiopia. *Ethiop. Pharm. J.* **2008,** *26*, 9–16.

Ghoulami, S.; Idrissi, A.; Fkih-Tetouani, S. Phytochemical Study of *Mentha longifolia* of Morocco. *Fitoterapia.* **2001,** *72*, 596–598.

Gulluce, M.; Sahin, F.; Sokmen, M.; Ozer, H.; Daferera, D.; Sokmen, A.; Polissiou, M.; Adiguzel, A.; Ozkan, H. Antimicrobial and Antioxidant Properties of the Essential Oils and Methanol Extract from *Mentha longifolia* L. ssp. *longifolia. Food Chem..* **2007,** *103*, 1449–1456. DOI: 10.1016/j.foodchem.2006.10.061.

Kee, L. A.; Shori, A. B.; Baba, A. S. Bioactivity and Health Effects of *Mentha spicata. Integr. Food Nutr. Metab.* **2017,** *5* (1), 1–2. DOI: 10.15761/IFNM.1000203.

Kumar, P.; Mishra, S.; Malik, A.; Satya. Insecticidal Properties of *Mentha* Species: A Review. *Ind. Crops Prod.* **2011,** *34*, 802–817. DOI: 10.1016/j.indcrop.2011.02.01

Mahboubi, M. *Mentha spicata* as Natural Analgesia for Treatment of Pain in Osteoarthritis Patients. *Complem. Therap Clin. Pract.* **2017,** *26*, 1–4. http://dx.doi.org/10.1016/j.ctcp.2016.11.001

Mahendran, G.; Rahman, L. U. Ethnomedicinal, Phytochemical and Pharmacological Updates on Peppermint (*Mentha × piperita* L.)—A Review. *Phytother. Res.* **2020,** 1–52. https://doi.org/10.1002/ptr.6664.

Makkar, M. K.; Sharma, S.; Kaur, H. Evaluation of *Mentha arvensis* Essential Oil and Its Major Constituents for Fungi Toxicity. *J. Food Sci. Technol.* **2018,** *55* (9), 3840–3844. https://doi.org/10.1007/s13197–018–3291-y

Malekmohammad, K.; Rafieian-Kopaei, M.; Sardari, S.; Sewell, R. D. E. Toxicological Effects of *Mentha x piperita* (Peppermint): A Review. *Toxin Rev.* **2019.** DOI: 10.1080/15569543.2019.1647545.

Marinova, E. M.; Yanishlieva, N. V. Antioxidative Activity of Extracts from Selected Species of the Family Lamiaceae in Sunflower Oil. *Food Chem.* **1997,** *58*, 245–248.

McKay, D. L.; Blumberg, J. B. A Review of the Bioactivity and Potential Health Benefits of Peppermint Tea (*Mentha piperita* L.). *Phytother Res.* **2006,** *20*, 619–633. DOI: 10.1002/ptr.1936.

Mikaili, P.; Mojaverrostami, S.; Moloudizargari, M.; Aghajanshakeri, S. Pharmacological and Therapeutic Effects of *Mentha longifolia* L. and Its Main Constituent, Menthol. *Ancient Sci Life.* **2013,** *33*, 129–36. DOI: 10.4103/0257–7941.139059.

Mimica-Dukić, N.; Popović, M.; Jakovljević, V; Szabo, A.; Gašić, O. Pharmacological Studies of *Mentha longifolia* Phenolic Extracts. II. Hepatoprotective Activity. *Pharma. Biol.* **1999,** *37* (30), 221–224. DOI: 10.1076/phbi.37.3.221.6306.

Mir, S. A.; Jeyabalan,G.; Parveen, P.; Rahman, M. U.; Hepatoprotective Activity of *Mentha arvensis* in Anti-Tubercular Drugs Induced Hepatotoxicity in Rats. *Res. Pharm. Health Sci.* **2020,** *6* (2), 115–120. https://doi.org/10.32463/RPHS.2020.v06i02.01

Mohammed, H. H. Repellency of Ethanolic Extract of Some Indigenous Plants Against *Tribolium confusum. IOSR Indian J. Agric. Veter. Sci.* **2013,** *2* (6), 27–31.

Nikšić, H.; Kovač-Bešović, E.; Makarević. E.; Durić, K. Chemical Composition, Antimicrobial and Antioxidant Properties of *Mentha longifolia* (L.) Huds. Essential Oil. *J. Health Sci.* **2012,** *2* (3), 192–200.

Oinonen, P. P.; Jokela, J. K.; Hatakka, A. I.; Vuorela, P. M. Linarin, a Selective Acetylcholinesterase Inhibitor from *Mentha arvensis*. *Fitoterapia*. **2006,** *77*, 429–434. DOI: 10.1016/j.fitote.2006.05.002.

Ojewumi, M. E.; Adedokun, S. O.; Omodara, O. J.; Oyeniy, E. A.; Taiwo, O. S.; Ojewumi, E. O. Phytochemical and Antimicrobial Activities of the Leaf Oil Extract of *Mentha spicata* and Its Efficacy in Repelling Mosquito. *Intern. J. Pharma. Res. Allied Sci.* **2017,** *6* (4), 17–27.

Okut, N.; Yagmur, M.; Selcuk, N.; Yildirim, B. Chemical Composition of Essential Oil of *Mentha longifolia* L. subsp. *longifolia* growing wild. *Pak. J. Bot.* **2017,** *49* (2), 525–529.

Olsen, H. T.; Stafford, G. I.; Stadenb, J. V.; Christensena, S. B.; Jager, A. K. Isolation of the MAO-Inhibitor Naringenin from *Mentha aquatica* L. *J. Ethnopharmacol.* **2008,** *117*, 500–502. doi:10.1016/j.jep.2008.02.015.

Patil, R. S.; Patil, S. R.; Godghate, A. G. *Mentha piperita* Linn: Phytochemical, Antibacterial and Dipterian Adulticidal Approach. *Int. J. Pharm. Pharm. Sci.* **2016,** *8* (3), 352–355.

Patti, F.; Palmioli, A.; Vitalini, S.; Bertazza. L.; Redaelli, M.; Zorzan, M.; Rubin, B.; Mian, C.; Bertolini, C.; Iacobone, M.; Armanini, D.; Barollo, S.; Airoldi, C.; Iriti, M.; Pezzani, R.

Pattnaik, S.; Subramanyam, V. R.; Kole, C. Antibacterial and Antifungal Activity of Ten Essential Oils In Vitro. *Microbios*. **1996,** *86* (349), 237–246.

Pazdzioch-Czochra, M.; Widenska, A. Spectrofluorimetric Determination of Hydrogen Peroxide Scavenging Activity. *Anal. Chim. Acta.* **2002,** *452*, 177–184.

Rita, P.; Animesh, D. K. An Updated Overview on Peppermint (*Mentha piperita* L.). *Intern. Res. J. Pharm.* **2011,** *2* (8), 1–10.

Saharkhiz, M. J.; Motamedi, M.; Zomorodian, K.; Pakshir, K.; Miri, R.; Hemyari, K. Chemical Composition, Antifungal and Antibiofilm Activities of the Essential Oil of *Mentha piperita* L. *ISRN Pharma.* **2012,** Article ID 718645, 1–6. DOI: 10.5402/2012/718645.

Saleem, M. N.; Idris, M. Podina (*Mentha arvensis*): Transformation from Food Additive to Multifunctional Medicine. *ARC J. Pharma. Sci.* **2016,** *2* (2), 6–15. http://dx.doi.org/10.20431/2455–1538.0202002

Salehi, B.; Stojanovi´c-Radic, Z.; Matejic, J.; Sharopov, F.; Antolak, H.; Kregiel, D.; Sen, S.; Sharifi-Rad, M.; Acharya, K.; Sharifi-Rad, R.; Martorell, M.; Sureda, A.; Martins, N.; Sevindik, M. Pharmacological Properties of *Mentha* Species. *J. Trad. Med. Clin. Nat.* **2018,** 7:1. DOI: 10.4172/2573–4555.1000259.

Sharifi-Rad, J. Plants of Genus *Mentha*: From Farm to Food Factory. *Plants* **2018,** *7*, 70. DOI: 10.3390/plants7030070.

Shah, P. P.; D'Mello, P. M. A Review of Medicinal Uses and Pharmacological Effects of *Mentha piperita. Nat. Prod. Radiance*. **2004,** *3* (4), 214–221.

Singh, P.; Pandey, A. K. Prospective of Essential Oils of the Genus *Mentha* as Biopesticides: A Review. *Front. Plant Sci.* **2018,** *9*, 1295. DOI: 10.3389/fpls.2018.01295

Singh, R. K.; Gautam, R.; Karchuli, M. S. Evaluation of Nephroprotective Activity of *Mentha arvensis*in Cisplatin-Induced Nephrotoxicity. *Asian J. Pharm. Clin. Res.* **2014,** *7* (4), 188–189.

Snoussi, M.; Noumi, E.; Trabelsi, N.; Flamini, G.; Papetti, A.; De Feo, V. *Mentha spicata* Essential Oil: Chemical Composition, Antioxidant and Antibacterial Activities Against Planktonic and Biofilm Cultures of *Vibrio* spp. Strains. *Molecules* **2015,** *20*, 14402–14424. DOI: 10.3390/molecules200814402.

Sulieman, A. M. E.; Abdelrahman, S. E.; Rahim, A. M. A. Phytochemical Analysis of Local Spearmint (*Mentha spicata*) Leaves and Detection of the Antimicrobial Activity of Its Oil. *J. Microbiol. Res.* 2011, *1* (1), 1–4. DOI: 10.5923/j.microbiology.20110101.01.

Thawkar, B. S.; Jawarkar, A. G.; Kalamkar, P. V.; Pawar, K. P.; Kale, M. K. Phytochemical and Pharmacological Review of *Mentha arvensis*. *Intern. J. Green Pharm.* **2016,** *10* (2), 71–76.

Ullah, N.; Khurram, M.; Amin, M. U.; Afridi, H. H.; Khan, F. A.; Khayam, S. M. U.; Ullah, S.; Najeeb, U.; Hussain, J.; Khan, M. A. Comparison of Phytochemical Constituents and Antimicrobial Activities of *Mentha spicata* from Four Northern Districts of Khyber Pakhtunkhwa. *J. Appl. Pharma. Sci.* **2011,** 1 (7), 72–76.

Venditti, A.; Frezza, C.; Celona, D.; Sciubba, F.; Foddai, S.; Delfini, M.; Serafini, M.; Bianco, A. Phytochemical Comparison with Quantitative Analysis Between Two Flower Phenotypes of *Mentha aquatica* L.: Pink-Violet and White. *AIMS Mol. Sci.* **2017,** *4* (3), 288–300. DOI: 10.3934/molsci.2017.3.288.

Verma, S. M.; Arora, H.; Dubey, R. Anti-Inflammatory and Sedative—Hypnotic Activity of the Methanolic Extract of the Leaves Of *Mentha arvensis*. *Ancient Sci. Life* **2003,** *23* (2), 95–99.

Zaidi, S.; Dahiya, P. In Vitro Antimicrobial Activity, Phytochemical Analysis and Total Phenolic Content of Essential Oil from *Mentha spicata* and *Mentha piperita*. *Intern. Food Res. J.* **2015,** *22* (6), 2440–2445.

Zangeneh, M. M.; Zangeneh, A.; Moradi, R.; Seydi, N. Pharmacological, Biochemical, and Histopathological Evaluations of Cutaneous Wound Healing Property of *Mentha piperita* Aqueous Extract Ointment. *Compar. Clin. Pathol.* **2019,** *28*, 1181–1190. https://doi.org/10.1007/s00580–019–02989-w

Zebelo, S. A.; Bertea, C. M.; Bossi, S.; Occhipinti, A.; Gnavi, G.; Maffei, M. E. *Chrysolina herbacea* Modulates Terpenoid Biosynthesis of *Mentha aquatica* L. *PLoS ONE* **2011,** *6* (3), e17195. https://doi.org/10.1371/journal.pone.0017195.

Zheng, W.; Wang, S. Y. Antioxidant Activity and Phenolic Compounds in Selected Herbs. *J. Agric. Food Chem.* **2001,** *49*, 5165–5170.

CHAPTER 12

THE PHARMACOLOGICAL ACTIONS OF THE HYBRID PLANT *MENTHA* × *VILLOSA* HUDS.

TAYHANA PRISCILA MEDEIROS SOUZA,
MARIA DANIELMA DOS SANTOS REIS,
FELIPE LIMA PORTO, RAFAEL VRIJDAGS CALADO,
EMILIANO BARRETO, and JAMYLLE NUNES DE SOUZA FERRO

Laboratory of Cell Biology, Federal University of Alagoas, Alagoas, Brazil

ABSTRACT

Mentha villosa Huds., popularly known as mint, is a hybrid of *M. spicata* and *M. suaveolens* used in folk medicine to treat pain, inflammation, and infections. In the essential oil and the aqueous extract of this plant was found a wide range of potential bioactives such as the monoterpenes eucalyptol and limonene, phenolic compounds, and flavonoid glycosides. These molecules may be responsible for the plant's pharmacological activity. In the literature, it was reported to have antinociceptive, antimicrobial, anti-helmintic, antioxidant and hypotensive effects; however, further studies are yet to be carried out to unveil the full potential of the mint.

12.1 INTRODUCTION

The species *Mentha villosa* Huds, or *Mentha crispa*, is popularly known as low mint, common mint, peppermint, peppermint, or mint (Radunz, 2013).

Bioactives and Pharmacology of Lamiaceae. T. Pullaiah, PhD (Ed.)

It is a hybrid of the species *Mentha spicata* and *Mentha suaveolens*. Its biological properties are poorly studied when compared to other species of the genus *Mentha*, but it has been considered a potential alternative for use as a flavoring agent in food and beverages (Kumar et al., 2011; Lahlou et al., 2002).

Various preparations using *M. villosa* leaves have already been reported in folk medicine, through decoction or infusion, to treat menstrual cramps, otalgia, flu, headache, migraine, inflammation in general, intestinal infection, indigestion, fever, ophthalmic problems, stroke, sinusitis, worms, and pain in general (Cartaxo et al., 2010). In northeastern Brazil, indigenous and rural communities use this plant for the treatment of bronchitis, helminthiasis, liver problems, and indigestion (Albuquerque et al., 2007).

12.2 BIOACTIVES

In the essential oil of *M. villosa* were identified through gas chromatography and mass spectrometry the compounds piperitenone oxide, eucalyptol, and limonene in greater quantities, followed by isopentyloxyethyl acetate, β-pinene, myrcene, α-pinene, sabinene, β-karyophylene, 6-methyl-5-hepten-2-ol, piperitone oxide, germacrene D, and *p*-chimene (Almeida et al., 2018; Passos-Braga et al., 2019).

Sixteen phenolic compounds were identified and quantified in the aqueous extract of the leaves of *M. villosa*, consisting of eight phenolic derivatives and eight flavonoid glycosides: quinic acid, chlorogenic acid, coumaroyl-hexoside, caffeic acid, coumaroylquinic acid, lithospheric acid, rosmarinic acid, salvianolic, luteolin-7-*O*-glucuronide, luteolin-7-*O*-glucoside, luteolin-7-*O*-rutinoside, eriodictiol-7-*O*-rutinoside, apigenin-7-*O*-glucuronide, kaempferol-3-*O*-glucuronide, cryoseriol-7-*O*-rutinoside, and hesperetin-7-*O*-rutinoside (Fialova et al., 2015).

12.3 PHARMACOLOGY

12.3.1 ANTINOCICEPTIVE ACTIVITY

The essential oil of *M. villosa* and its main constituent piperitenone oxide, when orally administrated in mice, could induce an antinociceptive effect in contortions induced by acetic acid (Sousa et al., 2009).

12.3.2 ANTIMICROBIAL ACTIVITY

The essential oil of *M. villosa* showed potential antibacterial activity against *Staphylococcus aureus* and *Escherichia coli* with a minimum inhibitory concentration value of 90% (MIC_{90}) of 1.0 and 3.0 mg/mL, respectively (Arruda et al., 2006). Moreover, the bactericide properties of the essential oil in the vapor phase were evaluated in vitro against five foodborne bacteria using the disc volatilization method. The compound showed antibacterial activity only against *S. aureus*, with a MIC of 0.53 $\mu L/cm^3$ (Nedorostova et al., 2009). Corroborating these results, the aqueous extract of this plant showed an MIC_{90} value = 3.0 mg/mL against *S. aureus* (Kapp et al., 2020).

The essential oil of *M. villosa* also presented antifungal effects. The compound was able to inhibit the growth of *Candida albicans*, *Candida tropicalis*, *Pichia anomala*, and *Saccharomyces cerevisiae*, with a MIC of 17.5 μL/mL. In addition, the incorporation of essential oil in Sabouraud Dextrose broth and fruit juices (cashew, guava, mango, and pineapple) decreased the counts of yeast strains at concentrations of 3.75, 7.5, and 15 μL/mL during 72 h of storage. This anti-yeast activity was dose-dependent, in which a reduction in yeast count was greater at higher concentrations of essential oil (Almeida et al., 2018).

12.3.3 ANTIOXIDANT EFFICACY

The aqueous extract of the leaves of *M. villosa* showed an antioxidant effect in vitro by the 2,2-diphenyl-1-picryl-hydrazyl-hydrate and 2,2′-azino-bis(3-ethylbenzothiazoline-6-sulfonic acid) tests and ex vivo in the mesenteric artery of rats injured by ischemia/reperfusion (Fialova et al., 2015).

12.3.4 HYPOTENSIVE EFFECT

Intravenous treatment with essential oil of *M. villosa* in different concentrations (1, 5, 10, and 20 mg/kg) showed hypotensive and bradycardic effects in a dose-dependent manner in Wistar rats (Lahlou et al., 2001; Lahlou et al., 2002; Guedes et al., 2004).

12.3.5 ANTHELMINTIC ACTIVITY

The hydrolate obtained from *M. villosa* was evaluated in heifers infected with gastrointestinal nematodes, where 90% of the larvae belonged to the genera *Cooperia*, *Haemonchus*, and to a lesser extent, larvae of the genera *Oesophagostomum* and *Trichostrongylus*. In vitro, the hydrolate showed efficacy of 91.88%, 94.15%, 98.40%, and 100% at concentrations of 40%, 60%, 80%, and 100%, respectively, using the quantitative co-culture method. In the in-vivo tests, no anthelmintic activity was observed at a dose of 0.1 mL/kg/day in the treated calves (Nascimento et al., 2009).

The essential oil of *M. villosa*, at a concentration of 500 μg/mL, showed schistosomicidal effects against *Schistosoma mansoni* through ultrastructural changes in the integument and vacuoles in the region of the adult worm syncytial matrix (Matos-Rocha et al., 2016).

KEYWORDS

- **mint**
- **folk medicine**
- **essential oil**
- **phenolic compounds**
- **flavonoids**

REFERENCES

Albuquerque, U. P. de; Medeiros, P. M. de; Almeida, A. L. S. de; Monteiro, J. M.; Freitas Lins Neto, E. M. de; Melo, J. G. de; Santos, J. P. dos. Medicinal Plants of the Caatinga (Semi-Arid) Vegetation of NE Brazil: A Quantitative Approach. *J. Ethnopharmacol.* **2007,** *114* (3), 325–354.

Almeida, E. T. D. C.; Barbosa, I. de M.; Tavares, J. F.; Barbosa-Filho, J. M.; Magnani, M.; Souza, E. L. de. Inactivation of Spoilage Yeasts by *Mentha spicata* L. and *M.* × *villosa* Huds. Essential Oils in Cashew, Guava, Mango, and Pineapple Juices. *Front. Microbiol* **2018,** *9*, 1–12.

Arruda, T. A.; Antunes, R. M. P.; Catão, R. M. R.; Lima, E. O.; Sousa, D. P.; Nunes, X. P.; Pereira, M. S. V.; Barbosa-Filho, J. M.; Cunha, E. V. L. da. Preliminary Study of the Antimicrobial Activity of *Mentha x villosa* Hudson Essential Oil, Rotundifolone and Its Analogues. *Revista Brasileira de Farmacognosia,* **2006,** *16* (3), 307–311.

Cartaxo, S. L.; Almeida Souza, M. M. de; Albuquerque, U. P. de. Medicinal Plants with Bioprospecting Potential Used in Semi-Arid Northeastern Brazil. *J. Ethnopharmacol.* **2010,** *131* (2), 326–342.

Fialova, S.; Veizerova, L.; Nosalova, V.; Drabikova, K.; Tekelova, D.; Grancai, D.; Sotnikova, R. Water Extract of *Mentha* × *villosa*: Phenolic Fingerprint and Effect on Ischemia-Reperfusion Injury. *Nat. Prod. Commun.* **2015,** *10* (6), 937–940.

Guedes, D.N.; Silva, D. F.; Barbosa-Filho, J. M.; Almeida de Medeiros, I. Endothelium-Dependent Hypotensive and Vasorelaxant Effects of the Essential Oil from Aerial Parts of *Mentha x villosa* in Rats. *Phytomedicine*, **2004,** *11* (6), 490–497.

Kapp, K.; Püssa, T.; Orav, A.; Roasto, M.; Raal, A.; Vuorela, P.; Vuorela, H.; Tammela, P. Chemical Composition and Antibacterial Effect of *Mentha* spp. Grown in Estonia. *Nat. Prod. Commun.* **2020,** *15* (12), 1–14.

Kumar, P.; Mishra, S.; Malik, A.; Satya, S. Insecticidal Properties of *Mentha* Species: A Review. *Ind. Crops Products*, **2011,** *34* (1), 802–817.

Lahlou, S.; Carneiro-Leão, R. F. L.; Leal-Cardoso, J. H.; Toscano, C. F.. Cardiovascular Effects of the Essential Oil of *Mentha x villosa* and Its Main Constituent, Piperitenone Oxide, in Normotensive Anaesthetised Rats: Role of the Autonomic Nervous System. *Planta Medica*, **2001,** *67* (7), 638–643.

Lahlou, S.; Lima Carneiro-Leão, R. F.; Leal-Cardoso, J. H. Cardiovascular Effects of the Essential Oil of *Mentha x villosa* in DOCA-Salt-Hypertensive Rats. *Phytomedicine* **2002,** *9* (8), 715–720.

Matos-Rocha, T. J.; Cavalcante, M. G. S.; Veras, D. L.; Feitosa, A. P. S.; Gonçalves, G. G. A.; Portela-Junior, N. C.; Lúcio, A. S. S. C.; Silva, A. L.; Padilha, R. J. R.; Marques, M. O. M.; Barbosa-Filho, J. M.; Alves, L. C.; Brayner, F. A. Ultrastructural Changes in *Schistosoma mansoni* Male Worms After In Vitro Incubation with the Essential Oil of *Mentha x villosa* Huds. *Revista do Instituto de Medicina Tropical de Sao Paulo* **2016,** *58* (2), 2–7.

Nascimento, É. M.; Furlong, J.; Pimenta, D. S.; Prata, M. C. de A. Efeito Anti-helmíntico do hidrolato de *Mentha villosa* Huds. (Lamiaceae) em nematóides gastrintestinais de bovinos. *Ciência Rural* **2009,** *39* (3), 817–824.

Nedorostova, L.; Kloucek, P.; Kokoska, L.; Stolcova, M.; Pulkrabek, J. Antimicrobial Properties of Selected Essential Oils in Vapour Phase Against Foodborne Bacteria. *Food Control* **2009,** *20* (2), 157–160.

Passos Braga, S. dos; Lundgren, G. A.; Macedo, S. A.; Tavares, J. F.; Santos Vieira, W. A. dos; Câmara, M. P. S.; Souza, E. L. de. Application of Coatings Formed by Chitosan and *Mentha* Essential Oils to Control Anthracnose Caused by *Colletotrichum gloesporioides* and *C. brevisporum* in Papaya (*Carica papaya* L.) Fruit. *Int. J. Biol. Macromol.* **2019,** *139*, 631–639.

Radunz, L. Efeito da temperatura do ar de secagem no teor e na composição dos óleos essenciais de guaco (*Mikania glomerata* Sprengel) e hortelã-comum (*Mentha x villosa* Huds.) *J. Chem. Inf. Model.* **2013,** *53* (9), 1689–1699.

Sousa, P. J. C.; Linard, C. F. B. M.; Azevedo-Batista, D.; Oliveira, A. C.; Coelho-de-Souza, A. N.; Leal-Cardoso, J. H. Antinociceptive Effects of the Essential Oil of *Mentha x villosa* Leaf and Its Major Constituent Piperitenone Oxide in Mice. *Braz. J. Med. Biol. Res.* **2009,** *42* (7), 655–659.

CHAPTER 13

CHEMICAL CONSTITUENTS AND PHARMACOLOGY OF *OCIMUM AMERICANUM* L.

THADIYAN PARAMBIL IJINU[1,2], IRIN KURIKESU[3], SASIDHARAN LATHIKAKUMARI SREEJITH[4], and PALPU PUSHPANGADAN[1]

[1]*Amity Institute for Herbal and Biotech Products Development, Thiruvananthapuram, Kerala, India*

[2]*Naturæ Scientific, Kerala University Business Innovation and Incubation Centre, Karyavattom Campus, Thiruvananthapuram, Kerala, India*

[3]*Inter University Centre for Evolutionary and Integrative Biology, University of Kerala, Thiruvananthapuram, Kerala, India*

[4]*Central Analytical Facility, Sree Chitra Tirunal Institute for Medical Sciences and Technology, Thiruvananthapuram, Kerala, India*

ABSTRACT

Ocimum americanum L. is a woody herb that belongs to the family Lamiaceae. It is widely distributed in tropical and South Africa, as well as from China and India to Brazil and Australia. *O. americanum* is used for various medicinal and culinary purposes and as an insect repellent. Leaf juice of *O. americanum* along with *Triphala* is used in Ayurvedic eye drop preparations against glaucoma, cataract, chronic conjunctivitis, and other painful eye diseases. Moreover, the fresh leaf juice is used to treat patients with chronic fever, dysentery, constipation, diabetes, haemorrhage, bronchial catarrh, and dyspepsia. The major

Bioactives and Pharmacology of Lamiaceae. T. Pullaiah, PhD (Ed.)

bioactive compounds reported from *O. americanum* include (*E*)-caryophyllene, (*E*)-γ-bisabolene, (*Z*)-methyl cinnamate, 1,8-cineole, camphor, eugenol, linalool, methyl chavicol, β-bisabolene, nevadensin, salvigenin, cirsimaritin, apigenin 7,4'-dimethyl ether, xanthomicrol, cirsilineol, 5-demethylnobiletin, etc. *O. americanum* possesses hepatoprotective, anti-inflammatory, antioxidant, antihypertensive, and antimicrobial activities.

13.1 INTRODUCTION

Ocimum americanum L. (Syn.: *Ocimum album* Roxb., *Ocimum brachiatum* Blume, *Ocimum canum* Sims, *Ocimum dichotomum* Hochst. ex Benth., *Ocimum dinteri* Briq., *Ocimum fluminense* Vell., *Ocimum fruticulosum* Burch., *Ocimum hispidulum* Schumach. & Thonn., *Ocimum incanescens* Mart., *Ocimum stamineum* Sims, and *Ocimum thymoides* Baker) belongs to the family Lamiaceae. Vernacular names of this plant include kali tulsi (Hindi), kshudraparna, gambhira (Sanskrit), kaattu tulasi (Malayalam), nai thulasi (Tamil), nayi tulasi (Kannada), kukka thulasi (Telugu), ram tulshi (Marathi), kalo tulashi (Bengali), mayangton (Manipuri), runhmui (Mizo), sari (Tangkhul), mang lahk nahand (Thai), e hoang (Vietnamese) and hoary basil, wild basil, lemon basil, and American basil (English). It is a woody herb and stem hirtus; leaves elliptic, acute to acuminate, base truncate or obtuse, subentire, sparsely hairy or glabrous; petiole to 2 cm long; racemes 13 cm long, densely pubescent; flowers 5–7 in each whorl; calyx 3 × 5 mm, upper lip 3 × 3 mm, orbicular, densely hairy, corolla white, 5 mm long, glabrous; filaments glabrous; nutlets 1.5 mm long, subtrigonous, minutely pitted, dark brown (Sasidharan, 2004, 2011).

O. americanum is used for various medicinal and culinary purposes and as an insect repellent. Leaf juice of *O. americanum* along with *triphala* is used in Ayurvedic eye drop preparations against glaucoma, cataract, chronic conjunctivitis, and other painful eye diseases. Moreover, the fresh leaf juice is used to treat patients with chronic fever, dysentery, constipation, diabetes, hemorrhage, bronchial catarrh, and dyspepsia (Kokwaro, 1993; Vidhya et al., 2020). *O. americanum* essential oil is used as a flavoring and fragrance agent. It is a good insecticide, and the aerial parts of the plant are burned to give the repellent effect against mosquitoes and other insects. Swahili people, an ethnic group inhabiting East Africa used *O. americanum* to treat ulcer, catarrh, high blood pressure, and stomach ache (Watt et al., 1962). *O. americanum* is widely distributed in tropical and South Africa, also from China and India to Brazil and in Australia.

13.2 PHYTOCHEMICAL CONSTITUENTS

Singh et al. (2013) found that the essential oil composition of *O. amercanum* collected from different locations in Uttarakhand, India, varies with soil properties and microclimatic conditions. The essential oil obtained from samples collected from different regions contains group (1) methyl chavicol, 1,8-cineole, (E)-γ-bisabolene, β-bisabolene, and eugenol; (2) (E)-γ-bisabolene, aliphatic hydrocarbons, eugenol, β-bisabolene, and methyl chavicol; (3) eugenol and (E)-caryophyllene; (4) linalool and methyl chavicol; (5) aliphatic hydrocarbons, eugenol, camphor, and 1,8-cineole; and (6) camphor and aliphatic hydrocarbons as major constituents. They also found that linalool was positively correlated with methyl chavicol, and eugenol with (E)-α-bisabolene and β-bisabolene with (E)-γ-bisabolene. Shadia et al. (2007) found that *O. americanum* var. *americanum* essential oil collected from Egypt contains eugenol and methyl chavicol as major components. Essential oil from *O. americanum* growing in Western Kenya contains mainly terpinen-4-ol (43.21%), 1,8-cineole (16.13%), α-terpineol (4.01%), *trans*-caryophyllene (3.06%), α-bergamotene (2.68%), and cis-sabinene hydrate (2.59%). Carović-Stanko et al. (2010) found that *O. americanum* contains linalool (12.15%), neral (20.16%), geranial (28.58%), nerol (7.15), and caryophyllene oxide (7.12) as major components. Yamada et al. (2013) found that the essential oil obtained from *O. amercanum* collected from Paraná, Brazil, contains linalool (19.63%) and 1,8-cineole (17.27%) as the main components. Vieira et al. (2014) found that the major constituents present in the essential oil of *O. americanum* collected from Ceará capital, Brazil, are 1,8-cineole (25.9%) and (*Z*)-methyl cinnamate (29.4%). Vieira et al. (2003) found that different flavone chemotypes of *O. americanum* contain nevadensin, salvigenin, cirsimaritin, cirsilineol, apigenin 7,4'-dimethyl ether, xanthomicrol, 5-desmethylnobiletin, and xanthomicrol.

13.3 PHARMACOLOGICAL STUDIES

13.3.1 HEPATOPROTECTIVE ACTIVITY

Studies carried in carbon tetrachloride and acetaminophen-induced toxicity in rats proved the hepatoprotective activity of *O. americanum* (Genfi et al., 2020). Oral administration of leaf and stem extracts, for 7 days at 250 mg/

kg, prevented carbon tetrachloride-induced elevation of serum biochemical parameters, prooxidants, as well as the expression of nuclear factor kappa B and interleukin-1.

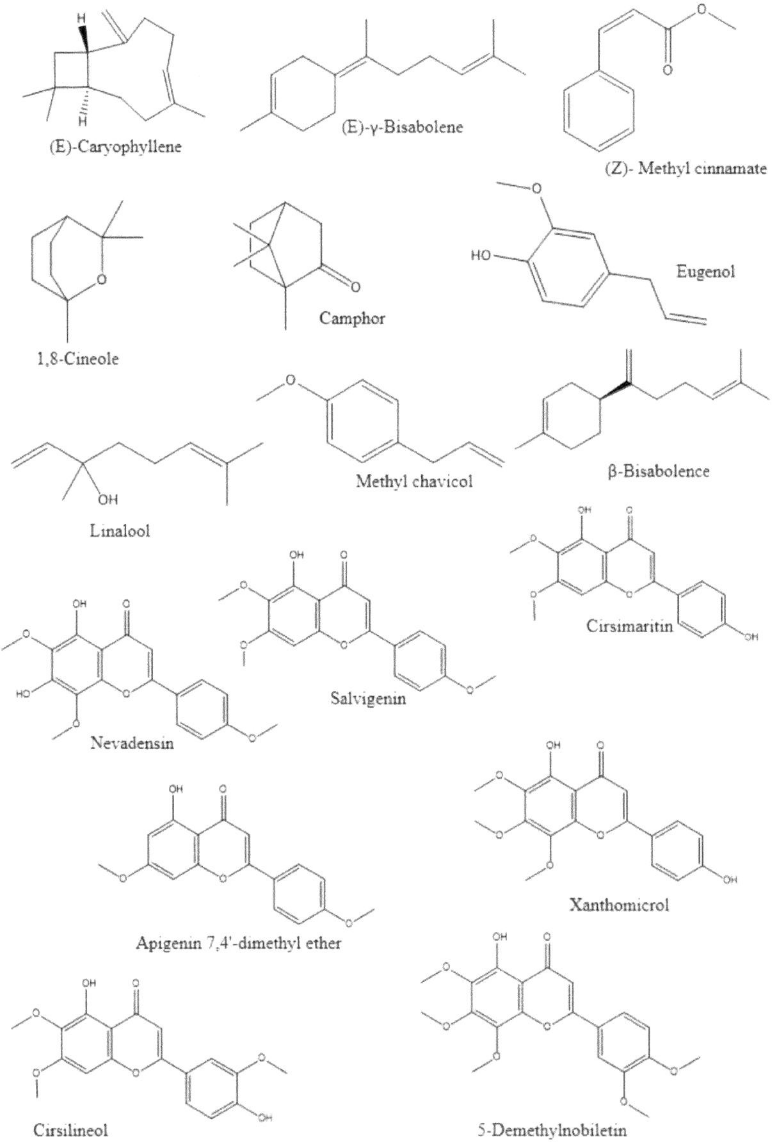

13.3.2 ANTI-INFLAMMATORY ACTIVITY

Yamada et al. (2013) found that the linalool and 1,8-cineole obtained from the fresh leaves of *O. americanum* reduced leukocyte migration at all of the doses tested (50, 150, and 300 mg/kg). Linalool and 1,8-cineole also inhibited paw edema formation at 12 h. Another study has shown that linalool reduces thermal hyperalgesia induced by carrageenan and causes antinociception in a formalin-induced nociception in mice (Peana et al., 2003). 1,8-Cineole exerted anti-inflammatory effect in a rat model (Santos and Rao, 2001). Juergens et al. (2003) observed the anti-inflammatory effect of 1,8-cineole in patients with bronchial asthma. The phenolic-rich water extract of *O. americanum* flower and leaf extracts (methanol, ethyl acetate, and water) showed good anti-inflammatory activity by blocking the upregulation of nitrite and 5-hydroxytryptamine levels in lipopolysaccharide-induced HCT116 cells. Also, the ethyl acetate extract of flower reduced the lipopolysaccharide-induced LDH level (Zengin et al., 2019).

13.3.3 ANTIOXIDANT ACTIVITY

O. americanum leaves contain biologically active components with effective antioxidant activity (Aluko et al., 2013). The phenolic-rich water extract of *O. americanum* flowers and leaves showed higher radical scavenging properties. Flowers showed 654.64 and 453.50 mg TE/g for CUPRAC and FRAP assays, respectively, and leaves showed 494.56 and 342.39 mg TE/g for CUPRAC and FRAP assays, respectively (Zengin et al., 2019).

13.3.4 ANTIHYPERTENSIVE ACTIVITY

AdjokÃ et al. (2019) reported that the administration of oral doses of *O. americanum* ethanolic extract (250 mg/kg) in rat significantly reduced the mean blood pressure from 146 ± 4.80 to 98.4 ± 9.44 mmHg.

13.3.5 ANTIMICROBIAL ACTIVITY

O. americanum inflorescences essential oil at the low concentrations (10 and 20 mg/L) added to the water, promoted the survival of fish experimentally infected with *Aeromonas hydrophila*, and also showed effective against

Gyrodactylus sp. (dose 50 mg/L) by reducing (60%) the number of parasites in the fish (Sutili et al., 2016). The methanol, ethyl acetate, and water extracts of *O. americanum* leaves and flowers showed significant inhibitory effects on bacterial strains, *Staphylococcus aureus* and *Staphylococcus typhimurium* and fungal strains, *Candida albicans* and *Candida tropicalis* (Zengin et al., 2019). Carović-Stanko et al. (2010) found that *O. americanum* essential oil showed good activity against *Enterococcus faecalis* (15.67 mm), *Enterococcus faecium* (15.67 mm), *Proteus vulgaris* (17.67 mm), *S. aureus* (33.33 mm), and *Staphylococcus epidermidis* (22.67 mm).

ACKNOWLEDGMENTS

The authors express their sincere thanks to Dr. Ashok K. Chauhan, Founder President, Ritnand Balved Education Foundation (RBEF) and Amity Group of Institutions, and Dr. Atul Chauhan, Chancellor, Amity University Uttar Pradesh (AUUP) for facilitating this work. Thadiyan Parambil Ijinu has received Young Scientist Fellowship from the Department of Science and Technology, Government of India (SP/YO/413/2018).

KEYWORDS

- **traditional medicinal use**
- **(E)-caryophyllene**
- **(Z)-methyl cinnamate**
- **hepatoprotective activity**
- **antioxidant activity**
- **antihypertensive activity**

REFERENCES

AdjokÃ¨, A. R. A.; OlatoundÃ, A. A. M.; Adamou, R.; Bonaventure, A.; Anatole, L.; Latifou, L. UHPLC-DAD Characterization of Bioactive Secondary Metabolites from *Ocimum americanum* and *Pupalia lappacea* Extracts: Antioxidant Activity and Antihypertensive Effects on L-NAME-Induced Hypertensive Rats. *J. Pharmacogn. Phytother.* **2019,** *1,* 17–27.

Aluko, B. T.; Oloyede, O. I.; Afolayan, A. J. Polyphenolic Contents and Free Radical Scavenging Potential of Extracts from Leaves of *Ocimum americanum* L. *Pak. J. Biol. Sci.* **2013,** *16*, 22–30.

Carović-Stanko, K.; Orlić, S.; Politeo, O.; Strikić, F.; Kolak, I.; Milos, M.; Satovic, Z. Composition and Antibacterial Activities of Essential Oils of Seven *Ocimum* taxa. *Food Chem.* **2010,** *119*, 196–201.

Genfi, A. K. A.; Larbie, C.; Emikpe, B. O.; Oyagbemi, A. A.; Firempong, C. K.; Adjei, C. O. Modulation of Oxidative Stress and Inflammatory Cytokines as Therapeutic Mechanisms of *Ocimum americanum* L. Extract in Carbon Tetrachloride and Acetaminophen-Induced Toxicity in Rats. *J. Evid. Based Integr. Med.* **2020,** *25*, 1–13.

Juergens, U. R.; Dethlefsen, U.; Steinkamp, G.; Gillissen, A.; Repges R.; Vetter, H. Anti-Inflammatory Activity of 1.8-cineol (Eucalyptol) in Bronchial Asthma: A Double-Blind Placebo-Controlled Trial. *Respir. Med.* **2003,** *97*, 250–256.

Kokwaro, J. O. *Medicinal Plants of East Africa*, 2nd ed.; Kenya Literature Bureau, 1993.

Matasyoh, J. C.; Bendera, M. M.; Ogendo, J. O.; Omollo, E. O.; Deng, A. L. Volatile Leaf Oil Constituents of *Ocimum americanum* L. Occuring in Western Kenya. *Bull. Chem. Soc. Ethiop.* **2006,** *20*, 177–180.

Peana, A. T.; D'Aquila, P. S.; Chessa, M. L.; Moretti, M. D.; Serra, G.; Pippia, P. (-)-Linalool Produces Antinociception in Two Experimental Models of Pain. *Eur. J. Pharmacol.* **2003,** *460*, 37–41.

Santos, F. A.; Rao, V. S. 1,8-Cineol, a Food Flavoring Agent, Prevents Ethanol-Induced Gastric Injury in Rats. *Dig. Dis. Sci.* **2001,** *46*, 331–337.

Sasidharan, N. *Biodiversity Documentation for Kerala, Part 6: Flowering Plants*; Kerala Forest Research Institute: Kerala, India, 2004.

Sasidharan, N. *Flowering Plants of Kerala*, Ver. 2.0 (DVD), Serial Number 698329520, Kerala Forest Research Institute, Thrissur 680653, Kerala, India, 2011.

Shadia, E.; El-Aziz, A.; Omer, E. A.; Sabra, A. S. Chemical Composition of *Ocimum americanum* Essential Oil and Its Biological Effects Against *Agrotis ipsilon* (Lepidoptera: Noctuidae). *Res. J. Agric. Biol. Sci.* **2007,** *3*, 740–747.

Singh, S.; Tewari, G.; Pande, C.; Singh, C. Variation in Essential Oil Composition of *Ocimum americanum* L. from North-Western Himalayan Region. *J. Essent. Oil Res.* **2013,** *25*, 278–290.

Sutili, F. J.; Velasquez, A.; Pinheiro, C. G.; Heinzmann, B. M.; Gatlin, D. M.; Baldisserotto, B. Evaluation of *Ocimum americanum* Essential Oil as an Additive in Red Drum (*Sciaenops ocellatus*) Diets. *Fish Shellfish Immunol.* **2016,** *56*, 155–161.

Vidhya, E.; Vijayakumar, S.; Rajalakshmi, S.; Kalaiselvi, S.; Pandiyan, P. Antimicrobial Activity and Phytochemical Screening of *Ocimum americanum* L. Extracts Against Pathogenic Microorganisms. *Acta Ecol. Sin.* **2020,** *40*, 214–220.

Vieira, P. R. N., de Morais, S. M.; Bezerra, F. H. Q.; Ferreira, P. A. T.; Oliveira, Í. R.; Silva, M. G. V. Chemical Composition and Antifungal Activity of Essential Oils from *Ocimum* Species. *Ind. Crops Prod.* **2014,** *55*, 267–271.

Vieira, R. F.; Grayer, R. J.; Paton, A. J. Chemical Profiling of *Ocimum americanum* Using External Flavonoids. *Phytochemistry* **2003,** *63*, 555–567.

Watt, J. M.; Gerdinat, M.; Breyer-Brandwij, K. *Medicinal and Poisonous Plants of Southern and Eastern Africa*, 2nd ed.; E&S Lwinstone Ltd., 1962.

Yamada, A. N.; Grespan, R.; Yamada, Á. T.; Silva, E. L.; Silva-Filho, S. E.; Damião, M. J.; Dalalio, M. M. D. O.; Amado, C. A. B.; Cuman, R. K. N. Anti-Inflammatory Activity of *Ocimum americanum* L. Essential Oil in Experimental Model of Zymosan-Induced Arthritis. *Am. J. Chin. Med.* **2013,** *41*, 913–926.

Zengin, G.; Ferrante, C.; Gnapi, D. E.; Sinan, K. I.; Orlando, G.; Recinella, L.; Diuzheva, A.; Jekő, J.; Cziáky, Z.; Chiavaroli, A.; Leone, S.; Brunetti, L.; Picot-Allain, C.; Mahomoodally, M. F.; Angelini, P.; Covino, S.; Venanzoni, R.; Tirillini, B.; Menghini, L. Comprehensive Approaches on the Chemical Constituents and Pharmacological Properties of Flowers and Leaves of American Basil (*Ocimum americanum* L.). *Food Res. Int.* **2019,** *125*, 108610.

CHAPTER 14

PHYTOCHEMICAL CONSTITUENTS AND PHARMACOLOGY OF *OCIMUM BASILICUM* L.

THADIYAN PARAMBIL IJINU[1,2], MAHESWARI PRIYA RANI[3], SHARAD SRIVASTAVA[4], VARUGHESE GEORGE[1], and PALPU PUSHPANGADAN[1]

[1]*Amity Institute for Herbal and Biotech Products Development, Thiruvananthapuram, Kerala, India*

[2]*Naturæ Scientific, Kerala University Business Innovation and Incubation Centre, Karyavattom Campus, Thiruvananthapuram, Kerala, India*

[3]*Phytochemistry and Phytopharmacology Division, Jawaharlal Nehru Tropical Botanic Garden and Research Institute, Thiruvananthapuram, Kerala, India*

[4]*Pharmacognosy Division, CSIR-National Botanical Research Institute, Council of Scientific and Industrial Research, Lucknow, Uttar Pradesh, India*

ABSTRACT

Ocimum basilicum L. or sweet basil, is an erect annual herb up to 60 cm tall and belongs to the family Lamiaceae. It is indigenous to Asian countries and can be observed growing wildly in tropical and subtropical regions. The ancient Ayurvedic texts such as *Charaka Samhita*, *Sushrutha Samhita*, *Bhava Prakasa*, *Dhanwantari Nighandu*, *Ashtanga Hridaya*, and *Kaiyadeva Nighandu* described the medicinal properties of *O. basilicum*. In *Ayurveda*,

Bioactives and Pharmacology of Lamiaceae. T. Pullaiah, PhD (Ed.)

the whole plant is used for cough, asthma, bronchitis, ophthalmia, giddiness, intermittent and malarial fever, catarrh, otalgia, cephalalgia, dyspepsia, and spasmodic affections. The major bioactive components of basil include rosmarinic acid, gentisic acid, vanillic acid, chicoric acid, caffeic acid, caftaric acid, quercetin, rutin, apigenin, etc. *O. basilicum* is rich in essential oil, and the major essential oil components include linalool, methyl chavicol, 1,8-cineole, eugenol, methyl eugenol, thymol, citral, camphor, bergamotene, linalyl acetate, and nerol. *O. basilicum* possesses cardioprotective, anti-inflammatory, analgesic, sedative, anxiolytic, bronchodilator, antidiabetic, antioxidant, hepatoprotective, gastroprotective, anticancer, antimicrobial, and larvicidal properties.

14.1 INTRODUCTION

Ocimum basilicum L. (Syn.: *Ocimum album* L., *O. anisatum* Benth., *O. barrelieri* Roth, *O. bullatum* Lam., *O. caryophyllatum* Roxb., *O. chevalieri* Briq., *O. ciliare* B.Heyne ex Hook.f., *O. ciliatum* Hornem., *O. citrodorum* Blanco, *O. cochleatum* Desf., *O. dentatum* Moench, *O. hispidum* Lam., *O. integerrimum* Willd., *O. lanceolatum* Schumach. & Thonn., *O. laxum* Vahl ex Benth., *O. majus* Garsault, *O. medium* Mill., *O. minus* Garsault, *O. nigrum* Thouars ex Benth., *O. odorum* Salisb., *O. scabrum* Wight ex Hook.f., *O. simile* N.E.Br., *O. thyrsiflorum* L., *O. urticifolium* Benth., *Plectranthus barrelieri* (Roth) Spreng.) belongs to the family Lamiaceae. Vernacular names of this plant include babuitulsi, babul, babuyitulsi, bahari, ban tulsi, barbar, kalitulsi, marua, niyakshbo, rihan, sabzah, tukhmerihan, tulsi, babestul, jangli tulsi, safed tulsi, gola-tulasi (Hindi), ajagandhika, ajaka, ambheram, apetarakhsasi, asurasa, barba, barbari, bisvatulasi, gandhapatra, jambirah, karahi, kharapushpa, kutheraka, manjariki, manjirika, marubaka, munjariki, pasanabheda, phanijjah, suklatulasi, surabhi, surasa, tulasidvesha, tungi, ugragandha, varavara, vishvatulasi (Sanskrit), pachcha, ramathulasi, soladi-tirtava, tirunitri, tirunitru, tirunitru-pach-cha, kattarama-tulasi (Malayalam), cankanirakkarantai, capcavitai, capja, civatocikappaccai, civatotikam, cunaikkarantai, kalimokam, kalimokappaccai, kamakasturi, kancankorai, karunthulasi, mancilikkan, nolatikam, paccai, paccaicapja, paccilai, pasanacceti, rudrajadai, ruttiracatai, thiruneetru patchi, tirnoot patchie tirunitru-pachai, tirunutpatchi, uruttiracatai, varuttiraccatai, viputippaccai, kiruttina parani, ganjankorai, kanchan korai, nay-tolashi, thiruneethupachai (Tamil), kamma gaggare, kaama kasturi, karavgagri, neeru thulasi, ramkasturi, sabjaa gida, sajjebiya, tulasi, nayi-tulasi, bhoothulasi (Kannada), bhu-tulasi, bhutulasi, kukkatulasi, rudra-jeda, vepudu-pach-cha, vibudipatri, gakarakau, komma gaggiri, sabja, vaepudupachha, viboodipathri (Telugu), bhumy-tulas, marva, bhoo tulasi, sabja, sabji, sabza, tukhamariya (Marathi), naosek lei tulsi (Manipuri), pha ni dza (Tibetan), burg faranjmushk, burg tulsi, sosambar, tukhm faranjmishk (Urdu), asaba-ul-feteyat, badruj, buklut-ul-zub, hebak, raihan, shahas-faram, tukhme rehan, badruje-abyaz (Arabic), daban-shab, daban-shah, firanj-mushk, nazbu, shahasparam, tureh-korasani, ungusht-kuni-zuckan, raihane-kohi, raihanekohi (Persian), and basil, common basil, lemon basil, and sweet basil (English).

O. basilicum is an erect annual herb up to 60 cm tall and stem obtusely four-angular. Leaves are 3.5–6 × 1.5–2.5 cm, ovate-elliptic, base cuneate, margin entire to distantly crenulate, apex acute, puberulous above, pubescent with sessile oil glands beneath; petiole up to 2 cm long. Racemes are up to 14 cm long; bracts oblanceolate and ciliate; pedicel up to 2 mm long. The calyx is up to 6 mm long; tube c. 3 mm long, sparsely strigose; upper lip c. 3 mm long, broadly ovate; lower lip to 4 mm long, hairy within. Corolla is cream-colored, 7–10 mm long; tube c. 4 mm long, inflated below; lobes 5, stamens 4; filaments of posterior pair with a transverse process of tufted hairs. Nutlets are 2–3 × 1–1.5 mm, ellipsoid, black, mucilaginous when wet. Fruiting calyx is 6–8 mm long (Sasidharan, 2004, 2011). It is indigenous to Asian countries (India, Pakistan, Iran, Thailand, etc.) and can be observed growing wildly in tropical and subtropical regions.

Several species of *Ocimum* are cultivated in India, and their medicinal and culinary uses have been known for several millennia. It is considered as a panacea for several ailments, the mother of all medicinal plants, pure and a provider of good health. Basil is invariably a constituent in all materials used for worship in Hindu temples. Just as in India, the ancient Greeks also used basil for the worship of Greek Gods and Goddesses. The great Ayurveda Acharya Sushruta classified basil among vegetables. The ancient Ayurvedic texts such as *Charaka Samhita*, *Sushrutha Samhita*, *Bhava Prakasa*, *Dhanwantari Nighandu*, *Ashtanga Hridaya*, *Kaiyadeva Nighandu*, etc. also described the medicinal properties of basil. Bhavamisra, the great Ayurvedic Acharya referred to *Ocimum basilicum* as '*Barbari*' (Sanskrit) (Pushpangadan et al., 1993). According to Ayurveda, the plant is used for diseases caused by aggravation of *Kapha* and *Vata*, while the seeds are used for pacifying aggravation of *Vata* and *Pitta*. In Ayurveda, the whole plant is used for cough, asthma, bronchitis, ophthalmia, giddiness, intermittent and malarial fever, catarrh, otalgia, cephalalgia, dyspepsia, and spasmodic affections (Udayan and Balachandran, 2009). The plant is stomachic, stimulant, carminative, diaphoretic, expectorant, diuretic, and antipyretic (Kirtikar and Basu, 1935; Pushpangadan and Bradu, 1995; Pushpangadan and George, 2012; George et al., 2017; Ijinu et al., 2021).

Basil is used as a medicine in Africa, Arabia, Myanmar, Sri Lanka, Philippines, etc. In the Mediterranean regions such as Eastern Morocco, the infusions of basil are used to decrease plasma lipid content (Zhang et al., 2009). It is also used for the treatment of diarrhea, constipation, warts, and worms (Simon et al., 1999). Externally, basil can be used as an ointment for insect bites, and its oil is applied directly to the skin to treat acne (Waltz, 1996). The leaves are used for inflammation, pain, bronchitis, colds, coughs, exhaustion, flatulence, flu, gout, insect bites, insect repellent, muscle aches, rheumatism, and sinusitis. The juice expressed from the leaves is used against cough and cold. The juice of leaves gives luster to eyes, good for toothache, earache, headache, and can be mixed with camphor to stop nasal hemorrhage. It also forms an excellent nostrum for the cure of ringworms, scorpion sting, and snake bite. The juice of leaves warmed with honey is useful in the treatment of croup (Pushpangadan and Bradu, 1995; Pushpangadan and George, 2012; George et al., 2017; Ijinu et al., 2021).

The consumption of seeds steeped in water is said to be cooling and very nourishing. Its infusion is given in gonorrhea, diarrhea, and chronic dysentery. The washed and pounded seeds are used as a poultice for unhealthy sores and sinuses. It is also used in sharbat in habitual constipation and in internal piles. The seeds are chewed in snake bite (Kirtikar and Basu, 1935). The aqueous extract of the seeds is used as a diuretic (Pushpangadan et al., 1993). A teaspoon full of seeds infused in a tumbler of water with little sugar, when taken daily, acts as a demulcent in genitourinary diseases. A cold infusion of seeds is said to relieve afterpains of child birth. An infusion of seed is also given in fever. The Santhal tribe (ethnic group native to India and Bangladesh) uses sweet basil for headache, earache, cough, cold, inflammation, snake bite, and rabies (Pushpangadan et al., 1993; Ijinu et al., 2021). The roots of basil are used for bowel complaints in children (Chopra et al., 1956).

14.2 PHYTOCHEMICAL CONSTITUENTS

O. basilicum contains several potent antioxidant compounds like caffeic acid, vanillic acid, rosmarinic acid, quercetin, rutin, apigenin, chlorogenic acid, and 4-hydroxybenzoic acid (Teofilovic et al., 2017). Phenolic acids such as rosmarinic acid, gentisic acid, chicoric acid, caffeic acid, and caftaric acid are reported in high quantities in various cultivars (Zgórka and Glowniak, 2001; Javanmardi et al., 2002; Lee and Scagel, 2009; Kwee and Niemeyer, 2011; Flanigan and Niemeyer, 2014). The purple cultivars of basil are rich in anthocyanins (Phippen and Simon, 1998; Kwee and Niemeyer, 2011). The major essential oil components of basil include linalool, methyl chavicol, 1,8-cineole, eugenol, methyl eugenol, methyl isoeugenol, thymol, methyl cinnamate, citral, and camphor (Kéita et al., 2000; Jirovetz et al., 2003; Bowes and Zheljazkov, 2004; Labra et al., 2004; Joshi, 2014; Farouk et al., 2016; Avetisyan et al., 2017; Perveen et al., 2020). The essential oil composition of basil is influenced by agroclimatic conditions, cultivars, harvesting, drying, and processing methods (Damalas, 2019; Milenkovic et al., 2019). The major fatty acid compositions of basil are linolenic, linoleic, oleic, and palmitic acids (Angers et al., 1996).

There are several other compounds reported from various laboratories including isoquercetin, quercetin-3-O-2″-gallate, quercetin-3-O-(2″-O-galloyl)-rutinoside, quercetin-3-O-α-L-rhamnoside, quercetin-3-O-β-D-galactoside, and syringin (Marwat et al., 2011); apigenin, genkwanin, ladanein, eupatorin, cirsimaritin, hispidulin, acacetin, apigenin 7,4′-dimethylether, salvigenin, nevadensin, and gardenin B (Grayer et al., 2001); kaempferol (de Lira Mota et al., 2009; Marwat et al., 2011); galuteolin (Hiltunen and Holm, 2003); (−) rabdosiin, shimobashiric acid A, 3-(3,4-dihydroxyphenyl)-2-hydroxypropanoic acid, and (E)-P-coumaroyl-4-O-β-D-glucoside (Flegkas et al., 2019); estragol, eugenol, and methyl eugenol (Vani et al., 2009); lithospermic acid (Tada et al., 1996); rosmarinic acid methyl ester (Tada et al., 1996; Bang et al., 2016; Singh and Chaudhuri, 2018); myristicin (Lee et al., 2007; Marwat et al., 2011); coumarin (Marwat et al., 2011; Jia et al., 2018); aesculetin (Chu et al., 2001; Marwat et al., 2011); aesculin (Chu et al., 2001; Lee et al., 2007); ocimarin (Singh and Chaudhuri, 2018); cyanidine-3,5-diglucoside, peonidin, peonidin-3,5-diglucoside, cyanidine-3-glucoside, cyanidin-3-(pcoumaryl glucoside), cyanidin 3-(pcoumaryl glucoside)-5-glucoside, and peonidin 3-(pcoumaryl glucoside)-5-glucoside (Phippen and Simon, 1998).

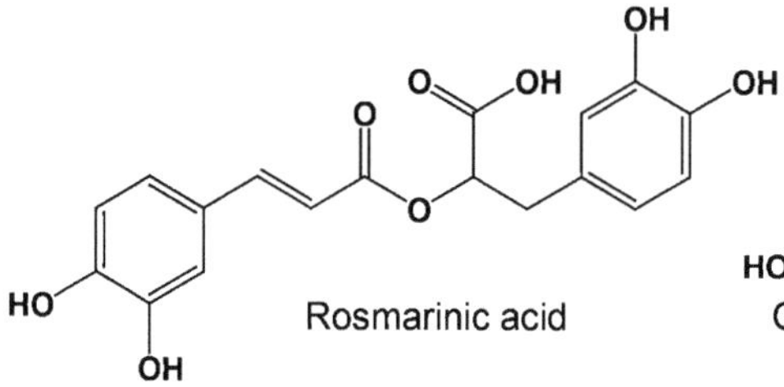
Rosmarinic acid

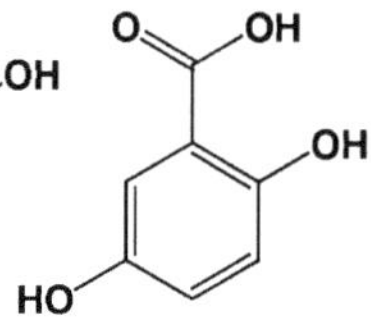
Gentisic acid

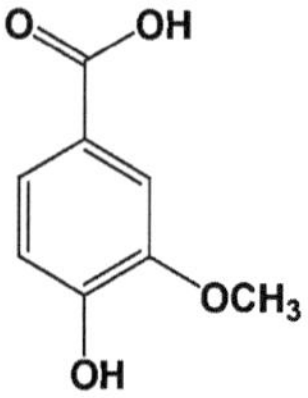
Vanillic acid

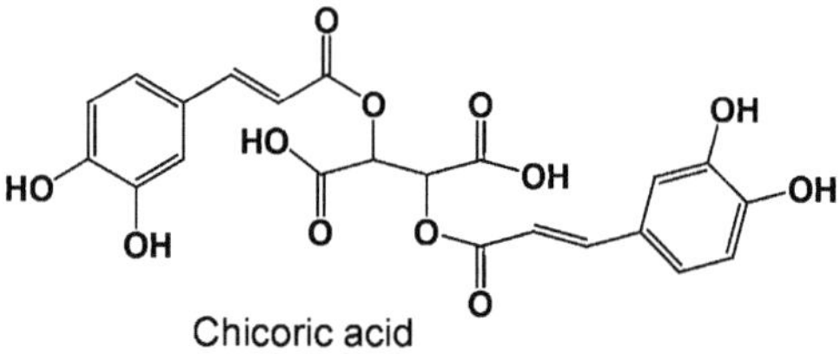
Chicoric acid

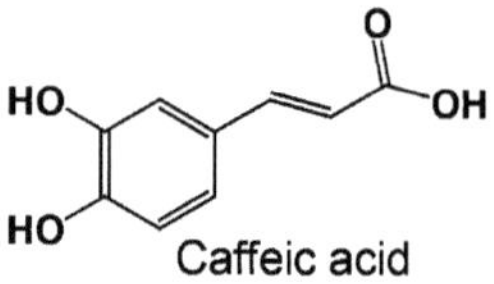
Caffeic acid

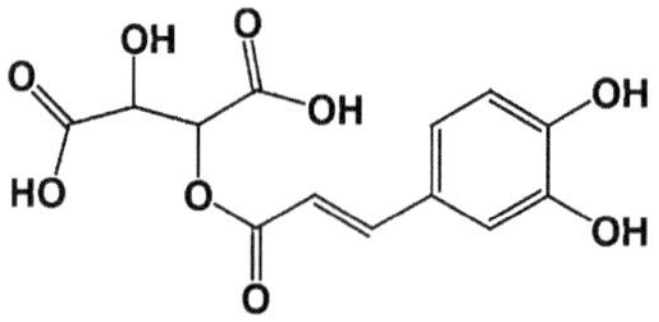
Caftaric acid

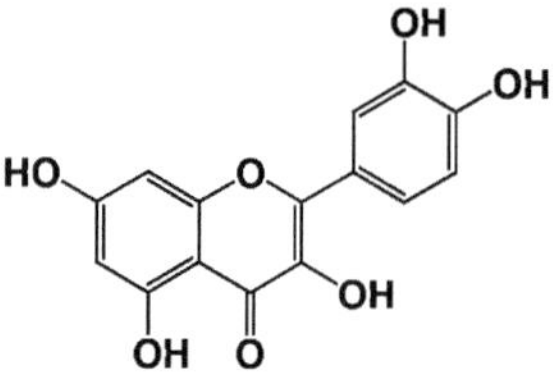
Quercetin

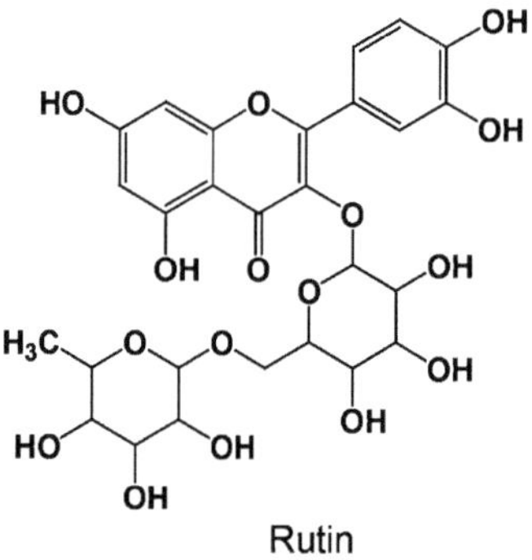
Rutin

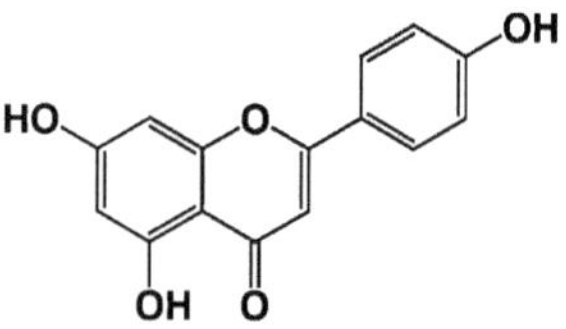
Apigenin

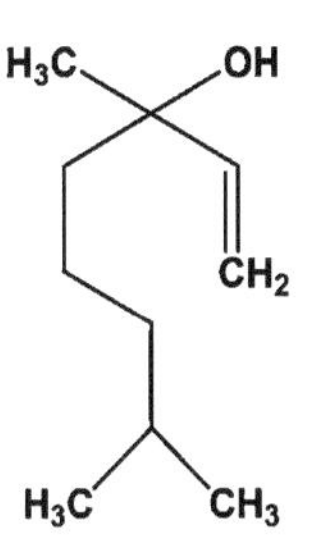

Linalool

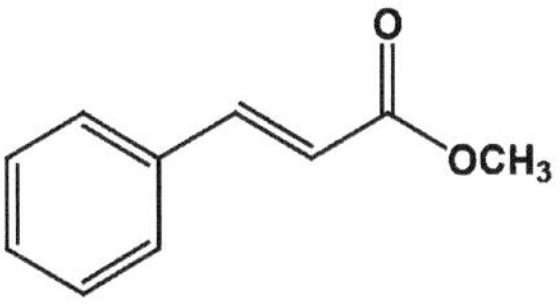

Methyl chavicol

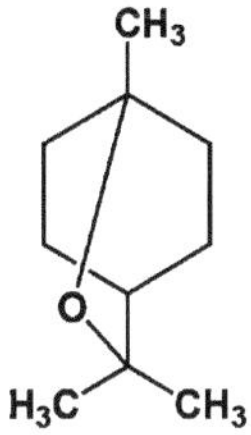

1,8-Cineole

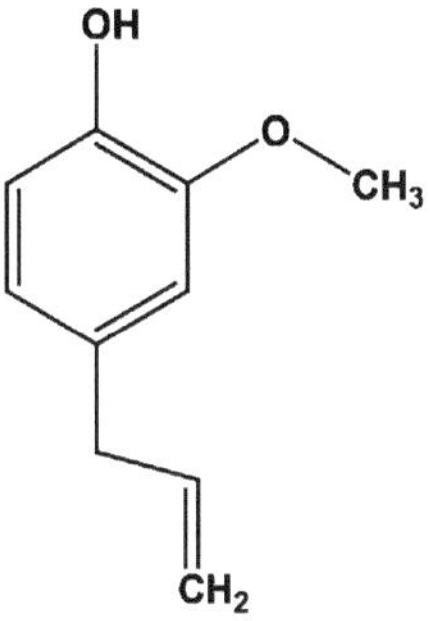

Eugenol

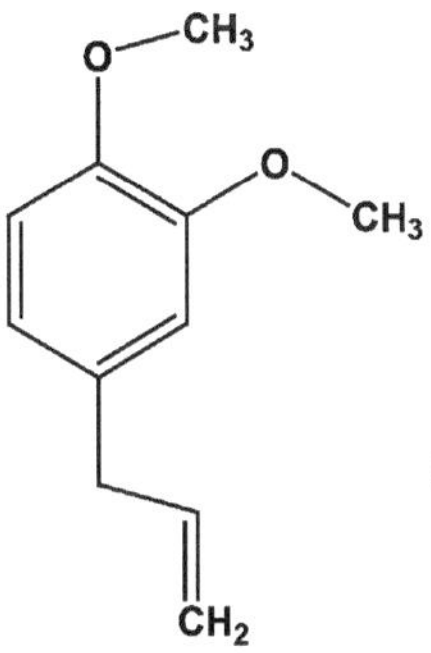

Methyl eugenol

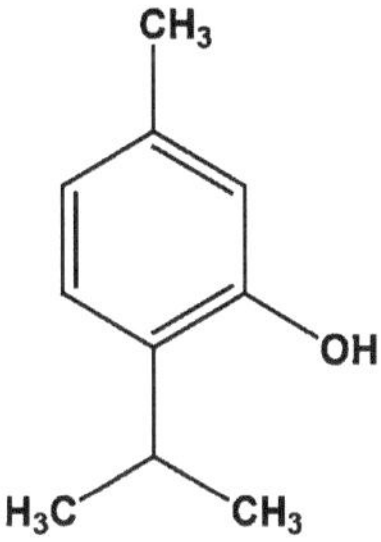

Thymol

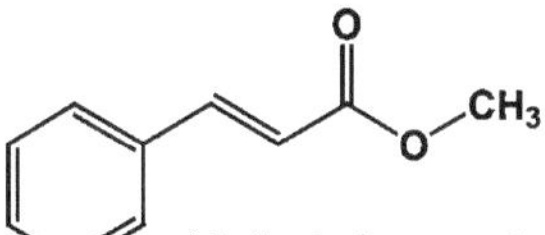

Methyl cinnamate

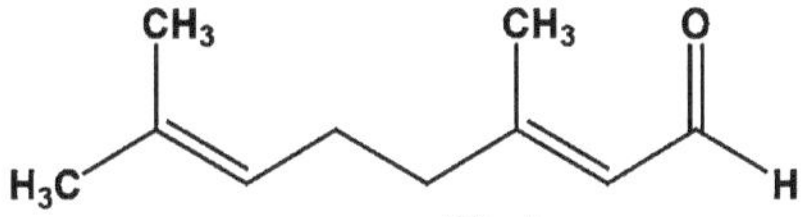

Citral

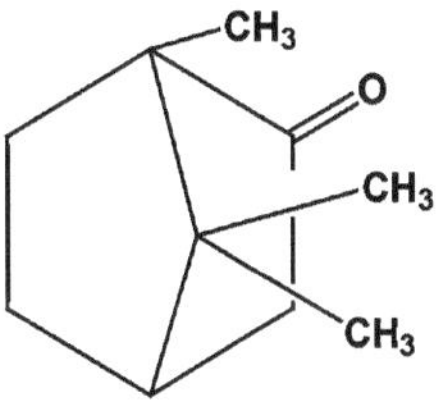

Camphor

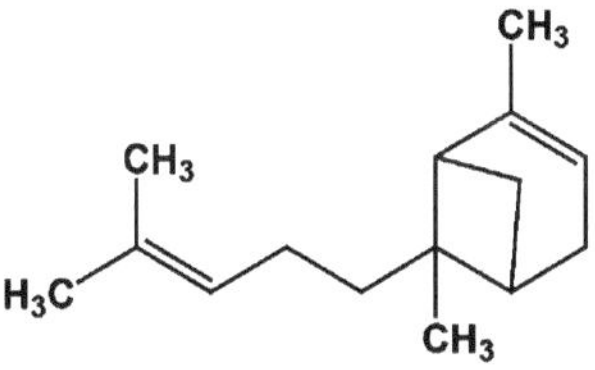

Bergamotene

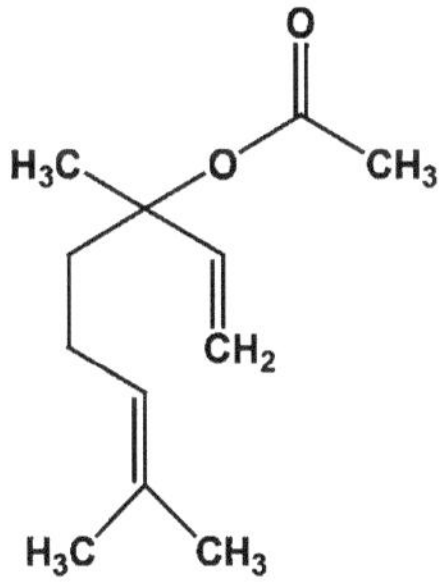

Linalyl acetate

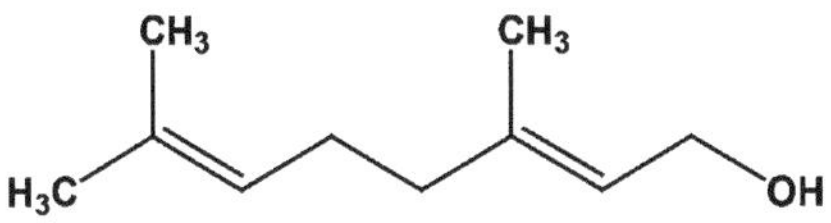

Nerol

14.3 PHARMACOLOGICAL STUDIES

14.3.1 CARDIOPROTECTIVE ACTIVITY

Short term exposure of human umbilical vein endothelial cells to *O. basilicum* leaf extract enhanced the production of prostacyclin metabolite 6-keto-PGF1α, a potent vasodilator (Umar et al., 2014). *O. basilicum* interferes with the mechanisms involved in the development of atherosclerotic diseases by acting as a vasodilator and reduces plaque aggregation (Benedec et al., 2007; Amrani et al., 2009). The alcoholic extract of *O. basilicum* showed cardioprotective and the aqueous extract showed ß-adrenergic effects in albino rats (Muralidharan and Dhananjayan, 2004). *O. basilicum* extract showed hypolipidemic effect stronger than of fenofibrate treatments (Amrani et al., 2006). At a dose of 100 mg/kg per day (4 weeks), *O. basilicum* extract reduced the systolic and diastolic blood pressure by 20 and 15 mm of Hg, respectively, in 2K1C Goldblatt renovascular hypertensive rats (Umar et al., 2010). Phenolic acid-rich leaf extract of *O. basilicum* at a dose of 10–40 mg/kg showed significant protection against isoproterenol-induced myocardial infarction in Wistar rats (Fathiazad et al., 2012). Administration of ethanolic extract of *O. basilicum* (100–400 mg/mL) activated the fibrinolytic system and exerted antithrombotic effects in carrageenan-induced mice model (Kuerban et al., 2017).

14.3.2 ANTI-INFLAMMATORY ACTIVITY

Rosmarinic acid-rich *O. basilicum* extract showed anti-inflammatory activity by inhibiting the level of proinflammatory cytokines (interleukin 6 and tumor necrosis factor α); mediators (cyclooxygenase-2, inducible nitric oxide synthase, and prostaglandin E_2); and increased anti-inflammatory cytokine levels (interleukin 10) (Mueller et al., 2010; Umar et al., 2014). The linolenic acid present in the fixed oil of *O. basilicum* seed showed significant inhibition in cyclooxygenases and lipoxygenases of arachidonate metabolism (Singh, 1998). *O. basilicum* fatty seed oil possesses potent anti-inflammatory activity against toxin-induced paw edema in rats (Singh, 1999a, 1999b). *O. basilicum* extract showed significant inhibition of inducible nitric oxide synthase and reactive nitrogen species in lipopolysaccharide-induced RAW264.7 cells (Tsai et al., 2007). *O. basilicum* tincture (1:10 in 70% ethanol) significantly reduced bone marrow acute phase response hampering phagocytes activity and decreased circulating polymorphonuclear neutrophils and macrophages in male Wistar rats treated with turpentine oil (Benedec et al., 2007). Estragole-rich *O. basilicum* essential oil was reported to show anti-inflammatory and antiedematogenic activity in acute and chronic inflammatory animal models (Rodrigues et al., 2016). Ethanol extract of leaf and leaf callus of *O. basilicum* reduced the nitric oxide production in lipopolysaccharide-induced RAW264.7 cells (Aye et al., 2019). The administration of *O. basilicum* essential oil along with β-cyclodextrin (10 mg/kg) improved the prevention of paw edema formation by decreasing vascular permeability, inhibited leukocyte recruitment to the peritoneal cavity, and inhibited granuloma formation in carrageenan-induced mice (Rodrigues et al., 2017).

14.3.3 ANALGESIC ACTIVITY

Venancio et al. (2011) and Nascimento et al. (2014) reported antinociceptive effect of *O. basilicum* essential oil in Swiss albino mice. Nascimento et al. (2014) ascribed analgesia not only to cyclooxygenase related mechanisms, but also to interactions with opioid receptors and to the activation of descending inhibitory pain pathways. Administration of *O. basilicum* essential oil with β-cyclodextrins produces more potent antinociceptive effect (Rodrigues et al., 2017). The methanolic extract of *O. basilicum* (200 mg/kg) showed a potent analgesic effect in mice (Mandoulakani et al., 2017).

14.3.4 SEDATIVE AND ANXIOLYTIC ACTIVITIES

The hydroethanolic extract of *O. basilicum* (250–500 mg/kg) increased the brain pool of enzymatic and nonenzymatic antioxidants (Muneefa et al., 2017). Intraperitoneal administration of essential oil (200 mg/kg b.w.) and hydroalcoholic extract (150–200 mg/kg) of *O. basilicum* exhibited relevant anxiolytic and sedative properties in male Syrian mice (Rabbani et al., 2015). Inhalation of *O. basilicum* essential oil significantly improved behavioral alterations, serum glucocorticoid levels, neurogenesis in the dentate gyrus, reduced neuronal and glial apoptosis, and downregulated the gene expression levels of glucocorticoid receptors and brain-derived neurotrophic factor in the hippocampus (Ayuob et al., 2017). The oral administration of leaf extract of *O. basilicum* (100 mg/kg) significantly improved neuromuscular coordination, object recognition ability, exploratory behavior, and short-term memory in male albino mice (Zahra et al., 2015). The administration of 25–100 mg/kg *O. basilicum* hydroalcoholic extracts to mice was found as effective as diazepam in increasing pentobarbital-induced sleep duration (Askari et al., 2016). Geraniol, linalool, and 1,8-cineole rich *O. basilicum* essential oil (100–400 mg/kg i.p.) significantly accelerated the onset of sleeping, prolonged duration, and reduced pentylenetetrazole- and picrotoxin-induced convulsions in mice (Oliveira et al., 2009).

14.3.5 BRONCHODILATOR ACTIVITY

O. basilicum aqueous methanolic extract produced relaxation in carbachol and K^+-induced tracheal contraction. The study proposed that the vasodilatory and bronchodilatory responses evoked by *O. basilicum* are likely to be mediated by verapamil-like Ca^{2+} channel-blocking features (Janbaz et al., 2014). *O. basilicum* extract significantly reduced the tracheal chain contractile tone stimulated by methacholine and airway inflammation (Boskabady et al., 2005).

14.3.6 ANTIDIABETIC ACTIVITY

In murine L6 skeletal muscle cells, *O. basilicum* extract (0.25 mg/mL) elevated the glucose transporter type-4 translocation by 3.5 and 7 folds in the absence or presence of insulin, respectively (Kadan et al., 2016). *O. basilicum* extract (100–400 mg/kg) significantly reduced the fasting blood glucose level, improved oral glucose tolerance, and increased liver glycogen reserves via the inhibition of α-glucosidase and α-amylase in alloxan-induced diabetic rats (Ezeani et al., 2017). El-Beshbishy and Bahashwan (2012) found that aqueous extract of *O. basilicum* was dose dependently inhibiting the enzymes responsible for carbohydrates hydrolysis.

14.3.7 ANTIOXIDANT ACTIVITY

The essential oils, crude extracts, and bioactive compounds isolated from *O. basilicum* have been reported to possess antioxidant activity (Javanmardi et al., 2003; Gulcin et al., 2007; Kaurinovic et al., 2011; Flanigan and Niemeyer, 2014). The methanol, ethanol, and water extracts of *O. basilicum* seeds showed potent antioxidant activity, and further found that the relative potency was linearly related to the total phenolic content in the extracts (Zakaria et al., 2008). Jayasinghe et al. (2003) found that the rosmarinic acid isolated from *O. basilicum* showed potent antioxidant activity in liposome system. They also found the synergistic antioxidant effect between alpha-tocopherol and rosmarinic acid. The ethanol extract of *O. basilicum* at a concentration of 500 μg/mL showed 75.87% antioxidant activity in vitro (Durga et al., 2009). Amrani et al. (2006) found that the aqueous extract of *O. basilicum* possessed a very high antioxidant power. Meera et al. (2009) found that the ethanolic leaf extract of *O. basilicum* significantly reduced lipid peroxidation in goat liver in vitro and also showed super oxide and nitric oxide radical scavenging activities.

14.3.8 HEPATOPROTECTIVE ACTIVITY

O. basilicum significantly reduced the liver hydroxyproline and increased the activity of hyaluronidase in carbon tetrachloride-induced liver toxicity in rats. The liver superoxide dismutase level was stimulated and the lipid peroxidation was significantly reduced. *O. basilicum* also significantly reduced the levels of liver function enzymes like aspartate aminotransferase, alanine aminotransferase, and alkaline phosphatase (Galila et al., 2012). Meera et al. (2009) reported hepatoprotective effect of ethanolic extract of leaves of *O. basilicum* in hydrogen peroxide- and carbon tetrachloride-induced hepatic damage. Ugwu et al. (2013) found that *O. basilicum* leaf extract (200 mg/kg) significantly prevented the lipid peroxidation, maintained normal levels of alanine and aspartate transferase, and increased superoxide dismutase and catalase activities in streptozotocin-induced rats.

14.3.9 GASTROPROTECTIVE ACTIVITY

The fixed oil of *O. basilicum* was found to possess significant inhibition in aspirin-induced gastric ulceration and secretion in pylorus-ligated rats (Singh, 1999a). The hydroalcoholic extract of *O. basilicum* ameliorated the clinical manifestations in patients with functional dyspepsia (Rafieian and Hosseini-asl, 2015). Rashidian et al. (2016) found that *O. basilicum* essential oil (200 and 400 μL/kg) showed significant reduction in ulcer severity, ulcer area, and ulcer index against acetic acid-induced colitis in rats. The hexane extract of *O. basilicum* (100 and 200 mg/kg) showed significant anxiolytic, antioxidant, and anti-inflammatory prophylactic effects on the aspirin-induced gastric ulcer (El-Ghffar et al., 2018).

14.3.10 ANTICANCER ACTIVITY

The ethanolic extract of dried *O. basilicum* showed a significant cytotoxic effect in human laryngeal epidermoid carcinoma cell line (Abirami and Nirmala, 2014). Eugenol isolated from *O. basilicum* showed cytotoxic activity against a broad range of different cancer cell lines through multiple mechanisms (Jaganathan and Supriyanto, 2012). Eugenol-activated caspases 3, 6, and 9 increased the levels of p53, Bax, Bid, Bad, release of cytochrome c, and decreased Bcl2 and Bcl-xL expression (Hussain et al., 2012). In animal model, eugenol (100 mg/kg, i.p.) inhibited the growth of solid Ehrlich tumors (Jaganathan et al., 2010). Ursolic acid (100 μM) isolated from *O. basilicum* induced a significant decrease in the percentage of cells in anaphase/telophase stages along with F-actin aggregation and mitotic spindle distortion in MCF-7 cells (Qamar et al., 2010). The hydroalcoholic extract of *O. basilicum* (200 mg/kg) was reported to reduce the volume and the incidence of B16F10 melanoma cell-induced metastases in C57BL mice (Monga et al., 2011). *O. basilicum* leaf powder (1% in basal diet) significantly reduced colon tumorigenesis in azoxymethane-treated Fisher 344 male rats, by increasing the cellular antioxidant status (Gajula et al., 2010).

14.3.11 ANTIMICROBIAL ACTIVITY

The essential oil obtained from *O. basilicum* inhibited *Bacillus subtilis*, *Staphylococcus aureus*, *Aspergillus niger*, *Candida albicans*, *Salmonella typhi*, and *Escherichia coli* growth with minimum inhibitory concentration ranging from 40 to 160 μg/mL (Shirazi et al., 2014). Linalool, eugenol, and τ-cadinol rich *O. basilicum* essential oil showed dose-dependent activity against *S. aureus* and *E. coli* (Stefan et al., 2013). Kadhim et al. (2016) showed that exposure of *S. aureus*, *E. coli*, *Klebsiella pneumoniae*, and *Pseudomonas aeruginosa* to ethanolic extract of *O. basilicum* caused significant inhibition (4.17–5.70 mm). Opalchenova and Obreshkova (2003) found that *O. basilicum* essential oil showed strong inhibitory activity against multi-drug resistant *Staphylococcus*, *Pseudomonas*, and *Enterococcus* strains, with MICs ranging from 0.003 to 0.007% (v/v). The methanol extract of *O. basilicum* showed inhibition against *P. aeruginosa*, *Shigella* sp., *S. aureus*, *E.*

coli, and *Listeria monocytogenes* (Kaya et al., 2008). The methanol extract of *O. basilicum* showed potent activity against *Mycobacterium tuberculosis* and *M. bovis*, with minimum inhibitory concentration ranging between 25–100 and 25–50 μg/mL, respectively (Gemechu et al., 2013). Bacilicin and (E)-3′-hydroxy-4′-(1′-hydroxyethyl)-phenyl-4-methoxycinnamate isolated from *O. basilicum* showed 38 and 49% of inhibition, respectively in *M. tuberculosis* H37Rv at a concentration of 6.25 μg/mL (Siddiqui et al., 2012). The methanol extract of *O. basilicum* showed activity against *B. cereus*, *B. subtilis*, *B. megaterium*, *S. aureus*, *L. monocytogenes*, *Vibrio parahaemolyticus*, and *V. mimicus* with diameters of inhibition zones 20.1, 17.2, 18.1, 16.1, 16.2, 15.4, and 15.2 mm, respectively (Hossain et al., 2010).

Ocimum basilicum aqueous alcoholic leaf extract is reported to cause a 95.8% growth inhibition of *Aspergillus flavus* NKD-235 and to abrogate aflatoxin B1 (AFB1) production at 3.1 μg/g concentration (Shukla et al., 2012). Concentration-dependent antifungal and AFB1-production inhibitory activities of *O. basilicum* essential oil were also reported by El-Soud et al. (2015). A complete inhibition of *A. flavus* growth along with a significant reduction of AFB1 production was found at 1000 ppm oil concentration. The aqueous and ethanolic extracts of *O. basilicum* and its purified components, namely ursolic acid, apigenin, and linalool exhibited significant antiviral activities against herpes virus, hepatitis virus, adenovirus, RNA virus, and enterovirus (Chiang et al., 2005). *Ocimum basilicum* essential oil was shown to inhibit the growth of *Trichomonas vaginalis* trophozoites, with a minimal lethal concentration of 10 μg/mL after 96 h (Eldin and Badawy, 2015). *O. basilicum* has found to be effective against drug-resistant strains of *Plasmodium falciparum* (Ntonga et al., 2014). Inbaneson et al. (2012) found that *O. basilicum* extract inhibited 50% of *Plasmodium* population at a concentration of 43.81 μg/mL. Chavan and Nikam (1982) found that methyl cinnamate (70%) and linalool (20%) rich *O. basilicum* essential oil showed 100% larvicidal activity against *Culex quinquefasciatus* larvae at 120 ppm. The essential oil of *O. basilicum* and its purified component linalool showed antigiardial activity. Linalool at a dose of 300 μg/mL showed 100% mortality of parasites after 1 h of incubation (de Almeida et al., 2007).

14.3.12 LARVICIDAL ACTIVITY

The petroleum ether extract of the leaves of *O. basilicum* was effective against the *Anopheles stephensi* and *Culex quinquefasciatus* larvae with LC_{50}

values of 8.29 and 4.57 ppm, and 10.06 and 6.06 ppm observed after 24 and 48 h of exposure, respectively (Prejwltta et al., 2009). The methanol extract of *O. basilicum* at concentrations of 2, 4, 6, 8, and 10% exhibited larvicidal activities against different instars (I, II, III, & IV) and pupa of *Aedes aegypti*. The LC_{50} value against I instar larvae was 10.059%, II instar 12.085%, III instar 13.233%, and IV instar 13.824%, respectively. The extract also showed considerable pupal mortality. The LC_{50} value of pupae was 16.925% (Murugan et al., 2007). Momen and Amer (2003) found that *O. basilicum* essential oil was highly toxic to phytoseiid mite *Euseius yousefi* female.

14.3.13 TOXICITY STUDIES

In an acute oral toxicity study, Parra et al. (2001) found that the medium lethal dose (LD_{50}) value of the hydroalcoholic extract of *O. basilicum* was 956.5 mg/kg body weight in mice. Rasekh et al. (2012) found that in acute toxicity study in Wistar rats, the hydroalcoholic extract of *O. basilicum* did not show any mortality or toxic signs up to a dose of 2 g/kg body weight. As per the European Food Safety Authority (2009), the maximum recommended dose of *O. basilicum* is dried leaf 2–4 g, 2–3 times/day or equivalent quantity in extract. *O. basilicum* essential oil showed good safety profile in acute and subacute toxicity studies in Wistar rats. Toxicity signs like erosions to stomach mucosa, muscular atrophy, and significant functional damage to liver tissues were observed only with doses more than 2000 mg/kg body weight (Fandohan et al., 2008; Venancio et al., 2011). Miller et al. (1983) reported increased incidence of hepatomas in estragole- (54 and 107 mg/kg body weight for 1 year) treated female rats. Chronic oral intake of high doses of methyleugenol (26, 54, and 107 mg/kg body weight for 2 years) resulted in an increased incidence of liver and stomach neoplasms in male and female F344/N rats and B6C3F1 mice (National Toxicology Program, 2000).

ACKNOWLEDGMENTS

The authors express their sincere thanks to Dr. Ashok K. Chauhan, Founder President, Ritnand Balved Education Foundation (RBEF) and Amity Group of Institutions, and Dr. Atul Chauhan, Chancellor, Amity University Uttar Pradesh (AUUP) for facilitating this work. Thadiyan Parambil Ijinu has received Young Scientist Fellowship from the Department of Science and Technology, Government of India (SP/YO/413/2018).

KEYWORDS

- **traditional medicinal use**
- **rosmarinic acid**
- **gentisic acid**
- **linalool**
- **cardioprotective activity**
- **anti-inflammatory activity**
- **bronchodilator**

REFERENCES

Abirami, S. G.; Nirmala, P. A Comparative In Vitro Study of Anticancer Effect of *Mentha piperita, Ocimum basilicum* and *Coleus aromaticus* Against Human Laryngeal Epidermoid Carcinoma (HEP-2) Cell Lines. *J. Med. Plants Stud.* **2014,** *2*, 6–9.

Amrani, S.; Harnafi, H.; Bouanani, N. E. H.; Aziz, M.; Caid, H. S.; Manfredini, S.; Besco, E.; Napolitano, M.; Bravo, E. Hypolipidaemic Activity of Aqueous *Ocimum basilicum* Extract in Acute Hyperlipidaemia Induced by Triton WR-1339 in Rats and Its Antioxidant Property. *Phytother. Res.* **2006,** *20*, 1040–1045.

Amrani, S.; Harnafi, H.; Gadi, D.; Mekhfi.; Legssyer, A.; Aziz, M.; Martin-Nizard, F.; Bosca, L. Vasorelaxant and Anti-Platelet Aggregation Effects of Aqueous *Ocimum basilicum* Extract. *J. Ethnopharmacol.* **2009,** *125*, 157–162.

Angers, P.; Morales, M. R.; Simon, J. E. Fatty Acid Variation in Seed Oil Among *Ocimum* Species. *J. Am. Oil Chem. Soc.* **1996,** *73*, 393–395.

Askari, V. R.; Rahimi, V. B.; Ghorbani, A.; Rakhshandeh, H. Hypnotic Effect of *Ocimum basilicum* on Pentobarbital-Induced Sleep in Mice. *Iran. Red Crescent Med. J.* **2016,** *18*, e24261.

Avetisyan, A.; Markosian, A.; Petrosyan, M.; Sahakyan, N.; Babayan, A.; Aloyan, S.; Trchounian, A. Chemical Composition and Some Biological Activities of the Essential Oils from Basil *Ocimum* Different Cultivars. *BMC Complement. Altern. Med.* **2017,** *17*, 60.

Aye, A.; Jeon, Y. D.; Lee, J. H.; Bang, K. S.; Jin, J. S. Anti-Inflammatory Activity of Ethanol Extract of Leaf and Leaf Callus of Basil (*Ocimum basilicum* L.) on RAW 264.7 Macrophage Cells. *Orient. Pharm. Exp. Med.* **2019,** *19*, 217–226.

Ayuob, N. N.; Firgany, A. E. L.; El-Mansy, A. A.; Ali, S. Can *Ocimum basilicum* Relieve Chronic Unpredictable Mild Stress-Induced Depression in Mice? *Exp. Mol. Pathol.* **2017,** *103*, 153–161.

Bang, S.; Ha, T. K. Q.; Lee, C.; Li, W.; Oh, W. K.; Shim, S. H. Antiviral Activities of Compounds from Aerial Parts of Salvia plebeia R. Br. *J. Ethnopharmacol* **2016,** *192*, 398–405.

Benedec, D.; Pârvu, A. E.; Oniga, I.; Toiu, A.; Tiperciuc, B. Effects of *Ocimum basilicum* L. Extract on Experimental Acute Inflammation. *Rev. Med. Chir. Soc. Med. Nat. Iasi.* **2007,** *111*, 1065–1069.

Boskabady, M. H.; Kiani, S.; Haghiri, B. Relaxant Effects of *Ocimum basilicum* on Guinea Pig Tracheal Chains and Its Possible Mechanism (s). *Daru.* **2005,** *13*, 28–33.

Bowes, K. M.; Zheljazkov, V. D. Factors Affecting Yields and Essential Oil Quality of *Ocimum sanctum* L. and *Ocimum basilicum* L. Cultivars. *J. Am. Soc. Hortic. Sci.* **2004,** *129*, 789–794.

Chavan, S. R.; Nikam, S. T. Mosquito Larvicidal Activity of *Ocimum basilicum* Linn. *Indian J. Med. Res.* **1982,** *75*, 220–222.

Chiang, L. C.; Ng, L. T.; Cheng, P. W.; Chiang, W.; Lin, C. C. Antiviral Activities of Extracts and Selected Pure Constituents of *Ocimum basilicum. Clin. Exp. Pharmacol. Physiol.* **2005,** *32*, 811–816.

Chopra, R. N.; Nayar, S. L.; Chopra, I. C. *Glossary of Indian Medicinal Plants*; Council of Scientific and Industrial Research: New Delhi, India, 1956; p 329.

Chu, C. Y.; Tsai, Y. Y.; Wang, C. J.; Lin, W. L.; Tseng, T. H. Induction of Apoptosis by Esculetin in Human Leukemia Cells. *Eur. J. Pharmacol.* **2001,** *416*, 25–32.

Damalas, C. A. Improving Drought Tolerance in Sweet Basil (*Ocimum basilicum*) with Salicylic Acid. *Sci. Hortic.* **2019,** *246*, 360–365.

de Almeida, I.; Alviano, D. S.; Vieira, D. P.; Alves, P. B.; Blank, A. F.; Lopes, A. H.; Alviano, C. S.; Rosa Mdo, S. Antigiardial Activity of *Ocimum basilicum* Essential Oil. *Parasitol. Res.* **2007,** *101*, 443–452.

de Lira Mota, K. S.; Dias, G. E. N.; Pinto, M. E. F.; Luiz-Ferreira, A.; Monteiro Souza-Brito, A. R.; Hiruma-Lima, C. A.; Barbosa-Filho, J. M.; Batista, L. M. Flavonoids with Gastroprotective Activity. *Molecules* **2009,** *14*, 979–1012.

Durga, K. R.; Karthikumar, S.; Jegatheesan, K. Isolation of Potential Antibacterial and Antioxidant Compounds from *Acalypha indica* and *Ocimum basilicum. J. Med. Plant Res.* **2009,** *3*, 703–706.

El-Beshbishy, H.; Bahashwan, S. Hypoglycemic Effect of Basil (*Ocimum basilicum*) Aqueous Extract Is Mediated Through Inhibition of Alpha-Glucosidase and Alpha-Amylase Activities: An In Vitro Study. *Toxicol. Ind. Health* **2012,** *28*, 42–50.

Eldin, H. M. E.; Badawy, A. F. In Vitro Anti-*Trichomonas vaginalis* Activity of *Pistacia lentiscus* Mastic and *Ocimum basilicum* Essential Oil. *J. Parasit. Dis.* **2015,** *39*, 465–473.

El-Ghffar, E. A. A.; Al-Sayed, E.; Shehata, S. M.; Eldahshan, O. A.; Efferthd, T. The Protective Role of *Ocimum basilicum* L. (Basil) Against Aspirin-Induced Gastric Ulcer in Mice: Impact on Oxidative Stress, Inflammation, Motor Deficits and Anxiety-Like Behavior. *Food Funct.* **2018,** *9*, 4457–4468.

El-Soud, N. H.; Deabes, M.; El-Kassem, L. A.; Khalil, M. Chemical Composition and Antifungal Activity of *Ocimum basilicum* L. Essential Oil. *Open Access Maced. J. Med. Sci.* **2015,** *3*, 374–379.

European Food Safety Authority. Scientific Opinion on the Substantiation of Health Claims Related to *Ocimum basilicum* L. and Improvement of Diuretic Function (ID 2314, 3465) Pursuant to Article 13 (1) of Regulation (EC) No 1924/2006. *EFSA J.* **2009,** *7*, 1298.

Ezeani, C.; Ezenyi, I.; Okoye, T.; Okoli, C. *Ocimum basilicum* Extract Exhibits Antidiabetic Effects via Inhibition of Hepatic Glucose Mobilization and Carbohydrate Metabolizing Enzymes. *J. Intercult. Ethnopharmacol.* **2017,** *6*, 22–28.

Fandohan, P.; Gnonlonfin, B.; Laleye, A.; Gbenou, J. D.; Darboux, R.; Moudachirou, M. Toxicity and Gastric Tolerance of Essential Oils from *Cymbopogon citratus*, *Ocimum gratissimum* and *Ocimum basilicum* in Wistar Rats. *Food Chem. Toxicol.* **2008,** *46*, 2493–2497.

Farouk, A.; Fikry, R.; Mohsen, M. Chemical Composition and Antioxidant Activity of *Ocimum basilicum* L. Essential Oil Cultivated in Madinah Monawara, Saudi Arabia and Its Comparison to the Egyptian Chemotype. *J. Essent. Oil-Bear. Plants*. **2016,** *19*, 1119–1128.

Fathiazad, F.; Matlobi, A.; Khorrami, A.; Hamedeyazdan, S.; Soraya, H.; Hammami, M.; Maleki-Dizaji, N.; Garjani, A. Phytochemical Screening and Evaluation of Cardioprotective Activity of Ethanolic Extract of *Ocimum basilicum* L. (Basil) Against Isoproterenol Induced Myocardial Infarction in Rats. *Daru*. **2012,** *20*, 87.

Flanigan, P. M.; Niemeyer, E. D. Effect of Cultivar on Phenolic Levels, Anthocyanin Composition, and Antioxidant Properties in Purple Basil (*Ocimum basilicum* L.). *Food Chem.* **2014,** *164*, 518–526.

Flegkas, A.; Ifantis, T. M.; Barda, C.; Samara, P.; Tsitsilonis, O.; Skaltsa, H. Antiproliferative Activity of (–)-Rabdosiin Isolated from *Ocimum sanctum* L. *Medicines.* **2019,** *6*, 1–10.

Gajula, D.; Verghese, M.; Boateng, J.; Shackelford, L.; Mentreddy, S. R.; Sims, C.; Asiamah, D.; Walker, L. T. Basil (*Ocimum basilicum* and *Ocimum tenuiflorum*) Reduces Azoxymethane Induced Colon Tumors in Fisher 344 Male Rats. *Res. J. Phytochem.* **2010,** *4*, 136–145.

Galila, A. Y.; Nihal, M. E.; Eman, F. E. A. Hepatoprotective Effect of Basil (*Ocimum basilicum* L.) on CCl_4-Induced Liver Fibrosis in Rats. *Afr. J. Biotechnol.* **2012,** *11*, 15702–15711.

Gemechu, A.; Giday, M.; Worku, A.; Ameni, G. In Vitro Anti-Mycobacterial Activity of Selected Medicinal Plants Against *Mycobacterium tuberculosis* and *Mycobacterium bovis* Strains. *BMC Complement. Altern. Med.* **2013,** *13*, 291.

George, V.; Pushpangadan, P.; Ijinu, T. P. Basil. In *Genesis and Evolution of Horticultural Crops*; Peter, K. V. Ed., Vol. 1; Grugger Brentt Publishers UK Ltd., 2017; pp 67–84.

Grayer, R. J.; Veitch, N. C.; Kite, G. C.; Price, A. M.; Kokubun, T. Distribution of 8-Xygenated Leaf-Surface Flavones in the Genus *Ocimum*. *Phytochem.* **2001,** *56*, 559–567.

Gulcin, I.; Elmastas, M.; Aboul-Enein, H. Y. Determination of Antioxidant and Radical Scavenging Activity of Basil (*Ocimum basilicum* L. Family Lamiaceae) Assayed by Different Methodologies. *Phytother. Res.* **2007,** *21*, 354–361.

Hiltunen, R.; Holm, Y. *Basil*: The Genus *Ocimum*; CRC Press: Amsterdam, 2003; pp 1–146.

Hossain, M. A.; Kabir, M. J.; Salehuddin, S. M.; Mizanur Rahman, S. M.; Das, A. K.; Singha, S. K.; Alam Md, K.; Rahman, A. Antibacterial Properties of Essential Oils and Methanol Extracts of Sweet Basil *Ocimum basilicum* Occurring in Bangladesh. *Pharm. Biol.* **2010,** *48*, 504–511.

Hussain, A.; Priyani, A.; Sadrieh, L.; Brahmbhatt, K.; Ahmed, M.; Sharma, C. Concurrent Sulforaphane and Eugenol Induces Differential Effects on Human Cervical Cancer Cells. *Integr. Cancer Ther.* **2012,** *11*, 154–165.

Ijinu, T. P.; Aswany, T.; George, V.; Pushpangadan, P. Basil. In *Greens, Sprouts and Edible Flowers*; Singh, B., Peter, K. V., Eds.; Brillion Publishing: New Delhi**,** India, 2021; pp.51–66.

Inbaneson, S. J.; Sundaram, R.; Suganthi, P. In Vitro Antiplasmodial Effect of Ethanolic Extracts of Traditional Medicinal Plant *Ocimum* Species Against *Plasmodium falciparum*. *Asian Pac. J. Trop. Med.* **2012,** *5*, 103–106.

Jaganathan, S. K.; Mondhe, D.; Wani, Z. A.; Pal, H. C.; Mandal, M. Effect of Honey and Eugenol on Ehrlich Ascites and Solid Carcinoma. *J. Biomed. Biotechnol.* **2010,** 989163.

Jaganathan, S. K.; Supriyanto, E. Antiproliferative and Molecular Mechanism of Eugenol Induced Apoptosis in Cancer Cells. *Molecules* **2012,** *17*, 6290–6304.

Janbaz, K. H.; Hamid, I.; Gilani, A. H.; Qadir, M. I. Spasmolytic, Bronchodilator and Vasodilator Activities of Aqueous-Methanolic Extract of *Ocimum basilicum. Int. J. Agric. Biol.* **2014,** *16*, 321–327.

Javanmardi, J.; Khalighi, A.; Kashi, A.; Bais, H. P.; Vivanco, J. M. Chemical Characterization of Basil (*Ocimum basilicum* L.) found in Local Accessions and Used in Traditional Medicines in Iran. *J. Agric. Food Chem.* **2002,** *50*, 5878–5883.

Javanmardi, J.; Stushnoff, C.; Locke, E.; Vivanco, J. M. Antioxidant Activity and Total Phenolic Content of Iranian *Ocimum* Accessions. *Food Chem.* **2003,** *83*, 547–550.

Jayasinghe, C.; Gotoh, N.; Aoki, T.; Wada, S. Phenolics Composition and Antioxidant Activity of Sweet Basil (*Ocimum basilicum* L.). *J. Agric. Food. Chem.* **2003,** *51*, 4442–4449.

Jia, C.; Zhang, J.; Yu, L.; Wang, C.; Yang, Y.; Rong, X.; Xu, K.; Chu, M. Antifungal Activity of Coumarin Against *Candida albicans* Is Related to Apoptosis. *Front. Cell Infect. Microbiol.* **2018,** *8*, 1–13.

Jirovetz, L.; Buchbauer, G.; Shafi, M. P.; Kaniampady, M. M. Chemotaxonomical Analysis of the Essential Oil Aroma Compounds of Four Different *Ocimum* Species from Southern India. *Eur. Food Res. Technol.* **2003,** *217*, 120–124.

Joshi, R. K. Chemical Composition and Antimicrobial Activity of the Essential Oil of *Ocimum basilicum* L. (Sweet Basil) from Western Ghats of North West Karnataka, India. *Anc. Sci. Life.* **2014,** *33*, 151–156.

Kadan, S.; Saad, B.; Sasson, Y.; Zaid, H. In Vitro Evaluation of Anti-Diabetic Activity and Cytotoxicity of Chemically Analysed *Ocimum basilicum* Extracts. *Food Chem.* **2016,** *196*, 1066–1074.

Kadhim, M. J.; Sosa, A. A.; Hameed, I. H. Evaluation of Anti-Bacterial Activity and Bioactive Chemical Analysis of *Ocimum basilicum* Using Fourier Transform Infrared (FT-IR) and Gas Chromatography-Mass Spectrometry (GC-MS) Techniques. *J. Pharmacognosy Phytother.* **2016,** *8*, 127–146.

Kaurinovic, B.; Popovic, M.; Vlaisavljevic, S.; Trivic, S. Antioxidant Capacity of *Ocimum basilicum* L. and *Origanum vulgare* L. Extracts. *Molecules.* **2011,** *16*, 7401–7414.

Kaya, I.; Yigit, N.; Benli, M. Antimicrobial Activity of Various Extracts of *Ocimum basilicum* L. and Observation of the Inhibition Effect on Bacterial Cells by Use of Scanning Electron Microscopy. *Afr. J. Tradit. Complement. Altern. Med.* **2008,** *5*, 363–369.

Kéita, S. M.; Vincent, C.; Schmit, J. P.; Bélanger, A. Essential Oil Composition of *Ocimum basilicum* L., *O. gratissimum* L. and *O. suave* L. in the Republic of Guinea. *Flavour Frag. J.* **2000,** *15*, 339–341.

Kirtikar, K. R.; Basu, B. D. *Indian Medicinal Plants*, Vol. III, 2nd ed.; Lalit Mohan Basu: Allahabad, 1935; pp. 1959–1968.

Kuerban, A.; Almulaiky, Y. Q.; Ryan, A. S.; Alzahrani, S. A.; Al-Talhi, H. A.; Nasrullah, M.; Moselhy, S. S.; Abulnaja, K. O. In Vivo Antithrombotic Activity of Ethanolic Extract from *Ocimum basilicum* L. *Annu. Res. Rev. Biol.* **2017,** *20*, 1–6.

Kwee, E. M.; Niemeyer, E. D. Variations in Phenolic Composition and Antioxidant Properties Among Fifteen Basil (*Ocimum basilicum* L.) Cultivars. *Food Chem.* **2011,** *128*, 1044–1050.

Labra, M.; Miele, M.; Ledda, B.; Grassi, F.; Mazzei, M.; Sala, F. Morphological Characterization, Essential Oil Composition and DNA Genotyping of *Ocimum basilicum* L. Cultivars. *Plant Sci.* **2004,** *167*, 725–731.

Lee, B. C.; Lee, S. Y.; Lee, H. J.; Sim, G. S.; Kim, J. H.; Kim, J. H.; Cho, Y. H.; Lee, D. H.; Pyo, H. B.; Choe, T. B. Anti-Oxidative and Photo-Protective Effects of Coumarins Isolated from *Fraxinus chinensis*. *Arch. Pharm. Res.* **2007,** *30*, 1293

Lee, J.; Scagel, C. F. Chicoric Acid Found in Basil (*Ocimum basilicum* L.) Leaves. *Food Chem.* **2009,** *115*, 650–656.

Mandoulakani, B. A.; Eyvazpour, E.; Ghadimzadeh, M. The Effect of Drought Stress on the Expression of Key Genes Involved in the Biosynthesis of Phenylpropanoids and Essential Oil Components in Basil (*Ocimum basilicum* L.). *Phytochemistry* **2017,** *139*, 1–7.

Marwat, S. K.; Khan, M. S.; Ghulam, S.; Anwar, N.; Mustafa, G.; Usman, K. Phytochemical Constituents and Pharmacological Activities of Sweet Basil-*Ocimum basilicum* L. (Lamiaceae). *Asian J. Chem.* **2011,** *23*, 3773–3782

Meera, R.; Devi, P.; Kameswari, B.; Madhumitha, B.; Merlin, N. J. Antioxidant and Hepatoprotective Activities of *Ocimum basilicum* Linn. and *Trigonella foenum-graecum* Linn. Against H_2O_2 and CCl_4 Induced Hepatotoxicity in Goat Liver. *Indian J. Exp. Biol.* **2009,** *47*, 584–590.

Milenkovic, L.; Stanojevic, J.; Cvetkovic, D.; Stanojevic, L.; Lalevic, D.; Sunic, L.; Fallik, E.; Ilic, Z. S. New Technology in Basil Production with High Essential Oil Yield and Quality. *Ind. Crops Prod.* **2019,** *140*, 111718.

Miller, E. C.; Swanson, A. B.; Phillips, D. H.; Fletcher, T. L.; Liem, A.; Miller, J. A. Structure-Activity Studies of the Carcinogenicities in the Mouse and Rat of Some Naturally Occurring and Synthetic Alkenylbenzene Derivatives Related to Safrole and Estragole. *Cancer Res.* **1983,** *43*, 1124–1134.

Momen, F. M.; Amer, S. A. A. Influence of the Sweet Basil *Ocimum basilicum* L. on Some Predacious Mites of the Family Phytoseiidae (Acari: Phytoseiidae). *Acta Phytopathol. Entomol. Hung.* **2003,** *38*, 137–143.

Monga, J.; Sharma, M.; Tailor, N.; Ganesh, N. Antimelanoma and Radioprotective Activity of Alcoholic Aqueous Extract of Different Species of *Ocimum* in C (57) BL Mice. *Pharm. Biol.* **2011,** *49*, 428–436.

Mueller, M.; Hobiger, S.; Jungbauer, A. Anti-Inflammatory Activity of Extracts from Fruits, Herbs and Spices. *Food Chem.* **2010,** *122*, 987–996.

Muneefa, K.; Doss, V.; Sowndarya, R. Beneficial Effect of Hydroethanolic Extract of *Ocimum basilicum* L on Enzymic and Non Enzymic Antioxidant in Depression Induced Rats. *J. Med. Plants Stud.* **2017,** *5*, 185–188.

Muralidharan, A.; Dhananjayan, R. Cardiac Stimulant Activity of *Ocimum basilicum* Linn. Extracts. *Indian J. Pharmacol.* **2004,** *36*, 163–166.

Murugan, K.; Murugan, P.; Noortheen, A. Larvicidal and Repellent Potential of *Albizzia amara* Boivin and *Ocimum basilicum* Linn. Against Dengue Vector, *Aedes aegypti* (Insecta: Diptera: Culicidae). *Bioresour. Technol.* **2007,** *98*, 198–201.

Nascimento, S. S.; Araújo, A. A. S.; Brito, R. G.; Serafini, M. R.; Menezes, P. P.; DeSantana, J. M.; Lucca Jr, W.; Alves, P. B.; Blank, A. F.; Oliveira, R. C. M.; Oliveira, A. P.; Albuquerque Jr, R. L. C.; Almeida, J. R. G. S.; Quintans Jr, L. J. Cyclodextrin-complexed *Ocimum basilicum* Leaves Essential Oil Increases Fos Protein Expression in the Central Nervous System and Produce an Antihyperalgesic Effect in Animal Models for fibromyalgia. *Int. J. Mol. Sci.* **2014,** *16*, 547–563.

National Toxicology Program. Toxicology and Carcinogenesis Studies of Methyleugenol (CAS NO. 93–15–2) in F344/N Rats and B6C3F1 Mice (Gavage Studies). US-National Institutes of Health, National Toxicology Program Technical Report Series. **2000,** *491*, 1.

Ntonga, P. A.; Baldovini, N.; Mouray, E.; Mambu, L.; Belong, P.; Grellier, P. Activity of *Ocimum basilicum*, *Ocimum canum*, and *Cymbopogon citratus* Essential Oils Against

Plasmodium falciparum and Mature-Stage Larvae of *Anopheles funestus* s.s. *Parasite* **2014,** *21*, 33.

Oliveira, J. S.; Porto, L. A.; Estevam, C. S.; Siqueira, R. S.; Alves, P. B.; Niculau, E. S.; Blank, A. F.; Almeida, R. N.; Marchioro, M.; Quintans Jr, L. J. Phytochemical Screening and Anticonvulsant Property of *Ocimum basilicum* Leaf Essential Oil. *B. Latinoam Caribe Pl.* **2009,** *8*, 195–202.

Opalchenova, G.; Obreshkova, D. Comparative Studies on the Activity of Basil—An Essential Oil from *Ocimum basilicum* L. Against Multidrug Resistant Clinical Isolates of the Genera *Staphylococcus, Enterococcus* and *Pseudomonas* by Using Different Test Methods. *J. Microbiol. Methods.* **2003,** *54*, 105–110.

Parra, A. L.; Yhebra, R. S.; Sardiñas, I. G.; Buela, I. G. Comparative Study of the Assay of *Artemia salina* L. and the Estimate of the Medium Lethal Dose (LD_{50} Value) in Mice, to Determine Oral Acute Toxicity of Plant Extracts. *Phytomedicine.* **2001,** *8*, 395–400.

Perveen, K.; Bokhari, N. A.; Al-Rashid, S. A. I.; Al-Humaid, L. A. Chemical Composition of Essential Oil of *Ocimum basilicum* L. and Its Potential in Managing the Alternaria Rot of Tomato. *J. Essent. Oil-Bear. Plants.* **2020,** *23*, 1428–1437.

Phippen, W. B.; Simon, J. E. Anthocyanins in Basil (*Ocimum basilicum* L.). *J. Agric. Food Chem.* **1998,** *46*, 1734–1738.

Prejwltta, M.; Preeti, S.; Lalit, M.; Lata, B.; Srivastava, C. N. Evaluation of the Toxicity of Different Phytoextracts of *Ocimum basilicum* against *Anopheles stephensi* and *Culex quinquefasciatus*. *J. Asia-Pacific Entom.* **2009,** *12*, 113–115.

Pushpangadan, P.; Bradu, B. L. Basil. In *Advances in Horticulture-Medicinal and Aromatic Plants*; Chadha, K. L., Gupta, R., Eds., Vol. 2; Malhotra Publishing House: New Delhi, 1995; pp 627–657.

Pushpangadan, P.; George, V. Basil. In *Handbook of Herbs and Spices*; Peter, K. V., Ed., Vol. 1, 2nd ed.; Woodhead Publishing Limited, 2012; pp 55–72.

Pushpangadan, P.; Rajasekharan, S.; Biju, S. D. *Tulasi* (in Malayalam). In: Serial Wise Publication *Primary Health Care and Medicinal Plants*; Jawaharlal Nehru Tropical Botanic Garden and Research Institute: Thiruvananthapuram, India, 1993; p 96.

Qamar, K. A.; Dar, A.; Siddiqui, B. S.; Kabir, N.; Aslam, H.; Ahmed, S.; Erum, S.; Habib, S.; Begum, S. Anticancer Activity of *Ocimum basilicum* and the Effect of Ursolic Acid on the Cytoskeleton of MCF-7 Human Breast Cancer Cells. *Lett. Drug Des. Discov.* **2010,** *7*, 726–736.

Rabbani, M.; Sajjadi, S. E.; Vaezi, A. Evaluation of Anxiolytic and Sedative Effect of Essential Oil and Hydroalcoholic Extract of *Ocimum basilicum* L. and Chemical Composition of Its Essential Oil. *Res. Pharm. Sci.* **2015,** *10*, 535–543.

Rafieian, K. M.; Hosseini-asl, K. Effects of *Ocimum basilicum* on Functional Dyspepsia: A Double-Blind Placebo-Controlled Study. *Iran J. Med. Sci.* **2015,** *30*, 134–137.

Rasekh, H. R.; Hosseinzadeh, L.; Mehri, S.; Kamli-Nejad, M.; Aslani, M.; Tanbakoosazan, F. Safety Assessment of *Ocimum basilicum* Hydroalcoholic Extract in Wistar Rats: Acute and Subchronic Toxicity Studies. *Iran. J. Basic Med. Sci.* **2012,** *15*, 645–653.

Rashidian, A.; Roohi, P.; Mehrzadi, S.; Ghannadi, A. R.; Minaiyan, M. Protective Effect of *Ocimum basilicum* Essential Oil Against Acetic Acid-Induced Colitis in Rats. *J. Evid. Based Complem Altern. Med.* **2016,** *21*, NP36–42.

Rodrigues, L. B.; Martins, A. O. B. P. B.; Cesário, F. R. A. S.; Castro, F. F. E.; de Albuquerque, T. R.; Fernandes, M. N. M.; da Silva, B. A. F.; Júnior, L. J. Q.; da Costa, J. G. M.; Coutinho, H. D. M.; Barbosa, R.; de Menezes, I. R. A. Anti-Inflammatory and Antiedematogenic

Activity of the *Ocimum basilicum* Essential Oil and Its Main Compound Estragole: In Vivo Mouse Models. *Chem. Biol. Interact.* **2016,** *257*, 14–25.

Rodrigues, L. B.; Martins, A. O. B. P. B.; Ribeiro-Filho, J.; Cesário, F. R. A. S.; E Castro, F. F.; de Albuquerque, T. R.; Fernandes, M. N. M.; da Silva, B. A. F.; Quintans Júnior, L. J.; Araújo, A. A. S.; Menezes, P. D. P.; Nunes, P. S.; Matos, I. G.; Coutinho, H. D. M.; Goncalves Wanderley, A.; de Menezes, I. R. A. Anti-Inflammatory Activity of the Essential Oil Obtained from *Ocimum basilicum* Complexed with β-cyclodextrin (β-CD) in Mice. *Food Chem. Toxicol.* **2017,** *109*, 836–846.

Sasidharan, N. *Biodiversity Documentation for Kerala, Part 6: Flowering Plants*; Kerala Forest Research Institute: Kerala, India, 2004.

Sasidharan, N. *Flowering plants of Kerala*, Ver. 2.0 (DVD), Serial Number 698329520, Kerala Forest Research Institute: Thrissur 680653, Kerala, India, 2011.

Shirazi, M. T.; Gholami, H.; Kavoosi, G.; Rowshan, V.; Tafsiry, A. Chemical Composition, Antioxidant, Antimicrobial and Cytotoxic Activities of *Tagetes minuta* and *Ocimum basilicum* Essential Oils. *Food Sci. Nutr.* **2014,** *2*, 146–55.

Shukla, R.; Sing, P.; Prakash, B.; Anuradha; Dubey, N. K. Antifungal, Aflatoxin Inhibitory and Free Radical-Scavenging Activities of Some Medicinal Plants Extracts. *J. Food Qual.* **2012,** *35*, 182–189.

Siddiqui, B. S.; Bhatti, H. A.; Begum, S.; Perwaiz, S. Evaluation of the Antimycobacterium Activity of the Constituents from *Ocimum basilicum* Against *Mycobacterium tuberculosis*. *J. Ethnopharmacol.* **2012,** *144*, 220–222.

Simon, J. E.; Morales, M. R.; Phippen, W. B.; Vieira, R. F.; Hao, Z. A source of aroma compounds and a Popular Culinary and Ornamental Herb. In *Perspectives on New Crops and New Uses*; Janick, J. Ed.; ASHS Press: Alexandria, VA, 1999, pp 499–505,

Singh, D.; Chaudhuri, P. K. A Review on Phytochemical and Pharmacological Properties of Holy Basil (*Ocimum sanctum* L.). *Ind. Crop. Prod.* **2018,** *118*, 367–382.

Singh, S. Comparative Evaluation of Anti-Inflammatory Potential of Fixed Oil of Different Species of *Ocimum* and Its Possible Mechanism of Action. *Indian J. Exp. Biol.* **1998,** *36*, 1028–1031.

Singh, S. Evaluation of Gastric Anti-Ulcer Activity of Fixed Oil of *Ocimum basilicum* Linn. and Its Possible Mechanism of Action. *Indian J. Exp. Biol.* **1999a,** *37*, 253–257.

Singh, S. Mechanism of Action of Anti-Inflammatory Effect of Fixed Oil of *Ocimum basilicum* Linn. *Indian J. Exp. Biol.* **1999b,** *37*, 248–252.

Stefan, M.; Zamfirache, M.; Padurariu, C.; Trută, E.; Gostin, I. The Composition and Antibacterial Activity of Essential Oils in Three *Ocimum* Species Growing in Romania. *Cent. Eur. J. Biol.* **2013,** *8*, 600–608.

Tada, H.; Murakami, Y.; Omoto, T.; Shimomura, K.; Ishimaru, K. Rosmarinic Acid and Related Phenolics in Hairy Root Cultures of *Ocimum basilicum*. *Phytochemistry*. **1996,** *42*, 431–434.

Teofilovic, B.; Grujic-Letic, N.; Golocorbin-Kon, S.; Stojanovic, S.; Vastag, G.; Gadzuric, S. Experimental and Chemometric Study of Antioxidant Capacity of Basil (*Ocimum basilicum*) Extracts. *Ind. Crops Prod.* **2017,** *100*, 176–182.

Tsai, P. J.; Tsai, T. H.; Yu, C. H.; Ho, S. C. Evaluation of No-Suppressing Activity of Several Mediterranean Culinary Spices. *Food Chem. Toxicol.* **2007,** *45*, 440–447.

Udayan, P. S.; Balachandran, I. *Medicinal plants of Arya Vaidya Sala herb garden*; Centre for Medicinal Plants Research, Arya Vaidya Sala: Kottakkal, India, 2009; p 362.

Ugwu, M. N.; Umar, I. A.; Utu-Baku, A. B.; Dasofunjo, K.; Ukpanukpong, R. U.; Yakubu, O. E.; Okafor, A. I. Antioxidant Status and Organ Function in Streptozotocin-Induced Diabetic Rats Treated with Aqueous, Methanolic and Petroleum Ether Extracts of *Ocimum basilicum* Leaf. *J. App. Pharm. Sci.* **2013,** *3,* S75–S79.

Umar, A.; Imam, G.; Yimin, W.; Kerim, P.; Tohti, I.; Berké, B.; Moore, N. Antihypertensive Effects of *Ocimum basilicum* L. (OBL) on Blood Pressure in Renovascular Hypertensive Rats. *Hypertens. Res.* **2010,** *33,* 727–730.

Umar, A.; Zhou, W.; Abdusalam, E.; Tursun, A.; Reyim, N.; Tohti, I.; Moore, N. Effect of *Ocimum basilicum* L. on Cyclooxygenase Isoforms and Prostaglandins Involved in Thrombosis. *J. Ethnopharmacol.* **2014,** *152,* 151–155.

Vani, S. R.; Cheng, S.; Chuah, C. Comparative Study of Volatile Compounds from Genus *Ocimum. Am. J. Appl. Sci.* **2009,** *6,* 523.

Venancio, A. M.; Onofre, A. S.; Lira, A. F.; Alves, P. B.; Blank, A. F.; Antoniolli, A. R.; Marchioro, M.; Estevam, C. D. S.; Araujo, B. S. D. Chemical Composition, Acute Toxicity, and Antinociceptive Activity of the Essential Oil of a Plant Breeding Cultivar of Basil (*Ocimum basilicum* L.). *Planta Med.* **2011,** *77,* 825–829.

Waltz, L. *The Herbal Encyclopedia*, 1996. http://www.wic.net/waltzark/herbenc.htm

Zahra, K.; Khan, M. A.; Iqbal, F. Oral Supplementation of *Ocimum basilicum* Has the Potential to Improves the Locomotory, Exploratory, Anxiolytic Behavior and Learning in Adult Male Albino Mice. *Neurol. Sci.* **2015,** *36,* 73–78.

Zakaria, Z.; Aziz, R.; Lachimanan, Y. L.; Sreenivasan, S.; Rathinam, X. Antioxidant Activity of *Coleus blumei*, *Orthosiphon stamineus*, *Ocimum basilicum* and *Mentha arvensis* from Lamiaceae Family. *Int. J. Nat. Eng. Sci.* **2008,** *2,* 93–95.

Zgórka, G.; Glowniak, K. Variation of Free Phenolic Acids in Medicinal Plants Belonging to the Lamiaceae Family. *J. Pharm. Biomed. Anal.* **2001,** *26,* 79–87.

Zhang, J.; Li, S.; Wu, W. The Main Chemical Composition and In Vitro Antifungal Activity of the Essential Oils of *Ocimum basilicum* Linn. var. *pilosum* (Willd.) Benth. *Molecules* **2009,** *14,* 273–278.

CHAPTER 15

PHYTOCHEMICAL CONSTITUENTS AND PHARMACOLOGY OF *OCIMUM KILIMANDSCHARICUM* GÜRKE

THADIYAN PARAMBIL IJINU[1,2], BALAN KALLULLATHIL SHAJIMA[3], MAHESWARI PRIYA RANI[4], MAYANK SHUKLA[5,6], and PALPU PUSHPANGADAN[1]

[1]*Amity Institute for Herbal and Biotech Products Development, Thiruvananthapuram, Kerala, India*

[2]*Naturæ Scientific, Kerala University Business Innovation and Incubation Centre, Karyavattom Campus, Thiruvananthapuram, Kerala, India*

[3]*Food Corporation of India, Food Storage Depot-Thikkodi, Kozhikode, Kerala, India*

[4]*Phytochemistry and Phytopharmacology Division, Jawaharlal Nehru Tropical Botanic Garden and Research Institute, Thiruvananthapuram, Kerala, India*

[5]*Indian Pharmacopoeia Commission, Ministry of Health and Family Welfare, Ghaziabad, Uttar Pradesh, India*

[6]*Division of Basic Medical Sciences, Indian Council of Medical Research, V. Ramalingaswami Bhawan, New Delhi, India*

ABSTRACT

Ocimum kilimandscharicum Gürke or camphor basil, is a perennial shrub up to 2 m tall and belongs to the family Lamiaceae. It is extensively grown in the tropics and distributed in East Africa, India, and Thailand. Traditionally, *O. kilimandscharicum* is used to treat cold, cough, abdominal pain, measles,

Bioactives and Pharmacology of Lamiaceae. T. Pullaiah, PhD (Ed.)

and diarrhea. The major essential oil components in *O. kilimandscharicum* include camphor, 1,8-cineole, linalool, limonene, etc. *O. kilimandscharicum* also contains apigenin-7-O-neohesperidoside, quercetin, turkesterone, fisetin, apigenin, chrysin, lupeol, stigmasterol, friedelin, α-amyrin acetate, and n-octacosonoic acid. *O. kilimandscharicum* showed antioxidant, anti-inflammatory, analgesic, neuroprotective, antidiarrheal, and antimicrobial properties.

15.1 INTRODUCTION

Ocimum kilimandscharicum Gürke (Syn.: *Ocimum johnstonii* Baker and *Ocimum tortuosum* Baker) belongs to the family Lamiaceae. The vernacular names of this plant include karpur tulsi (Hindi), karpur tulsi (Sanskrit), karpoorathulasi (Malayalam), karpur tulsi (Tamil), karupuratulasi (Kannada), karpuratulasi (Telugu), karpur tulsi (Manipuri), and camphor basil, kilimanjaro basil (English). It is a perennial shrub up to 2 m tall. The stems are rounded-quadrangular, much branched, and woody with epidermis sometimes peeling off in strips below. The leaves are often with fascicles of young leaves or young shoots in axils; blade ovate, 1.5–5.5 × 1–3 cm, saw toothed, tip blunt or rounded, and base wedge-shaped. The leaf stalks are 4–10 mm long. The flowers are borne in lax inflorescence with distant whorls 8–20 mm apart; bracts usually deciduous. The flower stalks are 3 mm long, erect, slightly curved, and flattened. Calyx is downward-pointing, 3–4 mm long at anthesis, and hairy. The fruiting calyx is 5 mm long. The flowers are white, pink or mauve, 6–7 mm long; tube straight, funnel-shaped, scarcely exceeding calyx tube; stamens protrude out, 3–5 mm. It is extensively grown in the tropics and distributed in East Africa, India, and Thailand (Kokwaro, 1976; Paton, 1991; Demissew and Asfaw, 1994).

Traditionally, *O. kilimandscharicum* is used to treat cold, cough, abdominal pain, measles, and diarrhea (Obeng-Ofori et al., 1998). The essential oil obtained from *O. kilimandscharicum* is used as a veterinary medicine (Baerts and Lehmann, 1991); antibacterial (Prasad et al., 1986); and antioxidant (Hakkim et al., 2008). *O. kilimandscharicum* also possesses carminative, stimulant, and antipyretic properties (Joshi, 2013; Kokwaro, 1976). The seed essential oil containing major component camphor is suitable for industrial applications (Jembere et al., 1994; Angers et al., 1996). In Mediterranean area, *O. kilimandscharicum* is used for decorative purposes (Simon and Reiss-Bubenheim, 1987; Morales et al., 1993).

15.2 PHYTOCHEMICAL CONSTITUENTS

Essoung et al. (2020) found that the major constituents in *O. kilimandscharicum* essential oil are camphor (47.1%) and 1,8-cineole (19.3%); camphene (5.2%), fenchone (4.9%), terpinen-4-ol (2.6%), myrtenol (2.1%), (E)-caryophyllene (1.9%), and α-terpineol (1.5%) are minor constituents. The essential oil of *O. kilimandscharicum* collected from Indiana contains linalool (41.94–58.85%), camphor (17.0–15.82%), and 1,8-cineole (10.18–6.38%) as major constituents (Charles and Simon, 1992). Vinutha and Srikar (2007) reported camphor (57.87%) and Ntezurubanza et al. (1984) reported 1,8-cineol (62%) as the major compounds in *O. kilimandscharicum* collected from South India and Rwanda, respectively. Garg et al. (2004) reported camphor (53.89%), limonene (10.5%), and camphene (4.5%) as the major constituents. Kumar et al. (2009) found the seasonal variation in percentage of oil components in *O. kilimandscharicum*. It is observed that camphor is rich in winter (48.9%) and summer (58.9%); 1,8-cineole (winter 22.2% and summer 14.8%); and limonene (winter 5.5% and summer 5.6%). Lima et al. (2014) found that the essential oil obtained from *O. kilimandscharicum* collected from the Federal University of Grande Dourados, Brazil is rich in camphor (51.81%), 1,8 cineole (20.13%), and limonene (11.23%). Makenzi et al. (2019) reported two new compounds namely, 2α-hydroxy-3-oxodammara-20,24-diene and 2α,3β-dihydroxy dammara-20, 24-diene, along with apigenin-7-O-neohesperidoside, quercetin, turkesterone, fisetin, apigenin, chrysin, lupeol, stigmasterol, friedelin, α-amyrin acetate, and n-octacosonoic acid.

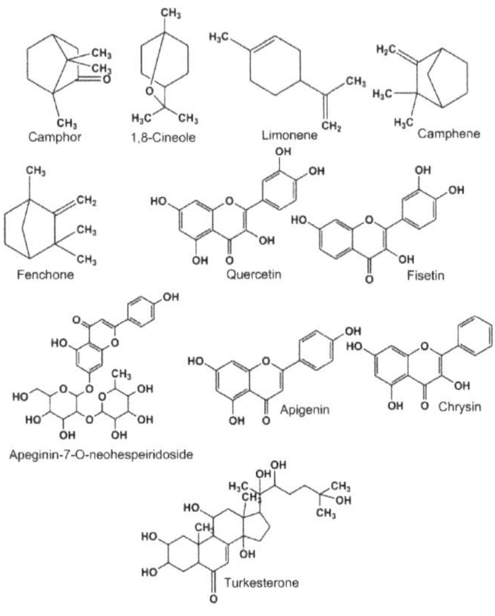

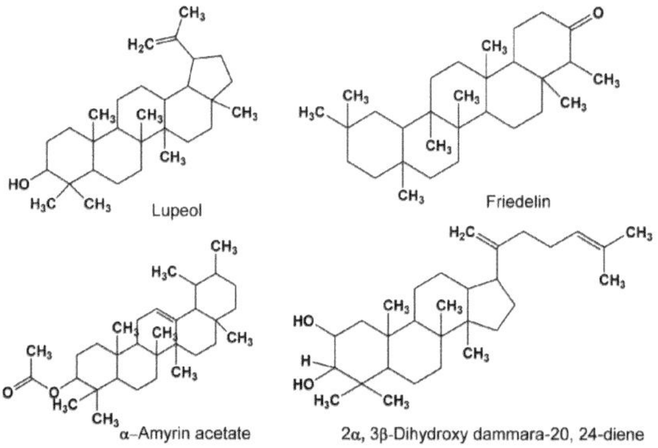

Lupeol, Friedelin, α-Amyrin acetate, 2α, 3β-Dihydroxy dammara-20, 24-diene

15.3 PHARMACOLOGICAL STUDIES

15.3.1 ANTIOXIDANT ACTIVITY

The methanolic extract of *O. kilimandscharicum* (100 μl of 2.5 mg/100 mL) was found to show good 2,2′-azino-bis(3-ethylbenzothiazoline-6-sulfonic acid) (79.2%) and 2,2-diphenyl-1-picrylhydrazyl (70.7%) free radicals scavenging activities (Gaurav et al., 2019). *O. kilimandscharicum* essential oil showed good free radical scavenging activity against 2,2-diphenyl-1-picrylhydrazyl radical (response parameter, GI_{50} 8.31 g/mL). Camphor and mixture of the limonene:1, 8 cineole (1:1) also showed potent 2,2-diphenyl-1-picrylhydrazyl radical scavenging activity with IC_{50} values 12.56 and 23.25 g/mL, respectively (Lima et al., 2014).

15.3.2 ANTI-INFLAMMATORY AND ANALGESIC ACTIVITIES

O. kilimandscharicum essential oil (30, 100, and 300 mg/kg) and camphor (30 mg/kg) showed a significant inhibition in carrageenan-induced edema and zymosan-induced articular inflammation/nociception in mice. Also, the essential oil (1, 3, 10, 30, and 60 μg/mL) significantly inhibited carrageenan-induced leukocyte rolling and adhesion (Dos Santos et al., 2021). The oral administration of *O. kilimandscharicum* essential oil (30 and 100 mg/kg) showed 82 and 95% inhibition, respectively in carrageenan-induced pleurisy. Camphor (15 mg/kg) showed 83% of inhibition and mixture of 1,8 cineole:limonene (1:1; 6 mg/kg) showed 80% of inhibition (Lima et al., 2014).

15.3.3 NEUROPROTECTIVE ACTIVITY

In mice, *O. kilimandscharicum* leaves extract (200 and 400 mg/kg), once daily for 7 days, showed protective activity against ischemia-reperfusion-induced brain injury, evidenced from histopathological and biochemical parameters (Singh et al., 2017).

15.3.4 ANTIDIARRHEAL ACTIVITY

The aqueous extract of leaves of *O. kilimandscharicum* at a dose of 200 mg/kg showed significant delay in the onset of defecation, reduction in the cumulative fecal weight, along with a change in the fecal consistency from watery to solid form in castor oil-induced diarrhea in rodent model (Sarin et al., 2013).

15.3.5 ANTIMICROBIAL ACTIVITY

O. kilimandscharicum essential oil showed significant activity against bacteria *Staphylococcus epidermidis*, *Enterococcus faecalis*, and *Staphylococcus aureus*, and fungi *Candida albicans* (Chaturvedi et al., 2018). The antimicrobial study of eugenol and cadinol isolated from *O. kilimandscharicum* in multidrug-resistant and methicillin-resistant *Staphylococcus aureus* strains showed minimum inhibitory concentrations range of 2–128 μg/mL (Shinde et al., 2010).

15.3.6 TOXICITY STUDIES

The acute oral toxicity study of methanolic extract of *O. kilimandscharicum* was conducted in female Sprague–Dawley rats. There was no mortality or toxic signs observed, and the LD_{50} value was found to be more than 2000 mg/kg body weight (Gaurav et al., 2019). *O. kilimandscharicum* essential oil showed cytotoxicity against human ovarian cancer cell line (response parameter, GI_{50} 31.90 mg/mL) (Lima et al., 2014).

ACKNOWLEDGMENTS

The authors express their sincere thanks to Dr. Ashok K. Chauhan, Founder President, Ritnand Balved Education Foundation (RBEF) and Amity Group of Institutions, and Dr. Atul Chauhan, Chancellor, Amity University Uttar Pradesh (AUUP) for facilitating this work. Thadiyan Parambil Ijinu has received Young Scientist Fellowship from the Department of Science and Technology, Government of India (SP/YO/413/2018).

KEYWORDS

- **traditional medicinal use**
- **camphor**
- **1,8-cineole**
- **antioxidant activity**
- **anti-inflammatory activity**
- **neuroprotective activity**

REFERENCES

Angers, P.; Morales, M. R.; Simon, J. E. Basil Seed Oils. In *Progress in New Crops*; Janick, J., Ed., Vol. 1; ASHS Press: Arlington, 1996, pp 598–601.

Baerts, M.; Lehmann, J. Veterinary Medicinal Plants of the Region of Cretes Zaire-Nil in Burundi. *Annalen Economic Wetenscbappen, Koninklijk Museum voor Midden, Africa.* **1991,** *21*, 133.

Charles, D. J.; Simon, J. E. Essential Oil Constituents of *Ocimum kilimandscharicum* Guerke. *J. Essent. Oil Res.* **1992,** *4*, 125–128.

Chaturvedi, T.; Kumar, A.; Kumar, A.; Verma, R. S.; Padalia, R. C.; Sundaresan, V.; Chauhan, A.; Saikia, D.; Singh, V. R.; Venkatesha, K. T. Chemical Composition, Genetic Diversity, Antibacterial, Antifungal and Antioxidant Activities of Camphor-Basil (*Ocimum kilimandscharicum* Guerke). *Ind. Crops Prod.* **2018,** *118*, 246–258.

Demissew, S.; Asfaw, N. Some Useful Indigenous Labiates from Ethiopia. *Lamiales Newsletter, Kew Gardens* **1994,** *3*, 5–6.

Dos Santos, E.; Leitão, M. M.; Aguero Ito, C. N.; Silva-Filho, S. E.; Arena, A. C.; Silva-Comar, F. M. S.; Nakamura Cuman, R. K.; Oliveira, R. J.; Nazari Formagio, A. S.; Leite Kassuya, C. A. Analgesic and Anti-Inflammatory Articular Effects of Essential Oil and Camphor Isolated from *Ocimum kilimandscharicum* Gürke Leaves. *J. Ethnopharmacol.* **2021,** *269*, 113697.

Essoung, F. R. E.; Tadjong, A. T.; Chhabra, S. C.; Mohamed, S. A.; Hassanali, A. Repellence and Fumigant Toxicity of Essential Oils of *Ocimum gratissimum* and *Ocimum kilimandscharicum* on Tuta absoluta (Lepidoptera: Gelechiidae). *Environ. Sci. Pollut. Res.* **2020,** *27*, 37963–37976.

Garg, S. N.; Naqvi, A. A.; Bahl J. R.; Khanuja, S. P. S. Composition of the Essential Oil of *Ocimum kilimandscharicum* Leaf. *Indian Perfum.* **2004,** *48*, 47–49.

Gaurav, S.; Abhilasha, M.; Wadood, S. A. Evaluation of Preliminary Phytochemical Screening, Acute Toxicity and Antioxidant Profile of *Ocimum kilimandscharicum*. *J. Drug Deliv. Ther.* **2019,** *9*, 372–375.

Hakkim, F. L.; Arivazhagan, G.; Boopathy, R. Antioxidant Property of Selected *Ocimum* Species and Their Secondary Metabolite Content. *J. Med. Plant Res.* **2008,** *2*, 250–257.

Joshi, R. K. Chemical Composition of the Essential Oil of Camphor Basil (*Ocimum kilimandscharicum* Guerke). *Glob. J. Med. Plant. Res.* **2013,** *1*, 207–209.

Kokwaro, J. O. *Medical Plants of East Africa*; East African Literature Bureau, General Printers Limited, Nairobi, 1976; p. 384.

Kumar, N.; Shukla, R.; Chanotiya, C. S.; Yadav, A.; Singh, A. K.; Bagchi, G. D. Effect of Seasons and Drying on the Essential Oil Composition of Leaf in *Ocimum kilimandscharicum* Guerke. *J. Essent. Oil Res.* **2009,** *21*, 400–402.

Lima, V. T. D.; Vieira, M. C.; Kassuya, C. A. L.; Cardoso, C. A. L.; Alves, J. M.; Foglio, M. A., de Carvalho, J. E.; Formagio, A. S. N. Chemical Composition and Free Radical-Scavenging, Anticancer and Anti-Inflammatory Activities of the Essential Oil from *Ocimum kilimandscharicum*. *Phytomedicine*, **2014,** *21*, 1298–1302.

Makenzi, A. M.; Manguro, L. O. A.; Owuor, P. O.; Opiyo, S. A. Chemical Constituents of *Ocimum kilimandscharicum* Guerke Acclimatized in Kakamega Forest, Kenya. *Bull. Chem. Soc. Ethiop.* **2019,** *33*, 527–539.

Morales, M. R.; Charles, D. J.; Simon, J. E. New Aromatic Lemon Basil Germplasm. In *New Crops*; Janick, J., Simon, J. E., Eds.; John Wiley and Sons: New York, 1993, pp 632–635.

Ntezurubanza, L.; Scheffer, J. J.; Looman A.; Svendsen, A. B. Composition of Essential Oil of *Ocimum kilimandscharicum* Grown in Rwanda. *Planta Med.* **1984,** *50*, 385–388.

Obeng-Ofori, D.; Reichmuth, C. H.; Bekele, A. J.; Hassanali, A. Toxicity and Protectant Potential of Camphor, a Major Component of Essential Oil of *Ocimum kilimandscharicum*, Against Four Stored Product Beetles. *Int. J. Pest. Manage.* **1998,** *44*, 203–209.

Paton, A. A synopsis of *Ocimum* (Labiatae) in Africa. *Kew Bull.* **1991,** *47*, 403–435.

Prasad, G.; Kumar, A.; Singh, A. K.; Bhattacharya, A. K.; Singh, K.; Harma, V. D. Antimicrobial Activity of Essential Oils of Some *Ocimum* Species and Clove Oil. *Fotpterapia*, **1986,** *57*, 429–432.

Sarin, R. V.; Narwal, S.; Bafna, P. A. Anti-Diarrhoeal Activity of Aqueous Extract of *Ocimum kilimandscharicum*. *J. Ethnopharmacol.* **2013,** *148*, 223–228.

Shinde, K.; Shinde, V.; Mahadik, K.; Gibbons, S. Phytochemical and Antibacterial Studies on *Ocimum kilimandscharicum*. *Planta Med.* **2010,** *76*, P412.

Simon, J. E.; Reiss-Bubenheim, D. Characteristics of American Basil Varieties. In *Proceedings of the Second National Herb Growing and Marketing Conference*; Simon, J. E., Grant, L., Eds.; Indianapolis, Indiana. 1987; pp 48–51.

Singh, V.; Krishan, P.; Singh, N.; Kumar, A.; Shri, R. Amelioration of Ischemia-Reperfusion Induced Functional and Biochemical Deficit in Mice by *Ocimum kilimandscharicum* Leaf Extract. *Biomed. Pharmacother.* **2017,** *85*, 556–563.

Vinutha, T.; Srikar, L. N. Essential Oil Composition of *Ocimum sanctum* and *Ocimum kilimandscharicum* Inoculated with Biofertilizers. *Indian Perfum.* **2007,** *51*, 60–62.

CHAPTER 16

PHYTOCHEMICAL AND PHARMACOLOGICAL PROFILE OF DIVINER'S SAGE (*SALVIA DIVINORUM* EPLING A JATIVA.)

CHINNADURAI IMMANUEL SELVARAJ[1],
VRUSHALI MANOJ HADKAR[2], and GADWAL SHAIK NISHAT ANJUM[1]

[1]*VIT School of Agricultural Innovations and Advanced Learning (VAIAL), Vellore Institute of Technology, Vellore, Tamil Nadu, India*

[2]*Department of Biotechnology, School of Biosciences and Technology, VIT, Vellore, Tamil Nadu, India*

ABSTRACT

Salvia divinorum, or the 'diviner's sage', is a member of the Lamiaceae family. it is a perennial shrub-like herb and grows to a height of 0.5–1.5 metres. *S. divinorum* is native to the Sierra Mazateca in Oaxaca, Mexico where the Mazatec people consume the plant's fresh leaves or leaf preparations for medicinal, therapeutic, and divinatory rites. The plant is a rich source of diterpenes and terpenoids. Divinatorins, salvinicins, salvidivins, neoclerodane diterpene and salvinorins are the predominant diterpenes isolated from the plant. The terpenoids such as divinorin A and divinorin B found in *S. divinorum* have psychotropic effect. Traditionally the plant is used to cure diarrhoea, anemia, headaches, rheumatism, and a semi-magical disease known as *panzón de borrego*, or a swollen belly, hemialgia and bursitis. The plant is believed to possess hallucinogenic effect which is due to the presence of Salvinorin A. The plant extracts have been studied to

Bioactives and Pharmacology of Lamiaceae. T. Pullaiah, PhD (Ed.)

evaluate its antinociceptive effects, anti-inflammatory effect, neuroprotective activity, psychoactive properties, antidepressant and anxiolytic activity. Salvinorin A can be used as a remedy for treatment of gastrointestinal dysfunction such as inflammatory bowel disease and diarrhoea. This chapter helps in understanding the phytochemistry and pharmacological aspect of *S. divinorum* and provides information about the potential pharmaceutical and psychotropic properties of the plant.

16.1 INTRODUCTION

The genus *Salvia* has a substantial diversity of species in the universe (Fragoso-Martínez et al., 2018). Mexico is recognized for higher frequency holding more than 300 endemic species (Cornejo-Tenorio and Ibarra-Manríquez, 2011). *Salvia divinorum* Epling & Jativa. (Family: Lamiaceae) is a precious perpetual herb and detailed as an endemic to the Oaxaca, Mazatec state, Mexico. This land is adjoined by the Sierra Mazateca mountain range in the north and the Pacific Ocean to the west. This peak range is covered by tropical montane cloud jungle (Reisfield, 1993; Ott, 1995, 1996). The whole registered indigenous populations of *Salvia divinorum* are found in Oaxaca, southern Mexico. *S. divinorum* is commonly grown for recreational use (Johnson, 1939). The name *divinorum* was christened by botanists Epling and Jativa for the plant's divinatory use (Wasson, 1962). Mazatecs have various names for *Salvia divinorum*; in their native language it is referred to as *Ska Maria Pastora, Ska Maria, Ska Pastora,* and in Spanish it is called *Hojas de la Pastora, Hojas de Maria Pastora, Hierba Maria, Hoja de adivinación,* or *La Maria* (Wasson, 1962; Schultes, 1972; Valdés et al., 1983; Valdés, 2001; Grundmann et al., 2007). Mazatecs correlate *Salvia divinorum* with the Christian saint, Mary (Valdés et al., 1983). The evidence of her as a shepherdess is not logical with Christian belief (Wasson, 1962).

The plant species, frequently spotted near streams in partial or complete shadow, thrive in damp, nutrient-rich loamy soil. In these circumstances, *Salvia divinorum* spreads and propagates principally vegetatively, blooming sporadically when enough sun infiltrates the woodland canopy (Reisfield, 1993). The hardwood cuttings are obtained from the parent plant and planted into the wet loamy soil; nevertheless, this plant may further root itself if a twig snaps off and drops on the earth (Beifuss, 1997).

Salvia divinorum is most nearly associated to the Columbian species *Salvia venulosa* (Jenks et al., 2011). The botanical description of *S. divinorum* describes that the plant grows up to a height of 1.5 m. The plant's stem is hollow, quadrangular, green, and translucent (Reisfield, 1993; Ott, 1996). The leaves are 10–25 cm lengthier, 5–10 cm broad, opposite on the stem, elliptic in form, and serrated edges (Epling and Jativa, 1962; Reisfield, 1993). Copious glandular and non-glandular trichomes are seen on the leaf cover (Siebert, 2004; Kowalczuk et al., 2014). The wildflowers have white corollas with purplish calyces. The flowers are 3–4 cm in the stretch and develop on panicles ranging from 20 to 30 flowers. Studies were done on natural populations and lab tests. *Salvia divinorum* seldom bears flowers on a periodical, seasonal basis (Valdés et al., 1987; Reisfield, 1993). This plant is observed to flower amid June and October in Oaxaca (Reisfield, 1993). The flowering initiates by fixed durations of unbroken darkness greater than 12 h, which is supposed to be a long-day plant (Reisfield, 1993).

Salvia divinorum is grown in hidden forests deep inside the jungle by traditional healers named *curandera* (female) or *curandero* (male) (Reisfield, 1993). *Salvia divinorum*, consumed by chewing the leaves or the extract from leaves, is mixed in proportions of water and ingested as a juice (Diaz, 1979, 2013; Valdés, 2001). The rituals are directed by a *curandera* or *curandero* and lasts about 2–3 h, by which the participants lead through various phases of cognizance (Ott, 1996; Schultes et al., 2001). In healing rituals, *S. divinorum* is used to prompt a delusory mood that enables the partakers to beat the situation of sickness or frailty (Valdés et al., 1983). The Mazatec Indians of Oaxaca use various parts of this plant for ethnopharmacological, spiritual, and traditional practices to create "hallucinogenic" or "mystical" events (Valdés et al., 1983). The usage of *Salvia divinorum* takes place principally in recuperating and prophecy rituals and in educating traditional therapeutic practitioners of Mazatec (Giovannini and Heinrich, 2009). There are four illnesses for which Mazatecs are known to have used *Salvia divinorum* (Johnson, 1939; Valdés et al., 1983; Ott, 1996; Prisinzano, 2005). First, this plant is frequently used to cure diarrhea. Second, the patients nearing death can be given a liquid of the plant's extracts as a palliative; the sufferer often recovers for a brief term. Third, *S. divinorum*, in petite portions, is used to cure hemialgia and bursitis. Lastly, it cures a Mazatec ailment known as *panzón de arrego* or bloated abdomen. A curse causes this Mazatec illness from a brujo, someone who exercises a black or malevolent spell (Johnson, 1939; Ott, 1996; Valdés et al., 1983; Prisinzano, 2005).

16.2 BIOACTIVES

The bioactive compounds found in *S. divinorum* were significant secondary metabolites. These secondary metabolites isolated from *S. divinorum* have primarily been diterpenoids (Kutrzeba et al., 2007). The diterpenoids obtained from the condensation of four isoprene units in *Salvia* species lead to bicyclic, tricyclic, and tetracyclic compounds (Bonito et al., 2011). The diterpene salvinorin A, a principal neoclerodane diterpenes is accountable for the bioactivity in *Salvia divinorum,* considered a potential lead compound in pharmaceutical research. Further additional diterpenoids include divinatorins, salvinicins, salvidivins, and nine additional salvinorins (Casselman et al., 2014). Ortega et al. (1982) isolated a novel neoclerodane diterpene, named as salvinorin A from *Salvia divinorum* (Valdés, 1994). Salvinorin A alone is recorded as a neoclerodane diterpene known for its psychoactive characteristics (Hanson, 2010). It was reported that salvinorin A is known to be a representative member of the subclass of the diterpenes structural class, the same configuration as the clerodin, which is isolated from *Clerodendron infortunatum* (Sim et al., 1961). (−)-Hardwickiic acid is a furanoid diterpene discovered in *Salvia divinorum* (Misra et al., 1968).

The psychotropic terpenoids reported in *S. divinorum* were divinorin A and divinorin B. These bioactive compounds were similar, based on the salvinorin structure (Ortega et al., 1982). Thus, divinorin A and B are also known as salvinorin A and B. Also, the isolation of another neoclerodane diterpene was done, named salvinorin C (Siebert, 1994). Additional neoclerodane diterpenes, salvinorin D–F and divinatorins A–C were known to be isolated (Bigham et al., 2003). Further phytochemicals investigations in *S. divinorum* identified two compounds, salvinicin A and B (Harding et al., 2005). The structures of various chemical compounds present in *S. divinorum* plant are given in Figures 16.1 and 16.2.

16.3 PHARMACOLOGY

16.3.1 ANTINOCICEPTIVE EFFECTS

A dose-dependent antinociceptive effect was produced by *S. divinorum* extract at the central and peripheral levels in pain models. The single extract's potential effect was seen due to bioactive compounds like salvinorins, of which the main compound was salvinorin A. The antinociceptive effect of *S. divinorum* was observed for both the receptors serotonin 5-HT1A and opioids, which inhibits neurotransmission. Thus it depended on the type of nociception (Tlacomulco-Flores et al., 2020). The investigations indicated that the nonsteroidal anti-inflammatory drugs induced antinociception by two known mechanisms, which inhibits cyclooxygenases activity (COX) and

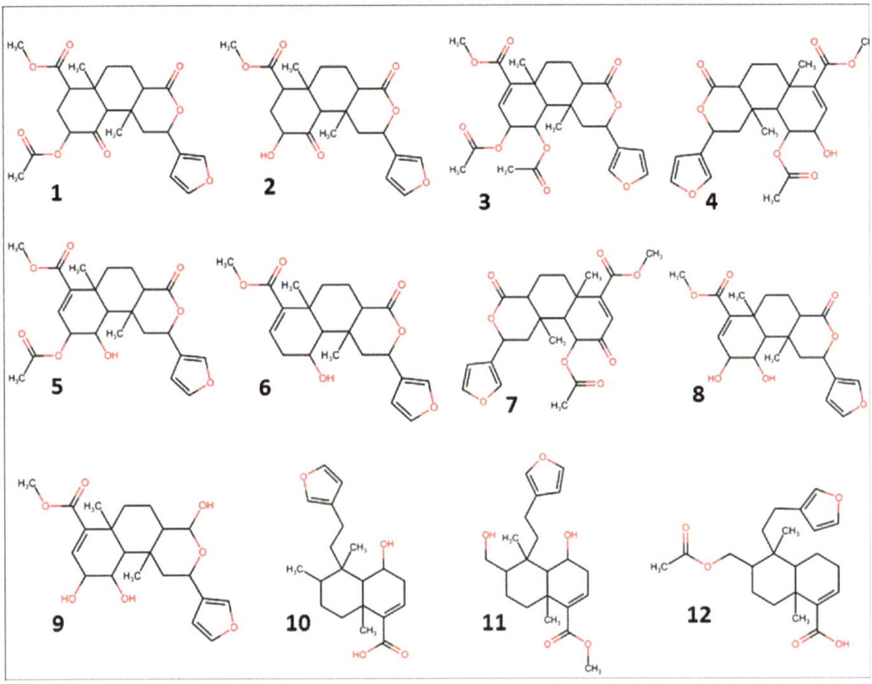

FIGURE 16.1 Chemical compounds present in *Salvia divinorum*. Salvinorin A (1), Salvinorin B (2), Salvinorin C (3), Salvinorin D (4), Salvinorin E (5), Salvinorin F (6), Salvinorin G (7), Salvinorin H (8), Salvinorin I (9), Divinatorin A (10), Divinatorin B (11) and Divinatorin C (12); [Marvin 17.21.0, ChemAxon tool was used for drawing chemical structures].

decreases the concentration of prostaglandins. Other studies reported COX-independent anti-inflammatory actions of nonsteroidal anti-inflammatory drugs-induced inhibition of the known nuclear factor kappa B. It was also observed that L-selectin's downregulation, inhibition of proinflammatory cytokines, and nitric oxide synthase activity inhibition was present (Miranda et al., 2019).

The investigations indicate that modifications in the expression of KOR and CB1 receptor were due to the effect of salvinorin A. The expression of KOR and CB1 receptor in macrophage protein lysates was followed by lipopolysaccharide stimulation and the related effect of salvinorin A. After 18-h exposure to lipopolysaccharides (1 μg/mL), KOR and the CB1 receptor expression was increased. Both KOR and CB1 receptor overexpression was reduced due to salvinorin A (10 pM). Besides, the statistically tabulated inhibitory effect was obtained only for the KOR receptor (Aviello et al., 2011).

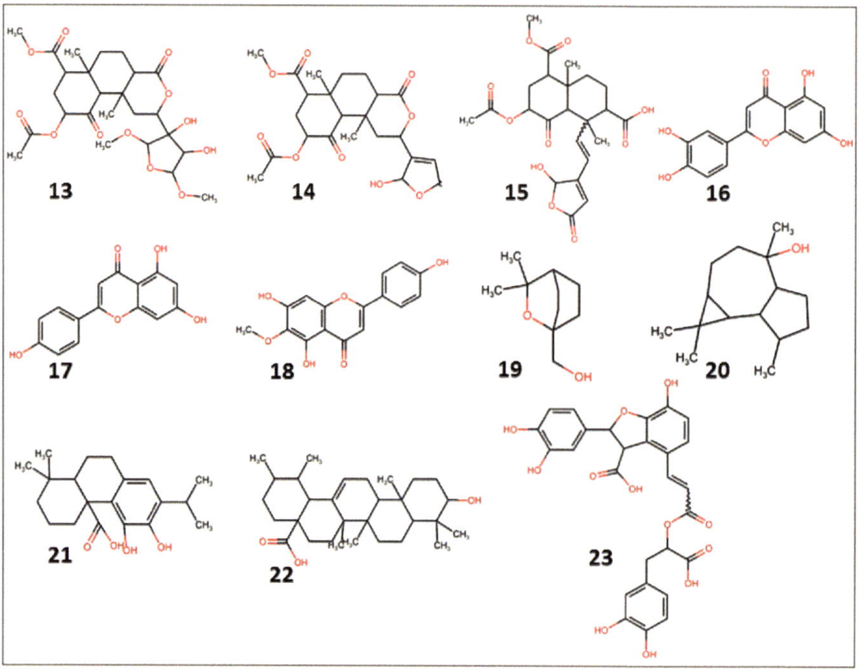

FIGURE 16.2 Chemical compounds present in *Salvia divinorum*; Salvinicin A (13), Salvidivin B (14), Salvidivin C (15), Luteolin (16), Apigenin (17), Hispidulin (18), 1,8-cineole (19), Viridiflorol (20), Carnosic acid (21), Ursolic acid (22) and Lithospermic acid (23); [Marvin 17.21.0, ChemAxon tool was used for drawing chemical structures].

The molecular target of diterpene salvinorin A is the κ-opioid receptor identified in vitro. It was investigated whether salvinorin B, salvinorin A, and the C2-substituted derivative salvinorinyl-2-propionate can behave as κ-opioid receptor agonists in vivo. The intracerebroventricular shot over a dose range of 1–30 μg of both salvinorinyl-2-propionate and salvinorin A produces antinociception in control mice and not in a κ-opioid receptor knockout mice. Salvinorinyl-2-propionate and salvinorin A decrease rectal temperature in a genotype-dependent manner, comparable to standard κ-opioid receptor agonists. Further, salvinorin A has a significant affinity for κ1 opioid receptors rather κ2 receptors. Subsequently, medication with the equivalent dosage span of salvinorin B, which is idle in vitro, exhibited neither hypothermic nor antinociceptive effects in wild mice. In conclusion, salvinorin A is the active component of *S. divinorum*, particularly for κ1 opioid receptors. Salvinorin A and specific structurally associated analogues which exhibit behavioral outcomes need κ-opioid receptor. The sedation and antinociceptive effects of salvinorin A in mice were short-lived maxima for 15 min. No effect was seen at 45 min, followed by an intracerebroventricular injection of 7.5 μg of salvinorin A, while hypothermia was long-lasting (maximum at 75 min and no effects at 120 min) but only with 50 μg of salvinorin A (Ansonoff et al., 2006).

16.3.2 ANTI-INFLAMMATORY EFFECT

S. divinorum can be an alternative therapy for inflammatory and neuropathic pain due to active compound salvinorin A, an allosteric modulator of CB1 receptors and a potent KOR agonist. The experimental confirmation supports the fact that *S. divinorum*, salvinorin A, and their analogues lower the pain induced by neuropathy and inflammation (Coffeen and Pellicer, 2019). *S. divinorum* extracts are of folkloric use in managing inflammatory ailments and in the knowledge of the study that primary spots of the salvinorin A action, identified as CB1 and KORs receptors, play an influential function during inflammation processes (Ji et al., 1995; Di Marzo, 2008; Pertwee, 2008). TNF-α and interleukins (IL-10 and IL-1β) are linked to LPS-evoked inflammatory responses in macrophages (Hawiger, 2001). The study reported that the anti-inflammatory effect of salvinorin A for specific lipopolysaccharide on carrageenan-induced paw edema represented probably an effective model for acute inflammation (Morris, 2003). TNF-α and IL-1 accomplish a crucial function in the acute stage of brain abscess formation and control the ensuing antibacterial inflammatory response. The levels of the interleukin-10 (IL-10), interleukin-1β (IL-1β), and tumor necrosis factor-α (TNF-α) in macrophages containing medium significantly raised following 18-h exposure to LPS. IL-1β levels remained unaltered when salvinorin A (0.1–10 pM) was given as a pretreatment for 30 min but it significantly, in a concentration-dependent fashion, reduced IL-10 and TNF-α levels (Aviello et al., 2011).

In the in vivo conditions, salvinorin A subdued the carrageenan-induced paw edema, formalin-induced inflammatory pain, and the LPS in rimonabant and norbinaltorphimine-sensitive mode. In conclusion, salvinorin A—through CB1 and KORs receptors exerts ultrapotent influences on macrophages and exhibits average anti-inflammatory effects. Salvinorin A repressed LPS-induced response in vitro, and was assessed for its influence on the inflammatory response provoked by LPS in vivo. The intraplantar dose of LPS evoked a rise in rear paw mass, manifested as edema. Salvinorin A, at the moderate dosage (0.5 mg/kg), significantly lessened LPS-influenced paw edema during the 2–6 h duration (Aviello et al., 2011). A dosage of salvinorin

A (0.5–2 mg/kg) in a dose-dependent manner significantly reduced the initial stage and effectively repressed the next stage of formalin-induced nociceptive function (Aviello et al., 2011).

The study of anti-inflammatory activity of *S. divinorum* revealed that the potent phytochemical salvinorin A induced the inhibition of leukotriene (LT) biosynthesis (potent mediators of inflammation) in macrophages in a dose-dependent manner both for in vitro (tested in activated rat peritoneal macrophages, at 0.01–1 μM) and in vivo tests (tested in zymosan-induced peritonitis mouse and carrageenan-induced pleurisy rat models, at 1–10 mg/kg). Besides inhibition of leukotriene biosynthesis, potent bioactive compound salvinorin A also restrained myeloperoxidase, cell infiltration, and vascular permeability in the peritoneal cavity of the peritonitis mouse model; as well as it also lowered inflammatory cells, exudate volume, and myeloperoxidase activity in the pleurisy rat model. Therefore, it was suggested that the possibility of using salvinorin A for the treatment of LT-related diseases (e.g., allergic rhinitis, asthma, and cardiovascular diseases) showed operational benefits in the control of allergic inflammation. Salvinorin A (10 mg/kg, i.p.) inhibited LT and IL-13 production, and pulmonary mast cell degranulation that reduced bronchial hyperreactivity in an air pouch mouse model (Rossi et al., 2017). *S. divinorum* showed significant antihyperalgesic effects at 100 and 200 mg/kg, and it increased paw withdrawal latency only at 100 mg/kg. Therefore, these results suggested that salvinorin A can be an alternative therapy for neuropathic pain (Mascarenhas et al., 2020).

16.3.3 ANTIMICROBIAL ACTIVITY

(−)-Hardwickiic acid (Misra et al., 1968), a furanoid diterpene discovered in *Salvia divinorum* has antimicrobial activity as evidenced from other studies. The ethanol extract with 1 mg/mL (−)-Hardwickiic acid has antibacterial action against *Staphylococcus aureus, Mycobacterium smegmatis,* and *Bacillus subtilis,* and antifungal activity against *Trichophyton mentagrophytes, Helminthosporium* spp., and *Candida albicans* (McChesney et al., 1991). The diterpene alcohol (E)-phytol (Bigham et al., 2003) separated from *Salvia divinorum* displays antimicrobial activity with a minimum inhibitory concentration (MIC) of 2 mg/mL (Rajab et al 1998). Nepetoidin B (Munro, 2006), from *Salvia divinorum,* a caffeic acid ester was shown to have antifungal properties against *Aspergillus niger* (Grayer et al., 2003; Hanson, 2010). (−)-Hardwickiic acid displays antimicrobial action (McChesney et al., 1991; Chaichantipyuth et al., 2004).

16.3.4 ANTIDEPRESSANT AND ANXIOLYTIC ACTIVITY

A 26-year old female administered with *S. divinorum* leaves at the concentration of 0.5–0.75 g sublingually three times in a week showed a complete reduction of depressive symptoms after 6 months, as estimated by the Hamilton Depression Scale (Hanes, 2001). Other studies have shown depressive-like symptoms (e.g., reduced locomotor activity and motivation) for high salvinorin A compound doses at the concentration of 0.125–2.0 mg/kg followed by reduced DA (Carlezon et al., 2006). Another study investigated the antidepressant and anxiolytic properties of the potent phytochemical compound salvinorin A from *S. divinorium* that was showed in rats (0.001–1000 μg/kg, s.c.) (Braida et al., 2009). *Salvia divinorum* when typically inhaled, intense influences are short and persistent unpropitious consequences are rare. Besides severe hallucinogenic effects, *Salvia divinorum* may result in subacute increase in personal well-being. Such a subacute consequence would be unique for a tranquillizer used non-medically, as a retreat from other sedatives typically either does not influence mood or produces dysphoria (Baggott et al., 2010). Chronic mild stress (CMS) was induced in female and male Long–Evans rats to model anhedonia in depression. The gradual decline in choice for a sucrose liquid over plain water, a pattern of anhedonia, and locomotor action observed for 49 days with 21 days of CMS exposure, began to lower sucrose preference. CMS was extended for 21 additional days, and rats were randomly allocated to treatment with 1 mg salvinorin A/kg body weight or a vehicle check group. The outcomes showed that salvinorin A changed anhedonia, whereas control animals persisted in exhibiting a subdued choice for the sucrose liquid. Thus, the outcomes show that salvinorin A is an efficient antidepressant compound when given chronically to rat models displaying signs of distress compared to those perceived in human being (Harden et al., 2012).

16.3.5 NEUROPROTECTIVE ACTIVITY

Salvinorin A signifies the very first naturally known nonnitrogenous full agonist of κ-opioid receptors (Chavkin et al., 2004). The neuroprotective properties of *S. divinorum* were demonstrated in animal models, namely a piglet (10 μg/g) and C57BL/6J mouse with a dose of 0.5 mg/kg, i.p. The results indicated therapeutic benefits for neonate brain protection due to a potent phytochemical compound salvinorin A isolated from *S. divinorum*. The drug caused conserved artery autoregulation, cerebral vasodilation, reduced mortality rate, and improved several short-term neurological developmental effects, leading to a reduction in neonatal hypoxic-ischemic-induced injury (Su et al., 2012; Chen et al., 2014; Wang et al., 2012). Newborn pigs were administered with 10 nM and 1 μM of salvinorin A and the effects on the cerebral vasculature were evaluated. In a dose-dependent manner, active phytochemical compound salvinorin A dilated the pial artery, both under normal and vessel-constricted conditions induced by endothelin and hypocarbia. This effect blocks a KATP channel inhibitor and a KOP-selective antagonist by a nitric oxide synthase (NOS) inhibitor. The dilation effects may be due to bioactive compound salvinorin A, which mediates ATP channels, NOS, and KOP activation (Su et al., 2011). A study was conducted in the mouse model with middle cerebral artery occlusion. The activation of specific KOP by salvinorin A with intranasal administration of 12.5, 25, and 50 μg/kg led to a decrease in cerebral infarct volume in a dose-dependent manner that protected the vascular integrity. Improvement in neurological development was observed (Chen et al., 2016).

16.3.6 PSYCHOACTIVE PROPERTIES

Salvinorin A is a primary potential diterpene bioactive compound in *S. divinorum*. Salvinorin A is the first known to be reported diterpene with psychoactive properties and exceedingly selective agonist for receptor named κ-opioid (Valdés, 1994). The κ-opioid receptor (KOR) is among the five associated receptors in individuals that connect opioid type compounds in the brain, connected in feeling, motor control, and pain discernment. *OPRK1* gene encodes κ-opioid receptor (KOR) (Butelman et al., 2007); this receptor is located in the pain neurons, brain, and spinal cord (Fine and Portenoy, 2004). A study revealed the time and dose-related effects of salvinorin A. In the study, participants were selected and then administered with the drug sessions accordingly. First, drug action was peaked at 2 min (first-time point) and then progressively observed to be reduced by preinhalation levels with a possible mild effect after 20 min inhalation. Two participants were provided with the maximal rating of 10 during at least one session. The observations indicated that both male participants were insensitive (scores imputed as 10). One participant was insensitive at the 2 min time point in 15, 16.5, 18, and 21 μg/kg dosages and at the 4 min time point in the 21 μg/kg dosage. Another male was insensitive at the 2 min time point in the 18 and 19.5 μg/kg conditions. After considering results from nonhuman animal research (Mowry et al., 2003), results indicated that salvinorin A's profile was physiologically safe at the studied doses.

The endocannabinoid system is fundamentally recognized for alleviating the psychoactive influences of cannabis and accentuating pain, appetite, memory, and mood (Pertwee, 2008). The interplay between the endocannabinoid receptor CB1 and salvinorin A was initially examined in zebrafish (Braida et al., 2007), and investigated and validated in rat models (Braida et al., 2008). *Salvia divinorum* is used traditionally to treat gastrointestinal discomfort, which associates with in vitro findings revealing that the activity of the CB1 receptor represses gastrointestinal flux in the inflamed stomach of rats and ex vivo investigation proving that salvinorin A hinders nerve cells in the small intestines of the guinea pig (Capasso et al., 2006).

Under controlled conditions, participants psychologically and physically had a healthy experience of hallucinogenic effect. No significant changes were seen in heart rate or blood pressure, no adverse events, and no tremor reported for bioactive compound salvinorin A (Johnson et al., 2011). *S. divinorum* is a hallucinogenic plant with an active component of neoclerodane diterpene salvinorin A. It was seen to be a potent and selective agonist by in vitro pharmacological studies. In in vivo studies, salvinorin A produced distinctive stimulus effects that were known to be similar to those of a high-valued agonist (Harding et al., 2005). Inhaling of the volatile compound, salvinorin A gets absorbed leading to an accelerated rush of the psychoactive effects. The influences of salvinorin A are quick acting, but brief in continuance (Siebert, 1994). When inhaled by humans, the recorded psychoactive dosage of purified compound is 0.375–21 mg/kg (Johnson et al., 2011; MacLean et al., 2013). If volatile salvinorin A is breathed, the influences reveal within 60 s following inhalation and persist up to 15 min (Valdés, 1994; Johnson et al., 2011).

Addy et al. (2015) investigated the overall psychical consequences of inebriation, aided by the commonly-occurring plant stimulant *Salvia divinorum* utilizing a randomized, placebo-controlled double-blind trial. The 30 healthful people were administered with *S. divinorum* via smoking and inhaling in a calm and pleasant experimentation environment. Laboratory sittings, post-session conversations, and 64 days follow-up conversations were audio-recorded and deciphered. Further, post-session replies to the Hallucinogen Rating Scale provided a quantitative basis for investigation. Three principal issues and 10 sub-issues of intense inebriation appeared, comprising the attributes of the sense, perceptual modifications, and cognitive changes. The encounter characterized as having an accelerated rush, deep, and freakish. The subjects described notable differences in hearing, optical, and interoceptive sensory input, missing ordinary consciousness of themselves and their surroundings, and a collection of delusional events. In conclusion, the investigation presented an essential skeleton for a more comprehensive perception of the psychic effects of *S. divinorum*, an emerging drug of misuse.

16.3.7 REMEDIATION OF GASTROINTESTINAL DYSFUNCTION BY SALVINORIN A

The study for remediating gastrointestinal dysfunction measures described that isolated mice smooth muscle strips were exposed to phytochemical compound salvinorin A (3 mg/kg), which led to reduced contractions of the ileum, colon, and stomach and also inhibited neurogenic ion transport. The Swiss albino mice when exposed to salvinorin A (3 mg/kg) enhanced the colonic expulsion time (Fichna et al., 2009). These phytochemical compound salvinorin A's actions were mediated by activation of KOP, CB1, and CB2 receptors (Fichna et al., 2009). In another study, a standardized extract from *S. divinorum* leaves (1–100 mg/kg) and isolated salvinorin A compound (0.01–10 mg/kg) were administered to mice, which demonstrated a weak (only at the highest doses of salvinorin A) or an absent effect in the healthy control mice when a mouse model of chronic small intestine inflammation was administered an intestinal motility normalization was observed that occurred through a KOP activation-dependent mechanism. These data investigated the active phytochemical compound salvinorin A's long-term potential to manage gastrointestinal illnesses, for example, inflammatory bowel disease and diarrhea (Capasso et al., 2008).

KEYWORDS

- **diviner's sage**
- **psychotropic**
- **anti-inflammatory**
- **salvinorin a**
- ***Salvia divinorum***
- **pharmacology**

REFERENCES

Addy, P. H.; Garcia-Romeu, A.; Metzger, M.; Wade, J. The Subjective Experience of Acute, Experimentally-Induced *Salvia divinorum* Inebriation. *J. Psychopharmacol.* **2015,** *29* (4), 426–435.

Ansonoff, M. A.; Zhang, J.; Czyzyk, T.; Rothman, R. B.; Stewart, J.; Xu, H.; Zjwiony, J.; Siebert, D. J.; Yang, F.; Roth, B. L.; Pintar, J. E. Antinociceptive and Hypothermic Effects of Salvinorin A Are Abolished in a Novel Strain of κ-Opioid Receptor-1 Knockout Mice. *J. Pharmacol. Exp. Ther.* **2006,** *318* (2), 641–648.

Aviello, G.; Borrelli, F.; Guida, F.; Romano, B.; Lewellyn, K.; De Chiaro, M.; Luongo, L.; Zjawiony, J. K.; Maione, S.; Izzo, A. A.; Capasso, R. Ultrapotent Effects of Salvinorin A, a Hallucinogenic Compound from *Salvia divinorum*, on LPS-Stimulated Murine Macrophages and Its Anti-Inflammatory Action In Vivo. *J. Mol. Med.* **2011,** *89* (9), 891–902.

Baggott, M. J.; Erowid, E.; Galloway, G. P.; Mendelson, J. Use Patterns and Self-Reported Effects of *Salvia divinorum*: An Internet-Based Survey. *Drug Alcohol Depend.* **2010,** *111* (3), 250–256.

Beifuss, W. Cultivating Diviner's Sage: A Step by Step Guide to Cultivation, Propagation, and Keeping Your Salvia Plants Happy. *Resonance Project* **1997,** *1*, 32.

Bigham, A. K.; Munro, T. A.; Rizzacasa, M. A.; Robins-Browne, R. M. Divinatorins A– C, New Neoclerodane Diterpenoids from the Controlled Sage *Salvia divinorum*. *J. Nat. Prod.* **2003,** *66* (9), 1242–1244.

Bonito, M. C.; Cicala, C.; Marcotullio, M. C.; Maione, F.; Mascolo, N. Biological Activity of Bicyclic and Tricyclic Diterpenoids from *Salvia* Species of Immediate Pharmacological and Pharmaceutical Interest. *Nat. Prod. Commun.* **2011,** *6* (8), 1205–1215.

Braida, D.; Capurro, V.; Zani, A.; Rubino, T.; Viganò, D.; Parolaro, D.; Sala, M. Potential Anxiolytic-and Antidepressant-Like Effects of Salvinorin A, the Main Active Ingredient of *Salvia divinorum*, in Rodents. *Br. J. Pharmacol.* **2009,** *157* (5), 844–853.

Braida, D.; Limonta, V.; Capurro, V.; Fadda, P.; Rubino, T.; Mascia, P.; Zani, A.; Gori, E.; Fratta, W.; Parolaro, D.; Sala, M. Involvement of κ-Opioid and Endocannabinoid System on Salvinorin A-Induced Reward. *Biol. Psichiatr.* **2008,** *63* (3), 286–292.

Braida, D.; Limonta, V.; Pegorini, S.; Zani, A.; Guerini-Rocco, C.; Gori, E.; Sala, M. Hallucinatory and Rewarding Effect of Salvinorin A in Zebrafish: κ-Opioid and CB 1-Cannabinoid Receptor Involvement. *Psychopharmacology* **2007,** *190* (4), 441–448.

Butelman, E. R.; Mandau, M.; Tidgewell, K.; Prisinzano, T. E.; Yuferov, V.; Kreek, M. J. Effects of Salvinorin A, a κ-Opioid Hallucinogen, on a Neuroendocrine Biomarker Assay in Nonhuman Primates with High κ-Receptor Homology to Humans. *J. Pharmacol. Exp. Ther.* **2007,** *320* (1), 300–306.

Capasso, R.; Borrelli, F.; Capasso, F.; Siebert, D. J.; Stewart, D. J.; Zjawiony, J. K.; Izzo, A. A. The Hallucinogenic Herb *Salvia divinorum* and Its Active Ingredient Salvinorin A Inhibit Enteric Cholinergic Transmission in the Guinea-Pig Ileum. *Neurogastroenterol. Motil.* **2006,** *18* (1), 69–75.

Capasso, R.; Borrelli, F.; Zjawiony, J.; Kutrzeba, L.; Aviello, G.; Sarnelli, G.; Capasso, F.; Izzo, A. A. The Hallucinogenic Herb *Salvia divinorum* and Its Active Ingredient Salvinorin A Reduce Inflammation-Induced Hypermotility in Mice. *Neurogastroenterol. Motil.* **2008,** *20* (2), 142–148.

Carlezon, W. A.; Beguin, C.; DiNieri, J. A.; Baumann, M. H.; Richards, M. R.; Todtenkopf, M. S.; Rothman, R. B.; Ma, Z.; Lee, D. Y. W.; Cohen, B. M. Depressive-Like Effects of the κ-Opioid Receptor Agonist Salvinorin A on Behavior and Neurochemistry in Rats. *J. Pharmacol. Exp. Ther.* **2006,** *316* (1), 440–447.

Casselman, I.; Nock, C. J.; Wohlmuth, H.; Weatherby, R. P.; Heinrich, M. From Local to Global- Fifty Years of Research on *Salvia divinorum*. *J. Ethnopharmacol.* **2014,** *151* (2), 768–783.

Chaichantipyuth, C.; Muangsin, N.; Chaichit, N.; Roengsumran, S.; Petsom, A.; Watanabe, T.; Ishikawa, T. Crystal Structure of (-)-Hardwickiic Acid, $C_{19}H_{27}OCOOH$. *Z. Kristallogr.: New Cryst. Struct.* **2004,** *219*, 111–113.

Chavkin, C.; Sud, S.; Jin, W.; Stewart, J.; Zjawiony, J. K.; Siebert, D. J.; Toth, B. A.; Hufeisen, S. J.; Roth, B. L. Salvinorin A, an Active Component of the Hallucinogenic Sage *Salvia divinorum* Is a Highly Efficacious κ-Opioid Receptor Agonist: Structural and Functional Considerations. *J. Pharmacol. Exp. Ther.* **2004,** *308* (3), 1197–1203.

Chen, C.; Cui, X.; Matsunaga, F.; Ma, J.; Ma, N.; Abel, T.; Liu, R., Salvinorin A Decreases Mortality and Improves Neurological Outcome in a Neonatal Mouse Hypoxia Model. *Transl. Perioper. Pain. Med.* **2014,** *1* (2), 9.

Chen, C.; Xi, C.; Liang, X.; Ma, J.; Su, D.; Abel, T.; Liu, R. The Role of Kappa Opioid Receptor in Brain Ischemia. *Crit. Care. Med.* **2016,** *44* (12), e1219.

Coffeen, U.; Pellicer, F. *Salvia divinorum*: From Recreational Hallucinogenic Use to Analgesic and Anti-Inflammatory Action. *J. Pain Res.* **2019,** *12*, 1069.

Cornejo-Tenorio, G.; Ibarra-Manríquez, G. Diversity and Distribution of the Genus *Salvia* (Lamiaceae) in Michoacán, México. *Mexican J. Biodivers.* **2011,** *82* (4), 1279–1296.

Di Marzo, V. Targeting the Endocannabinoid System: To Enhance or Reduce? *Nat. Rev. Drug. Discov.* **2008,** *7* (5), 438–455.

Díaz, J. L. Ethnopharmacology and Taxonomy of Mexican Psychodysleptic Plants. *J. Psychedelic Drugs* **1979,** *11* (1–2), 71–101.

Díaz, J. L. *Salvia divinorum*: A Psychopharmacological Riddle and a Mind-Body Prospect. *Curr. Drug Abuse Rev.* **2013,** *6* (1), 43–53.

Epling, C.; Jativa,C. D. A New Species of *Salvia* from Mexico. *Bot. Mus. Lealf. Harv. Univ.* **1962,** *20*, 297–299.

Fichna, J.; Schicho, R.; Andrews, C. N.; Bashashati, M.; Klompus, M.; McKay, D. M.; Sharkey, K. A.; Zjawiony, J. K.; Janecka, A.; Storr, M. A. Salvinorin A Inhibits Colonic Transit and Neurogenic Ion Transport in Mice by Activating κ-Opioid and Cannabinoid Receptors. *Neurogastroenterol. Motil.* **2009,** *21* (12), 1326–e128.

Fine, P. G.; Portenoy, R. K. Chapter 2: The Endogenous Opioid System. *Clin. Guide Opioid Analgesia*, 2004, 9–15.

Fragoso-Martínez, I.; Martínez-Gordillo, M.; Salazar, G. A.; Sazatornil, F.; Jenks, A. A.; Peña, M. D. R. G.; Barrera-Aveleida, G.; Benitez-Vieyra, S.; Magallón, S.; Cornejo-Tenorio, G.; Mendoza, C. G. Phylogeny of the Neotropical Sages (*Salvia* subg. *Calosphace*; Lamiaceae) and Insights into Pollinator and Area Shifts. *Plant Syst. Evol.* **2018,** *304* (1), 43–55.

Giovannini, P.; Heinrich, M. *Xki yoma'* (our medicine) and *xki tienda* (Patent Medicine) -Interface Between Traditional and Modern Medicine Among the Mazatecs of Oaxaca, Mexico. *J. Ethnopharmacol.* **2009,** *121* (3), 383–399.

Grayer, R. J.; Eckert, M. R.; Veitch, N. C.; Kite, G. C.; Marin, P. D.; Kokubun, T.; Simmonds, M. S.; Paton, A. J. The Chemotaxonomic Significance of Two Bioactive Caffeic Acid Esters, Nepetoidins A and B, in the Lamiaceae. *Phytochemistry*, **2003,** *64* (2), 519–528.

Grundmann, O.; Phipps, S. M.; Zadezensky, I.; Butterweck, V. *Salvia divinorum* and Salvinorin A: An Update on Pharmacology and Analytical Methodology. *Planta Medica*, **2007,** *73* (10), 1039–1046.
Hanes, K. R. Antidepressant Effects of the Herb *Salvia divinorum*: A Case Report. *J. Clin. Psychopharmacol.* **2001,** *21* (6), 634–635.
Hanson, J. R. Natural Products from the Hallucinogenic Sage. *Sci. Prog.* **2010,** *93*, 171–180.
Harden, M. T.; Smith, S. E.; Niehoff, J. A.; McCurdy, C. R.; Taylor, G. T. Antidepressive Effects of the κ-Opioid Receptor Agonist Salvinorin A in a Rat Model of Anhedonia. *Behav. Pharmacol.* **2012,** *23* (7), 710–715.
Harding, W. W.; Tidgewell, K.; Byrd, N.; Cobb, H.; Dersch, C. M.; Butelman, E. R.; Rothman, R. B.; Prisinzano, T. E.; Neoclerodane Diterpenes as a Novel Scaffold for μ Opioid Receptor Ligands. *J. Med. Chem.* **2005,** *48* (15), 4765–4771.
Hawiger, J., 2001. Innate Immunity and Inflammation: A Transcriptional Paradigm. *Immunol. Res.* **2001,** *23* (2), 99–109.
Jenks, A. A.; Walker, J. B.; Kim, S. C. Evolution and Origins of the Mazatec Hallucinogenic Sage, *Salvia divinorum* (Lamiaceae): A Molecular Phylogenetic Approach. *J. Plant Res.* **2011,** *124* (5), 593–600.
Ji, R. R.; Zhang, Q.; Law, P. Y.; Low, H. H.; Elde, R.; Hokfelt, T. Expression of Mu-, Delta-, and Kappa-Opioid Receptor-Like Immunoreactivities in Rat Dorsal Root Ganglia After Carrageenan-Induced Inflammation. *J. Neurosci.* **1995,** *15* (12), 8156–8166.
Johnson, J. B. 1939. *Elements of Mazatec Witchcraft. "Ethnological Studies" No. 9*; Gothenburg Ethnographical Museum: Gothenburg. Sweden, 1939.
Johnson, M. W.; MacLean, K. A.; Reissig, C. J.; Prisinzano, T. E.; Griffiths, R. R. Human Psychopharmacology and Dose-Effects of salvinorin A, a Kappa Opioid Agonist Hallucinogen Present in the Plant *Salvia divinorum*. *J. Alcohol Drug Depend.* **2011,** *115* (1–2), 150–155.
Kowalczuk, A. P.; Raman, V.; Galal, A. M.; Khan, I. A.; Siebert, D. J.; Zjawiony, J. K. Vegetative Anatomy and Micromorphology of *Salvia divinorum* (Lamiaceae) from Mexico, Combined with Chromatographic Analysis of Salvinorin A. *J. Nat. Med.* **2014,** *68* (1), 63–73.
Kutrzeba, L.; Dayan, F. E.; Feng, J.; Giner, J. L.; Zjawiony, J. K. Biosynthesis of Salvinorin A Proceeds *via* the Deoxyxylulose Phosphate Pathway. *Phytochemistry* **2007,** *68* (14), 1872–1881.
MacLean, K. A.; Johnson, M. W.; Reissig, C. J.; Prisinzano, T. E.; Griffiths, R. R. Dose-Related Effects of Salvinorin A in Humans: Dissociative, Hallucinogenic, and Memory Effects. *Psychopharmacology* **2013,** *226* (2), 381–392.
Mascarenhas, C. J.; Liu, R.; Barr, G. A. Effects of Plant-Derived Analgesic Compounds Sinomenine and Salvinorin A in Infant Rats. *J. Integr. Med.* **2020,** *18* (2), 174–180.
McChesney, J. D.; Clark, A. M.; Silveira, E. R. Antimicrobial Diterpenes of *Croton sonderianus*, 1. Hardwickic and 3,4-secotrachylobanoic Acids. *J. Nat. Prod.* **1991,** *54*, 1625–1633.
Miranda, H. F.; Noriega, V.; Sierralta, F.; Poblete, P.; Aranda, N.; Prieto, J. C. Non-steroidal Anti-Inflammatory Drugs in Tonic, Phasic and Inflammatory Mouse Models. *Drug Res.* **2019,** *69* (10), 572–578.
Misra, R.; Pandey, R. C.; Dev, S. The Absolute Stereochemistry of Hardwickiic Acid and Its Congeners. *Tetrahedron Lett.* **1968,** *9*, 2681–2684.

Morris, C. J. Carrageenan-Induced Paw Edema in the Rat and Mouse. *Inflamm. Protocols*, **2003,** 115–121.

Mowry, M.; Mosher, M.; Briner, W. Acute Physiologic and Chronic Histologic Changes in Rats and Mice Exposed to the Unique Hallucinogen Salvinorin A. *J. Psychoactive Drugs* **2003,** *35* (3), 379–382.

Munro, T. A. *The Chemistry of Salvia divinorum*; Doctoral Dissertation; Department of Chemistry, The University of Melbourne, Australia, 2006.

Ortega, A.; Blount, J. F.; Manchand, P. S. Salvinorin, a New Trans-Neoclerodane Diterpene from *Salvia divinorum* (Labiatae). *J. Chem. Soc. Perkin* **1982,** *1*, 2505–2508.

Ott, J. Ethnopharmacognosy and Human Pharmacology of *Salvia divinorum* and Salvinorin A. *Curare*, **1995,** *18* (1), 103–129.

Ott, J. Psychoactive Card IV. *Eleusis*, **1996,** *4*, 31–39.

Pertwee, R. G. The Diverse CB1 and CB2 Receptor Pharmacology of Three Plant Cannabinoids: Δ^9-Tetrahydrocannabinol, Cannabidiol and Δ^9-Tetrahydrocannabivarin. *Br. J. Pharmacol.* **2008,** *153* (2), 199–215.

Prisinzano, T. E. Psychopharmacology of the Hallucinogenic Sage *Salvia divinorum*. *Life Sci.* **2005,** *78* (5), 527–531.

Rajab, M. S.; Cantrell, C. L.; Franzblau, S. G.; Fischer, N. H. Anti-Mycobacterial Activity of (E)-Phytol and Derivatives: A Preliminary Structure-Activity Study. *Planta Med. Nat. Prod. Med. Plant Res.* **1998,** *64*, 2–4.

Reisfield, A. S. The Botany of *Salvia divinorum* (Labiatae). *Sida, Contr. Bot.* **1993,** 349–366.

Rossi, A.; Caiazzo, E.; Bilancia, R.; Riemma, M. A.; Pagano, E.; Cicala, C.; Ialenti, A.; Zjawiony, J. K.; Izzo, A. A.; Capasso, R.; Roviezzo, F. Salvinorin A Inhibits Airway Hyper Reactivity Induced by Ovalbumin Sensitization. *Front Pharmacol.* **2017,** *7*, 525.

Schultes, 1972. An Overview of Hallucinogens in the Western Hemisphere. In *Flesh of the Gods: The Ritual Use of Hallucinogens*; Furst, P. T., Ed.; pp 3–54.

Schultes.; Hofmann, A.; Rätsch, C. *Plants of the Gods—Their Sacred, Healing, and Hallucinogenic Powers*; Healing Arts Press: Rochester, 2001.

Siebert, D. J. Localization of Salvinorin A and Related Compounds in Glandular Trichomes of the Psychoactive Sage, *Salvia divinorum*. *Ann.Bot.* **2004,** *93* (6), 763–771.

Siebert, D. J. *Salvia divinorum* and Salvinorin A: New Pharmacologic Findings. *J. Ethnopharmacol.* **1994,** *43* (1), 53–56.

Sim, G. A.; Hamor, T. A.; Paul, I. C.; Robertson, J. M. The Structure of Clerodin. *Proc. Chem. Soc.* **1961,** 75–77.

Su, D.; Riley, J.; Armstead, W. M.; Liu, R. Salvinorin A Pretreatment Preserves Cerebrovascular Autoregulation After Brain Hypoxic/Ischemic Injury *via* Extracellular Signal-Regulated Kinase/Mitogen-Activated Protein Kinase in Piglets. *Anaesth. Analg.* **2012,** *114* (1), 200.

Su, D.; Riley, J.; Kiessling, W. J.; Armstead, W. M.; Liu, R. Salvinorin A Produces Cerebrovasodilation Through Activation of Nitric Oxide Synthase, κ Receptor, and Adenosine Triphosphate–Sensitive Potassium Channel. *J. Am. Soc. Anesthesiologists* **2011,** *114* (2), 374–379.

Tlacomulco-Flores, L. L.; Déciga-Campos, M.; González-Trujano, M. E.; Carballo-Villalobos, A. I.; Pellicer, F. Antinociceptive Effects of *Salvia divinorum* and Bioactive Salvinorins in Experimental Pain Models in Mice. *J. Ethnopharmacol.* **2020,** *248*, 112276.

Valdés, L. J. *Salvia divinorum* and the Unique Diterpene Hallucinogen, Salvinorin (Divinorin) A. *J. Psychoact. Drugs.* **1994,** *26* (3), 277–283.

Valdés, L. J. The Early History of *Salvia divinorum*. *Entheogen Rev.* **2001,** *10*, 73–75.

Valdés, L. J.; Hatheld, G. M.; Koreeda, M.; Paul, A. G. Studies of *Salvia divinorum* (Lamiaceae), an Hallucinogenic Mint from the Sierra Mazateca in Oaxaca, Central Mexico. *Econ. Bot*. **1987,** *41* (2), 283–291.
Valdés, L. J.; Díaz, J.; Paul, A. G. Ethnopharmacology of SKA Maria Pastora (*Salvia divinorum*, Epling and Játiva-M.). *J. Ethnopharmacol.* **1983,** *7* (3), 287–312.
Wang, Z.; Ma, N.; Riley, J.; Armstead, W. M.; Liu, R. Salvinorin A Administration After Global Cerebral Hypoxia/Ischemia Preserves Cerebrovascular Autoregulation *via* Kappa Opioid Receptor in Piglets. *PloS One*, **2012,** *7* (7), e41724.
Wasson, R. G. A New Mexican Psychotropic Drug from the Mint Family. *Botanical Museum Leaflets Harv. Univ.* **1962,** *20* (3), 77–84.

CHAPTER 17

AN UPDATED OVERVIEW ON GREEK SAGE (*SALVIA FRUTICOSA* MILL.) AND CHERRY SAGE (*SALVIA GREGGII* A. GRAY)

SHARMILA ARUNAGIRI[1], KALLIPUDI CHARISHMA REDDY[2], OISHIKA SAH[2], and CHINNADURAI IMMANUEL SELVARAJ[2]

[1]*Department of Biotechnology, School of Biosciences and Technology, Vellore Institute of Technology, Vellore, Tamil Nadu, India*

[2]*VIT School of Agricultural Innovations and Advanced Learning (VAIAL), Vellore Institute of Technology, Vellore, Tamil Nadu, India*

ABSTRACT

Greek sage, also known as *Salvia fruticosa* Mill., is a shrub or perennial herb native to the eastern Mediterranean, including the Canary Islands, North Africa, southern Italy, and Israel. The essential oil from *S. fruticosa* is applied topically to the stomach to treat dysmenorrhea and is used as a disinfectant on open wounds and bruises. The leaf infusion is used to treat different conditions, including colds, coughs, stomach aches, constipation, cardiovascular illness, diuretics, antiseptics, ulcers, sterility, respiratory diseases, and brain disorders. Autumn sage, also known as *Salvia greggii* A. Gray., is commonly found in the Mexican and Texan regions of the United States. Similar to *Salvia officinalis*, it is a soft, evergreen shrub. The antioxidant and inhibitory properties of *S. greggii* regulates obesity and diabetes. Flowers of *S. greggii* are edible, while leaves are utilized in teas and as seasonings in food. The plant species is used in traditional medicine as herbal antidiabetic and antiobesity medication, as well as to treat oral and

Bioactives and Pharmacology of Lamiaceae. T. Pullaiah, PhD (Ed.)

digestive issues. Significant levels of bioactive substances such as terpenoids, flavonoids, and phenolics are present in *S. fruticosa*. Aliphatic compounds, volatile monoterpenes, phenylpropanoids, and sesquiterpenes constitute the essential oil of *S. greggii*. They include cirsimaritin, ellagic acid, luteolin, genkwanin, α-thujone, caryophyllene, cineole, sabinene, α-cadinene, ursolic acid, quercetin, and rosmarinic acid. This chapter summarizes the bioactive components of both *Salvia fruticosa* and *Salvia greggii* and its various pharmacological activities.

17.1 INTRODUCTION

Lamiaceae is a flowering plant family with aromatic leaves, square stems, and a corolla segmented into two lip-like parts, similar to that in the snapdragon. *Salvia fruticosa* Mill, otherwise known as Greek sage, is a shrub or perennial herb inherited to the eastern Mediterranean, including the Canary Islands, southern Italy, North Africa, and the richest in Israel. Since 1400 BC, *Salvia fruticosa* has been utilized (Rivera et al., 2016). It is commonly known as *Adac-ayı*, *Almakeyik*, and *Almageyik* in Turkish (Gürdal and Kültür, 2013). The species is hermaphrodite, pollinated by bees. The leaf's shape varies widely; there has been a great deal of taxonomic confusion over the years; they have various names. These include *Salvia libanotica* Boiss & Gaill., *Salvia triloba* L.f., *Salvia lobryana* Azn., and *Salvia cypria* Unger & Kotschy, which are now considered "*Salvia fruticosa*." The leaf variation depends on geographical area, with plants growing on the western part of Crete having entire leaves with flat blades and margins and dark green upper sides. Plants growing on the island's eastern side have much smaller leaves, with deeply three-lobed yellowish–green blades and undulate margins. Leaves of certain *Salvia* spp. are predominantly used in meat preparations for their pleasant aroma, flavor, and preserved leaves to make flavored tea. Around 40 endemic *Salvia* species are thriving on the Mediterranean seashores of Turkey (Davis, 1982).

Salvia fruticosa, the essential oil (EO), is utilized as a disinfectant on exposed bruises and injuries and is rubbed over the stomach for dysmenorrhea. Its EO is administered to treat vaginitis (Gürdal and Kültür, 2013). The aqueous leaf extract of *Salvia fruticosa* can treat many ailments, particularly digestive complaints (Gürdal and Kültür, 2013). *Salvia fruticosa* treatment produces hypoglycemia mainly by decreasing the colonic absorption of glucose (Perfumi et al., 1991; Azevedo et al., 2011). The leaf infusion is used to treat various ailments, namely, cold (Honda et., 1996; Aslan et al., 2007; Ugurlu and Secman, 2008), cough (Honda et al., 1996), stomachache (Ertug, 2004), flatulence (Fakir et al., 2009), constipation (Honda et al., 1996), cardiovascular disease (Gonzalez-Tejero et al., 2008), diuretic, antiseptic (Fakir et al., 2009), prostatitis, digestive, ulcer, liver disorders, antiseptic, appetizing, antiaging and carminative (Everest and Ozturk, 2005), sterility, neuralgia, burn, herpes, (Ertuğ, 2004), respiratory disease, anti-inflammatory, wound (Akalın and Alpinar, 1994), muscular–skeletal illness, skin complaints (Gonza′lez-Tejero et al., 2008), and neural disorders (Akalın and Alpınar, 1994). *Salvia fruticosa* is an antiproliferative agent on cancer cells. It inhibits cholinesterase activity, protection from oxidative DNA damage, and antimicrobial and anti-inflammatory activities (Exarchou et al., 2015).

Salvia greggii A. Gray, commonly called "autumn sage", belongs to the Lamiaceae family. Other than autumn sage, it is also called Cherry Sage and Cherry Salvia. It originates from the region of Mexico and Texas, USA, and is spread worldwide. Contrary to its name, "autumn sage", *S. greggii* is a perennial plant. Flowering generally occurs during spring and fall times of the year, with more sporadic blooms during the summer heat. These flowers are usually characterized by their intense aroma and nectar present in abundant amount. It attracts a huge number of bees and hummingbirds, which in turn helps with pollination. This plant is heavily used for medicinal purposes and for producing EOs. *S. greggii* has been proven to be highly efficient in treating digestive and oral problems. Obesity and diabetes are controlled by the inhibitory activity and antioxidant properties of *S. greggii* toward critical metabolic enzymes. It is a soft, evergreen shrub with a height reaching anywhere from 1 to 4 ft and is quite similar to *Salvia officinalis*. The stem has multiple branches present with clumping, woody, and thin stems. The foliage is simple, elliptic-shaped, narrow, leathery, and aromatic, along with the opposite arrangement. Flowers are borne on racemes from spring to frost. Flowers are found in many colors, such as pink, purple, white, and orange, but red is predominantly found. The fruit is inconspicuous and not showy; unknown shape, color, and length (Gilman et al., 1999). The leaves are used for seasonings and teas (culinary use); flowers are edible. The plant species finds its usage as a traditional medicine to treat digestive and oral problems, as a natural antidiabetic and antiobesity agent (Frett, 1987; Kintzios, 2000).

17.2 BIOACTIVITES

Salvia fruticosa consists of significant amounts of bioactive compounds such as terpenoids and phenolics with antimicrobial, antioxidant, and anti-inflammatory properties (Torun et al., 2014). *Salvia fruticosa* contains numerous EOs in which many of which belong to the sesquiterpene hydrocarbons group. *Salvia fruticosa* has R-thujone, ranging from traces to 14.5%; 1,8-cineole, which ranges from 16.9% to 54.4%; α-thujone, ranges from 0.6% to 9.0%; €-caryophyllene, ranging from traces to 15.6% and camphor, ranging from 0.6% to 15.9%; (Pitarokili et al., 2003). *Salvia* species consists of biologically active diterpenoids, majorly with a skeleton of clerodane or abietane. Manool and isopimara9(11),15-diene have also been identified in the *Salvia fruticosa* EOs (Pitarokili et al., 2003). These oils are used extensively as an antifungal agents.

Salvia fruticosa was abundant in antioxidants source, in which the extracts of the root have potent antioxidant activity and high total phenolics concentration. The phenolic constituents like rutin, luteolin, and gallic acid were identified using column chromatography and quantified using HPLC. The plant extracts' polyphenolic content due to their radical scavenging activity may also have a part in a potential anti-inflammatory activity (Boukhary et al., 2016). The EOs of *Salvia fruticosa* are highly diverse; three-lobed sage oil with a high amount of 1,8-cineole (40–60%). The main compound present in *S. officinalis* are A and B-thujone but found only in trace amounts. Thujone acts as a convulsant and has emmenagogue properties based on its dose. Cineole is known as a tonic, antiseptic, and cardiovascular stimulant and treats respiratory tract disorders. *Salvia fruticosa* is used as vulnerary medicine and for treating vascular diseases and respiratory problems, but it has no use in treating menstrual disorders or childbirth difficulties (Rivera et al., 2016). The structures of major flavonoids and terpenes extracted from *Salvia fruticosa* (Fig. 17.1) and structures of EOs, terpenes, and terpenoids from *Salvia fruticosa* are shown in Figure 17.2.

The phytochemicals produced from the *Salvia* group of plants are monoterpenoids, mostly diterpenoids with a skeleton of clerodane or abietane, triterpenoids, and flavonoids. Triterpenoids and flavonoids are the main components in the medicinal history of *Salvia* species (Wu et al., 2012). The volatile profile of flowers was rich in terpene hydrocarbons, out of which monoterpenes were predominant, followed by sesquiterpenes. The most important components were limonene and β-pinene. Germacrene D and β-caryophyllene are the two highly abundant sesquiterpene hydrocarbons detected. The predominant compounds found in floral emission were monoterpenes (Giuliani et al., 2020).

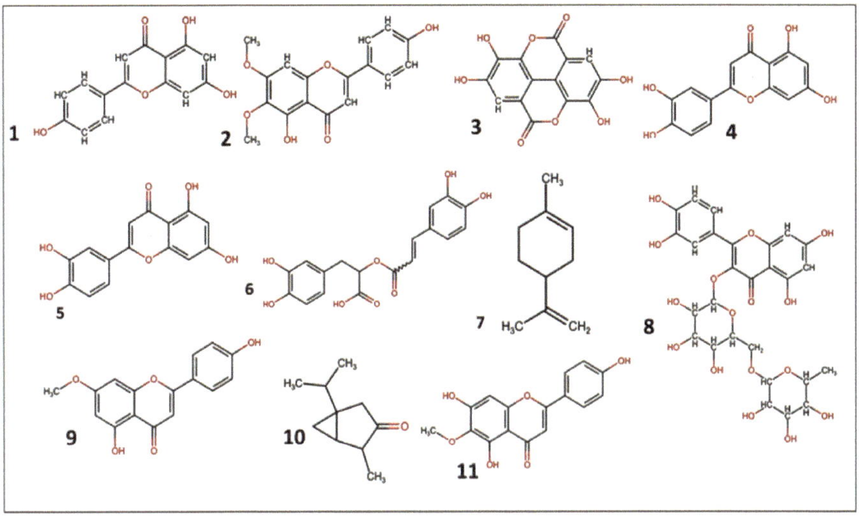

FIGURE 17.1 Structures of major flavonoids and terpenes extracted from *Salvia fruticosa.* Apigenin (**1**), cirsimaritin (**2**), ellagic acid (**3**), luteolin (**4**), quercetin (**5**), rosmarinic acid (**6**), limonene (**7**), rutin (**8**), genkwanin (**9**), α-thujone (**10**), and hispidulin (**11**). [Marvin 17.21.0, ChemAxon tool was used for drawing chemical structures].

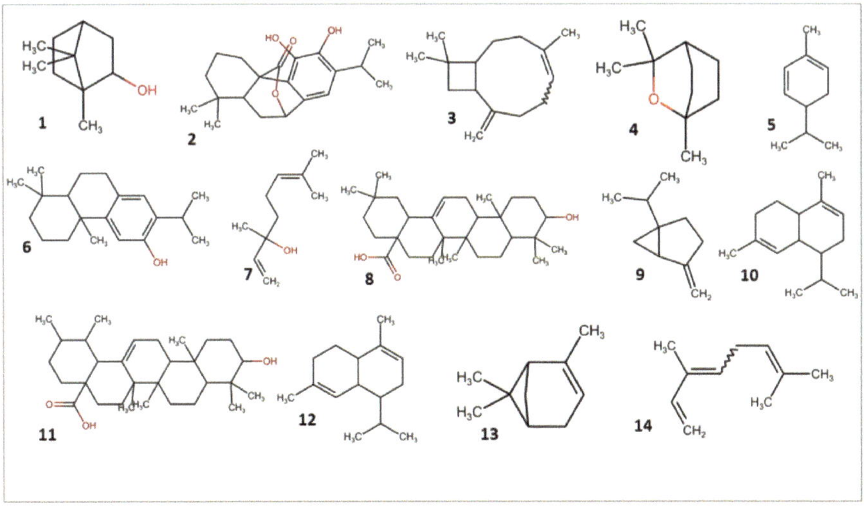

FIGURE 17.2 Structures of EOs, terpenes, and terpenoids from *Salvia fruticosa.* Borneol (**1**), carnosol (**2**), caryophyllene (**3**), cineole (**4**), phellandrene (**5**), ferruginol (**6**), linalool (**7**), oleanolic acid (**8**), sabinene (**9**), α-cadinene (**10**), ursolic acid (**11**), α-muurolene (**12**), α-pinene (**13**), and β-ocimene (**14**). [Marvin 17.21.0, ChemAxon tool was used for drawing chemical structures].

According to the profile of leaf samples, terpene hydrocarbons were abundantly present, out of which sesquiterpenes were dominant, followed by monoterpenes. The chief sesquiterpene hydrocarbons were muurolene, germacrene D, β-gurjunene, and β-caryophyllene; the natural bicyclic sesquiterpene, *β*-caryophyllene contains anti-inflammatory activity. Germacrene is a volatile hydrocarbon with antimicrobial and insecticidal properties. The chief compounds of monoterpenes are β-pinene and 1,8-cineole. There were twenty compounds exclusively present in the leaf profile, among which 1,8-cineole and camphor were the dominant ones (Giuliani et al., 2020). Diterpenoid compounds are found in the roots of *S. greggii* extensively which are listed as follows-salvigreside A, salvigreside B, salvigreside C, salvigreside D, salvigresin, 14α,18-dihydroxy-7,15-isopimaradiene, salviarin, 7,8β-dihydrosalviacoccin, 14α-hydroxy-7,15-isopimaradien-18-oic acid, 3β-hydroxy-7,15-isopimaradien-18-oic acid methyl ester, and isopimaric acid (Wu et al., 2012). Salviagresin is a diacetylated tetracyclic clerodane diterpenoid. Salvigreside A, B, C, and D are colorless amorphous, neoclerodane diterpenoid glucosides (Kawahara et al., 2004). Antibacterial activity was found in salvigreside D on *Bacillus subtilis* in an experiment conducted by Kawahara et al. (2003).

S. greggii EO consists of aliphatic substances, volatile monoterpenes, phenylpropanoid, and sesquiterpenes. In the analysis of EO from *S. greggii*, 49 different compounds were known with terpenes which account for 98.80% of the total, together with non-terpene derivatives and apocarotenoids. Monoterpene fraction, including oxygenated monoterpene, was more prominent (68.9%) and sesquiterpenes (24.5%). Significant constituents were *cis*-thujone and geranyl acetate, followed by camphor and geraniol in lesser amounts. The most abundant sesquiterpenes found are cadinene, α-muurolene, and biciclogermacrene (De Martino et al., 2010). Oxygenated sesquiterpenoids were also identified, including guaoil and caryophyllene oxide as the predominant ones. The monoterpene *β*-pinene was the most abundant compound found. 1,8-cineole was the predominant oxygenated monoterpene found. *S.greggii* aqueous extract was rich in phenolic compounds, mainly rosmarinic acid, apigenin-C-hexoside, luteolin-7-O-glucoside, and luteolin-8-C-glucoside. Luteolin-7-glucoside was found predominantly in the extract (Pereira et al., 2014). Rosmarinic acid is a hydroxycinnamic acid and it is 3,4-dihydroxyphenylacetic acid and caffeic acid esters. It is a major phenolic component in *Salvia* plants.

Fraternale et al. (2013) reported the phytochemical studies of the hybrid of *S. greggii* and *Salvia microphylla* called *Salvia* x *jamensis*. The EO analysis of the hybrid assisted in identifying 56 compounds, out of which 93.8% are volatile compounds. The most significant volatile compounds identified were sesquiterpenes, followed by oxygenated sesquiterpenes, monoterpenes, and oxygenated monoterpenes. The prominent compounds are *β*-caryophyllene, borneol, *β*-pinene, 1,8-cineole, caryophyllene oxide, α-copaene, α-pinene, and spathulenol. Hence, one can conclude that the oil showed similar composition to *S. greggii* and *S. microphylla.* The structures of important compounds present in *S. greggi* are given in Figure 17.3.

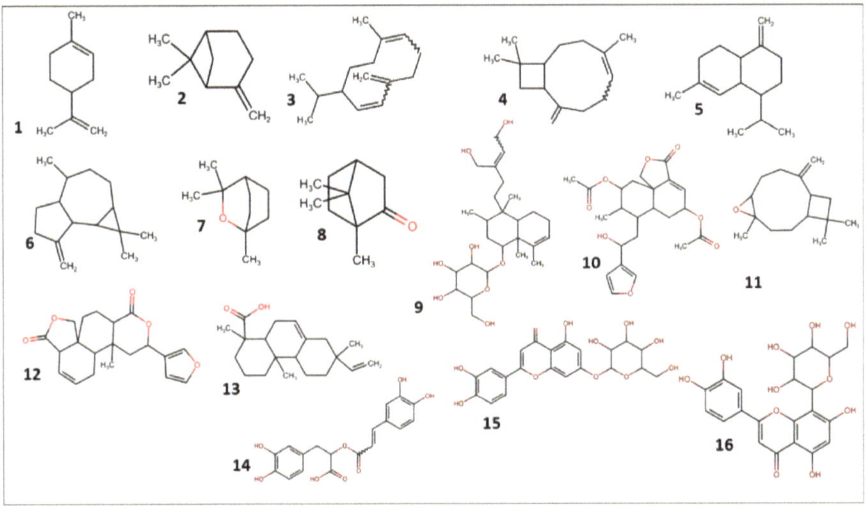

FIGURE 17.3 Structures of limonene (**1**), *β*-pinene (**2**), germacrene D (**3**), *β*-caryophyllene (**4**), γ-muurolene (**5**), *β*-gurjunene (**6**), 1,8-cineole (**7**), camphor (**8**), salvigreside A (**9**), salvigresin (**10**), caryophyllene oxide (**11**), salviarin (**12**), isopimaric acid (**13**), rosmarinic acid (**14**), luteolin-7-O-glucside (**15**), and luteolin-8-c-glucoside (**16**). [Marvin 17.21.0, ChemAxon tool was used for drawing chemical structures].

The chemical nature of *Salvia fruticosa* significantly correlates with the environment and developmental stages (Koliopoulos et al., 2010; Papageorgiou et al., 2008). Koliopoulos et al. (2010) described that *Salvia fruticosa* from Ikaria isles comprises 23.1% camphor as its principal ingredient. However, a lower borneol (12.6%), α-pinene (12.7%), 1,8-cineole (6.9%), camphene (9.0%), transcaryophyllene (5.3%), and β-pinene (5.8%) was observed, though the same researchers detailed on discrete chemical characterization for the *Salvia fruticosa* gathered from Kalymnos archipelago (camphene, 8.5%, α-pinene, 8.7%; camphor, 22.6%; 1,8-cineole, 31.4%). In contrast, Papageorgiou et al. (2008) observed that the volume of EO of the plants did associate with their quantum of 1.8-cineol. However, the antioxidant action of *Salvia fruticosa* was reliant on the unusual volume of sesquiterpenes and diterpenes.

17.3 PHARMACOLOGY

17.3.1 ANTIOXIDANT ACTIVITY

Salvia fruticosa oil essence and its alcoholic decoction, exhibited an intense antioxidant action (Papageorgiou et al., 2008; Şenol et al., 2011). The *S. frutitcosa* ethyl acetate root extract exhibited a strong 2,2-diphenyl-1-picryl-hydrazyl (DPPH) radical scavenging activity compared to the plant's aerial parts. The crocin test and Rancimat test displayed a fascinating antioxidant activity of *Salvia fruticosa* (Bonesi et al., 2017). The ethyl acetate root extract of *S. fruitcosa* exhibited the highest free radicals scavenging activity than their aerial parts (Boukhary et al., 2016). The *Salvia fruticosa* extracts of methanol, petroleum ether, and dichloromethane significantly reduce reactive oxygen species (ROS) present in SH-SY5Y cells when exposed to *tert*-butyl hydro-peroxide (TBHP). Among these, the *Salvia fruticosa* methanol extract showed efficient antioxidant activity. The optimal concentrations of *Salvia fruticosa* for its antioxidant activity were 50 and 20 μg/mL (Ververis et al., 2020).

The dry leaves of *S. fruticosa* were extracted in PBS and filtered utilizing vacuum and high-pressure filtrations. Every milliliter of the filtration carried (7.0 ± 1.0) mg of the extract. HEK-293 cells that were incubated with a defined concentration of *Salvia fruticosa* extract in the presence of 0.1 mmol/L of H_2O_2 and another set with the extract only. Fluorescence involving fluorescein isothiocyanate derivatization of 8-oxaguanine moieties indicates the level of DNA oxidation; fluorescence was measured using the flow cytometry technique. Cells incubated for three hours with 150 mL of extract and exposed to 0.1 mmol/L hydrogen peroxide exhibited fainter intensity of fluorescence and feebler DNA oxidation. Besides, cells incubated for 3 hours with 100 mL of the extract manifested more muffled strength of fluo-rescence and lowered inherent cellular DNA oxidation than control (in the absence of *S. fruticosa*). The outcomes from this investigation imply that the aqueous extract of *S. fruticosa* leaves shields against both H_2O_2 provoked and intrinsic cellular DNA oxidation in Human Embryonic Kidney 293 cells (Hani and Bayachou, 2014).

Aqueous extracts from *S. greggii* were examined for their antioxidant ability concerning the ability for scavenging free radicals such as $O_2^{\bullet -}$ (super-oxide), DPPH$^\bullet$, NO$^\bullet$ (nitric oxide), and $RO_2^\bullet$ (peroxyl), and its ability to reduce Fe^{3+} to Fe^{2+}, then the extracts were screened for its potency to inhibit xanthine oxidase. The decoction (flavones and flavonols, caffeic acid, and coumaric acid derivatives) showed a promising ability for scavenging free radicals and reducing Fe^{3+} to Fe^{2+}. *S. greggii* decoctions have more potential for scavenging DPPH• and RO_2•. The decoction also effectively inhibits xanthine oxidase activity. The enzyme xanthine oxidase catalyzes the hypoxanthine to xanthine by oxidation process and catalyzes xanthine to uric acid further, accompanied by the production of $O_2^{\bullet -}$ (Pereira et al., 2018).

17.3.2 ANTITUMOR ACTIVITY

S. fruticosa incites DNA repairing mechanism in HeLa cells (Ramos et al., 2010). The ethanolic essence of *S. fruticosa* has antiproliferative properties on breast cancer cells (Abu-Dahab et al., 2012). Liquid chromatography and tandem mass spectrometry methods are involved in identifying the phenolic components found in the methanolic extracts of *S. fruticosa*. The most abundant terpene phenolic compound present in *S. fruticosa* is carnosic acid and its metabolite carnosol. *S. fruticosa* extract is highly efficient compared to *Salvia pomifera* extract as it potentially decreases the multiplication of human melanoma cells. Carnosic acid exhibited the most significant antitumor activity on cancer cells. It was evident that microtubule dynamics were affected by carnosic acid, and it is also responsible for arresting the cell cycle in the G2/M phase (Koutsoulas et al., 2019).

There are not many studies available on *S. fruticosa* about its anticancer activity. It contains rosmarinic acid, which is the main phenolic component in their water extracts. Of all the phenolic components in *S. fruticosa* water extract, rosmarinic acid is 58%. This phenolic compound is known for potent anti-inflammatory and antioxidant activities but has little effect on cancer cells, particularly colorectal carcinoma (Xavier et al., 2009). The extracts of methanol, petroleum ether, and dichloromethane, showed a significant cytotoxic effect at relatively high concentrations. This cytotoxic effect of *S. fruticosa* was potent against different human cancer cell lines compared to noncancer cells. Hence, extracts of *S. fruticosa* can be efficient for the treatment of cancer (Ververis et al., 2020).

The antiproliferative activity of the EO from leaves was studied on HCT116 and MCF-7 cell lines. The EO exhibited promising dose-dependent repression of cell growth at low IC_{50}'s (HCT116: 23.6 μg/mL; MCF-7: 35.3 μg/mL) with a reasonable safety profile against the standard Vero cell line as opposed to the positive standard doxorubicin (Abaza et al., 2021).

17.3.3 CYTOTOXIC AND APOPTOTIC ACTIVITY OF S. GREGGI X JAMENSIS

A study on the surface exudate of *S. greggii* x *jamensis* revealed good platelet anticlumping activity. The profiled compounds from the extract, namely, 14-α-hydroxy-isopimaric acid, isopimaric acid, 7,8β-dihydrosalviacoccin, 3β-hydroxy-isopimaric acid, ursolic acid, and betulinic acid, were isolated collectively with a unique diterpene X. The structure of X was ascertained as 15,16-epoxy-cleroda-3-en-7α,10β-dihydroxy-12,17; 19,18-diolide using spectroscopic data investigation. With all examined compounds, isopimaric acid exhibited a notable concentration-dependent anticlumping activity when adenosine di-phosphate (ADP) with a concentration of 3 μM, was used as an agonist on rat platelets. In contrast, X enhanced ADP-influenced platelet aggregation (Bisio et al., 2008). U937 cell line culture was used for determining the cytotoxic ability of the hybrid's EO. The cell viability, cell growth, and apoptosis were measured every hour; apoptosis initiation and cell growth inhibition were noted every 4 hours. Volatile compounds such as β-caryophyllene, α-pinene, and β-pinene were tested separately and in combination. According to the results, β-caryophyllene had the most potent anti-inflammatory and anticarcinogenic effects. In contrast, α-pinene stimulates apoptosis of tumor cell lines and confers the protection of the antimetastatic state in a melanoma model. Another compound, limonene, was found to have antiproliferative activity and it also induces apoptosis ability. In DPPH radical assay, the EO did not exhibit any antioxidant activity (Fraternale et al., 2013).

17.3.4 ANTI-INFLAMMATORY ACTIVITY

In in vivo condition, *S. fruticosa* showed anti-inflammatory activity with the aerial part of the plant extract and crude extract of roots showed substantial protection at four hours against mouse paw edema induced by carrageenan to the extent of 50% and 44%, respectively. With that of the standard antibiotic diclofenac, the roots' anti-inflammatory abilities and the aerial part extracts were 0.9 and 0.8, respectively. The aerial parts of *S. fruticosa* exhibit intense anti-inflammatory activity (Boukhary et al., 2016).

17.3.5 ANTIMICROBIAL EFFECTS

The ethanol and water extracts of *S. fruticosa* were used to assess the antimicrobial capacity against various Gram-negative and Gram-positive bacteria and fungi by using microdilution. Ethanol extracts of *S. fruticosa* displayed higher activity than its water extracts on Gram-positive bacteria (Duletic-Lausevic et al., 2018). High antibacterial performance against common Gram-positive and -negative bacteria and fungi were shown by the methanol extract of the bark of *S. fruticosa*. The extract can treat infections caused by *B. subtilis*, *Escherichia coli*, and *Staphylococcus aureus*, inhibiting the infection with a MIC of 50 mg/mL. *S. fruticosa* bark will be more potent in treating diseases caused by *Candida albicans* and *Pseudomonas aeruginosa.* Both were inhibited by a lower minimal inhibitory concentration (MIC) of 12.5 mg/mL and 25 mg/mL, respectively (Abdulla-Eltawaty et al., 2021).

Zaccardelli et al. (2020) studied the antibacterial activity using gas chromatography–mass spectrometry (GC–MS) for volatile compounds isolated from the *S. greggii* EO on two phytopathogens, *i.e.*, *Pectobacterium carotovorum* subsp. *carotovorum* and *Xanthomonas campestris* pv. *campestris*. In vitro plate tests were performed for evaluating the ability to inhibit bacterial growth. They concluded that *S. greggii* showed partial inhibition against *X. campestris* pv. *campestris* and no inhibition in case of *P. carotovorum* subsp. *carotovorum*.

17.3.5.1 ANTIFUNGAL ACTIVITY

GC–MS was used to evaluate the volatile metabolites extracted from *S. fruticosa*. The presence of EO ranges from 0.69% to 4.68% and it exhibited an effective antifungal activity (Pitarokili et al., 2003). Among the extracts of *S. fruticosa* prepared through different solvent polarities such as methanol, ethyl acetate, hexane, and water, ethyl acetate extract of *S. fruticosa* had the potent antifungal activity. The flavonoids such as hispidulin, salvigenin, and cirsimaritin and diterpenes such as carnosol, 12-methoxycarnosic acid, and carnosic acid considered to be the significant components of the ethyl acetate extract characterized by HPLC-SPENMR and q-NMR experiments. *S. fruticosa* antifungal activity was primarily evaluated qualitatively against *Botrytis cinerea*, *Penicillium digitatum*, and *Aspergillus tubingensis*. At the concentration of 400 μg mL^{-1}, all the verified compounds' inhibitory activity was found to be improved, but *S. fruticosa* extract has potent fungal activity against fungal species (Exarchou et al., 2015).

Ultrastructural studies on *B. cinerea* cells, when exposed to 1,8-cineole showed harmful effects on cell organelles, concluding that 1,8-cineole leads to the antifungal activity of *S. greggii* (Giuliani et al., 2017).

17.3.5.2 ANTIVIRAL ACTIVITY

In a study against HSV-1 (an abundant human pathogen), *S. fruticosa* EO extract and its major compounds were tested for its virucidal action. At 0.2% concentration, the oil inactivates infectious viruses up to 80% in 30 min. The effect enhanced at a higher concentration of 0.4% by nearly 4log10 values. Thujone displayed more virucidal activity due to its most active nature among all the components. At 0.1% dilution, thujone inactivates 95% of viral pathogens. The EO of *S. fruticosa* contains some monoterpenes such as bornyl acetate, isoborneol, and borneol, which showed a reasonable amount of cytotoxic activity against Vero cells and antiviral activity against HSV-1 (Sivropoulou et al., 1997).

17.3.6 COGNITIVE AND MEMORY ENHANCING EFFECTS

Orhan et al. (2008) observed that *S. fruticosa* has a notable anticholinesterase activity. The extracts of *S. fruticosa* are known to have cognitive and memory enhancing effects. The ethanolic extract of *S. fruticosa* helps to increase retention of memory of passive avoidance learning in rats. *S. fruticosa* hydroalcoholic extract major flavonoid is known to be rosmarinic acid, which improves healthy rats' cognition and diabetes-induced memory deficits. *S. fruticosa* hydroalcoholic extract also weakens memory loss induced by morphine. *S. fruticosa* improves cognitive performance in normal healthy individuals and dementia or cognitive impairment patients (Lopresti, 2016).

17.3.7 METABOLIC ENZYME ACTIVITY

Administration of *S. fruticosa* infusion at a single dose per day cannot alter glucose levels in the blood of either diabetic or normoglycemic rabbits. Instead, *S. fruticosa* infusion on daily treatment, recurring for seven successive days, reduces blood glucose levels in the blood by alloxan treatment of hyperglycemic rabbits. *S. fruticosa* leaf extract induces hypoglycemic effects by influencing the intestinal transport of carbohydrates mechanisms and not influencing glucose metabolism. They might function by lowering glucose absorption in the intestine (Perfumi et al., 1991). After Streptozotocin (STZ) treatment, there is an adaptive rise of SGLT1 levels in enterocytes of rat brush border membrane (BBM), which *S. fruticosa* then inhibits due to the stimulation of a rich carbohydrate diet. The active principle present is rosmarinic acid. Rosmarinic acid and *Salvia* extracts help treat colon cancer by preventing DNA damage, inducing apoptosis, and inhibiting cell proliferation (Mahdizadeh et al., 2018).

The necessary digestive enzymes involved in carbohydrates and lipids metabolisms are *α*-amylase, pancreatic lipase, and α-glucosidase, which are the primary targets for the therapeutic control of diabetes and obesity. The enzymes α-Amylase and α-glucosidase catalyze the hydrolysis of carbohydrates into simple sugars. There is a retardation of starch and oligosaccharides digestion due to their inhibitory activity, which further leads to a decreased postprandial rise of glucose levels in plasma. On the other hand, lipase inhibition decreases the digestion of dietary triglycerides, reducing free fatty acids (FFA) levels and monoacylglycerols in the intestinal lumen. The ability of *S. greggii* decoction to have inhibitory activity on three digestive enzymes, as mentioned above, was assessed with the help of chemical models. The decoction holds effective inhibitory activity against α-glucosidase. Hence, *S. greggii* can be used as an ordinary anti-obesity and antidiabetic agent to control the glucose level by controlling the movement of α-glucosidase. It also showed more than 10% antilipase activity, probably due to the higher content of luteolin-7-O-glucoside. This phenolic component is considered an efficient inhibitor of lipase (Pereira et al., 2018).

17.3.8 NEUROPROTECTIVE ACTIVITY

The *S. fruticosa* petroleum ether (SFP) extract displayed intense neuroprotection activity against amyloid-β toxicity. A similar effect was found in the *S. fruticosa*'s dichloromethane (SFD) extract but in smaller quantities. The *S. fruticosa* methanol (SFM) extract is not able to reverse the consequences of A*β*25–35. *S. fruticosa* extracts with lesser polarity have more ability to influence neuroprotective activity. SFD extract exhibited a statistically significant neuroprotective effect at 20–50 μg/mL and SFP extract at 20 μg/mL (Ververis et al., 2020).

17.3.9 GENOTOXIC AND ANTIGENOTOXIC EFFECTS ON HPBLS OF SF

Human peripheral blood lymphocytes (HPBLs) subjected to mutagenicity tests could reveal the genotoxicity of substances or plant extracts. It is the proper method by which genotoxicity should get investigated in HPBLs (Madle et al., 1993). Three tests, namely, sister chromatid exchange (SCE), micronucleus (MN), and chromosome aberration (CA) tests, in HPBLs are amongst the cytogenetic markers extensively adopted for exposure of early physiological effects related to DNA-damaging agents (Kocaman and Topaktaş, 2010; Norppa and Falck, 2003). With these tests, the architectural and numerical chromosomal aberrations can be examined.

Sevindik and Rencuzogullari (2014) assessed the genotoxic and antigenotoxic effects of *S. fruticosa* extract by in vitro examination methods employing HPBLs with the absence and presence (-S9 mix). Maron and Ames (1983) used 3-Methylcolanthrene-influenced rodent liver microsomal portion as the metabolic activator (S9 mix) in their studies. HPBLs are exposed to 1.5, 3.0, and 6.0 mL concentrations for 1 and 2 days. The cytotoxicity of *S. fruticosa* leaf extract was investigated by measuring the nuclear division index (NDI), mitotic index (MI), and proliferation index (PI). In the absence of the S9 mix, *S. fruticosa* leaf extract solely increased the SCE rate at the 48 h exposure duration; nevertheless, it provoked the MN and CA with all concentrations and at all exposure phases.

S. fruticosa with MMC (mitomycin C) synergically influenced CA and SCE, besides the highest concentration of *S. fruticosa* leaf extract and MMC on initiation of SCE. Further, *S. fruticosa* leaf extract effected the influence of MMC on MN rate for 24 h. However, it significantly attenuated the impact of MMC on MN recurrence for the 48 h exposure period. *S. fruticosa* leaf extract exhibited a cytotoxic impact by minimizing the MI; However, it did not lower the PI and NDI. In the presence of S9 mix, *S. fruticosa* extract did not raise the SCE associated with solvent control. In contrast, it lessened the impact of cyclophosphamide (Cyp). *S. fruticosa* leaf extract induced the MN and CA but could not enhance the effect of Cyp on CA and MN development. *S. fruticosa* leaf extract did not show a cytotoxic effect (Sevindik and Rencuzogullari, 2014).

KEYWORDS

- ***Salvia fruticosa***
- ***Salvia greggii***
- **rosmarinic acid**
- **genkwanin**
- **bioactives**
- **Cirsimaritin**

REFERENCES

Abaza, I.; Aboalhaija, N.; Alsalman, A.; Talib, W.; Afifi, F. Aroma Profile, Chemical Composition and Antiproliferative Activity of the Hydrodistilled Essential Oil of a Rare Salvia Species (*Salvia greggii*). *J. Biol. Active Prod. Nat.* **2021,** *11*(2), 129–137.

Abdulla-Eltawaty, S. I.; Busba, A. O.; Omare, M. E.; Almagboul, A. Z.; Yagoub, S. O.; Alramli, A.; Ahmed, A.; Abdalbagi-Dafalla, M. Antimicrobial Activity of Libyan *Salvia fruticosa* MIL and Multi-drug Resistant Bacteria. *Eur. J. Pharm. Med. Res.* **2021,** *8*(1), 416–423.

Abu-Dahab, P.; Afifi, F.; Kasabri, V.; Majdalawi,L.; Naffa, R. Comparison of the Antiproliferative Activity of Crude Ethanol Extracts of Nine *Salvia* Species Grown in Jordan Against Breast Cancer Cell Line Models. *Pharmacognosy Mag.* **2012,** *8*(32), 319–324.

Akalın, E.; Alpınar, K. An Investigation on Medicinal and Edible Wild Plants of Tekirdağ. *Ege Üniv. Ecz. Fak. Der.* **1994,** *2*, 1–11.

Aslan, A.; Mat, A.; Ozhatay, N.; Sarıyar, G. A Contribution to Traditional Medicine in West Anatolia. *J. Fac. Pharm. Istanbul Univ.* **2007,** *39*, 73–83.

Azevedo, M. F.; Lima, C. F.; Fernandes-Ferreira, M.; Almeida, M. J.; Wilson, J. M.; Pereira-Wilson, C. Rosmarinic Acid, Major Phenolic Constituent of Greek Sage Herbal Tea, Modulates Rat Intestinal SGLT1 Levels with Effects on Blood Glucose. *Mol. Nutr. Food Res.* **2011,** *55*(S1), S15–S25.

Bisio, A.; Romussi, G.; Russo, E.; De Tommasi, N.; Mascolo, N.; Alfieri, A.; Bonito, M. C.; Cicala, C. Platelet Anti-aggregating Activity and Chemical Constituents of *Salvia* x *Jamensis* J. Compton. *Nat. Prod. Commun.* **2008,** *3*(6), 881–884.

Bonesi, M.; Loizzo, M. R.; Acquaviva, R.; Malfa, G. A.; Aiello, F.; Tundis, R. Anti-inflammatory and Antioxidant Agents from *Salvia* Genus (Lamiaceae): An Assessment of the Current State of Knowledge. *Antiinflamm. Antiallergy Agents Med. Chem.* (Formerly *Curr. Med. Chem. Anti. Inflamm. Anti. Allergy Agents*). **2017,** *16*(2), 70–86.

Boukhary, R.; Raafat, K.; Ghoneim, A. I.; Aboul-Ela, M.; El-Lakany, A. Anti-inflammatory and Antioxidant Activities of *Salvia fruticosa*: An HPLC Determination of Phenolic Contents. *Evid. Based Complement. Alternat. Med.* **2016,** *2016*, 1–6, Article ID 7178105, 1–6. https://doi.org/10.1155/2016/7178105

Davis, P. H. Ed. *Flora of Turkey and the East Aegean Islands*, Vol. 7; Edinburgh University Press: Edinburgh, UK, 1982.

De Martino, L.; Roscigno, G.; Mancini, E.; De Falco, E.; De Feo, V. Chemical Composition and Antigerminative Activity of the Essential Oils from Five *Salvia* Species. *Molecules* **2010,** *15*(2), 735–746.

Duletic-Lausevic, S,; Aradski, A.A,; Šavikin, K,; Knežević, A,; Milutinović, M,; Stević, T,; Vukojević, J,; Marković, S,; Marin, P. D. Composition and Biological Activities of Libyan *Salvia fruticosa* Mill. and *S. Lanigera* Poir. Extracts. *S. Afr. J. Bot.* **2018,** *117*, 101–109.

Ertuğ, F. *Bodrum Yöresinde Halk Tibbinda Yararlanilan Bitkiler. 14 Bitkisel İlaç Hammaddeleri Toplantısı (29–31 Mayıs, 2002, Eskişehir) Bildiri Kitabı*, **2004,** 76–93.

Everest, A.; Ozturk, E. Focusing on the Ethnobotanical uses of Plants in Mersin and Adana Provinces (Turkey). *J. Ethnobiol. Ethnomed* **2005,** *1*(1), 1–6.

Exarchou, V.; Kanetis, L.; Charalambous, Z.; Apers, S.; Pieters, L.; Gekas, V.; Goulas, V. HPLC-SPE-NMR Characterization of Major Metabolites in *Salvia fruticosa* Mill. Extract with Antifungal Potential: Relevance of Carnosic Acid, Carnosol, and Hispidulin. *J. Agric. Food Chem.* **2015,** *63*(2), 457–463.

Fakir, H.; Korkmaz, M.; Güller, B. Medicinal Plant Diversity of Western Mediterranean Region in Turkey. *J. Appl. Biol. Sci.* **2009,** *3*(2), 33–43.

Fraternale, D.; Albertini, M. C.; Rudov, A.; Flamini, G.; Ricci, D.; Bisio, A.; Battistelli, M.; Accorsi, A. *Salvia* x *jamensis* J. Compton: Trichomes, Essential Oil Constituents, and Cytotoxic-apoptotic Activity. *Nat. Prod. Res.* **2013,** *27*(17), 1583–1588.

Frett, J. F. Influence of Nutrient Salts, Auxins and Cytokinins on the In Vitro Growth of *Salvia greggii*. *Plant Cell Tissue Organ Cult.* **1987,** *9*, 89–93.

Gilman, E. F.; Marshal, D. Report on *Salvia greggii.* University of Florida, Cooperative Extension Service, Institute of Food and Agricultural Sciences, Fact Sheet, 1999, FPS-524, 1–3. https://hort.ifas.ufl.edu/database/documents/pdf/shrub_fact_sheets/salgrea.pdf

Giuliani, C.; Ascrizzi, R.; Corrà, S.; Bini, L. M.; Flamini, G.; Fico, G. Ultrastructural Insight Into Terpene-Producing Trichomes and Essential Oil Profile in *Salvia greggii* A. Gray. *Flora* **2017,** *236*, 107–114.

Giuliani, C.; Giovanetti, M.; Lupi, D.; Mesiano, M. P.; Barilli, R.; Ascrizzi, R;, Flamini, G.; Fico, G. Tools to Tie: Flower Characteristics, VOC Emission Profile, and Glandular Trichomes of Two Mexican *Salvia* Species to Attract Bees. *Plants* **2020,** *9*(12), 1645.

Gonzalez-Tejero, M. R.; Casares-Porcel, M.; Sanchez-Rojas, C. P.; Ramiro Gutierrez, J. M.; Molero-Mesa, J.; Pieroni, A.; Giusti, M. E.; Censorii, E.; de Pasquale, C.; Della, A.; Paraskeva-Hadijchambi, D.; Hadjichambis, A.; Houmani, Z.; EL-Demerdash, M.; El-Zayat, M.; Hmamouchi, M.; Eljohriq, S. Medicinal Plants in the Mediterranean Area: Synthesis of the Results of the Project Rubia. *J. Ethnopharmacol.* **2008,** *116*, 341–357.

Gürdal, B.; Kültür, Ş. An Ethnobotanical Study of Medicinal Plants in Marmaris (Muğla, Turkey). *J. Ethnopharmacol.* **2013,** *146*(1), 113-126.

Hani, S. B.; Bayachou, M. *Salvia fruticosa* Reduces Intrinsic Cellular and H_2O_2-induced DNA Oxidation in HEK 293 Cells; Assessment Using Flow Cytometry. *Asian Pac. J. Trop. Biomed.* **2014,** *4*(5), 399–403.

Honda, G.; Yeşilada, E.; Tabata, M.; Sezik, E.; Fujita, T.; Takeda, Y.; Takaishi, Y.; Tanaka, T. Traditional Medicine in Turkey VI. Folk Medicine in West Anatolia: Afyon, Kütahya, Denizli, Muğla, Aydin provinces. *J. Ethnopharmacol.* **1996,** *53*(2), 75–87.

Kawahara, N.; Inoue, M.; Kawai, K. I.; Sekita, S., Satake, M.; Goda, Y. Diterpenoid from *Salvia greggii*. *Phytochemistry* **2003,** *63*(8), 859–862.

Kawahara, N.; Tamura, T.; Inoue, M.; Hosoe, T.; Kawai, K. I.; Sekita, S.; Satake, M.; Goda, Y. Diterpenoid Glucosides from *Salvia greggii. Phytochemistry* **2004,** *65*(18), 2577–2581.

Kintzios, S. E. Sage The Genus *Salvia*; Harwood Academic Publisher, Amsterdam, The Netherlands, 2000.

Kocaman, A. Y.; Topaktaş M. In Vitro Evaluation of the Genotoxicity of Acetamiprid in Human Peripheral Blood Lymphocytes. *Environ. Mol. Mutagen.* **2007,** 48(6), 483–490.

Koliopoulos, G.; Pitarokili, D.; Kioulos, E.; Michaelakis, A.; Tzakou, O. Chemical Composition and Larvicidal Evaluation of *Mentha, Salvia,* and *Melissa* Essential Oils Against the West Nile Virus Mosquito *Culex Pipiens. Parasitol. Res.* **2010,** *107*(2), 327–335.

Koutsoulas, A.; Čarnecká, M.; Slanina, J.; Tóth, J.; Slaninová, I. Characterization of Phenolic Compounds and Antiproliferative Effects of *Salvia pomifera* and *Salvia fruticosa* Extracts. *Molecules* **2019,** *24*(16), 2921.

Lopresti, A. L. *Salvia* (Sage): A Review of its Potential Cognitive-enhancing and Protective Effects. *Drugs R D* **2017,** *17*(1), 53–64.

Madle, S.; Beek, B.; Nowak, C. Zum Verstandnis Von Chromo-somen Mutations Tests an Somazellen. (The Understanding of Chromosome and Somatic Cell Mutation test). In *Mutationsforschung und Genetische Toxicologie*; Fahrig, R., Ed. (Mutation research and genetic toxicology); Wissenschaftliche Buchgesellschaft, Dramstad, Germany, 1993; pp 224–242.

Mahdizadeh, R.; Moein, S.; Soltani, N.; Malekzadeh, K.; Mahmoodreza, M. Study the Molecular Mechanism of *Salvia* Species in Prevention of Diabetes. *Int. J. Pharm. Sci. Res.* **2018,** *9*(11), 4512–4521.

Maron, D. M.; Ames, B. N. Revised Method for the *Salmonella* Mutagenicity Test. *Mutat. Res.* **1983,** *113*, 173–215.

Norppa, H.; Falck, G. C. M. What Do Human Micronuclei Contain? *Mutagenesis* **2003,** *18*, 221–233.

Orhan, I.; Kartal, M.; Kan, Y.; Şener, B. Activity of Essential Oils and Individual Components Against Acetyl and Butyrylcholinesterase. *Zeitschrift fuer Naturforschung C* **2008,** *63*(7–8), 547–553.

Papageorgiou, V.; Gardeli, C.; Mallouchos, A.; Papaioannou, M.; Komaitis, M. Variation of the Chemical Profile and Antioxidant Behaviour of *Rosmarinus Officinalis* L. and *Salvia fruticosa* Miller Grown in Greece. *J. Agric. Food. Chem.* **2008,** *56*(16), 7254–7264.

Pereira, O. R.; Catarino, M. D.; Afonso, A. F.; Silva, A.; Cardoso, S. M. *Salvia Elegans, Salvia Greggii* and *Salvia officinalis* Decoctions: Antioxidant Activities and Inhibition of Carbohydrate and Lipid Metabolic Enzymes. *Molecules* **2018,** *23*(12), 3169.

Pereira, O.R;, Afonso, A. F.; Silva, J. D. A.; Sobral, A. J.; Cardoso, S. M. *Characterization of Phenols in Salvia Elegans and Salvia greggii.* In 62nd International Congress and Annual Meeting of the Society-of-Medicinal-Plant-and-Natural-Product-Research; Georg Thieme Verlag Kg, 2014, 1471.

Perfumi, M.; Arnold, N.; Tacconi, R. Hypoglycemic Activity of *Salvia fruticosa* Mill. from Cyprus. *J. Ethnopharmacol.* **1991,** *34*(2–3), 135–140.

Pitarokili, D.; Tzakou, O.; Loukis, A.; Harvala, C. Volatile Metabolites from *Salvia fruticosa* as Antifungal Agents in Soilborne Pathogens. *J. Agric. Food Chem.* **2003,** *51*(11), 3294–3301.

Ramos, A. A.; Azqueta, A.; Pereira-Wilson, C.; Collins, A. R. Polyphenolic Compounds from *Salvia* Species Protect Cellular DNA from Oxidation and Stimulate DNA Repair in Cultured Human Cells. *J. Agric. Food Chem.* **2010,** *58*(12), 7465–7471.

Rivera, D.; Obon, C.; Cano, F. The Botany, History and Traditional Uses of Three-lobed Sage (*Salvia fruticosa* Miller) (Labiatae). *Econ. Bot.* **2016,** *48*(2), 190–195.

Şenol, F. S.; Orhan, I. E.; Erdem, S. A.; Kartal, M.; Şener, B.; Kan, Y.; Celep, F.; Kahraman, A.; Dogan, M. Evaluation of Cholinesterase Inhibitory and Antioxidant Activities of Wild and Cultivated Samples of Sage (*Salvia fruticosa*) by Activity-guided Fractionation. *J. Med. Food* **2011,** *14*(11), 1476–1483.

Sevindik, N.; Rencuzogullari, E. The Genotoxic and Antigenotoxic Effects of *Salvia fruticosa* Leaf Extract in Human Blood Lymphocytes. *Drug and Chem. Toxicol.* **2014,** *37*(3), 295–302.

Sivropoulou, A.; Nikolaou, C.; Papanikolaou, E.; Kokkini, S.; Lanaras, T.; Arsenakis, M. Antimicrobial, Cytotoxic, and Antiviral Activities of *Salvia fructicosa* Essential Oil. *J. Agric. Food. Chem.* **1997,** *45*(8), 3197–3201.

Torun, M.; Dincer, C.; Topuz, A.; Sahin–Nadeem, H.; Ozdemir, F. Aqueous Extraction Kinetics of Soluble Solids, Phenolics and Flavonoids from Sage (*Salvia fruticosa* Miller) Leaves. *J. Food Sci. Tech.* **2015,** *52*(5), 2797–2805.

Ugurlu, E.; Secmen, O. Medicinal Plants Popularly Used in the Villages of Yunt Mountain (Manisa-Turkey). *Fitoterapia* **2008,** *79*(2), 126–131.

Ververis, A.; Savvidou, G.; Ioannou, K.; Nicolaou, P.; Christodoulou, K.; Plioukas, M. Greek Sage Exhibits Neuroprotective Activity Against Amyloid Beta-Induced Toxicity. *Evid. Based Complement. Alternat. Med.* **2020,** *2020*, 1–10, Article ID 2975284, https://doi.org/10.1155/2020/2975284

Wu, Y. B.; Ni, Z. Y.; Shi, Q. W.; Dong, M.; Kiyota, H.; Gu, Y. C.; Cong, B. Constituents from *Salvia* Species and Their Biological Activities. *Chem. Rev.* **2012,** *112*(11), 5967–6026.

Xavier, C. P.; Lima, C. F.; Fernandes-Ferreira, M.; Pereira-Wilson, C. *Salvia fruticosa, Salvia officinalis,* and Rosmarinic Acid Induce Apoptosis and Inhibit Proliferation of Human Colorectal Cell Lines: the Role in MAPK/ERK Pathway. *Nutr. Cancer* **2009,** *61*(4), 564–571.

Zaccardelli, M.; Pane, C.; Caputo, M.; Durazzo, A.; Lucarini, M.; Silva, A. M.; Severino, P.; Souto, E. B.; Santini, A.; De Feo, V. Sage Species Case Study on a Spontaneous Mediterranean Plant to Control Phytopathogenic Fungi and Bacteria. *Forests* **2020,** *11*(6), 704.

CHAPTER 18

BIOACTIVES AND PHARMACOLOGY OF TRADITIONAL CHINESE HERB *SALVIA MILTIORRHIZA* BUNGE

DEVIKA R. NATH[2], VRUSHALI MANOJ HADKAR[1], BABU SUBRAMANIAN[2], and CHINNADURAI IMMANUEL SELVARAJ[1]

[1]*School of Biosciences and Technology, Vellore Institute of Technology, Vellore, Tamil Nadu, India*

[2]*VIT School of Agricultural Innovations and Advanced Learning, Vellore Institute of Technology, Vellore, Tamil Nadu, India*

ABSTRACT

Salvia miltiorrhiza Bunge is a perennial plant in the Lamiaceae family of mints. It is also known as red sage or Danshen. It is a common medicinal herb in traditional Chinese medicine. The entire plant has been used for the treatment of several diseases for thousands of years. The plant is an erect perennial herb that bears purple or blue flowers. *S. miltiorrhiza* generally grows in central, north-eastern, west, and southwest China. It has been used for treating various diseases due to the presence of its active constituents. The bioactive constituents present in *S. miltiorrhiza* are of two groups (a) lipophilic chemicals such as diterpenoid, tanshinones, and (b) hydrophilic compounds such as salvianolic acids. Lipophilic phenanthrene compounds known as tanshinones are of two types, namely, ortho-quinones and para-quinones with various medicinal properties. The prime bioactive constituents present in roots are tanshinone I, tanshinone IIA, and cryptotanshinone. Major phenolic compounds present in *S. miltiorrhiza* are salvianolic acid A,

Bioactives and Pharmacology of Lamiaceae. T. Pullaiah, PhD (Ed.)

salvianolic acid B, and Danshensu. The extract of plant mainly from roots possesses various pharmacological activities. The plant has good antioxidant activity, antimicrobial activity, antiviral activity, anticancer, anti-inflammatory and can be used to treat cardiovascular diseases. It also has other pharmacological activities like antihypertensive activity, antiurolithiasis activity and anticoagulation effect. This chapter presents comprehensive analysis of its bioactives and pharmacology.

18.1 INTRODUCTION

Salvia miltiorrhiza Bunge belongs to the family Lamiaceae, one of the commonly used herbs in Traditional Chinese Medicine (TCM) (Chen et al., 2019b). *S. miltiorrhiza* is native to China and Japan. It generally grows in central, north-eastern, west, and southwest China (Peng et al., 2014). The plant is an erect perennial herb that bears purple or blue flowers. The plant grows between 90 and 1200 m altitudes and prefers grassy areas in forests, along stream banks and hillsides. The plant grows up to 30–60 cm high. Depending on the position of leaves on the stem, they are either simple or divided. The petals of the flower are purple or blue, having a dark purple calyx. Specifically, the name *miltiorrhiza* means "red juice extracted from a root" of *S. miltiorrhiza* (Clebsch et al., 2003).

The plant has various medicinal properties. It is used as a traditional natural medicine in clinics in different parts of China (Zhou et al., 2017; Cao et al., 2018). The plant is also known as Danshen. *Danshensu* (Salvianic acid A) is a product extracted from Danshen (Bao et al., 2018). It is a composite mixture of many chemical components (Pang et al., 2016). It structurally comprises catechol and lactic acid. A study indicates that catechol's antioxidant properties can protect cardiac injuries (Zhang et al., 2019). Apart from treating cardiovascular diseases and neurasthenic insomnia, TCM says that this plant functions to stimulate blood circulation, discharge blood stasis, balance the menstrual cycle, reduce pain, and improve blood production (Li et al., 2020).

18.2 BIOACTIVES

The phytochemical components of *S. miltiorrhiza* are of two groups: (a) lipophilic chemicals such as diterpenoid, tanshinones and (b) hydrophilic compounds such as salvianolic acids (Li et al., 2009). Phytochemical screening of *S. miltiorrizha* revealed tanshinones that are also known as total tanshinones. They come under the class of lipophilic phenanthrene compounds relatively found in the roots of *S. miltiorrhiza* (Ma et al., 2015). Tanshinones have significant pharmacological properties like antibacterial, anti-cancer, and anti-viral effects. More than 40 tanshinones are known so far from the plant *S. miltiorrhiza* (Pang et al., 2016).

The tanshinones are of two types, (a) ortho-quinones, that is., tanshinone I (tan I, $C_{18}H_{12}O_3$), tanshinone IIA (tan IIA, $C_{19}H_{18}O_3$), cryptotanshinone, and dihydrotanshinone. (b) the para-quinones that is., isotanshinone I, isotanshinone IIA, cryptotanshinone (CT), and isocryptotanshinone (Li et al., 2020); isocryptotanshinone II, tanshinlactone, and neo-tanshinlactone (Guo et al., 2014). Besides, diterpenoids and triterpenoids compounds reported from Danshen, namely danshenxinkun A/B/C/D, miltirone, and neo-tanshinlactone (Wu et al., 2012; Gao et al., 2012). The prime bioactive constituents of tanshinones are tanshinone I, tanshinone IIA, and cryptotanshinone in which 1% of the total tanishone content comprises tanshinone I and II, are present in the roots (Wang et al., 2010).

Major phenolic compounds known to be present in *S. miltiorrhiza* are salvianolic acid A, salvianolic acid B, Danshensu, lithospermic acid, caffeic acid, protocatechuic acid, and rosmarinic acid (Ho and Hong, 2011; Ma et al., 2015; Shi et al., 2019). In general, water, methanol, ethanol, or aqueous acetone can isolate the plant's phenolic acids (Wang et al., 2017). Danshensu is a combination of protocatechuic, aldehyde, protocatechuic acid, 3,4-dihydroxy phenyl lactic acid, and caffeic acid present in the form of single phenolic acid or also known as phenolic acid compound (Wang et al., 2010). *Dashensu* has a well-defined structure, and it comprises catechol and lactic acid. The catechol is the chief active group that exerts an antioxidation effect to protect against cardiac injury (Zhang et al., 2019). Simultaneously, polyphenolic acids include rosmarinic acid, lithospermic acid, salvianolic acid A, salvianolic acid B, and other salvianolic acids (Wang et al., 2010).

Most of the salvianolic acids are colorless or tan amorphous powders. *S. miltiorrhiza* has a main active constituent as tetramer known as salvianolic acid B. The major antioxidant components, salvianolic acid A and salvianolic acid B, present in *Salvia miltiorrhiza* can get extracted using high speed counter-current chromatography and ultrasound-assisted extraction (Wang et al., 2010). Salvianolic acid B is a potential anti-atherosclerotic agent with an anti-inflammatory activity that regulates lipid metabolism. The chemical structures of few important compounds of *S. miltiorrhiza* are given in Fig. 18.1.

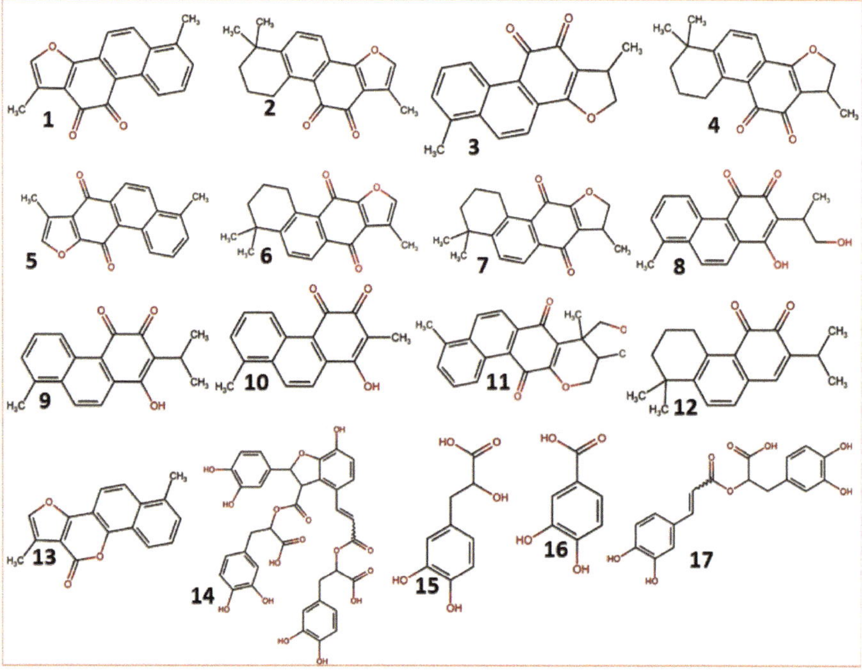

FIGURE 18.1 Structures of tanshinone I (1), tanshinone IIA (2), dihydrotanshinone (3), cryptotanshinone (4), isothanshinone I (5), isotanshinone IIA (6), isocryptotanshinone (7), danshenxinkun A (8), danshenxinkun B (9), danshenxinkun C (10), danshenxinkun D (11), miltirone (12), neo-tanshinlactone (13), salvianolic acid B (14), danshensu (15), protocatechuic acid (16), and rosmarinic acid (17); [Marvin 17.21.0, ChemAxon tool was used for drawing chemical structures].

18.3 PHARMACOLOGY

18.3.1 ANTI-TUMOR ACTIVITY

Blood stasis syndrome explains tumor development duration while cancer disease occurs, states Chinese medicine theory. Chinese medicine prescribes Danshen as anti-tumor due to blood-activating and stasis-resolving property. *Salvia miltiorrhiza* possesses anti-tumor mechanisms due to the presence of phenolic acids, which is involved in tumor cell growth inhibition, metastasis, angiogenesis, apoptosis, and cell cycle arrest. The phenolic acids are the major contributors to antioxidant activities (Wang et al., 2020).

A study investigated a phytochemical compound tanshinone IIA in *S. miltiorrhiza* regulated oxidative stress and reduced the mitochondrial membrane potential. This effect actively induced apoptosis in cancer cells such as human lung cancer A549 cells (Chiu and Su, 2010). Tanshinones inhibited human prostate cancer cell lines by arresting the cell cycle and inducing apoptosis *in vitro*. Besides, tanshinone I, has significant activity with an IC_{50} value of approximately 3–6 μM. Thus, it showed relatively fewer side effects on normal prostate epithelial cells (Gong et al., 2011). In an investigation, the anti-tumor activity of aqueous extract of *S. miltiorrhiza* actively inhibited human hepatoma HepG2 cells' growth. Reports show that apoptotic cell death was due to crude extract treatment (Liu et al., 2000).

In a Chinese patent, salvianolic acid A was administered as an anticancer drug. It showed potential results combined with other anti-tumor agents, for example, mitomycin C, 5-flurouracil (5-FU), and methotrexate (Jiang et al., 2005). Salvianolic acid A improved the anti-tumor effects with 5-FU without raising its toxicity in an animal study. Investigations indicate that salvianolic acid A was also known to protect rat heart mitochondria against toxicities induced by the drug adriamycin (Lin et al., 1992). Cancer is being one of the leading causes of death. The chemical constituents present in *S. miltiorrhiza* are capable of inhibiting various cancer cells. Several chemical screening of 76 medicinal plants recognized tanshinone IIA as one of the promising cytotoxic compounds in human leukemia cells. Tanshinone IIA is a potential anticancer component obtained from the dried root of *S. miltiorrhiza.* Specific experiments show that tanshinone IIA exerts a cytotoxic effect on seven human tumor cell lines. Induction of apoptosis is a crucial factor in contributing to the cytotoxic property of tanshinone IIA (Wang et al., 2010).

18.3.2 CARDIOVASCULAR EFFECTS

Experimental results indicated that phytochemical compound sodium tanshinone IIA silate controlled the growth and migration of high glucose-induced vascular smooth muscle cells (VSMC), probably by activation of adenosine 50-monophosphate (AMP)-activated protein kinase (AMPK) (Chen, 1984). Besides, sodium tanshinone IIA silate has shielding properties on myocardial apoptosis, produced by acute myocardial infarction (MI) in adult rats. Thus, it inhibited cardiac fibrosis after pathological stimuli to the cardiovascular system (Yang et al., 2009).

18.3.3 ANTI-BACTERIAL ACTIVITY AND ANTIVIRAL ACTIVITY

Micro-organisms cause various diseases and result in the deterioration of various products. Scientists conclude that the components (cryptotanshinone and dihydrotanshinone) present in *Salvia miltiorrhiza* exhibits antibacterial activity against a wide range of Gram-positive and Gram-negative bacteria. The crucial agents in antibacterial actions are superoxide radicals. Clinical utilization of *S. miltiorrhiza* can effectively prevent inflammatory diseases caused by Gram-negative bacteria. Lipopolysaccharide (LPS) has been, so far, embroiled as one of the significant reasons for Gram-negative micro-organisms instigated sepsis. These are reality undermining symptoms happening in the emergency unit. Additionally, ethanol extracts of *S. miltiorrhiza* have anti-microbial activity on *Porphyromonas gingivalis*, A. *actinomycetemcomitans*, *Streptococcus mutans*, and *Lactobacillus*.

The hexane and chloroform portion of *Salvia miltiorrhiza* hinders resistant gene expression hostile to *Staphylococcus aureus* and methicillin-resistant *Staphylococcus aureus* (Wang et al., 2010). The main lipophilic components in *S. miltiorrhiza* that play antibacterial roles are bioactive compounds like cryptotanshinone and dihydrotanshinone I, tanshinone IIB, and hydroxyl tanshinone (Honda et al., 1988). Water-soluble extract of *S. miltiorrhiza* showed a strong effect against HIV-1 integrase activity *in vitro* and *in vivo* viral replication. The isolated bioactive components from *S. miltiorrhiza* were effective against HIV. The IC_{50} value ranges from 0.83 μM for lithospermic acid and 0.48 μM for lithospermic acid B present in *S. miltiorrhiza* (Abd-Elazem et al., 2002).

18.3.4 ANTIOXIDANT ACTIVITY

Scientific studies have shown that phenolic components present in *S. miltiorrhiza* prevent several diseases. These components are potent scavengers of free radicals. A phytoconstituent, Tanshinone IIA, present in *S. miltiorrhiza,* efficiently regulates the interaction between DNA and the lipid peroxidation products intracellularly (Zhao et al., 2008). It also blocks the chain reaction of lipid peroxidation and shields DNA by eliminating lipid free radicals (Tang et al., 2007). Anti-oxidant studies of *S. miltiorrhiza* involved in the treatment of two-kidney, two-clip renovascular hypertensive rats with tanshinone IIA (35 and 70 $mg \cdot kg^{-1} \cdot d^{-1}$) for 6 weeks had restrained ROS production and inhibited maximum NAD(P)H oxidase activity and its expression (Wang et al., 2011). In another study, Dimethyl lithospermate

(DML) (5 and 10 $mg \cdot kg^{-1} \cdot d^{-1}$), intraperitoneally injected to 20-month-old rats; with 6-month-old rats as young control animals. The results showed that dimethyl lithospermate was known to reduce an age-associated increase in cyclooxygenase-2 and inducible nitric oxide synthase (iNOS) expression. They restrained NO metabolites and reactive species formation (Choi et al., 2014).

In China, Danshen is a commercial herbal medicine extracted from the roots of *S. miltiorrhiza,* and its leaf biomass has a considerable proportion of the entire plant. The leaf extract has considerable amounts of phenolics and possesses significant anti-oxidant activities. Salvianolic acid B and rosmarinic acid are the major contributors to phenolic compounds. They have higher anti-oxidant activities (Wang et al., 2010; Zhao et al., 2011). More than 20 phenolic acids have been extracted and examined by the researchers. Water-soluble salvianolic acid A and salvianolic acid B were isolated and purified from the crude extract of *S. miltiorrhiza*. The results indicated that salvianolic acid A and salvianolic acid B exhibited high anti-oxidant activities. Moreover, the aqueous fraction of *Salvia miltiorrhiza* contains salvianolic acid B and rosmarinic acid that exhibits great anti-lipo-peroxidant activity by scavenging superoxide anion radical.

18.3.5 ANTI-INFLAMMATORY ACTIVITY

Inflammation is concerned with the pathological system of many illnesses associated with the generation of inflammatory cytokines, such as interleukins and tumor necrosis factor-alpha (TNF-α). The extended cytokine stage in all possibility contributes to the widespread occurrence of inflammatory illnesses, neurodegenerative diseases, menopause, and cardiovascular illnesses (Wang et al., 2010). The anti-inflammatory activity studies reported that the phytochemical compound Sodium tanshinone IIA silate from *S. miltiorrhiza* decreases the expression of TNF-α, interleukin (IL)-6, and vascular cell adhesion molecule 1 (VCAM-1). It also reduced intercellular adhesion molecule 1 (ICAM-1) in atherosclerotic mice (Zhu et al., 2017).

In atherosclerosis pathogenesis, it is believed that inflammation plays a critical role in the initiation, progression, and destabilization of atherosclerotic plaques due to activation of innate and adaptive immunity (Shah et al., 2019). Thus atherosclerotic cardiovascular disease pathogenesis can be lowered and prevented due to a practical anti-inflammatory approach. Inflammatory cells and mediators are known to be an essential component of the inflammatory response. The immune system gets modified by producing several inflammatory factors and chemokines by macrophages (Siouti and Andreakos, 2019). The effects of tanishone I and cryptotanishone on the cell growth of RAW264.7 were detected by CCK-8 (Cholecystokinin Octapeptide) after 24 and 48 h treatment. Tanishone I inhibited macrophages' growth after 24 and 48 h, compared with the DMSO control group, 50 and 100 μM. A study concludes that 25.0–100.0 μM of cryptotanishone effectively restrained the macrophage proliferation after 24 and 48 h. Moreover, the results depicted that Cryptotanishone exhibited a higher effect than Tanishone I (Cui et al., 2020).

In the study of nitric oxide assay in cell culture supernatants, the NO was resolved by estimating nitrite accumulation (NO^{2-}) (Cui et al., 2019). Active effects of Tanishone I and Cryptotanishone on the nitric oxide secretion of RAW264.7 cells produced by lipopolysaccharides examined by using Griess reagent. Results indicate that macrophages did not produce nitric oxide without lipopolysaccharides stimulation. Thus secretion of nitric oxide strongly increased after 24 h of lipopolysaccharides treatment. Therefore, it was reported that phytocompounds Tanishone I and cryptotanishone (12.5–100.0 μM) from *S. miltiorrhiza* effectively inhibited lipopolysaccharides-induced nitric oxide secretion of RAW264.7 cells (Cui et al., 2020). Salvianolic acid B inhibits the development of inflammatory cytokine and infiltration of inflammatory cells (Wang et al., 2020). Tanshinone IIA separated from *S. miltiorrhiza* has shown anti-inflammatory activity. Tanshinone IIA can be an antagonist for cell infiltration. It is used toward the inflammatory disorders under cytokine production during menopause (Wang et al., 2010).

18.3.6 ENDOTHELIAL PROTECTIVE EFFECT

Sodium tanshinone IIA silate in *S. miltiorrhiza* produces an influential function in vascular endothelial cells. It repressed the apoptosis of human umbilical vein endothelial cells (HUVECs) induced by heat stress through the phosphoinositide-3-kinase/protein kinase B/eNOS signaling pathway (Cheng et al., 2017).

18.3.7 ANTICOAGULATION EFFECT AND VASORELAXANT EFFECT

Pharmacological studies revealed a bioactive compound tanshinone IIA in *S. miltiorrhiza* restrained platelet accumulation produced by adenosine diphosphate and collagen. Thus, the compound tanshinone IIA is believed to be a promising drug that can lower blood viscosity (Maione et al., 2014). The study indicates that tanshinone IIA improves endothelium-dependent vasodilation by enhancing eNOS expression and activity, which gets impaired due to diabetes. Thus *S. miltiorrhiza* shows significant vasorelaxant effect (Li et al., 2015).

18.3.8 ANTI UROLITHIASIS ACTIVITY

It has been reported that *S. miltiorrhiza* has shown both preventive and remedial properties on Calcium oxalate crystals that were produced in a lithogenic fly model. The study revealed that high-dose plant extract (60 μg/mL) was reported to be more active than K-citrate (Chen et al., 2019a).

18.3.9 ANTIHYPERTENSION ACTIVITY

Hypertension is a condition caused when the heart pumps blood through the blood vessels abnormally at a high rate. *Salvia miltiorrhiza* decreases blood pressure and hypertension through various mechanisms (Orgah et al., 2020). The study of an anti-hypertension activity of potent phytochemical compound Sodium danshensu from *S. miltiorrhiza* had represented biphasic effects on vessel tension. They are involved in producing small contractions due to low dosage (0.1–0.3 $g \cdot L^{-1}$) of sodium danshensu probably through the transient improvement of Ca^{2+} influx. Studies indicated that high dosage (1–3 $g \cdot L^{-1}$) produced effective vasodilation, which improved the opening of non-selective K+ channels and small-conductance calcium-sensitive K^+ channels in Vascular smooth muscle cells (VSMCs) (Zhang et al., 2010). Bioactive compound sodium tanshinone IIA sulfonate initiates high conductance Ca^{2+} activated K^+ channels in coronary artery smooth muscle cells in porcine (Yang et al., 2008). Another study reported increased BKCa (Large-conductance Ca2+-activated K+ channels) macroscopic as "Channels" by 43.6 and 42.1%, respectively, due to extracellular application of bioactive compound DS from *S. miltiorrhiza* (40 and 80 $\mu mol \cdot L^{-1}$), as well as it increases spontaneous transient of outward K^+ channels by 48.7 and 47.4%. Thus these results indicated that the activation of BK_{Ca} takes place due to vasodilatation by the DS compound. Another study had revealed that pretreatment with DS (10 $mg \cdot kg^{-1} \cdot d^{-1}$) for 3 weeks decreased right ventricle weight to the left ventricle with septum weight and also decreased the higher mean pulmonary arterial pressure in an animal model rats with hypoxic pulmonary hypertension but was not effective on normal rats (Huang et al., 2009).

KEYWORDS

- ***Salvia miltiorrhiza***
- **cryptotanshinone**
- **tanshinones**
- **salvianolic acid A**
- **Salvianolic acid B**
- **danshensu**

REFERENCES

Abd-Elazem, I. S.; Chen, H. S.; Bates, R. B.; Huang, R. C. C. Isolation of Two Highly Potent and Non-toxic Inhibitors of Human Immunodeficiency Virus Type 1 (HIV-1) Integrase from *Salvia miltiorrhiza. Antiviral Res.* **2002,** *55* (1), 91–106.

Bao, X. Y.; Zheng, Q.; Tong, Q.; Zhu, P. C.; Zhuang, Z.; Zheng, G. Q.; Wang, Y. Danshensu for Myocardial Ischemic Injury: Preclinical Evidence and Novel Methodology of Quality Assessment Tool. *Front. Pharmacol.* **2018,** *9*, 1445.

Cao, W.; Wang, Y.; Shi, M.; Hao, X.; Zhao, W.; Wang, Y.; Ren, J.; Kai, G. Transcription Factor SmWRKY1 Positively Promotes the Biosynthesis of Tanshinones in *Salvia miltiorrhiza. Front. Plant Sci.* **2018,** *9*, 554.

Chen, W. C.; Chou, T. Y.; Chen, H. Y.; Yang, Y. R.; Man, K. M.; Tsai, M. Y.; Chen, Y. H. *Salvia miltiorrhiza* Bunge (Danshen) for Treatment and Prevention of Urolithiasis: A Drosophila Animal Study. *Evidence-Based Complem. Altern. Med.* **2019a,** 1–5

Chen, W. Z. Pharmacology of *Salvia Miltiorrhiza. Acta Pharmacol. Sin.* **1984,** *19*, 876–880.

Chen, X.; Yu, J.; Zhong, B.; Lu, J.; Lu, J. J.; Li, S.; Lu, Y. Pharmacological Activities of Dihydrotanshinone I, a Natural Product from *Salvia miltiorrhiza* Bunge. *Pharmacol. Res.* **2019b,** *145*, 104254.

Cheng, Q.; Zhao, Y.; Li, J. Sodium Tanshinone IIA Sulfonate Suppresses Heat Stress-Induced Endothelial Cell Apoptosis by Promoting NO Production Through Upregulating the PI3K/AKT/eNOS Pathway. *Mol. Med. Rep.* **2017,** *16* (2), 1612–1618.

Chiu, T. L.; Su, C. C. Tanshinone IIA Induces Apoptosis in Human Lung Cancer A549 Cells Through the Induction of Reactive Oxygen Species and Decreasing the Mitochondrial Membrane Potential. *Int. J. Mol. Med.* **2010,** *25* (2), 231–236.

Choi, Y. J.; Kim, H. S.; Lee, J.; Chung, J.; Lee, J. S.; Choi, J. S.; Yoon, T. R.; Kim, H. K.; Chung, H. Y. Down-regulation of Oxidative Stress and COX-2 and iNOS Expressions by Dimethyl Lithospermate in Aged Rat Kidney. *Arch. Pharmacol. Res.* **2014,** *37* (8), 1032–1038.

Clebsch, B. *New Book of Salvias*; Timber Press, 2003.
Cui, S.; Chen, S.; Wu, Q.; Chen, T.; Li, S. A Network Pharmacology Approach to Investigate the Anti-Inflammatory Mechanism of Effective Ingredients from *Salvia miltiorrhiza. Intern. Immunopharmacol.* **2020,** *81*, 106040.
Cui, S.; Wu, Q.; Wang, J.; Li, M.; Qian, J.; Li, S. Quercetin Inhibits LPS-Induced Macrophage Migration by Suppressing the iNOS/FAK/Paxillin Pathway and Modulating the Cytoskeleton. *Cell Adh. Migr.* **2019,** *13* (1), 1–12.
Gao, S.; Liu, Z.; Li, H.; Little, P.J.; Liu, P.; Xu, S. Cardiovascular Actions and Therapeutic Potential of Tanshinone IIA. *Atherosclerosis* **2012,** *220* (1), 3–10.
Gong, Y.; Li, Y.; Lu, Y.; Li, L.; Abdolmaleky, H.; Blackburn, G. L.; Zhou, J. R. Bioactive Tanshinones in *Salvia miltiorrhiza* Inhibit the Growth of Prostate Cancer Cells In Vitro and In Mice. *Int. J. Cancer* **2011,** *129* (5), 1042–1052.
Guo, Y.; Li, Y.; Xue, L.; Severino, R. P.; Gao, S.; Niu, J.; Qin, L. P.; Zhang, D.; Brömme, D. *Salvia miltiorrhiza*: an Ancient Chinese Herbal Medicine as a Source for Anti-osteoporotic Drugs. *J. Ethnopharmacol.* **2014,** *155* (3), 1401–1416.
Ho, J. H. C.; Hong, C. Y. Salvianolic Acids: Small Compounds with Multiple Mechanisms for Cardiovascular Protection. *J. Biomed. Sci.* **2011,** *18* (1), 1–5.
Honda, G.; Koezuka, Y.; Tabata, M. Isolation of an Antidermatophytic Substrance from the Root of *Salvia miltiorrhiza. Chem. Pharmaceut. Bull.* **1988,** *36* (1), 408–411.
Huang, Y. F.; Liu, M. L.; Dong, M. Q.; Yang, W. C.; Zhang, B.; Luan, L. L.; Dong, H. Y.; Xu, M.; Wang, Y. X.; Liu, L. L.; Gao, Y. Q. Effects of Sodium Tanshinone II A Sulphonate on Hypoxic Pulmonary Hypertension in Rats In Vivo and on Kv2. 1 Expression in Pulmonary Artery Smooth Muscle Cells In Vitro. *J. Ethnopharmacol.* **2009,** *125* (3), 436–443.
Jiang, R. W.; Lau, K. M.; Hon, P. M.; Mak, T. C.; Woo, K. S.; Fung, K. P. Chemistry and Biological Activities of Caffeic Acid Derivatives from *Salvia miltiorrhiza. Curr. Med. Chem.* **2005,** *12* (2), 237–246.
Li, Y. G.; Song, L.; Liu, M.; Wang, Z. T. Advancement in Analysis of *Salviae miltiorrhizae* Radix et Rhizoma (Danshen). *J. Chromatogr. A* **2009,** *1216* (11), 1941–1953.
Li, Y. H.; Xu, Q.; Xu, W. H.; Guo, X. H.; Zhang, S.; Chen, Y. D. Mechanisms of Protection Against Diabetes-Induced Impairment of Endothelium-dependent Vasorelaxation by Tanshinone IIA. *Biochim. Biophys. Acta Gen. Subj.* **2015,** *1850* (4), 813–823.
Li, Z.; Zou, J.; Cao, D.; Ma, X. Pharmacological Basis of Tanshinone and New Insights into Tanshinone as a Multitarget Natural Product for Multifaceted Diseases. *Biomed. Pharmacother.* **2020,** *130*, 110599.
Lin, T. J.; Liu, G. T.; Liu, Y.; Xu, G. Z. Protection by Salvianolic Acid A Against Adriamycin Toxicity on Rat Heart Mitochondria. *Free Radical Biol. Med.* **1992,** *12* (5), 347–351.
Liu, J.; Shen, H. M.; Ong, C. N. *Salvia miltiorrhiza* Inhibits Cell Growth and Induces Apoptosis in Human Hepatoma HepG2 Cells. *Cancer Lett.* **2000,** *153* (1–2), 85–93.
Ma, X. H.; Ma, Y.; Tang, J. F.; He, Y. L.; Liu, Y. C.; Ma, X. J.; Shen, Y.; Cui, G. H.; Lin, H. X.; Rong, Q. X.; Guo, J. The Biosynthetic Pathways of Tanshinones and Phenolic Acids in *Salvia miltiorrhiza. Molecules* **2015,** *20* (9), 16235–16254.
Maione, F.; De Feo, V.; Caiazzo, E.; De Martino, L.; Cicala, C.; Mascolo, N. Tanshinone IIA, A Major Component of *Salvia milthorriza* Bunge, Inhibits Platelet Activation Via Erk-2 Signaling Pathway. *J. Ethnopharmacol.* **2014,** *155* (2), 1236–1242.
Orgah, J. O.; He, S.; Wang, Y.; Jiang, M.; Wang, Y.; Orgah, E. A.; Duan, Y.; Zhao, B.; Zhang, B.; Han, J.; Zhu, Y. Pharmacological Potential of the Combination of *Salvia*

miltiorrhiza (Danshen) and *Carthamus Tinctorius* (Honghua) for Diabetes Mellitus and its Cardiovascular Complications. *Pharmacol. Res.* **2020,** *153*, 104654.

Pang, H.; Wu, L.; Tang, Y.; Zhou, G.; Qu, C.; Duan, J. A. Chemical Analysis of the Herbal Medicine *Salviae Miltiorrhizae* Radix et Rhizoma (Danshen). *Molecules* **2016,** *21* (1), 51.

Peng, L.; Ru, M.; Wang, K.; Li, B.; Wang, Y.; Xia, P.; Liang, Z. Spaceflight Environment-induced Variation in Root Yield and Active Constituents of *Salvia miltiorrhiza. Planta Med.* **2014,** *80* (12), 1029–1035.

Shah, P. K. Inflammation, Infection and Atherosclerosis. *Trends Cardiovasc. Med.* **2019,** *29* (8), 468–472.

Shi, M.; Huang, F.; Deng, C.; Wang, Y.; Kai, G. Bioactivities, Biosynthesis and Biotechnological Production of Phenolic Acids in *Salvia miltiorrhiza. Critical Rev. Food Sci. Nutrition* **2019,** *59* (6), 953–964.

Siouti, E.; Andreakos, E. The Many Facets of Macrophages in Rheumatoid Arthritis. *Biochem. Pharmacol.* **2019,** *165*, 152–169.

Tang, F;Wu, X.; Wang, T.; Wang, P.; Li, R.; Zhang, H.; Gao, J.; Chen, S.; Bao, L.; Huang, H.; Liu, P., 2007. Tanshinone II A Attenuates Atherosclerotic Calcification in Rat Model by Inhibition of Oxidative Stress. *Vascular Pharmacol.* **2007,** *46* (6), 427–438.

Wang, X.; Yang, Y.; Liu, X.; Gao, X. Pharmacological Properties of Tanshinones, the Natural Products from *Salvia miltiorrhiza. Adv. Pharmacol.* **2020,** *87*, 43–70.

Wang, B. Q. *Salvia miltiorrhiza*: Chemical and Pharmacological Review of A Medicinal Plant. *J. Med. Plants Res.* **2010,** *4* (25), 2813–2820.

Wang, L.; Ma, R.; Liu, C.; Liu, H.; Zhu, R.; Guo, S.; Tang, M.; Li, Y.; Niu, J.; Fu, M.; Gao, S. *Salvia miltiorrhiza*: A Potential Red Light to the Development of Cardiovascular Diseases. *Curr. Pharm. Des.* **2017,** *23* (7), 1077–1097.

Wang, P.; Wu, X.; Bao, Y.; Fang, J.; Zhou, S.; Gao, J.; Pi, R.; Mou, Y. G.; Liu, P. Tanshinone IIA Prevents Cardiac Remodeling Through Attenuating NAD (P) H Oxidase-derived Reactive Oxygen Species Production in Hypertensive Rats. *Pharmazie.* **2011,** *66* (7), 517–524.

Wu, Y. B.; Ni, Z. Y.; Shi, Q. W.; Dong, M.; Kiyota, H.; Gu, Y. C.; Cong, B. Constituents from *Salvia* Species and their Biological Activities. *Chem. Rev.* **2012,** *112* (11), 5967–6026.

Yang, L.; Zou, X. J.; Gao, X.; Chen, H.; Luo, J. L.; Wang, Z. H.; Liang, Q. S.; Yang, G. T. Sodium Tanshinone IIA Sulfonate Attenuates Angiotensin II-induced Collagen Type I Expression in Cardiac Fibroblasts In Vitro. *Exp. Mol. Med.* **2009,** *41* (7), 508–516.

Yang, Y.; Cai, F.; Li, P. Y.; Li, M. L.; Chen, J.; Chen, G. L.; Liu, Z. F.; Zeng, X. R., Activation of High Conductance Ca^{2+}-activated K^+ Channels by Sodium Tanshinoneii-A Sulfonate (DS-201) in Porcine Coronary Artery Smooth Muscle Cells. *European J. Pharmacol.* **2008,** *598* (1–3), 9–15.

Zhang, J.; Zhang, Q.; Liu, G.; Zhang, N. Therapeutic Potentials and Mechanisms of the Chinese Traditional Medicine Danshensu. *European J. Pharmacol.* **2019, 864**, 172710.

Zhang, N.; Zou, H.; Jin, L.; Wang, J.; Zhong, M. F.; Huang, P.; Gu, B. Q.; Mao, S. L.; Zhang, C.; Chen, H. Biphasic Effects of Sodium Danshensu on Vessel Function in Isolated Rat Aorta. *Acta Pharmacologica Sinica* **2010,** *31* (4), 421–428.

Zhao, G. R.; Zhang, H. M.; Ye, T. X.; Xiang, Z. J.; Yuan, Y. J.; Guo, Z. X.; Zhao, L. B. Characterization of the Radical Scavenging and Antioxidant Activities of Danshensu and Salvianolic Acid B. *Food Chem. Toxicol.* **2008,** *46* (1), 73–81.

Zhao, J.; Lou, J.; Mou, Y.; Li, P.; Wu, J.; Zhou, L. Diterpenoid Tanshinones and Phenolic Acids from Cultured Hairy Roots of *Salvia miltiorrhiza* Bunge and their Antimicrobial Activities. *Molecules* **2011,** *16* (3), 2259–2267.

Zhou, W.; Huang, Q.; Wu, X.; Zhou, Z.; Ding, M.; Shi, M.; Huang, F.; Li, S.; Wang, Y.; Kai, G. Comprehensive Transcriptome Profiling of *Salvia miltiorrhiza* for Discovery of Genes Associated with the Biosynthesis of Tanshinones and Phenolic Acids. *Scientific Rep.* **2017,** *7* (1), 1–12.

Zhu, J.; Xu, Y.; Ren, G.; Hu, X.; Wang, C.; Yang, Z.; Li, Z.; Mao, W.; Lu, D. Tanshinone IIA Sodium Sulfonate Regulates Antioxidant System, Inflammation, and Endothelial Dysfunction in Atherosclerosis by Downregulation of CLIC1. *European J. Pharmacol.* **2017,** *815*, 427–436.

CHAPTER 19

AN OVERVIEW OF BIOACTIVE CONSTITUENTS AND PHARMACOLOGICAL ACTIONS OF IRANIAN SAGE, *SALVIA MIRZAYANII* RECH. F. & ESFAND.

VRUSHALI MANOJ HADKAR[1] and
CHINNADURAI IMMANUEL SELVARAJ[2]

[1]*Department of Biotechnology, School of Biosciences and Technology, Vellore Institute of Technology, Vellore, Tamil Nadu, India*

[2]*VIT School of Agricultural Innovations and Advanced Learning (VAIAL), Vellore Institute of Technology, Vellore, Tamil Nadu, India*

ABSTRACT

Salvia mirzayanii Rech. f. and Esfand is an endemic, herbaceous, medicinal and aromatic plant belonging to the Lamiaceae family. The plant grows in the central and southern parts of Iran as a biennial or perennial flowering herb and is broadly used in folk medicine. *S. mirzayanii* is also known as Iranian Sage, *Moor-e-Talkh*, or "*Marv-e-Talkh*" in local areas. It is a traditional medicinal plant used in management of gastrointestinal diseases, skin infections, spasms, inflammations, and weakness by people from the native southern regions of Iran. *S. mirzayanii* is used in food preparations, cosmetics, drugs, and perfumery industries mainly due to presence of active essential oils. The essential oils in *Salvia* are classified into oxygenated monoterpenes, monoterpene hydrocarbons, diterpenes, and sesquiterpene hydrocarbons.

Bioactives and Pharmacology of Lamiaceae. T. Pullaiah, PhD (Ed.)

Linalool, spathulenol, δ-cadinene, cubenol, β-eudesmol, α-cadinol, linalyl acetate, and α-terpinyl acetate are the significant components found in essential oil of the plant. The sesquiterpenes present in *S. mirzayanii* are spathulenol, teuclatriol and bicyclogermacrene and flavonoids such as chrysoeriol, cirsimaritin and salvigenin. The plant has many pharmacological activities such as immunomodulatory activity, antimicrobial activity, antidepressant activity, antioxidant activity, cytotoxic activity, anti-inflammatory activity, and neuroprotective activity.

19.1 INTRODUCTION

The genus *Salvia* belongs to the family Lamiaceae, having around 900 species. *Salvia mirzayanii* Rech. f. and Esfand is a popularly known endemic species of Iran. In Hormozgan, Kerman, and Fars provinces in Southern Iran, this species is found growing wild on mountains and rocky slopes (Mozaffarian, 1996). *S. mirzayanii* is an aromatic flowering herb generally 25–40 cm high, biennial or a perennial. Flowers of this plant are blue and also commonly known as *Moor-e-Talkh*, "*Marve-Talkh*," or Iranian Sage in local areas of Iran (Soltanipour, 2005). It has great importance as a folk medicine; the aerial parts of *Salvia mirzayanii* are used to treat various diseases such as stomach ache, diarrhea, headache, hypercholesteremia, and diabetes in the south of Iran (Sadeghi and Alizadeh, 2013). In Iranian folk medicine, *S. mirzayanii* is used as a tonic and an astringent (Zarshenas and Krenn, 2015). It is a medicinal plant that is also used as herbal remedy for treatment and management of some other diseases such as gastrointestinal diseases, spasms, skin infections, inflammations, and weakness by people from the native southern regions of Iran (Javidnia et al., 2002; Sadeghi and Alizadeh, 2013). *Salvia* species are a prolific source of essential oils. The essential oils from *S. mirzayanii* are used in food preparations, cosmetics, drugs, and perfumery industries. For flavorings or fragrances and as a medicine, *Salvia* species are widely used in many different parts of the world (Miliauskas et al., 2004).

19.2 BIOACTIVES

Essential oils in *Salvia* spp. have a great significance for medicinal purpose. Genus *Salvia* has more than 100 active elements in essential oils that show various pharmacological effects. The essential oils in *Salvia* are classified into oxygenated monoterpenes, monoterpene hydrocarbons, diterpenes, and sesquiterpene hydrocarbons. In *S. mirzayanii*, the main components reported are 1,8-cineole, camphor, borneol, α-pinene, β-pinene, α-thujene, and camphene (Fu et al., 2013). Also, linalool, spathulenol, δ-cadinene, cubenol, β-eudesmol, α-cadinol, linalyl acetate, and α-terpinyl acetate were found to be the significant components of *S. mirzayanii* essential oil (Javidnia et al. 2002). Another study found that 1,8-cineole, 8-acetoxy linalool, linalyl acetate, and linalool were the chief components in *S. mirzayanii* essential oil (Yamini et al. 2008). *S. mirzayanii* has also shown the presence of sesquiterpenes, among which tricyclic sesquiterpenes alcohol, spathulenol, teuclatriol as a type of guaiane sesquiterpene and bicyclogermacrene have been reported (Ziaei et al., 2011a). In addition, phytoconstituents belonging to class flavonoids also have been reported in *S. mirzayanii*. Methoxyflavonoid named chrysoeriol was reported to have muscle relaxant, antioxidant, anti-inflammatory, and chemopreventive properties (Choi et al., 2005). Also, flavonoid cirsimaritin was analyzed that shown a positive effect against acute renal failure. Additionally, the presence of eupatorium, salvigenin, and the 6′, 7′, 3′, 4′-tetramethyl ether of 6-hydroxyluteolin was reported, and methanolic extract has shown the presence of rutin and luteolin (Wollenweber et al., 1992; Lu and Foo, 2002). In addition, the aerial parts of *S. mirzayanii* shown the presence of some phenol compounds known as catechin and rosmarinic acid (Asadi et al., 2011). The chemical structures of essential oil constituents of *S. mirzayanii* are given in Figure 19.1 and the important bioactive compounds are given in Figure 19.2.

19.3 PHARMACOLOGY

19.3.1 IMMUNOMODULATORY ACTIVITY

19.3.1.1 LYMPHOCYTE PROLIFERATION ASSAY

Lymphocyte proliferation assay was assessed using *S. mirzayanii* extract, which at low concentrations of 0.01–10 mg/mL showed a notable stimulatory effect on the proliferation of lymphocytes. The highest stimulation index was found in cultures treated with 0.1 and 1 mg/mL extract concentration. It was observed that increasing the concentration of extract led to the reduced stimulatory effect of lymphocytes. After adding 100 and 200 mg/mL of the extract, PHA-treated cells indicated the inhibitory effect of

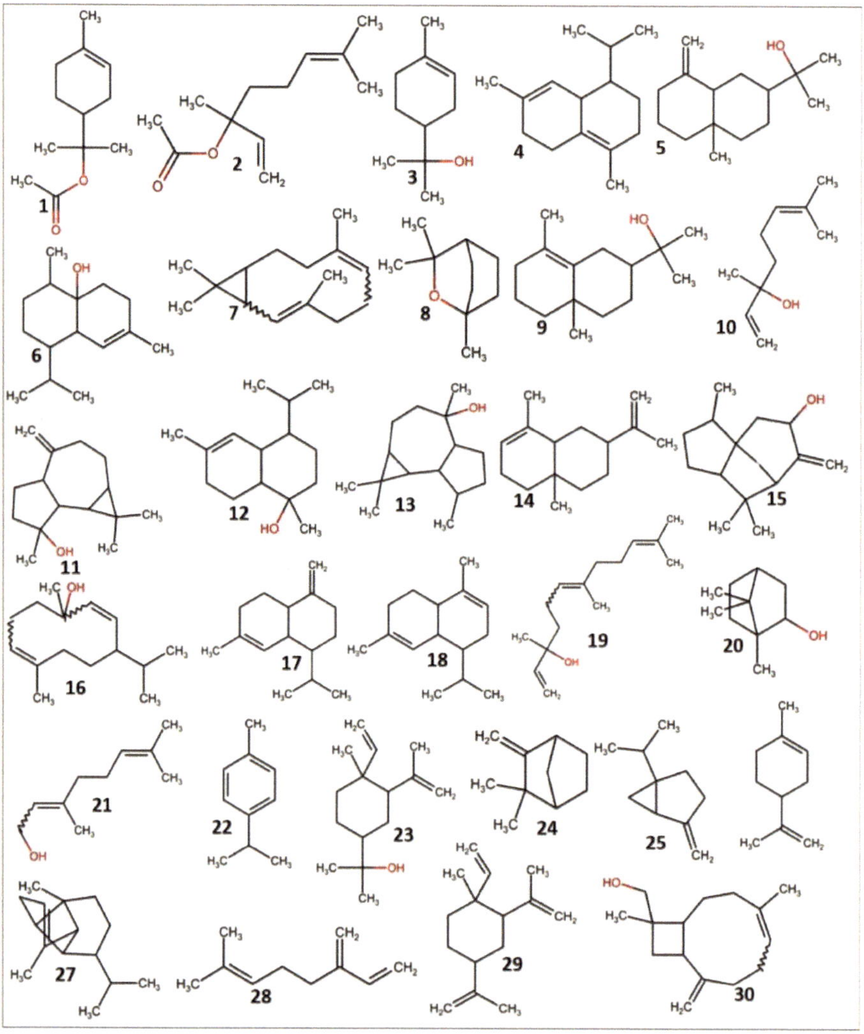

FIGURE 19.1 Chemical structures of the essential oil constituents of *Salvia mirzayanii*. Linalyl acetate (1), *α*-terpinyl acetate (2), *α*-terpineol (3), *δ*-cadinene (4), *β*-eudesmol (5), Cubenol (6), Biocyclogermacrene (7), 1,8-cineol (8), γ-eudesmol (9), Linalool (10), Spathulenol (11), *α*-cadinol (12), Globulol (13), 7-epi-*α*-selinene (14), 5-neo-cedranol (15), Germacrene D-4-ol (16), γ-cadinene (17), *α*-cadinene (18),(Z)-nerolidol (19), borneol (20), geraniol (21), *p*-cymene (22), elemol (23), camphene (24), sabinene (25), limonene (26), *α*-copaene (27), myrcene (28), *β*-elemene (29), and (E)-caryophyllene (30); [Marvin 17.21.0, ChemAxon tool was used for drawing chemical structures].

lymphocyte activation by actively decreasing 3H-thymidine uptake with SI 0.60.02 and 0.20.04, respectively (Amirghofran et al., 2010). Lymphocyte proliferation assay of *S. mirzayanii* was studied using phyto-hemagglutinin mitogen for T-cells activation in the presence of three concentrations 1, 10, and 100 µg/mL of compounds 5,7-dihydroxy,6,4′-dimethoxyflavone, 5-hydroxy,6,7,3′,4′-tetramethoxyflavone, and 5,3′-dihydroxy,6,7,4′-trimethoxyflavone. Active suppression of proliferation of human blood lymphocytes was done by compounds 5-hydroxy,6,7,3′,4′-tetramethoxyflavone and 5,3′-dihydroxy,6,7,4′-trimethoxyflavone in a dose-dependent manner. The reported IC_{50} values were 1.3 ± 0.04 µg/mL and 1.3 ± 0.21 µg/mL compared with drug prednisolone having IC_{50} 1.45 ± 0.60 µg/mL. The results were related to the activity of other methoxylated flavonoids like casting from *Vitex agnus-castus* that suggested *S. mirzayanii* bioactive compounds have shown an immunoinhibitory activity (Mesaik et al., 2009).

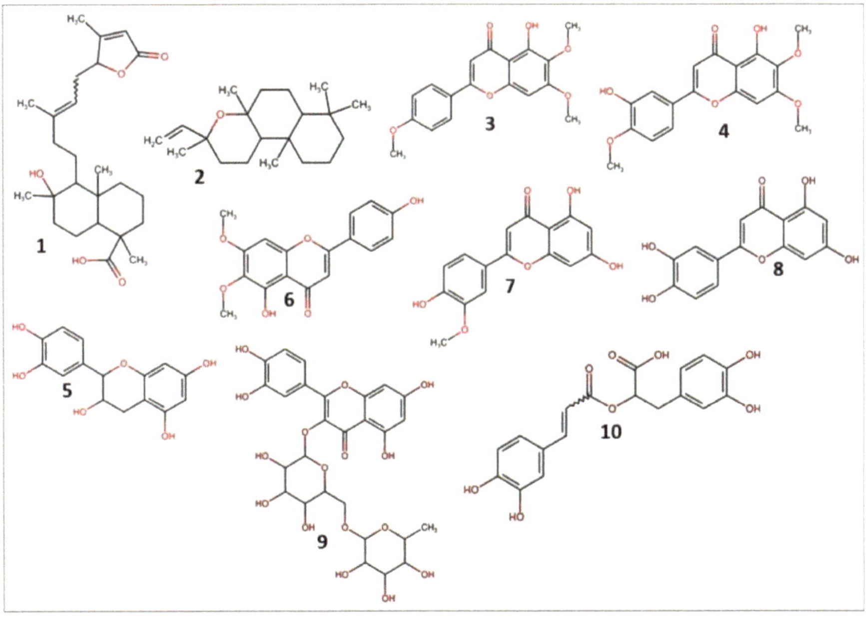

FIGURE 19.2 Chemical structures of bioactive compounds of *Salvia mirzayanii*. Salvimirzacolide (1), manoyloxide sesquiterpene (2), salvigenin (3), eupatorin (4), catechin (5), cirsimaritin (6), chrysoeriol (7), luteolin (8), rutin (9), and rosemarinic acid (10); [Marvin 17.21.0, ChemAxon tool was used for drawing chemical structures].

19.3.1.2 DETERMINATION OF IL-2 PRODUCTION

Studies of IL-2 production using *S. mirzayanii* extract reported that in the presence of 0.1 mg/mL of the extract, activated lymphocytes displayed 1568 μg/mL of IL-2-production compared to the non-treated cells 322.9 μg/mL. The IL-2 production at 50 and 100 mg/mL of the extract, reduced to 301.5 and 18 4.2 μg/mL. A higher level of IL-2 production was released from cells treated with less extract, and less production of IL-2 was found in cells treated with a higher dose of the *S. mirzayanii* extract (Amirghofran et al., 2010).

19.3.1.3 PHAGOCYTE CHEMILUMINESCENCE ASSAY

S. mirzayanii extract was used to study phagocyte chemiluminescence assay in which compounds 5,7-dihydroxy,6,4′-dimethoxyflavone and 5,3′-dihydroxy,6,7,4′-trimethoxyflavone were used on PMNCs (peripheral mononuclear cells) in human blood containing luminol and zymosan as an intracellular oxidative burst. These compounds have shown suppressive activities against reactive oxygen species. IC_{50} values were found to be 55.3 ± 0.4 and 36.2 ± 0.7 μg/mL. The compound 5-hydroxy,6,7,3′,4′-tetramethoxyflavone showed low activity, having an IC_{50} value of more than 100 μg/mL (Ayatollahi et al., 2015).

19.3.1.4 HUMORAL ANTIBODY RESPONSE

The study reported that *S. mirzayanii* extract was injected into mice to determine its effect on humoral immunity. At the lower concentration of the extract, activation of the response at primary and secondary contact to the antigen was observed. The mean antibody titre was recorded according to log 2 for 1 mg/kg of the extract was 9.20.6 and 10.40.5 as compared to 8.30.5, and 90.4 was for both the response in non-treated mice. The results suggested that depending on the concentration of the extract, *S. mirzayanii* had different effects on the humoral immune response (Amirghofran et al., 2010).

Another study was carried out in which phytohemagglutinin (PHA)-activated human peripheral blood lymphocytes were treated with the varying volume of concentrated fractions of spathulenol from *S. mirzayanii*. Decrease in the proliferation of the lymphocytes was observed at concentrations of 50 mg/mL (71.1 ± 6.1%), 100 mg/mL (25.5 ± 4.3%), and 200 mg/mL (28.2 ± 6.8%) that was compared with untreated cells (100%), which indicated that spathulenol downregulated proliferation of lymphocytes having IC_{50}, 85.4 ± 11.08 mg/mL. It was observed that Annexin V-FITC and PI staining flow cytometry results showed a significant apoptotic inducing effect of compound spathulenol on Human peripheral blood lymphocytes. The treated cultures with 50, 100, and 200 mg/mL of compound spathulenol reported percentage of apoptosis cells that were 30.5 ± 1.5%, 71.57 ± 0.48%, and 90.54 ± 3.7%, respectively. The observed IC_{50} value was 77.2 ± 5.31 mg/mL (Ziaei et al., 2011b).

19.3.2 ANTIMICROBIAL ACTIVITY

Essential oils of *S. mirzayanii* have shown significant antimicrobial activity assessed by using the disc diffusion method against a standard strain of *Shigella dysenteriae*, *Staphylococcus aureus*, and *Escherichia coli*. The formation of the inhibition zone on the growth of microorganisms was used to determine the effectiveness of the essential leaf oil of *S.mirzayanii* (Priya et al., 2014). Different inhibition zone resulted against *E. coli* (8.67, 8.67, 10, 11.33, 12.33, and 14.33 mm), *S. aureus* (8, 10.33, 11.33, 12.67, 14, and 17.67 mm), and *S. dysenteriae* (8, 9, 10, 12, 13.33, and 15.67 mm) with different concentrations of *S. mirzayanii* essential oil (0.625, 1.25, 2.5, 5, 10, and 20 mm) were observed, respectively. The study reported that the most resistant microbe was *E. coli* and the most sensitive one was *S. dysenteriae*. Minimum inhibitory concentrations (MICs) of essential oils of *S. mirzayanii* were determined. The reported MIC value was 0.625 µl/mL in all microorganisms (Haghighat et al., 2012). Another study reported the antimicrobial activity of *S. mirzayanii* ecotypes against standard strains of *Escherichia coli* (ATCC 25922), *Staphylococcus aureus* (ATCC 6538), *Staphylococcus epidermidis* (ATCC 1435), and fungal strain *Candida albicans* (ATCC 10231) by measuring the inhibition zone, minimum bactericidal/fungicidal (MBC/MFC) concentration and MIC values. The results suggested that the highest antimicrobial activity was seen against *S. aureus* with the highest inhibition zones of 62.5 mm. The ranges of MIC and MBC/MFC values were 0.16–2.5 and 0.32–2.5 µg/mL, each. The results suggested that essential oils from *S. mirzayanii* have significant antimicrobial activity (Ghasemi et al., 2020).

Antimicrobial activity of extracts of 12 medicinal plants, including leaves of *Salvia mirzayanii* and *Nigella sativa* (seeds) from Iran, were examined for their antibacterial activity against *Helicobacter pylori* using two different methods. Compared with metronidazole and amoxicillin, the disc inhibition assay of the extracts has shown inhibitory effects at higher concentrations on the growth of *H. pylori*. The disc diffusion assay for the plants tested reported that *S. mirzayanii*, *Zataria multiflora*, *Teucrium polium*, and *Bunium persicum* showed potential anti-*H. pylori* activities having a zone of inhibition 20–30 mm. *S. mirzayanii* was found to be the most effective plant with great antibacterial activity with MIC (Minimum Inhibitory concentration) of 32–64 μg/mL among the plant extracts examined for MIC, against *H. pylori* (Atapour et al., 2009). Essential oils of *Salvia* species were evaluated for their antimicrobial activity by the disc diffusion method. It was observed that *S. mirzayanii* has shown significant antimicrobial activity of oil against the bacterial strains and produced significant zones of inhibition against the fungal strains. *Staphylococcus epidermidis* and *Bacillus subtilis* having MIC values of 1.25 mg/mL were the most susceptible microbial strains, followed by *Candida albicans* and *Aspergillus niger* with MIC of 2.5 mg/mL. Less susceptibility was observed in *Klebsiella pneumoniae* with essential oils of *S. santolinifolia* and *S. mirzayanii*. *Pseudomonas aeruginosa* was resistant to the essential oil of *S. mirzayanii* with the MIC value of 20.0 mg/mL (Sonboli et al., 2006). It was reported that *S. mirzayanii* essential oil such as linalool, α-terpinyl acetate, linalyl acetate, and 1,8-cineol was found to have a significant antimicrobial effect (Park et al., 2012).

19.3.3 ANTIDEPRESSANT ACTIVITY

Antidepressant activity of *S. mirzayanii* extracts was evaluated, exhibiting a notable effect of the *S. mirzayanii* extracts on locomotor activity. At doses of 1,800 mg/kg (35.5%) and 2,700 mg/kg (80.1%), *S. mirzayanii* aqueous extract effectively reduced the locomotor activity compared to the control group. The hydroalcoholic extract of *S. mirzayanii* at doses of 1,800 mg/kg (52.2%), 2,700 mg/kg (49.9%), and 5,400 mg/kg (94.1%) were noted to show significant decreases in locomotor activity compared to the control group (Sarkoohi et al., 2020). Effects of *S. mirzayanii* extracts on Forced Swimming Test (FST) were reported in which depression-like behavior was assessed in mice. The antidepressant activity was considered based on the decrease in immobility and development in climbing or swimming behavior. Dosage concentrations of 150, 300, 600, and 900 mg/kg of *S. mirzayanii* extract were used in this study. *S. mirzayanii* aqueous extract effectively reduced immobility at doses of 900 and 1,800 mg/kg that was 11.1 and 10.8%, respectively; and enhanced swimming behavior at a dose of 1,800 mg/kg that was reported to be 68.6%. Drugs used as standards, fluoxetine and imipramine, notably decreased immobility compared to the control group in mice. Aqueous extract *S. mirzayanii* at a dose of 300 mg/kg reported decreased immobility to 7.6% and increased swimming by 48.5%. The results suggested that *S. mirzayanii* plant extract has an antidepressant activity (Sarkoohi et al., 2020).

19.3.4 ANTIOXIDANT ACTIVITY

Antioxidant activity of *S. mirzayanii* was estimated in butanol fraction, ethyl acetate fraction, including the crude extract concentration. A concentration of 12.5–200 mg/mL was used to perform the assay. The observed activity of the crude extract was 24 and 18 mg/mL compared to butylated hydroxytoluene and gallic acid (Moein et al., 2007). A study of antioxidant and antiglycation activities of *S. mirzayanii* has shown the lowest antioxidant effect. *S. mirzayanii* effectively reduced glycation and end-product formation at the doses of 50 and 100 mg/mL. Also, a significant decrease in the formation of BSA fibrillar structure was observed (Asadi et al., 2011). In another study of antioxidant activity, the extracts from cultivated and wild *S. mirzayanii* using DPPH assay were performed. There was no difference in the activity for cultivated plants with an IC_{50} value of 34.64 mg/mL and wild ones with an IC_{50} value of 38.44 mg/mL (Sadeghi and Alizadeh, 2013). *S. mirzayanii* extracts were used to determine antioxidant activity in which two different fractions of *S. mirzayanii* have shown radical scavenging activity using DPPH assay in which IC_{50} values were 37.9 ± 0.85 and 40.05 ± 1.4 mg/mL that was compared with standard quercetin having an IC_{50} value of 38.84 ± 0.86 mg/mL (Moein et al., 2008).

19.3.5 CYTOTOXIC ACTIVITY

The H_2O_2 cytotoxic activity of *S. mirzayanii* in fibroblast cells was assessed using the MTT test. In H_2O_2-treated cells, the growth was strongly suppressed. The crude extract used was 12.5 and 25 mg/mL. Considering the same conditions, 25 mM of butylated hydroxytoluene decreased the cytotoxic effects of H_2O_2. It was reported that *S. mirzayanii* extract 12.5 and 25 mg/mL was not cytotoxic. Therefore, it was noted that *S. mirzayanii* had shown reduced H_2O_2-induced cytotoxicity in human non-immortalized fibroblast cells (Moein et al., 2007). Cytotoxic activity of *S. mirzayanii* essential oil was examined on human fetal skin fibroblast and human fetal liver fibroblast cell lines. The reported IC_{50} value for human fetal skin fibroblast was 625 mg/mL, and it was 1,250 mg/mL for human fetal liver fibroblast. Thus, the results suggested that *S. mirzayanii* essential oil has cytotoxic properties (Armana et al., 2012).

19.3.6 ANTI-INFLAMMATORY ACTIVITY

Amirghofran et al. (2011) reported an anti-inflammatory activity *S. mirzayanii* methanolic extract, on nitric oxide production in LPS-stimulated mouse macrophages. The low inhibition observed at 50 mg/mL concentration was 66.2 ± 8%. There was no inhibition observed in interleukin-1 β and tumor necrosis factor-α induced with LPS.

19.3.7 NEUROPROTECTIVE AND ANTI-APOPTOTIC ACTIVITY

S. mirzayanii and other species of *Salvia* were evaluated for the neuroprotective activity in rat pheochromocytoma cells. The cells were treated with varying concentrations of methanolic extracts 10, 25, 50, and 100 mg/mL and with hydrogen peroxide as oxidative stress. Cells were incubated with 10, 25, 50, and 100 mg/mL of *S. mirzayanii* extract that effectively renewed cell viability by 2.84, 3.01, 3.29, and 3.45-fold. The *S. mirzayanii* extract protected genomic DNA from nerve growth factor differentiated pheochromocytoma cells at 10 mg/mL concentration. The neuroprotective effect of *S. mirzayanii* in pheochromocytoma cells was examined by morphological observations and their resistance against the damaging effects of hydrogen peroxide. Hydrogen peroxide decreased the cell viability to 28%. Pretreatment with the extract of *S. mirzayanii* at 10, 25, 50, and 100 mg/mL decreased the malondialdehyde release from neuronal phospholipid bilayers by 1.29, 1.38, 1.53, and 1.59 fold each (Asadi et al., 2011).

The anti-apoptotic effects of *S. mirzayanii* were evaluated against neuron-like pheochromocytoma cells by inducing hydrogen peroxide cytotoxicity. Cell growth was observed by measuring some parameters: cell body area, average neurite width, and length. The methanolic extract with 10–100 mg/mL concentration effectively improved the changes in the average cell body and neurite length of cells exposed to hydrogen peroxide. The decreased caspase-3 level in pheochromocytoma cells was noticed using 10, 25, 50, and 100 mg/mL concentrations of extract for 24 h. In the presence of hydrogen peroxide, the concentration of glutathione was decreased nearly 2.23 fold. Pre-treatment with the extract at 10, 25, 50, and 100 mg/mL concentrations for 24 h were found to have 6.80 ± 0.38, 7.04 ± 0.14, 7.72 ± 0.32, and 7.83 ± 0.33 μL/mg protein, each (Alamdary et al., 2012).

19.3.8 ANTI-HYPERGLYCEMIC ACTIVITY

The methanolic extract of *S. mirzayanii* was used to evaluate anti-hyperglycemic activity in streptozocin-induced diabetic rats. The single doses of extract at concentrations of 50, 100, 200, and 400 mg/kg body weight were used to treat animals *via* intraperitoneal injection. Afterward, a daily dose of 400 mg/kg body weight was administered for 12 days. After the last dose, reduced serum glucose concentration was reported at 400 mg/kg extract. The extract of *S. mirzayanii* was administered in normal and diabetic animals in which the histopathological study of the organs like liver, spleen, and pancreas showed no progressive changes (Mehrabani and Heidary, 2010).

19.3.9 HYPOGLYCEMIC EFFECT

The hypoglycemic effect of *S. mirzayanii* extracts in patients who have diabetes were studied. The treatment with the herbal drug, $P = 0.473$, $P = 0.641$, and $P = 0.987$, noted no difference in body mass index, diastolic, and systolic pressures for trial and control groups. There were no differences reported before and after treatment between fasting blood sugar and hemoglobin A1C levels in trial and control groups. The low-density lipid cholesterol level of the trial group was found to be 89.6 ± 31.5 mg/dL, which was more than the control group 108.8 ± 30.9 mg/dL. A decrease in low-density lipid cholesterol level was reported after having *S. mirzayanii* capsule, and aspartate aminotransferase level was improved in the control group from 21.3 ± 4.2 mg/dL to 24.5 ± 6.2 mg/dL. The results suggested that the herbal drug *S. mirzayanii* improved insulin secretion and low-density lipid cholesterol levels in patients having diabetes (Moein et al., 2020).

KEYWORDS

- ***Salvia mirzayanii***
- **neuroprotective**
- **bicyclogermacrene**
- **spathulenol**
- **chrysoeriol**
- **cirsimaritin**

REFERENCES

Alamdary, S. Z.; Khodagholi, F.; Shaerzadeh, F.; Ansari, N.; Sonboli, A.; Tusi, S. K. S. Choloroleuca, *S. mirzayanii* and *S. santolinifolia* Protect PC12 Cells from H_2O_2-induced Apoptosis by Blocking the Intrinsic Pathway. *Cytotechnology* **2012,** *64* (4), 403–419.

Amirghofran, Z.; Malek-Hosseini, S.; Golmoghaddam, H.; Kalantar, F.; Shabani, M. Inhibition of Nitric Oxide Production and Proinflammatory Cytokines by Several Medicinal Plants. *Iran. J. Immunol.* **2011,** *8* (3), 159–169.

Amirghofran, Z.; Bahmani, M.; Azadmehr, A.; Javidnia, K.; Ramazani, M.; Ziaei, A. Effect of *Salvia mirzayanii* on the Immune System and Induction of Apoptosis in Peripheral Blood Lymphocytes. *Nat. Prod. Res.* **2010,** *24* (6), 500–508.

Armana, M.; Azizi, N.; Yousefzadi, M. Cytotoxicity, Antimicrobial Activity and Composition of the Essential Oil from *Salvia mirzayanii* Rech. f. & Esfand from Iran. *J. Biol. Act. Prod. Nat.* **2012,** *2* (1), 54–58.

Asadi, S.; Khodagholi, F.; Esmaeili, M. A.; Tusi, S. K.; Ansari, N.; Shaerzadeh, F.; Sonboli, A.; Ghahremanzamaneh, M. Chemical Composition Analysis, Antioxidant, Antiglycating Activities and Neuroprotective Effects of *S. choloroleuca*, *S. mirzayanii* and *S. santolinifolia* from Iran. *Am. J. Chin. Med.* **2011,** *39* (3), 615–638.

Atapour, M.; Zahedi, M. J.; Mehrabani, M.; Safavi, M.; Keyvanfard, V.; Foroughi, A.; Siavoshi, F.; Foroumadi, A. *In Vitro* Susceptibility of the Gram-negative Bacterium *Helicobacter pylori* to Extracts of Iranian Medicinal Plants. *Pharm. Biol.* **2009,** *47* (1), 77–80.

Ayatollahi, A. M.; Ghanadian, M.; Att-ur-Rahman, R.; Mesaik, M. A.; Khalid, A. S.; Adeli, F. Methoxylated Flavones from *Salvia mirzayanii* Rech. f. and Esfand with Immunosuppressive Properties. *Iran J. Pharm. Res.* **2015,** *14* (3), 955.

Choi, D. Y.; Lee, J. Y.; Kim, M. R.; Woo, E. R.; Kim, Y. G.; Kang, K. W. Chrysoeriol Potently Inhibits the Induction of Nitric Oxide Synthase by Blocking AP-1 Activation. *J. Biomed. Sci.* **2005,** *12* (6), 949–959.

Fu, Z.; Wang, H.; Hu, X.; Sun, Z.; Han, C. The Pharmacological Properties of *Salvia* Essential Oils. *J. Appl. Pharm. Sci.* **2013,** *3* (7), 122.

Ghasemi, E.; Sharafzadeh, S.; Amiri, B.; Alizadeh, A.; Bazrafshan, F. Variation in Essential Oil Constituents and Antimicrobial Activity of the Flowering Aerial Parts of *Salvia mirzayanii* Rech. & Esfand. Ecotypes as a Folkloric Herbal Remedy in Southwestern Iran. *J. Essent. Oil-Bear. Plants.* **2020,** *23* (1), 51–64.

Haghighat, M. H.; Alizadeh, A.; Nouroznejadfard, M. J. Essential Oil Composition and Antimicrobial Activity in Iranian *Salvia mirzayanii* Rech. & Esfand. *Adv. Environ. Biol.* **2012,** 1985–1990.

Javidnia, K.; Miri, R.; Kamalinejad, M.; Nasiri, A. Composition of the Essential Oil of *Salvia mirzayanii* Rech. f. & Esfand from Iran. *Flavour Fragr. J.* **2002,** *17* (6), 465–467.

Lu, Y.; Foo, L. Y. Polyphenolics of *Salvia*—A review. *Phytochem.* **2002,** *59* (2), 117–40.

Mehrabani, M. I.; Heidary, M. R. Study of Acute and Chronic Antihyperglycemic Activity of Methanolic Extract of *Salvia mirzayanii* Rech. & Esfand. in Rats. *J. Med. Plant.* **2010,** *9* (36), 106–206.

Mesaik, M. A.; Murad, S.; Khan, K. M.; Tareen, R. B.; Ahmed, A.; Choudhary, M. I. Isolation and Immunomodulatory Properties of a Flavonoid, Casticin from *Vitex agnus-castus. Phytother. Res.* **2009,** *23* (11), 1516–1520.

Miliauskas, G.; Venskutonis, P. R.; Van Beek, T. A. Screening of Radical Scavenging Activity of Some Medicinal and Aromatic Plant Extracts. *Food Chem.* **2004,** *85* (2), 231–237.

Moein, M. R.; Moein, S.; Ahmadizadeh, S. Radical Scavenging and Reducing Power of *Salvia mirzayanii* Subfractions. *Molecules* **2008,** *13* (11), 2804–2813.

Moein, S.; Farzami, B.; Khaghani, S.; Moein, M. R.; Larijani, B. A. Antioxidant Properties and Protective Effect on Cell Cytotoxicity of *Salvia mirzayani. Pharm. Biol.* **2007,** *45* (6), 458–463.

Moein, S.; Saberi, P.; Moein, M.; Mehdizadeh, R.; Zarshenas, M. Hypoglycemic Effects of Aqueous Extract of *Salvia mirzayanii* Rech. F & Esfand in Diabetic Patients; A Randomized Controlled Trial Study. *J. Nephropathol.* **2020,** *9* (1), 1–7.

Mozaffarian, V. *A Dictionary of Iranian Plant Names.* Tehran: Farhang Moaser. Mazda Publishers, Costa Mesa, California, USA, 1996, 396.

Park, S. N.; Lim, Y. K.; Freire, M. O.; Cho, E.; Jin, D.; Kook, J. K. Antimicrobial Effect of Linalool and α-Terpineol Against Periodontopathic and Cariogenic Bacteria. *Anaerobe* **2012,** *18* (3), 369–372.

Priya, B.; Gupta, V. K.; Pathania, D.; Singha, A. S. Synthesis, Characterization and Antibacterial Activity of Biodegradable Starch/PVA Composite Films Reinforced with Cellulosic Fibre. *Carbohydr. Polym.* **2014,** *109*, 171–179.

Sadeghi, F.; Alizadeh A. Phytochemical Composition of the Essential Oil, Total Phenolic Content and Antioxidant Activity in *Salvia mirzayanii* Rech. & Esfand. Grown Wild and cultivated in Iran. *Int. Res. J. Appl. Basic Sci.* **2013,** *6* (7), 977–982.

Sarkoohi, P.; Fathalipour, M.; Ghasemi, F.; Javidnia, K.; Emamghoreishi, M. Antidepressant Effects of the Aqueous and Hydroalcoholic Extracts of *Salvia mirzayanii* and *Salvia macrosiphon* in Male Mice. *Shiraz E Med. J.* **2020,** *2 1*(2), e91276.

Soltanipour M. A. Phenology of *Salvia mirzayanii* Rech. F. & Esfand in Different Elevations of Hormozgan Province. *Pajouhesh-Va-Sazandegi.* **2005,** *17* (65), 34–38. Available from: https://www.sid.ir/en/journal/ViewPaper.aspx?id=33140

Sonboli, A.; Babakhani, B.; Mehrabian, A. R. Antimicrobial Activity of Six Constituents of Essential Oil from *Salvia. Z. Naturforsch. C.* **2006,** *61* (3–4), 160–164.

Wollenweber, E.; Dörr, M.; Rustaiyan, A.; Roitman, J. N.; Graven, E. H. Exudate Flavonoids of Some *Salvia* and a *Trichostema* Species. *Z. Naturforsch. C.* **1992,** *47* (9–10), 782–784.

Yamini, Y.; Khajeh, M.; Ghasemi, E.; Mirza, M.; Javidnia, K. Comparison of Essential Oil Compositions of *Salvia mirzayanii* Obtained by Supercritical Carbon Dioxide Extraction and Hydrodistillation Methods. *Food Chem.* **2008,** *108* (1), 341–346.

Zarshenas, M. M.; Krenn, L. Phytochemical and Pharmacological Aspects of *Salvia mirzayanii* Rech. f. & Esfand. *J. Evid. Based Complem. Altern. Med.* **2015,** *20* (1), 65–72.

Ziaei, A.; Amirghofran, Z.; Zapp, J.; Ramezani, M. Immunoinhibitory Effect of Teuclatriol a Guaiane Sesquiterpene from *Salvia mirzayanii. Iran. J. Immunol.* **2011a**, *8* (4), 226–35.

Ziaei, A.; Ramezani, M.; Wright, L.; Paetz, C.; Schneider, B.; Amirghofran Z. Identification of Spathulenol in *Salvia mirzayanii* and the Immunomodulatory Effects. *Phytother. Res.* **2011b**, *25* (4), 557–62.

CHAPTER 20

PHYTOCHEMISTRY AND PHARMACOLOGY OF VIOLET SAGE (*SALVIA NEMOROSA* L.)

CHINNADURAI IMMANUEL SELVARAJ[1], SANKARALINGAM NIVITHA[2], SHARMILA ARUNAGIRI[1], NAYAN KUMAR SISHU[1], VRUSHALI MANOJ HADKAR[1], PARASURAMA DEEPA SANKAR, BABU SUBRAMANIAN[2], and PARTHASARATHI THEIVASIGAMANI[2]

[1]*Department of Biotechnology, School of Biosciences and Technology, Vellore Institute of Technology, Vellore, Tamil Nadu, India*

[2]*VIT School of Agricultural Innovations and Advanced Learning, Vellore Institute of Technology, Vellore, Tamil Nadu, India*

ABSTRACT

Salvia nemorosa L. is a perennial herb which is believed to be native to Europe and west-central Asia. Many cultivars of *S. nemorosa* are cultivated for ornamental purposes. It has a long history of use in many countries for a variety of purposes, including antihemorrhagic agent and gastrointestinal disturbances. The plant is believed to possess astringent, antiseptic, carminative, and disinfecting properties. It is an aromatic herb because of its powerful scent, which is nearly camphor-like. It has a little bitter flavour and is energizingly warming. The plant is rich in phenols, flavonoids, terpenoids, and essential oils. Caffeic acid, luteolin, apigenin, ferulic acid, quercetin, kaempferol, p-hydroxybenzoic acid, rosmarinic acid, germacrene D, β-caryophyllene, spathulenol and sabinene are predominantly reported

Bioactives and Pharmacology of Lamiaceae. T. Pullaiah, PhD (Ed.)

in the plant. *S. nemorosa* extracts have been studied for their antioxidant, antimicrobial, antifungal, antidiabetic and analgesic properties. The plant is a good source of antioxidants and it exhibits acetylcholinesterase inhibitory activity and possess potential antidiabetic effect.

20.1 INTRODUCTION

The genus *Salvia* is well known, and it is the biggest genus of the family Lamiaceae, consists of more than a thousand species; approximately 60 species are scattered in Iran; 17 of them remain endemic (Bahadori et al., 2017). The species *Salvia* are generally utilized worldwide, from European countries to China and from South Africa to Mexico. *Salvia nemorosa,* a valued natural product source is utilized for formulating pharmaceutical ingredients, new functional foods, and cosmetics (Bahadori et al., 2017). *Salvia* is a Latin word "*Salvare,*" which indicates "health," known for its extended scope of therapeutic outcomes (Ulubelen, 2003).

S. nemorosa is a valuable and well-known herb in pharmacology and medicine, generally referred to as wood sage due to its prevalence in the woods and is spread across the interior of Europe and the Western regions of the Asian Continent (Skala and Wysokinska, 2004). *S. nemorosa* L. (syn. *S. sylvestris*) grows in Asia Minor, southern Siberia, and southeastern Europe (Hedge, 1972; WCSP, 2016). It thrives in open spaces, including farming areas and deciduous forests. The foliage of *S. nemorosa,* when applied for cuts and wounds, arrests blood loss as reported to be used as the folkloric medication in Turkey (Josifovic et al., 1970; Betsy and Barner, 2003). *S. nemorosa* is used for treating several similar illnesses as that of *S. officinalis* in Russia. Customary medication in Bulgaria uses *S. nemorosa* primarily for the treatment of stomach pain, hemorrhagic diarrhea, and abscesses (Bahadori et al., 2017, Daskalova, 2004).

Currently, *S. nemorosa* is grown as an ornamental with several cultivars with unique disease and pest-resistance properties. There are numerous reports on the characterization and separation of bioactive components from various species of *Salvia.* The genus *Salvia* is abundant in phenolic and terpenoid compounds (Lu and Foo, 2002). In this context, numerous views centered on the biological activities of plant essences, filtered mixtures, and plant's essential oils; such as toxicity toward cells, antimicrobial, antitumor, anti-malarial, and cardioprotective activity (Bahadori et al., 2017, Ulubelen, 2003). The volatile oil of *S. nemorosa* is abundant in monoterpenes, possessing carminative, antiseptic (bactericide), diuretic, and antispastic properties (Coisin et al., 2010).

20.2 BIOACTIVES

The aerial portion of *S. nemorosa*, when distilled with water, results in a release of essential oil, a pale yellow liquid. Based on dry weight, it yields 0.62% (w/w). More than 98.24% of the total oil was from the identified 42 components. The main components of the essential oil are germacrene-D (15.34%), E-caryophyllene (26.37%), caryophyllene oxide (6.33%), phytol (16.92%), bicyclogermacrene (4.71%), n-tetradecane (2.45%), and sabinene (12.86%). Oxygenated sesquiterpenes (25.07%), oxygenated monoterpenes (3.39%), sesquiterpene hydrocarbons (51.91%), monoterpene hydrocarbons (17.87%) were the most abundant fractions present in the oil (Tabei and Alizadeh, 2018). The studies on essential oils of *S. nemorosa* from Iran have germacrene B, b-caryophyllene, cis-b-farnesene, caryophyllene oxide, and germacrene D as their main components (Hatipoğlu et al., 2016).

Phenolic and flavonoid compounds have drawn significant attention because of their extensive biological actions, such as anticancer influences, cardioprotective, and antimicrobial activities. Besides, specific phenolic components have potent toxicity against disease-causing microbes. The cumulative phenolic content was 47.48 milligrams of gallic acid equivalent/g dry weight. Hydroextracted essential oil from the aerial parts of *S. nemorosa* was perceived to be pale yellow fluid with a yield of 0.62% (w/w) based on the dry weight basis (Tabei and Alizadeh, 2018). A study conducted by Bahadori et al. (2017) revealed the total phenolic contents (TPCs) of the *S. nemorosa* extracts. Eleven phenolic compounds were detected altogether. The first three compounds, namely, rosmarinic acid, caffeic acid, and quercetin, remained the principal compounds. The three compounds are familiar with numerous biological characteristics, such as antimicrobial, antioxidant, anticancer, and aldehyde oxidase inhibitory activities. The vital biological characteristics recorded for rosmarinic acid comprise anti-inflammatory, astringency, antibacterial, antioxidative, antiviral, and antimutagenic activities (Petersen and Simmonds, 2003). The other compounds are protocatechuic acid, p-coumaric acid, syringic acid, ferulic acid, quercetin, kaempferol, *p*-hydroxybenzoic acid, and apigenin. The plant's methanolic extract showed the maximum TPCs (295.0 ± 10.8

mg GAE/ g). The phenolic compound is higher when compared with eight other *Salvia* species (Tosun et al., 2009). Apigenin 7-methyl ether, apigenin, apigenin scutellarein 7,4′-dimethyl ether, 7,4′- dimethyl ether, scutellarein 6,7,4′- trimethyl ether, scutellarein 6,7-dimethyl ether, 6-methyl 6,7-dimethyl 6,3′,4′-trimethyl and 6,7,3′-trimethyl ethers of 6-hydroxyluteolin, luteolin, quercetin, and 6-methyllutcolin (2, 5,7,3′,4′- tetrahydroxy-6-methyl-flavone are the flavonoids reported (Skala and Wysokinska, 2004; Bahadori et al., 2017; Nikolova et al., 2006). The 50% aqueous methanolic leaf extract of *S. nemorosa* consists following compounds; rosmarinic acid (1.72%), caffeic acid (0.05%), total flavonoid (0.38%), and total phenol (11.05%) (Zupkó et al., 2001).

The plant extract had the maximum total flavonoid content when compared with nine other *Salvia* species. Therefore, *S. nemorosa* seems to hold comprehensive flavonoid content when compared with the earlier investigated *Salvia* species. The study further indicated that extracted flavonoid and phenolic compounds yield is dependent on the polarity of the solvents used. Therefore, flowers and leaves of *S. nemorosa* are known to be high reservoirs of precious phenolic components with tremendous nutritional and well-being benefits for end-users. The studies indicate that flavonoids are a substantial proportion of phenolic compounds (Bahadori et al., 2017). Three new crystalline compounds were isolated from the aerial parts of *S. nemorosa* from the hydroethanol infusion, examined with chromatography, yielded flavones, namely, 6-methyllutcolin (2, 5,7,3′,4′-tetrahydroxy-6-methyl-flavone), cinaroside (3, lutcolin-7-O-glucoside), and salvigenin (1, 5-hydroxy-6,7,4′-trimethoxy-flavone), have been reported for the first time (Milovanović et al., 1996).

Sixteen ingredients are present in the oil derived from flowers; eight ingredients are reported within the oil obtained from leaves. The separated oil had a slightly yellow color and rich fragrance. Therefore, the major and significant compounds can be grouped under the categories of oxygenated sesquiterpenes, sesquiterpenes hydrocarbon, monoterpene hydrocarbon, and oxygenated monoterpene. The essential oil constituents are not always the same among a range of populations. Each population of *S. nemorosa* contains a different amount of each compound. For example, the thymol content of the polar variety is as high as 17.59%. In comparison, the Tehran variety has only around 0.88% of thymol. Studies reveal that the synthesis of phytochemicals directly impacts the biotic and abiotic factors of a particular geographical location (Mahdieh et al., 2018). Essential oil extraction from the shoot and the root parts of *S. nemorosa* discloses many terpene compounds, especially sesquiterpenes and their oxides. Monoterpene hydrocarbons, such as thujene, sabinene, and cymene are present in the essential oil extracted from the floral parts of *S. nemorosa.* Spathulenol and caryophyllene oxide is the abundantly present compounds in the foliage and flowers of *S. nemorosa,* respectively. Other compounds include E-caryophyllene, ledene oxide, hexahydrofarnesyl acetone (Bahadori et al., 2017). The essential oil (EO) of *S. nemorosa* sample consists of germacrene D (18.62%), β-caryophyllene (19.03%), and sabinene (38.2%). The other major EO compounds profiled include α-thujene (2.71%), caryophyllene oxide (3.53%), α-caryophyllene (0.72%), Elixene (2.98%), β-myrcene (0.84%), β-pinene (0.48%), p-cymene (0.71%), γ-terpinene (1.28%), z-β-farnesene (0.44%), β-cadinene (0.72%), spatulenol (0.81%), τ-cadinol (0.96%), and δ-cadinol (0.63%) (Coisin et al., 2012). Kováts et al. (2011) reported that the toxicity of *S. nemorosa* with relevance to the EC_{50} was 1.854g dried plant/l.

The percentage yield of woodland sage essential oil (EO) was significantly lesser than that generally obtained from Dalmatian sage (*Salvia officinalis*) and different wild-growing or cultivated *Salvia* species (Mimica-Dukic et al., 2002; Bozin et al., 2007; Anačkov et al., 2009). The total number of chemical constituents identified in the essential oil was 48 (88.37%). Oxygenated sesquiterpenes (26.15%) and sesquiterpene hydrocarbons (39.19%) covered 65.34% of all ingredients, while monoterpene hydrocarbons contribute 16.61% of the plant sample; insignificant ingredients comprised aliphatic compounds (1.39%) and oxygenated monoterpenes (5.29%). Sesquiterpenes, caryophyllene oxide (17.98%), and E-caryophyllene (22.95%) were recognized as the principal ingredients, accompanied by monoterpene hydrocarbons, limonene (3.07%), and sabinene (7.69%). In addition, notable quantities of sesquiterpene hydrocarbons; alpha-humulene (2.17%), gamma-muurolene (5.23%), and beta-cubebene (2.22%) were profiled along with spathulenol, 3.39% and caryophyllenol, 2.97% as oxygenated sesquiterpene alcohols (Božin et al., 2012). Romanova et al. (1971) detailed the separation of compounds from the root extract of *S. nemorosa* diterpene quinones hydroxyroyleanone (horminone) $C_{20}H_{28}O_4$ (mp. 173–174°C), acetoxyroyleanone $C_{22}H_{30}O_5$ (mp. 211.5–214°C) and a yellowish crystalline compound with the structure $C_{22}H_{28}O_6$ (mp. 192–194°C), confirmed to be distinct and separated from violet sage, named as nemorone (7-acetoxy-10-formyl 12-hydroxy- 13-isopropyl-4,4-dimethyloctahydrophenanthrene-11,14-quinone). The important compounds present in *Salvia nemorosa* are given in Fig. 20.1.

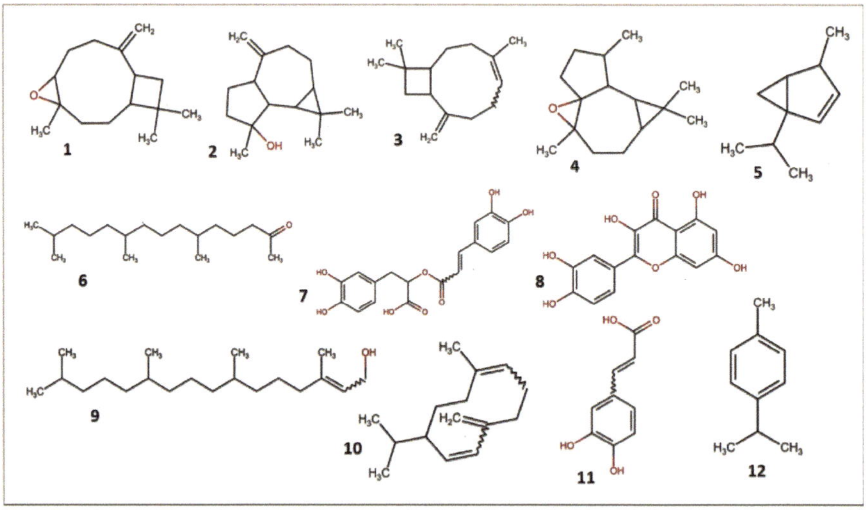

FIGURE 20.1 Chemical structures of spathulenol (1), caryophyllene oxide (2), caryophyllene (3), ledene oxide (4), hexahydrofarnesyl acetone (5), rosemarinic acid (6), quercitin (7), caffeic acid (8), thujene (9), cymene (10), phytol (11) and germacrene-D (12); [Marvin 17.21.0, ChemAxon tool was used for drawing chemical structures].

20.3 PHARMACOLOGY

20.3.1 ANTIOXIDANT EFFICACY

The DPPH free radical scavenging method is used to assess the antioxidant activity of the methanolic extract of *S. nemorosa.* The IC_{50} value was 40.22 μg/mL with moderate antioxidant activity (Tabei and Alizadeh, 2018). The DPPH assay estimates the scavenging of reactive oxygen species or free radicals in the essential oils of *S. nemorosa* and its plant extracts. DPPH assay indicated that the methanolic plant extract (IC_{50} = 82.1 milligrams/milliliter) conferred excellent scavenging activity. The IC_{50} values of essential oils and additional infusions were more than 500 mg/mL and were not resolved. The antioxidant capacity of the biological compounds correlates with the reducing potential of such compounds. FRAP (The ferric reducing antioxidant power) test done with essential oils and the methanolic leaf extracts of *S. nemorosa* showed the most substantial reducing activity with SRP =0.99 with context to the ascorbic acid control with SRP =1.02, succeeded by the essential oil of leaves with SRP = 0.77. Phenolic compounds and flavonoids are responsible for donating hydrogen elements in the extracts. The essential oils (spathulenol) accountable for the observed reducing power of *S. nemorosa* (Bahadori et al., 2017). Şenol et al. (2010) affirmed that ethyl acetate and dichloromethane extracts of *S. nemorosa* had not shown inhibitory activity upon BchE and AchE enzymes with 0.1 mg/mL concentration. The extracts exhibited potent (ethyl acetate and methanol extracts) to low (dichloromethane extract) activity on DPPH radicals. The extracts revealed average to high acetylcholinesterase, xanthine oxidase, and aldehyde oxidase activity. It also exhibited moderate cytotoxicity and alpha-glucosidase inhibitory activity.

The antioxidant activity of *S. nemorosa* indicates as the most potent radical mitigators of DPPH radicals (DPPH IC_{50}; 138.43 ± 4.6 μg plant extracted or μg quercetin/ 1 mL 10^{-4} M DPPH), with higher whole phenolic content (30.36±1.08 mg eq. gallic acid in 1 g dried plant) (Firuzi et al., 2013). The antioxidant activity of *S. nemorosa* was assessed by estimating the lipid peroxidation (LP) effects and radical oxygen species scavenging capacity (RSC). Particularly, RSC was estimated by scaling the scavenging activity of essential oils (EOs) on DPPH and hydroxyl radicals. At the same time, LP effects were estimated by observing EO activities in and Fe^{2+}/H^2O^2 and Fe^{2+}/ascorbate induction systems. The examined EO of *S. nemorosa* decreased DPPH (IC_{50} = 0.98 μL/mL) and OH radical formation (IC_{50} = 0.11 μL/mL) in a dose-dependent mode. Powerful LP inhibition was observed in both induction systems (Božin et al., 2012). The antioxidant potential was resolved by quantifying the repression of the conjugated diene hydroperoxides and unstable organic compounds occurring from linoleic acid oxidation. In β-Carotene/linoleic acid assay, oxidation of linoleic acid was restrained moderately by *S. nemorosa* extract (68.6%). The free radical-scavenging potential of *S. nemorosa* reported was 32.0 μg/ mL. The quantity of the cumulative phenolics present in *S. nemorosa* methanolic extract was 68.6 mg GAE/g DW (Tosun et al., 2009).

Malencic et al. (2002) examined the antioxidant characteristics of a sage species from the Serbian vegetation, *S. nemorosa*. They found that the examined species displayed tremendous peroxidase and superoxide dismutase activities and has a reasonable quantity of cumulative flavonoids, and the principal commonly occurring compound in *S. nemorosa* was rosmarinic acid. In a study conducted by Janicsak et al. (2011), the water and methanolic extract of *S. nemorosa* demonstrated lipid peroxidation-inhibiting activities more than ascorbic acids. The antioxidant potential of *S. nemorosa* (EC_{50} 9.6 μg/mL) was moderate. Concerning the white-flowered types, the EC_{50} values taken on the rat-brain homogenate were substantially identical to the values found in earlier research with *S. nemorosa* on ox-brain homogenate (Zupkó et al., 2001). They also authenticated a close association between the antioxidant actions and the cumulative phenol contents of the *S. nemorosa* extract. The correlation was much weaker for caffeic and rosmarinic acids, intimating the critical role of polyphenols in antioxidant activity.

20.3.2 ANTIMICROBIAL ACTIVITY

S. nemorosa essential oils' antimicrobial efficacy was estimated using disc diffusion and serial dilution methods in the in vitro condition against *Staphylococcus epidermidis, Escherichia coli, Staphylococcus aureus,* and *Candida albicans* for antimicrobial activity. Diameter of the inhibition zones (DIZ), minimum bactericidal/fungicidal (MBC /MFC) concentration, and minimum inhibitory concentration (MIC) values are the indicators used to express the antimicrobial activity. The disc diffusion and serial dilution methods were used to evaluate the in vitro antimicrobial activity of the essential oils of *S. nemorosa* upon microbial strains, including *Staphylococcus epidermidis*, *Staphylococcus aureus, Candida albicans*, and *Escherichia coli.* The Gram-negative bacteria, *E. coli*, were more resistant to the essential oil than Gram-positive bacteria, and *Candida albicans* were more sensitive. The results reveal that *S. nemorosa* essential oils possess modest antimicrobial activities on four microorganisms studied, associated with the positive control antibiotics (Tabei and Alizadeh, 2018). Extended-spectrum beta-lactamase (ESBL)-producing *Escherichia coli* strains can result in several diseases, especially urinary tract diseases. The inhibitory effects of the EOs were tested against ESBL-producing *E. coli*, secluded from the urinary tract of infectious patients. Results revealed that EOs from *S. nemorosa* had significant antibacterial effects against *E. coli* isolates with a high MIC of 8 mg/mL (Sharifi-Rad et al., 2016).

The essential oil of *S. nemorosa* was tested against 13 bacterial strains and six fungi for antimicrobial activity, including *Candida albicans* and five dermatomycetes; an ineffective antifungal activity with significant antibacterial activity was observed against *Salmonella typhi, Shigella sonei,* and *Staphylococcus epidermidis* (Božin et al., 2012). The antimicrobial outcomes of alcoholic (methanolic and ethanolic) and aqueous infusions taken from the leaf, flower, root, and stem of *S. nemorosa*, tested on various species of bacteria, namely, *Staphylococcus aureus, Staphylococcus epidermidis, Staphylococcus saprophyticus, Escherichia coli, Bacillus serus, Salmonella typhi*, and *Shigella flexeneri.* The fungi— *Aspergillus flavus* and *Candida albicans* were investigated. The evaluation of the diameter of the inhibitory zone was assessed by the diffusion method. The results showed that alcoholic infusions of different plant parts have the most potent antimicrobial effect on the microorganisms relevant to aqueous extracts (Chalabian et al., 2006).

Sesquiterpene compounds do possess an antifungal property. Due to the presence of a high amount of sesquiterpenes, *S. nemorosa* has a high degree of antifungal properties compared with other members of the Lamiaceae family. *S. nemorosa* exhibited the ability against the following fungi species: *Epidermophyton floccosum, Trichophyton rubrum, Trichophyton mentagrophytes*, and *Candida albicans*. A high amount of sesquiterpene hydrocarbons and oxygenated sesquiterpenes are present in *S. nemorosa.* Due to this property, *S. nemorosa* essential oil has the highest antifungal activity and the lowest antibacterial activity for *Streptomyces enteritidis* and *E. coli*, respectively (Djakovic Sekulic et al., 2016). Essential oils from *S. nemorosa* exhibited inhibitory activity against the examined bacteria to a differing degree, higher on *Staphylococcus aureus* than *Escherichia coli.* The MIC values of EOs observed for these wild-growing species registered encouraging activities over *S. aureus* during incubation at 37 deg. C for 24 h. The bioassay authenticates that Gram-positive bacteria are more susceptible than Gram-negative bacteria, *E. coli* remaining the most defiant strain. Particularly in moderate concentrations, the essential oils of *S. nemorosa* failed in their inhibitory response toward the pathogen mentioned above (Coisin et al., 2012).

20.3.3 ANALGESIC ACTIVITY

The water and ethanolic extracts of the aerial plant sections revealed antinociceptive action with an effectiveness comparable to established analgesic medications. The plant parts of *S. nemorosa* comprised sterols, diterpenoid compounds, triterpenes, and megastigmane glycosides (Nikolova and Aneva, 2017). The extracts of *S. nemorosa* have anti-nociceptive properties with effectiveness comparable to traditional analgesic medicines. The antinociceptive properties of *S. nemorosa* were analyzed adopting a hot plate and writhing experiments on the aqueous and ethanolic homogenized extracts of the aerial parts of this plant. In contradiction to the ethanolic extract, the intraperitoneal dose of the aqueous extract exhibited anti-nociceptive activity in the hot-plate test hindered by naloxone. In the writhing test, significant anti-nociceptive activity was noted in the case of both aqueous and ethanolic extracts. Naloxone had an inhibitory influence on the anti-nociceptive attributes of only the aqueous extract. This study shows that the aqueous extract of *S. nemorosa* might possess opioid receptor-mediated central and peripheral anti-nociceptive activity. In contrast, the ethanolic extract has only peripheral anti-nociceptive activity (Hosseinzadeh and Amel, 2000).

20.3.4 ANTICANCER ACTIVITY

Janicsák et al. (2011) investigated the cytotoxic activities of the portions made from the methanol extract of *S. nemorosa* upon MCF7, A431, and HeLa cell lines; the n-hexane fractions of *S. nemorosa* induced higher than 50% growth repression of the skin carcinoma cell line (A431). None of the examined portions displayed principal antiproliferative outcomes toward MCF7 and HeLa cells. In the brine shrimp toxicity assay, the leaf and root extract showed significantly less cytotoxic effect with LC_{50} values greater than 500 μg/mL (Eskandani et al., 2016). The cytotoxic activity of methanol (80% methanol) and dichloromethane extracts of *S. nemorosa* were evaluated on HL60, K562, and MCF-7 human cancer cell lines. Results indicated that the methanol (80%) extract alone showed moderate cytotoxic activity against K562 cell lines (87.0 ± 6.7 μg/mL) (Firuzi et al., 2013). Clerodermic acid, a clerodone diterpene isolated from *S. nemorosa* is an anticancer agent that inhibits a hypoxic condition by suppressing Hypoxia-inducible factor1–α (HIF1α) (Bahadori et al., 2018). The cancer tumors require a hypoxic microenvironment condition to multiply and replenish. HIF1- α and HIF1-β are two sub-units of heterodimeric transcription factors that present more than the required amount in cancer conditions (Bos et al., 2003).

20.3.5 ACETYLCHOLINESTERASE ACTIVITY

Alzheimer's disease (AD) is a neurodegenerative disorder that results in mental disorders and finally leads to death. The species of *Salvia* are active on the central nervous system (CNS) disorders. The most substantial acetylcholinesterase enzyme's inhibitory potential was shown by the dichloromethane (DCM) extract of the plant. Essential oils and other extracts were usually less effective than dichloromethane (Bahadori et al., 2017).

20.3.6 ANTIDIABETIC ACTIVITY

Alpha-glucosidase inhibitor drugs reduce the level of glucose in the blood for type 2 diabetes treatment. The methanolic leaf extract of *S. nemorosa* exhibited moderate alpha-amylase inhibitory activity with (0.170 ± 0.006), roots (0.270 ± 0.014 mmol ACAEs/g extracts) mmol acarbose equivalents/g extract; manifested moderate alpha-glucosidase activity; leaves extract (5.401 ± 0.099) followed by roots with (2.724 ± 0.128) mmol ACAEs/g (Eskandani et al., 2016). The methanolic extract of *S. nemorosa* exhibited a potent α-glucosidase inhibitory activity with IC_{50} 19 mg/mL. Compared with acarbose, a standard antidiabetic drug with IC_{50}, 16 mg/mL, the methanolic extract of *S. nemorosa* is considered highly active. The detected activity was due to more flavonoid and phenolic compounds present in the methanolic extract of *S. nemorosa*. Thus, the phenolic compounds perform as potent a-glucosidase inhibitors (Bahadori et al., 2017). The hypoglycemic effect was studied in streptozotocin-induced diabetic rat models with the ethanolic extract of *S. nemorosa* aerial parts. The dosage of ethanolic extract 100, 200, and 400 mg/kg body weight, injected intraperitoneally. The serum glucose was estimated by the enzymatic method of glucose oxidase. The results showed that plant extracts significantly reduced serum glucose in diabetic rodents but not in control rodents (Sadeghzadeh et al., 2008).

KEYWORDS

- **violet sage**
- **rosmarinic acid**
- **germacrene d**
- **antidiabetic**
- **β-caryophyllene**
- **antidiabetic**
- **analgesic**

REFERENCES

Anačkov, G.; Božin, B.; Zorić, L.; Vukov, D.; Mimica-Dukić, N.; Merkulov, L. J.; Igić, R.; Jovanović, M.; Boža, P. Chemical Composition of Essential Oil and Leaf Anatomy of

Salvia bertolonii Vis. and *Salvia pratensis* L. (Sect. Plethiosphaceae, Lamiaceae. *Molecules* **2009,** *14*, 1–9.

Bahadori, M. B.; Asghari, B.; Dinparast, L.; Zengin, G.; Sarikurkcu, C.; Abbas-Mohammadi, M.; Bahadori, S. *Salvia nemorosa L.*: A Novel Source of Bioactive Agents with Functional Connections. *Lebensm. Wiss. Tech.* **2017,** *75*, 42–50.

Bahadori, M. B.; Eskandani, M.; De Mieri, M.; Hamburger, M.; Nazemiyeh, H. Anti-Proliferative Activity-Guided Isolation of Clerodermic Acid from *Salvia nemorosa L.*: Geno/Cytotoxicity and Hypoxia-Mediated Mechanism of Action. *Food. Chem. Toxicol.* **2018,** *120*, 155–163.

Betsy, C.; Barner, C. D. *The New Book of Salvias*; Timber Press, 2003; p 216.

Bos, R.; Van der Groep, P.; Greijer, A. E.; Shvarts, A.; Meijer, S.; Pinedo, H. M.; Semenza, G. L.; Van Diest, P. J.; Van der Wall, E. Levels of Hypoxia-Inducible Factor-1α Independently Predict Prognosis in Patients with Lymph Node Negative Breast Carcinoma. *Cancer* **2003,** *97* (6), 1573–1581.

Božin, B.; Lakić, N.; Čonić, B. S.; Kladar, N.; Orčić, D.; Mimica-Dukić, N. Antioxidant and Antimicrobial Properties of a New Chemotype of Woodland Sage (*Salvia nemorosa* L. subsp. *nemorosa*, Lamiaceae) Essential Oil. *Biologia Serbica* **2012,** *34* (1–2), 51–60.

Chalabian, F.; Ourmazdi, P.; Arbabian, S. Study of Anatomical Properties and Antimicrobial Effects of *Salvia nemorosa. J. Sci. (Islamic Azad University)*, [Biology, Online] **2006,** *15* (58/1), 440–451. https://www.sid.ir/en/journal/ViewPaper.aspx?id=94833

Coisin, M.; Burzo, I.; Stefan, M.; Rosenhech, E.; Zamfirache, M. M. Chemical Composition and Antibacterial Activity of Essential Oils of Three *Salvia* Species, Widespread in Eastern Romania. *Analele Stiintifice ale Universitatii" Al. I. Cuza" din Iasi* **2012,** *58* (1), 51.

Coisin, M.; Padurariu, C.; Andro, A. R.; Boz, I.; Zamfirache, M. M.; Burzo, I. Biochemical and Physyological Researches in *Salvia nemorosa* L. *Analele Stiintifice ale Universitatii" Al. I. Cuza" din Iasi* **2010,** *56* (2), 31–37.

Daskalova, T. On Some Specificities of Seed Formation in *Salvia nemorosa* (Lamiaceae). *Phytologia Balcanica* **2004,** *10* (1), 79–84.

Djakovic Sekulic, T.; Bozin, B.; Smolinski, A. Chemometric Study of Biological Activities of 10 Aromatic Lamiaceae Species' Essential Oils. *J. Chemom.* **2016,** *30* (4), 188–196.

Eskandani, M.; Bahadori, M. B.; Zengin, G.; Dinparast, L.; Bahadori, S. Novel Natural Agents from Lamiaceae Family: An Evaluation on Toxicity and Enzyme Inhibitory Potential Linked to Diabetes Mellitus. *Curr. Bioact. Compd.* **2016,** *12* (1), 34–38.

Firuzi, O.; Miri, R.; Asadollahi, M.; Eslami, S.; Jassbi, A. R. Cytotoxic, Antioxidant and Antimicrobial Activities and Phenolic Contents of Eleven *Salvia* species from Iran. *Iran. J. Pharm. Res.* **2013,** *12* (4), 801.

Hatipoğlu, S. D.; Zorlu, N.; Dirmenci, T.; Gören, A. C.; Öztürk, T.; Topçu, G. Determination of Volatile Organic Compounds in Forty Five *Salvia* Species by Thermal Desorption-GC-MS Technique. *Rec. Nat. Prod.* **2016,** *10* (6), 659–700.

Hedge, I. C. *Salvia* L. In *Flora Europaea*; Tutin, T. G., Heywood, V. H., Burges, N. A., Valentine, D. H., Walters, S. M., Webb, D. A., Eds., Vol. 3. Cambridge University Press: Cambridge, 1972; pp 188–192.

Hosseinzadeh, H.; Amel, S. Anti-Nociceptive Effects of the Aerial Parts of Salvia *nemorosa* L. Extracts in Mice. *Arch. Iran. Med.* **2000,** *3* (2), 1–4.

Janicsak G.; Zupko I.; Nikolova M. T.; Forgo P.; Vasas A.; Mathe I.; Blunden G.; Hohmann J. Bioactivity-Guided Study of Antiproliferative Activities of *Salvia* Extracts. *Nat. Prod. Commun.* **2011,** *6* (5), 1934578X1100600501.

Janicsák, G.; Zupkó, I.; Máthé, I.; Hohmann, J. Comparative Study of the Antioxidant Activities of Eleven *Salvia* Species. *Nat. Prod. Commun.* **2010,** *5* (2), 227–230.

Josifovic, M.; Stjepanovic, L.; Jankovic, M. M.; Gajic, M.; Kojic, M.; Diklic, N. *Flora of SR Serbia*; Beograd: SANU, 1970; pp 1–9.

Kováts, N.; Ács, A.; Gölöncsér, F.; Barabás, A. Quantifying of Bactericide Properties of Medicinal Plants. *Plant Signal. Behav.* **2011,** *6* (6), 777–779.

Lu, Y.; Foo, L. Y. Polyphenolics of *Salvia*—A Review. *Phytochemistry* **2002,** *59* (2), 117–40.

Mahdieh, M.; Talebi, S. M.; Akhani, M. Infraspecific Essential Oil and Anatomical Variations of *Salvia nemorosa* L. (Labiatae) Populations in Iran. *Ind. Crops Prod.* **2018,** *123*, 35–45.

Malencic, D. J.; Popovic, M.; Stajner, D.; Mimica-Dukic, N.; Boza, P.; Mathe, I. Screening for Antioxidant Properties of *Salvia nemorosa* L. and *Salvia glutinosa* L. *Oxidation Commun.* **2002,** *25* (4), 613–619.

Milovanović, M.; Djermanović, V.; Dermanović, M.; Stefanović, M. Chemical Constituents of *Salvia nemorosa* L. and Its Antioxidant Effect in Lard. *J. Serb. Chem. Soc.* **1996,** *61* (6), 423–429.

Mimica-Dukic, N.; Boza, P.; Igic, R.; Spasic-Adanski, L. J.; Stajner, D. Volatile Constituents of Wild-Growing *Salvia* Species in Province Vojvodina in Serbia. *J. Essen. Oil Bearing Plants.* **2002,** 5 (1), 19–29.

Nikolova, M.; Aneva, I. European Species of Genus *Salvia*: Distribution, Chemodiversity and Biological Activity. *Salvia Biotech.* **2017,** 1–30.

Nikolova, M.; Janicsák, G.; Genova, E.; Máthé, I. Comparative Analysis of External Flavonoids of Bulgarian and Hungarian Samples of *Salvia* Species. *Acta. Bot. Hung.* **2006,** *48* (3–4), 361–367.

Petersen, M.; Simmonds, M. S. Rosmarinic acid. *Phytochemistry* **2003,** *62* (2), 121–125.

Romanova, A. S.; Pribylova, G. F.; Zakharov, P. I.; Sheichenko, V. I.; Ban'kovskii, A. I. "A New Quinone from *Salvia nemorosa*." *Chem. Nat. Comp.* **1971,** *7* (2), 190–191.

Sadeghzadeh, F.; Eidi, A.; Parivar, K. A. Z. E. M.; Mazooji, A. Hypoglycemic Effect of Alcoholic Extract of *Salvia nemorosa* in Normal and Diabetic Male Rats. *Res. Med.* **2008,** *32* (3), 233–238.

Şenol, F. S.; Orhan, I.; Celep, F. A.; Kahraman, M.; Doğan, G.; Yılmaz; Şener, B. Survey of 55 Turkish *Salvia* Taxa for Their Acetylcholinesterase Inhibitory and Antioxidant Activities. *Food Chem.* **2010,** *120*, 34–43.

Sharifi-Rad, J.; Mnayer, D.; Roointan, A.; Shahri, F.; Ayatollahi, S. A. M.; Sharifi-Rad, M.; Molaee, N. Antibacterial Activities of Essential Oils from Iranian Medicinal Plants on Extended-Spectrum β-Lactamase-Producing *Escherichia coli*. *Cell. Mol. Biol.* **2016,** *62* (9), 75–82.

Skala, E.; Wysokinska, H. In Vitro Regeneration of *Salvia nemorosa L.* from Shoot Tips and Leaf Explants. *In Vitro Cell Dev. Biol. Plant* **2004,** *40* (6), 596–602.

Tabei, S. M.; Alizadeh, A. Phytochemical Constituents and Antimicrobial activity of *Salvia. Bangladesh J. Bot.* **2018,** *47* (4), 847–854.

Tosun, M.; Ercisli, S.; Sengul, M.; Ozer, H.; Polat, T.; Ozturk, E. Antioxidant Properties and Total Phenolic Content of Eight *Salvia* Species from Turkey. *Biol. Res.* **2009,** *42* (2), 175–181.

Ulubelen, A. Cardioactive and Antibacterial Terpenoids from Some *Salvia* Species. *Phytochemistry* **2003,** *64* (2), 395–399.

World Checklist of Selected Plant Families (WCSP), Facilitated by the Royal Botanic Gardens, Kew, 2016. Published on the Internet. http://apps.kew.org/wcsp/

Zupkó, I.; Hohmann, J.; Rédei, D.; Falkay, G. Y.; Janicsák, G.; Máthé, I. Antioxidant Activity of Leaves of *Salvia* Species in Enzyme-Dependent and Enzyme-Independent Systems of Lipid Peroxidation and Their Phenolic Constituents. *Planta Medica* **2001,** *67*, 366–368.

CHAPTER 21

PHARMACOLOGY AND THERAPEUTIC POTENTIAL OF ROSEMARY [*SALVIA ROSMARINUS* (L.) SCHLEID.]

SHARMILA ARUNAGIRI[1] and
CHINNADURAI IMMANUEL SELVARAJ[2]

[1]*School of Biosciences and Technology, Vellore Institute of Technology, Vellore, Tamil Nadu, India*

[2]*School of Agricultural Innovations and Advanced Learning, Vellore Institute of Technology, Vellore, Tamil Nadu, India*

ABSTRACT

Salvia rosmarinus (L.) Schleid, commonly referred to as rosemary, is a perennial evergreen shrub that is utilized for both ornamental and medicinal purposes. Rosemary is commonly found in the Mediterranean region of Asia, Africa, and Europe. It contains major alkaloids, flavonoids, terpenoids, essential oils, and phenolic compounds. The leaves of the rosemary plant are used both as a spice and for numerous medical conditions. A popular household plant with strong pharmacological activity, *S. rosmarinus* contains more essential oils and bioactive components. Chlorogenic acid, caffeic acid, carnosic acid, oleanolic acid, alpha-pinene, camphor, carnosol, eucalyptol, luteolin, monomeric acid, rosmadial, rosmarinic acid, rosmaquinones A and B, rosmanol, and ursolic acid are the main bioactive components of *S. rosmarinus*. Rosemary has various pharmacological activities, including antibacterial, antiviral, antifungal, antioxidant, anti-inflammatory, anticancer, antidiabetic, antinociceptive, antidepressant and antiulcerogenic activities. It is also known to possess hypolipidemic effects, cytotoxic

Bioactives and Pharmacology of Lamiaceae. T. Pullaiah, PhD (Ed.)

activity and cognitive and cerebrovascular activity. Moreover, the rosemary extract also promotes hair growth. The *S. romarinus* extract regulates the process of aging. In addition to its therapeutic use, rosemary may be used in cosmetic formulations to treat pathological and non-pathological conditions such baldness, cellulite, aging, and UV damage. Manufacturers of food and cosmetics have employed rosemary extensively.

21.1 INTRODUCTION

Salvia rosmarinus (L.) Schleid. is generally known as rosemary, which is a perennial, aromatic evergreen shrub that grows majorly in the Mediterranean region of Asia, Africa, and Europe, mainly in the islands like Sicily, Elba, Sardinia, and the Balearic Islands and in other regions of the World (Borrás-Linares et al., 2014; De Oliveira et al., 2019; Sharifi-Rad et al., 2020). It belongs to the Family Lamiaceae. The name "Rosemary" is derived from the Latin word "*Ros Marines,*" which means dew of the sea. Until 2017, the plant *S. rosmarinus* is known as *Rosmarinus officinalis* L. The genus *Rosmarinus* has been consolidated into the genus *Salvia* in the modern phylogenetic investigation. *R. officinalis* is no longer the exact name of the species. Considering *Salvia officinalis* was previously occupied when the consolidation was done, this species required a new precise epithet in *Salvia,* which is now recognized as *S. rosmarinus* (Uniprot, 2021; Garden, 2021; Science, 2021). The leaves of *S. rosmarinus* are needle-like in shape, 1–5 mm wide, and 15–40 mm long. It is coriaceous, linear with revolute margins, sessile, and bright green. The calyx is green or purple, meagrely tomentose with 3–4 mm in young and grows up to 5–7 mm and veined distinctly. The corolla is big, pale blue, rarely white or pink, with 10–12 mm, producing brown color nutlets.

Rosemary is a thick bush having condensed foliage with stems reaching up to a height of 1–2 m (Valones et al., 2016). It is also a woody perennial herb, used as an aromatic and ornamental plant. The rosemary leaves are used as flavoring agents in foods as a seasoning and used for various medicinal purposes (Rašković et al., 2014). *S. rosmarinus* is a common household plant containing more essential oils and bioactive compounds, thereby exhibiting high pharmacological activity. Rosemary has curative characteristics and has been used in the traditional medication, cosmetics industries, and pharmaceuticals mainly for its anti-inflammatory and antioxidant characteristics associated with the presence of carnosic/carnosol and ursolic acids. The curative usage of rosemary has been investigated to treat inflammatory diseases; nevertheless, additional applications have been examined, such as mycoses, wound healing, and skin cancer therapies. Besides its curative uses, rosemary has possible utilization in cosmetics formulations, treating non-pathological and pathological circumstances, such as alopecia, cellulite, aging, and ultraviolet damage (de Macedo et al., 2020).

Rosemary has remedial characteristics, used in community medicine as an oral mixture to relieve muscle spasms, renal colic, and dysmenorrhea (Al-Sereiti et al., 1999; Begum et al., 2013; Ribeiro-Santos et al., 2015). Rosemary has antiviral, antifungal, anti-inflammatory, antibacterial, antithrombotic, antitumor, antidepressant, antinociceptive, antioxidant, and antiulcerogenic activities (Begum et al., 2013; Ribeiro-Santos et al., 2015; Ojeda-Sana et al., 2013). Numerous therapeutic uses for *S. rosemarinus* have been recognized viz., managing ailments correlated with the cardiovascular, nervous, genitourinary, gastrointestinal, hepatic, menstrual, respiratory, and reproductive systems and skin maladies (Begum et al., 2013). Rosemary has been used extensively by cosmetics and food manufacturers (Stefanovits-Bányai et al., 2003).

21.2 BIOACTIVES

S. rosmarinus generally has a wide variety of secondary metabolites and a high amount of bioactive compounds or phytochemicals, such as alkaloids, terpenoids, and flavonoids. The different rosemary extracts on chemical analysis exhibited highly potent compounds of phenolic diterpenes, terpenes, and phenolic acids (Alavi et al., 2021). The major bioactive compounds of *S. rosmarinus* include chlorogenic acid, ursolic acid, caffeic acid, rosmarinic acid, monomeric acid, carnosic acid, oleanolic acid, alpha-pinene, camphor, secohinokio, carnosol, luteolin, eucalyptol, rosmadial, rosmaquinones A and B, rosmanol (also its isomers epirosmanol and epiisorosmanol), other derivative compounds, such as epirosmanolmethylether, 5,6,7,10-tetrahydro-7-hydroxyrosmaquinone and methylcarnosate, and some eugenol derivatives (Rašković et al., 2014; Borrás-Linares et al., 2014; De Oliveira et al., 2019; Sharifi-Rad et al., 2020; Ahmed et al., 2020). The aerial parts of *S. rosmarinus* contain five terpenoid glycosides, such as officino terpenosides A1, A2, B, C, and D (Sharifi-Rad et al., 2020).

The phenolic compounds isolated from rosemary powder include hydroxycinnamic acids: rosmarinic acid, p-coumaric acid, ellagic acid, chlorogenic acid, ferulic acid, pyrogallol, benzoic acid, o-coumaric acid, m-hydroxybenzoic acid, coumarin, coumaroyl quinic acid, and isoferulic acid. The hydroxybenzoic acids, such as vanillic acid, caffeic acid, syringic acid, protocatechuic acid, dicaffeoylquinic acid, and hydroxyphenyl acetic acids which include p-hydroxybenzoic acid and homovanillic acid. The flavonoid and flavonol compounds consist of luteolin, luteolin 7-glucoside, naringin, hispidulin, naringenin, naringenin-C-hexoside, rosmarinic acid, rutin, hesperidin, kaempferol 3,7-dirhamnoside, kaempferol-3-O-rutinoside, isohamnetin-lutelion, isorhamnetin-3-O-hexoside, quercitrin, rhamnetin, quercetin, apigenin 7-O glucoside, genkwanin, apigenin, and hesperitin (Naimi et al., 2017; Zappalà et al., 2021; Hassanen et al., 2020).

The chemical composition of rosemary essential oil constitutes about 98.87%, in which monocyclic terpenes form 3.61%, bicyclic terpenes form 22.62%, aromatic hydrocarbons form 1.81%, aliphatic hydrocarbons are present about 2.58%, oxides 17.62%, ketones 27.44%, alcohols 10.86%, sesquiterpenes 6.92%, phenol/phenol ethers 0.53% and esters 4.88%. The monocyclic terpenes consist of α-phellandrene, α-terpinolene, α-terpinene, and γ-terpinene. The bicyclic terpenes are camphene, α-pinene, and β-pinene, and triterpene acids which include betulinic acid, ursolic acid, and oleanolic acid. The phenolic diterpenes of rosemary include carnosic acid, carnosol, rosmanol, isorosmanol, epirosmanol, rosmadial, rosemaridiphenol, methoxycarnosate, and methoxycarnosol (Naimi et al., 2017; Hassanen et al., 2020). The main compounds of essential oil are eucalyptol, α-pinene, and 1,8-Cineole involved in anti-inflammatory activity (Bogavac et al., 2017; Ahmed et al., 2020). The essential oil obtained from the *S. rosmarinus* leaves exhibits the principal component 1,8-cineole. The extracted oil of the stem has mostly caryophyllene, and the predominant compound present in the flower-extracted oil is caryophyllene oxide. The frequently reported terpenes are linear, oxygenated, and bicyclic monoterpenes, such as α-pinene, 1,8-cineol, camphene, camphor, b-pinene, borneol, p-cymene, bornyl acetate, b-myrcene, verbenone, limonene, a-terpinene, terpineol, terpinen-4-ol, and b-linalool. Among the various extracts of essential oil from *S. rosmarinus* 1,8-cineole, α-pinene, and camphor are considered to have more pharmacological activities (Borges et al., 2019).

In rosemary, the compounds, such as sagerinic acid, salvianolic acid A, and pentacyclic triterpenoid corosolic acid were also detected. The potential method to determine bioactive molecules of *S. rosmarinus* is Ultra-High-performance Liquid chromatography, Electrospray Ionization coupled with quadrupole time of Flight Mass Spectrometry (UHPLC-ESI-QTOF-MS)

(Sharma et al., 2020). The efficient method is known to be soxhlet extraction. The essential oils, such as citronellol, eudesmol, estragole, geraniol, eugenol, phytol, and linoleic acid were extracted from *S. rosmarinus* is also shown to have a significant amount of antimicrobial activity (Lorenzo-Leal et al., 2019). The new compounds identified from the rosemary plant are rosmarinones A and B, miltionone I, 3-epimaslinic acid, atriplicone B, 3α-dihydroxyurs-12-en-28-oic acid, 2α-hydroxymicromeric acid, and corosolic acid (Zhong et al., 2021). The important bioactive compounds present in rosemary are given in Fig. 21.1.

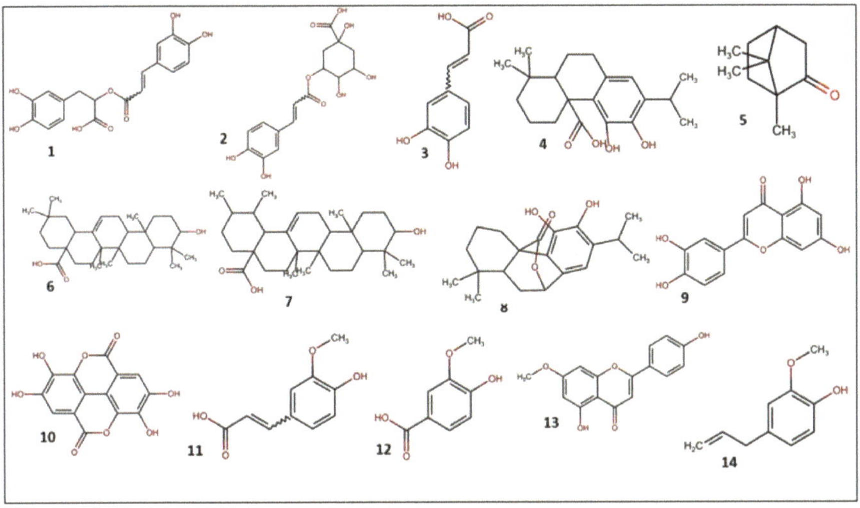

FIGURE 21.1 The important bioactive compounds of Rosemary. (1), chlorogenic acid (2), caffeic acid (3), carnosic acid (4), camphor (5), oleanolic acid (6), ursolic acid (7), carnosol (8), luteolin (9), ellagic acid (10), ferulic acid (11), vanillic acid (12), genkwanin (13) and eugenol (14); [Marvin 17.21.0, ChemAxon tool was used for drawing chemical structures].

21.3 PHARMACOLOGY

21.3.1 ANTIBACTERIAL ACTIVITY

The essential oil of *S. rosmarinus* combined with calcium hydroxide is effective against the bacteria *Enterococcus faecalis*, which shows bacteriostatic effects against microorganisms (Silva et al., 2019). Among the isolated bioactive compounds, carnosic acid, a major metabolite, exhibited weak activity against Gram-negative bacteria and strong antibacterial action against Gram-positive bacteria (Zhong et al., 2021). The *S. rosmarinus* alcoholic extract exhibited antimicrobial activity on oral bacteria of a dentifrice than the other herbal dentifrice. The toothpaste with *S. rosmarinus* alcoholic extract could prevent microbial growth of *Streptococcus mutans*, *Lactobacillus rhamnosus*, and *Streptococcus oralis*. This antimicrobial activity is similar to that of the commercial toothpaste used against the microbes, such as *S. oralis* and *S. mutans* (Valones et al., 2016). The essential oil of *S. rosmarinus* has excellent potential against Gram-positive and Gram-negative bacteria and Candida strains obtained from pregnant women having genital infections. It is another proof that the *S. rosmarinus* plant can be effectively used as an antimicrobial agent against various microorganisms (Bogavac et al., 2017). The essential oil of rosemary can prevent the growth of *Acinetobacter baumannii* and show a small amount of antibacterial action against *Escherichia coli*, *Pseudomonas aeruginosa*, and *Staphylococcus aureus* (Lorenzo-Leal et al., 2019).

21.3.2 ANTIFUNGAL, ANTI-ALFA-TOXIGENIC AND ANTI-DERMATOPHYTIC EFFECT

The *S. rosmarinus* essential oil from dried leaves efficiently inhibited *Aspergillus flavus* growth by creating structural modifications. The essential oil of rosemary has anti-alfa-toxigenic properties, which inhibits aflatoxin production. *S. rosmarinus* essential oil causes severe damage to mycotoxin-producing fungi (da Silva Bomfim et al., 2020). In a study conducted by Endo et al. (2015), the hydroalcoholic extracts from *S. rosmarinus* and *Tetradenia riparia* contain major bioactive components against fungi. These compounds can be used against dermatophytes by inhibiting the growth of fungi and producing hyphae with morphological variations. The hyphal growth inhibition by the extracts is effectively determined by scanning electron microscopy (SEM) and fluorescence microscopy, which exhibited more significant inhibition of hyphal growth and abnormal growth patterns. Hence, this facilitates using *S. rosmainus* to treat various skin problems and as an anti-dermatophytic agent/drug.

Dermatophytes are the general factors producing topical mycoses (Achterman and White, 2012). The World Health Organization predicts that 20% of the global community is hit by dermatomycoses (Negi et al., 2017); the pervasiveness of these complications rises with lifetime and depends on the location and climate (Negi et al., 2017). *S. rosmarinus* was stated to be effective on dermatophytes in vivo (Andrade et al., 2018). The essential oil from rosemary was tested for its antifungal activity on *Candida dubliniensis, Candida albicans, Candida krusei*, and *Candida parapsilosis* (Gauch et al., 2014). It was observed that a concentration of 8% oil could inhibit the growth of *Candida* sp. A similar study evaluated the effect of rosemary hydroalcoholic extract upon two dermatophytes, *Trichophyton rubrum* and *Microsporum gypseum*. It revealed inhibition of fungal growth (86%) with rosemary extract concentration of 10% (Sudan and Singh, 2019).

21.3.3 ANTICANCER ACTIVITY

S. rosmarinus anticancer activities include its antiangiogenic property, antioxidant activity, immune response regulation, anti-inflammatory action, hormone signaling alteration, epigenetic action, tumor suppressor genes increased expression, and specific metabolic pathways modifications. The phytochemical carnosic acid and carnosol in rosemary inhibit endothelial cells' angiogenic actions, such as proliferation, migration, differentiation, and differentiation capacity. The in vivo study of chick chorioallantoic membrane assay confirmed the potential of rosemary to inhibit in vitro angiogenesis. Rosemary also has a role in reducing the metastasizing process. It inhibits various types of cancers, such as oral cancer, colon cancer, skin cancer, lung cancer, ovarian cancer, cerebral neoplasms, hepatocellular carcinoma, gastric cancer, and pancreatic cancer (Allegra et al., 2020). Investigations revealed that carnosic acid presented the most potent inhibitory effect among different compounds obtained from *S. rosemarinus* (including carnosol) on HIV-1 protease in cell-free assays (Pariš et al., 1993). Carnosol repressed skin tumorigenesis, obtained from rosemary plants (Huang et al., 1994). The rosemary extract showed various anticancer effects for different types of cancer like breast, colon, lung, skin cancer, and various other cancers. The antitumor activity of rosemary is majorly due to the bioactive compounds of its extract, such as carnosol, ursolic acid, rosmarinic acid, and carnosic acid. Supercritical fluid technology is a method to extract rosemary extract containing bioactive compounds with more concentration than the other extraction methodologies (González-Vallinas et al., 2015).

21.3.4 ANTIOXIDANT AND HEPATOPROTECTIVE ACTIVITY

The presence of antioxidants in various spices and herbs, particularly rosemary (*S. rosmarinus*), is adequately recognized. The detection of antioxidants to enhance the storage of foods has created the feasibility of marketing numerous food products and has served the consumers. The food and pharmaceutical industries widely use antioxidants in food (Madsen and Bertelsen, 1995). Phytochemicals, such as carnosol, carnosic acid, and rosmarinic acid are responsible for *S. rosmarinus* antioxidant activity. Its cytotoxic activity is majorly due to the compounds, such as carnosic acid and triterpenoids, such as ursolic and betulinic acid. The antioxidant property and cytotoxic activity cause the damage of cancer cells by protecting healthy cells in cancer treatment, majorly due to its antioxidant activity. Carnosol and carnosic acid, among numerous additional antioxidants, can provide approximately 90% of the antioxidant action of rosemary plants (*S. rosmarinus*) (Aruoma et al., 1992). In comparison with sage extract, *S. rosmarinus* has carnosic acid in fourfold greater concentration, and it contains a significantly more quantity of betulinic acid (Kontogianni et al., 2013). The rosemary essential oil contains α-pinene of 11.51%, Camphor of 12.53%, and 1,8-cineole of 43.77%. These phytochemicals effectively showed antioxidant activity determined by the DPPH assay. It inhibited carbon tetrachloride (CCl_4)-induced lipid peroxidation increase in liver homogenate. In a particular 10 mg/kg dose, the essential oil pretreatment efficiently alters the free radical scavenging enzymes, such as catalase, glutathione reductase, glutathione peroxidase, and peroxidase. Thus, rosemary essential oil is known to have great hepatoprotective potential (Rašković et al., 2014).

21.3.5 ANTIDIABETIC ACTIVITY

The extract of rosemary and its polyphenols, such as rosmarinic acid and carnosic acid showed insulin-like performance in insulin's target cells in vitro condition. It exhibited a practical antidiabetic effect in various animal models of Type 2 diabetes mellitus (T2DM) in vivo conditions. The extract of *S. rosmarinus* and its polyphenols displayed protective action against hyperglycemia and hyperlipidemia in chemically induced, genetic, and alimentary animal models with type 2 diabetes mellitus (T2DM) and obesity. These in vivo animal studies' findings suggest the effectiveness of rosemary extracts and their polyphenol contents toward blood glucose level management and diabetes (Naimi et al., 2017).

Healing is a complex progressive process that occurs in reviving the anatomical limitations of the skin, jeopardized by disease attacks or wounds (Lazarus et al., 1994). Diabetes was provoked in BALB/c mice (male) by the intraperitoneal dose of alloxan (Abu-Al-Basal, 2010). Following verification of hyperglycemia, 4 mm slashes of 4 mm in width were made on their backs. The mice were earmarked for one of the four treatments: essential oil, vehicle control, and aqueous extract. The BALB/c mice of the essential oil group conferred improvements in granulation tissue, healing, and angiogenesis, at various grades associated with mice in the hydro extract group (Abu-Al-Basal, 2010). A study with diabetic-induced rats treated topically with *S. rosmarinus* oil extracts showed accelerated injury healing in diabetic and nondiabetic animals (Park et al., 2014).

21.3.6 ANTISTRESS ACTIVITY

The *S. rosmarinus* leaf extract (RLE) on restraint-stressed mice lowers the number of global cells of the intestine and hepatic triglycerides (TG) content. The leaf extract also decreases the forced swimming test (FST) immobility time and also activates the microglia present in the brain. These findings reported that the leaf extract of rosemary has positive actions on stress-induced dysfunctions. Thus, rosemary is considered an effective antistress agent (Kayashima et al., 2020).

21.3.7 ANTIVIRAL ACTIVITY

The extract of rosemary is used against Herpes Simplex Virus 1 and 2 in vitro conditions. The antioxidant activity of the rosemary extract was determined by using superoxide anion and DPPH assay. At 50 μg/mL, the extract of *S. rosmarinus* completely prevented the plaque formation of HSV 1 and HSV 2. At 30 μg/mL, it causes 55% inhibition of HSV 1 plaques, whereas 65% prevention of HSV 2 plaques occurs at 40 μg/ml. The radical superoxide anion scavenging activity occurred at 65.74 mg/mL, and 50% DPPH radical scavenging activity was detected at 67.34 mg/mL. Thus, the *S. rosmarinus* extract acts as a suitable therapeutic agent for the infection caused by the herpes virus, thereby acting as an antiviral component (Al-Megrin et al., 2020).

21.3.8 COGNITIVE AND CEREBROVASCULAR ACTIVITY

The oral administration of rosemary water enhances the act of cognition. Rosemary water consumption also increases the deoxygenated hemoglobin levels during task performance of cognition monitored by near-infrared spectroscopy, thereby exhibiting cerebrovascular effect (Moss et al., 2018).

21.3.9 CYTOTOXIC, GENOTOXIC, AND ANTIESTROGENIC ACTIVITY

Cytochalasin B-blocked Micronucleus assay (CBMN) is used to determine the cytotoxic activity in the cell line of human ovarian carcinoma (A2780) and genotoxic activity in peripheral lymphocytes of humans. The antiestrogenic or estrogenic activity is determined by using human estrogen receptor alpha (ERα) expression in yeast strains. The administration of essential oil of *S. rosmarinus* exhibited high cytotoxic activity in the ovarian carcinoma cell line (A2780) and moderate genotoxic activity in peripheral lymphocytes of humans. The essential oil of rosemary also showed a significant amount of antiestrogenic activity in yeast strains which effectively disturbs the estrogen-dependent tissues (Contini et al., 2020). The higher cytotoxic activity exhibited by the rosemary extract is due to the presence of carnosic acid. The rosemary extract exerts an efficient synergistic effect with betulinic acid, which may cause differential cytotoxic activity (Kontogianni et al., 2013).

21.3.10 ANTI-INFLAMMATORY ACTIVITY

The anti-inflammatory activity of the rosemary extract essential oils is majorly due to the presence of a high abundance of α-pinene and 1,8-cineole, which are often more abundant, and whose mechanisms are better understood. These essential oils of the *S. rosmarinus* major monoterpenes exhibit anti-inflammatory activity by decreasing the function of the NK-κB transcription factor. The activity by α-pinene, 1,8-cineol, myrcene, and limonene hinders pro-inflammatory mediators, such as IL-1β, TNF-α synthesis, thereby preventing the arachidonic acid (AA) metabolites formation by inhibiting the AA enzymes like Cyclooxygenase (COX-2), which transcription is based on NK-κB (Borges et al., 2019).

21.3.11 IMPROVING HAIR GROWTH

The *S. rosmarinus* extract showed a potential hair regrowth effect in vivo condition in a male mouse model of C57BL/6NCrSlc, which is sensitive to testosterone. The reformation of hair is slowed in the testosterone-treated group than in the untreated group. In these conditions, the rosemary extract of 2 mg/day per mouse administration enhanced hair growth in the group administered with testosterone. The findings suggest that the rosemary extract is mainly due to its anti-androgenic activity, which can be a potential treatment against androgenic alopecia (AGA). The rosemary extract promotes hair growth in C3H/He mice whose dorsal areas are shaved. The anti-androgenic action of rosemary extract is due to the testosterone 5a-reductase (5aR) inhibition which is responsible for the androgenic alopecia (AGA) treatment and dihydrotestosterone (DHA) binding to androgen receptors (AR). The active component to inhibit 5a-reductase and dihydrotestosterone (DHA) binding to the androgen receptor (AR) is 12-methoxycarnosic acid (12-MCA) which is determined by activity-guided fractionation. Thus, the extract of rosemary is potentially used to treat hair regrowth (Murata et al., 2013).

21.3.12 HYPOLIPIDEMIC EFFECT

The bioactive compounds of *S. rosmarinus* reveal the existence of rosmarinic acid and carnosic acid, which significantly lower the amount of triglyceride present in the liver, free fatty acids (FFA), total cholesterol (TC) and enhances vacuolation, cell hypertrophy, and liver cell necrosis in the model rats having an orotic acid-induced non-alcoholic fatty liver disease (NAFLD). Hence, bioactive compounds, such as carnosic acid and rosmarinic acid act as potential hypolipidemic agents for treating non-alcoholic fatty liver disease (NAFLD) (Wang et al., 2019).

21.3.13 ANTIAGING PROPERTY

Aging is a skin-associated process that transpires owing to inherent and extraneous determinants. Inherent factors function at the cellular level, whereas extraneous agents are dictated by human activity, such as inadequate nourishment, prolonged sun appearance, lavish alcohol intake, and smoking (Baumann, 2007). These external and internal factors influence ROS generation (Stojiljković et al., 2014); when ROS levels surpass the cell's neutralization potential, injury to cell components happens, eventually commencing cell mortality (Liochev, 2015). A new compound separated from the hydrophilic portion of hydromethanol extract from Rosemary and was designated Rosm1. Rosm1 had a substantial antioxidant potential to counterbalance ROS, comparable to tocopherol. It inhibited free radical-mediated responses in vitro and in vivo, defending cell components and lipids from oxidative stress (Calabrese et al., 2000). An increase in the cutaneous diffusion of drugs was witnessed when lipid nanoparticles were used (Junyaprasert et al., 2009; Souto and Muller, 2008). These nanoparticles were primed with rosemary essential oil combined into a gel. An in vivo skin test was conducted to evaluate any increase in flexibility and the moisturizing capability of this formulation. The gel comprising nanoparticles packed with rosemary oil exhibited a more comprehensive ability to hydrate and improve skin flexibility than the gel comprising only free essential oil (Montenegro et al., 2017). Rosmarinic acid was encapsulated inside chitosan microparticles. The discharge profile of the primed particles was investigated in two different media: coconut oil and water. A more gradual release profile recognized in coconut oil was associated with the chitosan particles' lower solubility (Casanova et al., 2016). Rosemary extract has potent antioxidant action, which is principally connected to the presence of phenolic compounds. Antioxidant activity is frequently associated with scavenging free radicals. However, secondary metabolites may perform a physiological role in regulating xenobiotic metabolism in the liver, cell signal transduction, and apoptosis (Lamaison et al., 1991).

21.3.14 IMMUNOSUPPRESSIVE EFFECT

The phytochemical trans-caffeic acid present in *S. rosmarinus* exhibits the immunosuppressive effect. It is due to the inhibition of STAT3 activity by inducing apoptosis and preventing T-lymphocytes proliferation. Hence, the presence of trans-caffeic acid enables rosemary to act as a potential immunosuppressive agent (von Schönfeld et al., 2018).

KEYWORDS

- **rosemary**
- **rosmarinic acid**
- **antiulcerogenic**
- **anti-aging**
- **rosmaquinone A**
- **rosmaquinone B**

REFERENCES

Abu-Al-Basal, M. A. Healing Potential of *Rosmarinus officinalis* L. on Full-Thickness Excision Cutaneous Wounds in Alloxan-Induced-Diabetic BALB/c Mice. *J. Ethnopharmacol.* **2010,** *131*, 443–450.

Achterman, R. R.; White, T. C. Dermatophyte Virulence Factors: Identifying and Analyzing Genes That May Contribute to Chronic or Acute Skin Infections. *Int. J. Microbiol.* **2012,** *2012*, Article ID 358305.

Ahmed, H. M.; Babakir-Mina, M. Investigation of Rosemary Herbal Extracts (*Rosmarinus officinalis*) and Their Potential Effects on Immunity. *Phytother. Res.* **2020,** *34* (8), 1829–1837.

Alavi, M. S.; Fanoudi, S.; Ghasemzadeh Rahbardar, M.; Mehri, S.; Hosseinzadeh, H. An Updated Review of Protective Effects of Rosemary and Its Active Constituents Against Natural and Chemical Toxicities. *Phytother. Res.* **2021,** *35* (3), 1313–1328.

Allegra, A.; Tonacci, A.; Pioggia, G.; Musolino, C.; Gangemi, S. Anticancer Activity of *Rosmarinus officinalis* L.: Mechanisms of Action and Therapeutic Potentials. *Nutrients* **2020,** *12* (6), 1739.

Al-Megrin, W. A.; AlSadhan, N. A.; Metwally, D. M.; Al-Talhi, R. A.; El-Khadragy, M. F.; Abdel-Hafez, L. J. Potential Antiviral Agents of *Rosmarinus officinalis* Extract Against Herpesviruses 1 and 2. *Biosci. Rep.* **2020,** *40* (6), BSR20200992.

Al-Sereiti, M. R.; Abu-Amerb, K. M.; Sena, P. Pharmacology of Rosemary (*Rosmarinus officinalis* Linn.) and Its Therapeutic Potentials. *Indian J. Exp. Biol.* **1999,** *37*, 124–130.

Andrade, J. M.; Faustino, C.; Garcia, C.; Ladeiras, D.; Reis, C. P.; Rijo, P. *Rosmarinus officinalis* L.: An Update Review of Its Phytochemistry and Biological Activity. *Future Sci. OA.* **2018,** *4*, FSO283.

Aruoma, O. I.; Halliwell, B.; Aeschbach, R.; Löligers, J. Antioxidant and Pro-Oxidant Properties of Active Rosemary Constituents: Carnosol and Carnosic Acid. *Xenobiotica* **1992,** *22* (2), 257–268.

Baumann, L. Skin Ageing and Its Treatment. *J. Pathol.* **2007,** *211*, 241–251.

Begum, A.; Sandhya, S.; Vinod, K. R.; Reddy, S.; Banji, D. An In-Depth Review on the Medicinal Flora *Rosmarinus officinalis* (Lamiaceae). *Acta Sci. Pol. Technol. Aliment.* **2013,** *12*, 61–74.

Bogavac, M. A.; Karaman, M. A.; Suđi, J. J.; Radovanović, B. B.; Janjušević, L. N.; Ćetković, N. B.; Tešanović, K. D. Antimicrobial Potential of *Rosmarinus officinalis* Commercial Essential Oil in the Treatment of Vaginal Infections in Pregnant Women. *Nat. Prod. Commun.* **2017,** *12* (1), 1934578X1701200136.

Borges, R. S.; Ortiz, B. L.; Pereira, A. C.; Keita, H.; Carvalho, J. C. *Rosmarinus officinalis* Essential Oil: A Review of Its Phytochemistry, Anti-Inflammatory Activity, and Mechanisms of Action Involved. *J. Ethnopharmacol.* **2019,** *229*, 29–45.

Borrás-Linares, I.; Stojanović, Z.; Quirantes-Piné, R.; Arráez-Román, D.; Švarc-Gajić, J.; Fernández-Gutiérrez, A.; Segura-Carretero, A. *Rosmarinus officinalis* Leaves as a Natural Source of Bioactive Compounds. *Int. J. Mol. Sci.* **2014,** *15* (11), 20585–20606.

Calabrese, V.; Scapagnini, G.; Catalano, C.; Dinotta, F.; Geraci, D.; Morganti, P. Biochemical Studies of a Natural Antioxidant Isolated from Rosemary and Its Application in Cosmetic Dermatology. *Int. J. Tissue React.* **2000,** *22*, 5–13.

Casanova, F.; Estevinho, B. N.; Santos, L. Preliminary Studies of Rosmarinic Acid Microencapsulation with Chitosan and Modified Chitosan for Topical Delivery. *Powder Technol.* **2016,** *297*, 44–49.

Contini, A.; Di Bello, D.; Azzarà, A.; Giovanelli, S.; D'Urso, G.; Piaggi, S.; Pinto, B.; Pistelli, L.; Scarpato, R.; Testi, S. Assessing the Cytotoxic/Genotoxic Activity and Estrogenic/Antiestrogenic Potential of Essential Oils from Seven Aromatic Plants. *Food Chem Toxicol.* **2020,** *138*, 111205.

da Silva Bomfim, N.; Kohiyama, C. Y.; Nakasugi, L. P.; Nerilo, S. B.; Mossini, S. A.; Romoli, J. C.; Graton Mikcha, J. M.; Abreu Filho, B. A.; Machinski, Jr M. Antifungal and Antiaflatoxigenic Activity of Rosemary Essential Oil (*Rosmarinus officinalis* L.) against *Aspergillus flavus*. *Food Addit. Contam. A.***2020,** *37* (1), 153–161.

de Macedo, L. M.; Santos, É. M. D.; Militão, L.; Tundisi, L. L.; Ataide, J. A.; Souto, E. B.; Mazzola, P. G. Rosemary (*Rosmarinus officinalis* L., syn *Salvia rosmarinus* Spenn.) and Its Topical Applications: A Review. *Plants* **2020,** *9* (5), 651.

De Oliveira, J. R.; Camargo, S. E.; De Oliveira, L. D. *Rosmarinus officinalis* L. (Rosemary) as Therapeutic and Prophylactic Agent. *J. Biomed. Sci.* **2019,** *26* (1), 1–22.

Endo, E. H.; Costa, G. M.; Nakamura, T. U.; Nakamura, C. V.; Dias Filho, B. P. Antidermatophytic Activity of Hydroalcoholic Extracts from *Rosmarinus officinalis* and *Tetradenia riparia*. *J. Mycol. Med.* **2015,** *25* (4), 274–279.

Garden, M. B. *Salvia rosmarinus.* http://www.missouribotanicalgarden.org/PlantFinder/PlantFinderDetails.aspx?kempercode=b968 (accessed on 03 Aug 2021).

Gauch, L. M. R.; Pedrosa, S. S.; Esteves, R. A.; Silveira-Gomes, F.; Gurgel, E. S. C.; Arruda, A. C.; Marques-da-Silva, S. H. Antifungal Activity of *Rosmarinus officinalis* Linn. Essential Oil Against *Candida albicans*, *Candida dubliniensis*, *Candida parapsilosis* and *Candida krusei*. *Rev. Pan-Amaz. Saúde.* **2014,** *5*, 61–66.

González-Vallinas, M.; Reglero, G.; Ramírez de Molina A. Rosemary (*Rosmarinus officinalis* L.) Extract as a Potential Complementary Agent in Anticancer Therapy. *Nutr. Cancer*. **2015,** *67* (8), 1223–31.

Hassanen, N. H.; Fahmi, A.; Shams-Eldin, E.; Abdur-Rahman, M. Protective Effect of Rosemary (*Rosmarinus officinalis*) Against Diethylnitrosamine-Induced Renal Injury in Rats. *Biomarkers*. **2020,** *25* (3), 281–289.

Huang, M. T.; Ho, C. T.; Wang, Z. Y.; Ferraro, T.; Lou, Y. R.; Stauber, K.; Ma, W.; Georgiadis, C.; Laskin, J. D.; Conney, A. H. Inhibition of Skin Tumorigenesis by Rosemary and Its Constituents Carnosol and Ursolic Acid. *Cancer Res.* **1994,** *54* (3), 701–708.

Junyaprasert, V. B.; Teeranachaideekul, V.; Souto, E. B.; Boonme, P.; Muller, R. H. Q10-Loaded NLC Versus Nanoemulsions: Stability, Rheology and In Vitro Skin Permeation. *Int. J. Pharm.* **2009,** *377*, 207–214.

Kayashima, T.; Nagao, K.; Umino, M.; Kaikiri, H.; Shibata, S.; Matsubara, K. Anti-Stress Effects of Rosemary (*Rosmarinus officinalis* L.) Leaf Extract on Intestinal Goblet Cells and Immobility of Forced-Swimming Test in BALB/c Mice. *Biosci. Biotechnol. Biochem.* **2020,** *84* (11), 2385–2389.

Kontogianni, V. G.; Tomic, G.; Nikolic, I.; Nerantzaki, A. A.; Sayyad, N.; Stosic-Grujicic, S.; Stojanovic, I.; Gerothanassis, I. P.; Tzakos, A. G. Phytochemical Profile of *Rosmarinus officinalis* and *Salvia officinalis* Extracts and Correlation to Their Antioxidant and Anti-Proliferative Activity. *Food Chem.* **2013,** *136* (1), 120–129.

Lamaison, J. L.; Petitjean-Freytet, C.; Carnat, A. Medicinal Lamiaceae with Antioxidant Properties, a Potential Source of Rosmarinic Acid. *Pharm. Acta Helv.* **1991,** *66*, 185–188.

Lazarus, G. S.; Cooper, D. M.; Knighton, D. R.; Margolis, D. J.; Pecoraro, R. E.; Rodeheaver, G.; Robson, M. C. Definitions and Guidelines for Assessment of Wounds and Evaluation of Healing. *JAMA Dermatol.* **1994,** *130*, 489–493.

Liochev, S. I. Reflections on the Theories of Aging, of Oxidative Stress, and of Science in General. Is It Time to Abandon the Free Radical (Oxidative Stress) Theory of Aging? *Antioxidants Redox Signal*, **2015,** *23* (3), 187–207.

Lorenzo-Leal, A. C.; Palou, E.; López-Malo, A.; Bach, H. Antimicrobial, Cytotoxic, and Anti-Inflammatory Activities of *Pimenta dioica* and *Rosmarinus officinalis* Essential Oils. *Biomed. Res. Int.* **2019,** *2019*, 1–8. Article ID 1639726.

Madsen, H. L.; Bertelsen, G. Spices as Antioxidants. *Trends Food Sci. Tech.* **1995,** *6* (8), 271–277.

Montenegro, L.; Pasquinucci, L.; Zappalà, A.; Chiechio, S.; Turnaturi, R.; Parenti, C. Rosemary Essential Oil-Loaded Lipid Nanoparticles: In Vivo Topical Activity from Gel Vehicles. *Pharmaceutics*, **2017,** *9*, 48.

Moss, M.; Smith, E.; Milner, M.; McCready, J. Acute Ingestion of Rosemary Water: Evidence of Cognitive and Cerebrovascular Effects in Healthy Adults. *J. Psychopharmacol.* **2018,** *32* (12), 1319–1329.

Murata, K.; Noguchi, K.; Kondo, M.; Onishi, M.; Watanabe, N.; Okamura, K.; Matsuda, H. Promotion of Hair Growth by *Rosmarinus officinalis* Leaf Extract. *Phytother. Res.* **2013,** *27* (2), 212–217.

Naimi, M.; Vlavcheski, F.; Shamshoum, H.; Tsiani, E. Rosemary Extract as a Potential Anti-Hyperglycemic Agent: Current Evidence and Future Perspectives. *Nutrients.* **2017,** *9* (9), 968.

Negi, N.; Tripathi, V.; Choudhury, R.; Bist, J.; Kumari, N.; Chandola, I. Clinicomycological Profile of Superficial Fungal Infections Caused by Dermatophytes in a Tertiary Care Centre of North India. *Int. J. Curr. Microbiol. Appl. Sci.* **2017,** *6*, 3220–3227.

Ojeda-Sana, A. M.; van Baren, C. M.; Elechosa, M. A.; Juárez, M. A.; Moreno, S. New insights into antibacterial and Antioxidant Activities of Rosemary Essential Oils and Their Main Components. *Food Control* **2013,** *31*, 189–195.

Pariš, A.; Štrukelj, B.; Renko, M.; Turk, V.; Pukl, M.; Umek, A.; Korant, B. D. Inhibitory Effect of Carnosolic Acid on HIV-1 Protease in Cell-Free Assays. *J. Nat. Prod.* **1993,** *56* (8), 1426–1430.

Park, S. Y.; Song, H.; Sung, M.-K.; Kang, Y.-H.; Lee, K. W.; Park, J. H. Y. Carnosic Acid Inhibits the Epithelial-Mesenchymal Transition in B16F10 Melanoma Cells: A Possible Mechanism for the Inhibition of Cell Migration. *Int. J. Mol. Sci.* **2014,** *15*, 12698–12713.

Rašković, A.; Milanović, I.; Pavlović, N.; Ćebović, T.; Vukmirović, S.; Mikov, M. Antioxidant Activity of Rosemary (*Rosmarinus officinalis* L.) Essential Oil and Its Hepatoprotective Potential. *BMC Complem. Altern. Med.* **2014,** *14* (1), 1–9.

Ribeiro-Santos, R.; Carvalho-Costa, D.; Cavaleiro, C.; Costa, H. S.; Albuquerque, T. G.; Castilho, M. C.; Ramos, F.; Melo, N. R.; Sanches-Silva, A. A Novel Insight on an Ancient Aromatic Plant: The Rosemary (*Rosmarinus officinalis* L.). *Trends Food Sci. Technol.* **2015,** *45*, 355–368.

Science, R. B. G. K. *Salvia rosmarinus* Spenn. http://plantsoftheworldonline.org/taxon/urn:lsid:ipni.org:names:457138–1 (accessed on 03 Aug 2021).

Sharifi-Rad, J.; Ezzat, S. M.; El Bishbishy, M. H.; Mnayer, D.; Sharopov, F.; Kılıç, C. S.; Neagu, M.; Constantin, C.; Sharifi-Rad, M.; Atanassova, M.; Nicola, S. *S. rosmarinus*

Plants: Key Farm Concepts Towards Food Applications. *Phytother. Res.* **2020,** *34* (7), 1474–1518.

Sharma, Y.; Velamuri, R.; Fagan, J.; Schaefer, J. Full-Spectrum Analysis of Bioactive Compounds in Rosemary (*Rosmarinus officinalis* L.) as Influenced by Different Extraction Methods. *Molecules* **2020,** *25* (20), 4599.

Silva, S.; Alves, N.; Silva, P.; Vieira, T.; Maciel, P.; Castellano, L. R.; Bonan, P.; Velozo, C.; Albuquerque, D. Antibacterial Activity of *Rosmarinus officinalis*, *Zingiber officinale*, *Citrus aurantium bergamia*, and *Copaifera officinalis* Alone and in Combination with Calcium Hydroxide Against *Enterococcus faecalis*. *Biomed Res. Int.* **2019,** *2019*, 1–7. Article ID 8129439.

Souto, E. B.; Muller, R. H. Cosmetic Features and Applications of Lipid Nanoparticles (SLN, NLC). *Int. J. Cosmet. Sci.* **2008,** *30*, 157–165.

Stefanovits-Bányai, É.; Tulok, M.; Hegedus, A.; Renner, C.; Szollosi Varga, I. Antioxidant effect of various rosemary (*Rosmarinus officinalis* L.) clones. *Acta Biol. Szeged.* **2003,** *47*, 111–113.

Stojiljković, D.; Pavlović, D.; Arsić, I. Oxidative Stress, Skin Aging and Antioxidant Therapy/ Oksidacioni Stress, *Starenje Kože I Antioksidaciona Terapija. Acta Fac. Med. Naissensis* **2014,** *31*, 207.

Sudan, P.; Singh, J. Antifungal Potential of *Rosmarinus Officinalis* Against *Microsporum gypseum* and *Trichophyton rubrum*. *Int. Res. J. Pharm.* **2019,** *10*, 205–207.

UniProt. Taxonomy—*Rosmarinus Officinalis* (Rosemary) *(Salvia Rosmarinus)*. https://www.uniprot.org/taxonomy/39367 (accessed on 03 Aug 2021).

Valones, M. A.; Higino, J. S.; Souza, P. R.; Crovella, S.; Caldas, A. D.; Carvalho, A. D. Dentifrice Containing Extract of *Rosmarinus officinalis* Linn.: An Antimicrobial Evaluation. *Braz. Dent. J.* **2016,** *27*, 497–501.

von Schönfeld, C.; Huber, R.; Trittler, R.; Kammerer, B.; Garcia-Käufer, M.; Gründemann, C. Rosemary Has Immunosuppressant Activity Mediated Through the STAT3 Pathway. *Complement. Ther. Med.* **2018,** *40*, 165–170.

Wang, S. J.; Chen, Q.; Liu, M. Y.; Yu, H. Y.; Xu, J. Q.; Wu, J. Q.; Zhang, Y.; Wang, T. Regulation Effects of Rosemary (*Rosmarinus officinalis* Linn.) on Hepatic Lipid Metabolism in OA Induced NAFLD Rats. *Food Funct.* **2019,** *10* (11), 7356–7365.

Zappalà, A.; Vicario, N.; Calabrese, G.; Turnaturi, R.; Pasquinucci, L.; Montenegro, L.; Spadaro, A.; Parenti, R.; Parenti, C. Neuroprotective Effects of *Rosmarinus officinalis* L. Extract in Oxygen Glucose Deprivation (OGD)-Injured Human Neural-Like Cells. *Nat. Prod. Res.* **2021,** *35* (4), 669–675.

Zhong, X.; Wang, X.; Zhou, N.; Li, J.; Liu, J.; Yue, J.; Hao, X.; Gan, M.; Lin, P.; Shang, X. Chemical Characterization of the Polar Antibacterial Fraction of the Ethanol Extract from *Rosmarinus officinalis*. *Food Chem.* **2021,** *344*, 128674.

CHAPTER 22

BIOACTIVE COMPOUNDS AND PHARMACOLOGICAL VALUES OF *SALVIA SCLAREA* L.

SARIKA SINGH[1], SOBIA AMAN CHISHTI[1], SANGRAM SINGH[1], UROOJ HASSAN BHATT[1], M. MASROOR A. KHAN[1], and MOINUDDIN[2]

[1]*Department of Botany, Faculty of Life Sciences, Aligarh Muslim University, Aligarh, Uttar Pradesh, India*

[2]*Botany section, Women's College, Aligarh Muslim University, Aligarh, Uttar Pradesh, India*

ABSTRACT

Salvia sclarea, a member of Lamiaceae family, commonly called clary sage and is native to Southern Europe. *S. sclarea* is an ornamental and essential oil-bearing medicinal plant which is used in perfumery. In *S. sclarea*, essential oil contains several bioactive compounds like linalyl acetate, linalool, germacrene D, α-terpineol, geranyl acetate, (E)-caryophyllene. In roots of *S. sclarea*, several diterpenoids are found like ferruginol, salvipisone, sclareol, aethiopinone, 1-oxoaethiopinone. Essential oil of *S. sclarea* possesses several pharmacological properties including antimicrobial, cytotoxic activity, anti-depressant, anti-inflammatory, antidiabetic activity, insecticidal, antianxiety, antioxidant.

22.1 INTRODUCTION

Salvia sclarea L. (commonly called clary sage) is a member of the family Lamiaceae. The sage plant is an ornamental and essential oil-bearing plant,

Bioactives and Pharmacology of Lamiaceae. T. Pullaiah, PhD (Ed.)

basically used in perfumery, and is native to southern Europe. However, the essential oil is obtained from the fresh spikes during flowering stage. Besides this, the essential oil of *S. sclarea* possesses several multi-pharmacological effects, such as antimicrobial, insecticidal, antidiabetic, and also used as a stress-buster and in curing depression and insomnia (Abdelkader et al., 2014; Küçük et al.., 2019; Seol et al., 2010). There are several bioactive compounds, such as terpenoids (diterpenoids, triterpenes, sesquiterpenes, sesquiterpenoids), polyphenols (flavonoids, phenolic acids) present in the sage plant that might induce the above pharmacological effects (Coisin et al., 2012; Abdelkader et al., 2014). Apart from this, the oil of *S. sclarea* is rich in omega 3-linoleic acid, and can be used as edible oil (Aćimović et al., 2018).

22.2 BIOACTIVE COMPOUNDS

Diterpenes are a diverse class of plant metabolites, that are basically derived from GGPP (geranylgeranyl pyrophosphate) synthesized in a MEP pathway. In a recent study, GC-MS analysis reveals the presence of terpenoids in the essential oil of *S. sclarea* like oxygenated monoterpenes in the highest quantity approximately 66%, monoterpenes hydrocarbon (7–8%), sesquiterpenes (5–6%), and oxygenated sesquiterpenes (4–5%) (Raffat and Habib, 2018). From the roots of *S. sclarea*, four diterpenoids, such as ferruginol, salvipisone, aethiopinone, and 1-oxoaethiopinone are extracted (Fig. 22.1). In addition to this, a few triterpenoids (ursine-type), such as 2α, 3α-dihydroxy-urs-12-en-28-oic acid and 2α, 3α, 24-trihydroxy-urs-12-en-28-oic acid are also found (Kuźma et al., 2006). GC-MS analysis reveals the presence of several strolls, such as β-sitosterol, campesterol, and stigmasterol, and the presence of oleanolic acid and ursolic acids are also found (Kuźma et al., 2006). Sclareol is also a labdane type diterpenoid that is widely used in the perfume industry. Sclareol is extracted from the trichomes that protect the flower calices. However, another diterpene, manool is also found in plants of *Salvia* species, but in *S. sclarea*, this diterpene is found in minor quantity (Caniard et al., 2012).

Essential oil from the aerial part of *S. sclarea* reveals the presence of 59 active compounds and the highest amount of compounds found are linalyl acetate, linalool, germacrene D, α-terpineol, geranyl acetate, (E)-caryophyllene (Fig. 22.2) (Sharopov and Setzer, 2012). Several polyunsaturated fatty acids are present in the oil of *Salvia* seeds. In *S. sclarea*, α-linolenic acid is found predominantly. In addition, various tocopherols, tocotrienols, several carotenoids, as well as fatty acids are also present.

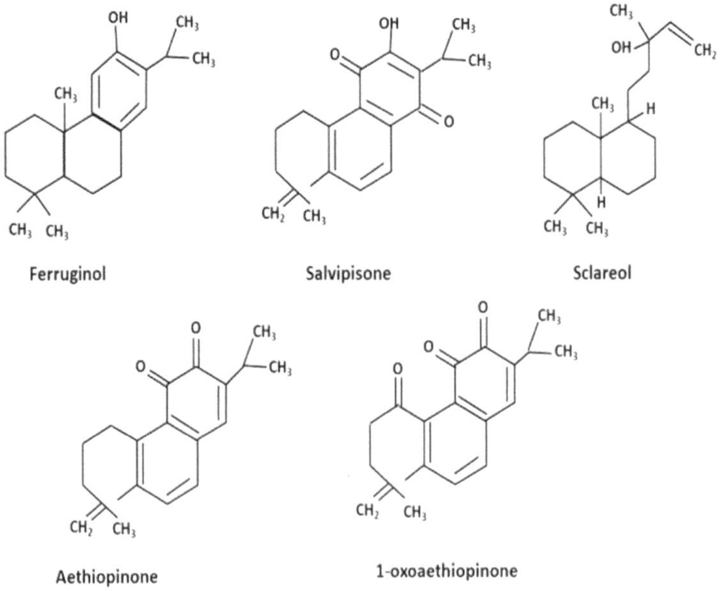

FIGURE 22.1 Chemical structure of diterpenoids present in roots of *S. sclarea.*

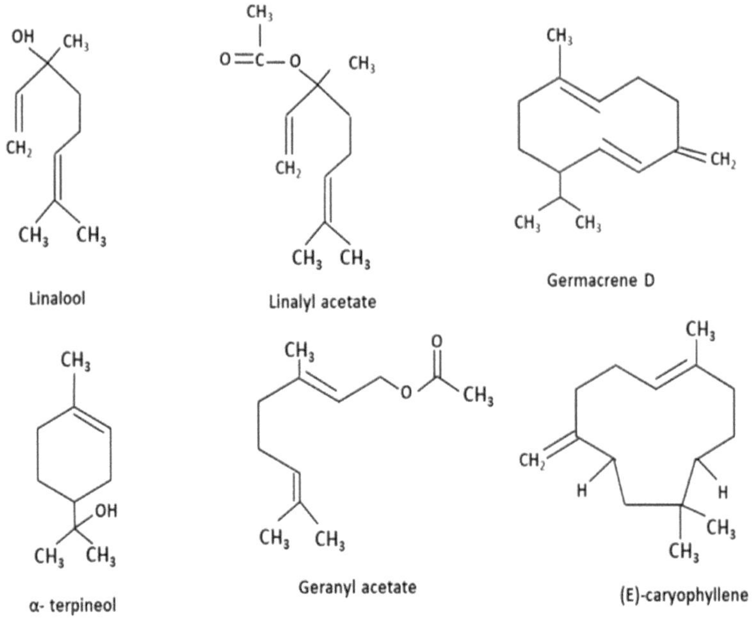

FIGURE 22.2 Chemical structure of bioactive compounds in the essential oil of *S. sclarea.*

The sage plant is a rich source of antioxidants, and this property added medicinal value to this plant. The action of phenolic compounds provides antioxidant properties to this plant. However, HPLC-DAD analysis reveals the presence of 18 phenolic compounds including 11 phenolic acids and 8 flavonoids in the extract of the sage plant. Phenolic acids include 3-hydroxybenzoic acid, 3, 4-dihydroxybenzoic acid, vanillic acid, caffeic acid, syringic acid, p-coumaric acid, rosmarinic acid, ferulic acid, ellagic acid, p-hydroxybenzoic acid, and chlorogenic acid, whereas different flavonoid compounds identified are galangin, chrysin, kaempferol, pinocembrin, luteolin, and myricetin (Jasicka-Misiak et al., 2018). In addition to this, reverse-phase HPLC with diode array detector analysis reveals that rosmarinic acid is also present in rich amounts in sage plants, and methanolic extract shows the presence of highest amount of phenolic and flavonoid compounds (Zengin et al., 2018).

22.3 PHARMACOLOGICAL VALUES

22.3.1 ANTIMICROBIAL

S. sclarea extract possesses antimicrobial and antibiofilm activities that inhibit the growth of many microorganisms, such as *Staphylococcus aureus, S. epidermidis, Candida albicans*, and *C. krusei* by degrading the membrane integrity of these microorganisms (Küçük et al.., 2019). Local application of *S. clarea* essential oil on wound infections works as antimicrobial or anti-staphylococcal agent and provides resistance to pathogenic microorganisms (Sienkiewicz et al., 2015). This might happen because the application of essential oil of *S. sclarea* generally damages the cell membrane of microbes as well as changes the cell membrane permeability due to which macromolecules like ATP and DNA get released from the cell. These events might take place in a series inside the cell or on the cell surface that leads to microbial death (Cui et al., 2015).

22.3.2 CYTOTOXIC ACTIVITY

Several diterpenes found in sage root extract (salvipisone, ferruginol, aethiopinone, 1-oxoaethiopinone) contain cytotoxic activity. Salvipisone and aethiopinone show high cytotoxic activity against different strains of leukemia cells (HL-60 and NAML-6) (Różalski et al., 2006). In another experiment, aethiopinone is an abietane diterpene, found in the sage roots, is used as a cytotoxic agent that can induce apoptosis in human melanoma cells (Vaccaro et al., 2017). Cytotoxic activity is also seen against *Artemia salina nauplli* via Brine Shrimp Lethality Assay. This could be due to the presence of various alkaloids, tannins, flavonoids, such as secondary metabolites in the extract of *S. sclarea* (Küçük et al., 2019). Therefore, diterpenoids of *S. sclarea* might be used against drug resistance and in cancer treatment.

22.3.3 ANTIDEPRESSANT ACTIVITY

There are various researches held to find out the uses of essential oil of various medicinal plants in reducing depression among patients. A study by Seol et al. (2010) revealed that 5% essential oil of *S. sclarea* is found best as a therapeutic agent for depressant patients and act as a stress-burster. In addition, with this, the effect of clary sage essential oil takes place through the DAnergic pathway. Therefore, the essential oil of *S. sclarea* might be used in pschycopharmacology and act as an anxiolytic and antidepressant agent.

22.3.4 ANTI-INFLAMMATORY ACTIVITY

S. sclarea extract act as a potential anti-inflammatory agent in the treatment of various diseases like periodontal disease. The essential oil of *S. sclarea* used in mucoadhesive preparations might be useful in the treatment related to the inflammation of mucosal tissues (Peana and Moretti, 2002). Oral invasion of sage extract in mice is found effective as anti-inflammatory agent because of the presence of rosmarinic acid which acts as an antioxidant agent (Kostić et al., 2017).

22.3.5 ANTIDIABETIC ACTIVITY

Diabetes mellitus is a chronic disorder in which the elevation of glucose in blood levels leads to several complications in the life of a healthy individual. A study by Rafaat and Habib (2018) reveals that the presence of linalool and linalyl acetate in the essential oil of *S. sclarea* shows the presence of acute and subchronic antidiabetic activities and might be used as a potential alternative for medicine for curing diabetes and related disorders. However, in another study, the aqueous extract of *S. sclarea* inhibited the action of α-amylase, α-glucosidase, and tyrosinase (Zengin et al., 2018). Therefore, bioactive compounds (flavonoids and phenolic acids) present in *S. sclarea* inhibited the action of active enzymes involved in the regulation of glucose level in blood.

22.3.6 PHYTOREMEDIATION

S. sclarea plant can be grown in contaminated soil containing heavy metals. These heavy metals, such as Pb, Zn, Cd, and Cu accumulate in the above-ground part like the vegetative organ (stem, leaves, and the inflorescence) of the sage plant through the root system (Angelova et al., 2016).

22.3.7 INSECTICIDAL ACTIVITY

Plant synthesizes several secondary metabolites that are beneficial in plant–insect interaction. In a study by Šućur et al. (2015, 2016), it was revealed that the aqueous extract of *S. sclarea* possesses toxic effects on the growth of *Rhyzopertha dominicana* (lesser grain borer) with a mortality rate of 95% and above and *Trialeurodes vaporariorum* (whitefly) with a mortality rate of 56.66%. Therefore, the extract of this plant might be beneficial as insecticidal products.

22.3.8 ANTIOXIDANT ACTIVITY

Phenolic compounds are basically responsible for antioxidant properties that are being synthesized in the sage plant. Several phenolic compounds and phenolic acids found in the sage plant, such as a-Tocopherol, quercetin, butylated hydroxyanisole, and butylated hydroxytoluene. Chloroform extract of *S. sclarea* constitutes stronger antioxidant activity and inhibits the linoleic acid peroxidation (Gülçin et al., 2004). Hydroxyl group in phenolic compounds is responsible for the antioxidant properties. In addition, phenolic compounds and their activity via DPPH were determined spectrophotometrically and the methanolic extract of *S. sclarea* shows the highest antiradical activity (Jasicka-Misiak et al., 2018).

22.3.9 ANTIANXIETY ACTIVITY

S. sclarea is an important medicinal plant that reduces anxiety levels by reducing the blood corticosone levels. Therefore, the essential oil of this plant might possess anxiolytic effects (Gross et al., 2013). Application of essential oil of *S. sclarea* might reduce the oxidative stress and increases the production of nitric oxide (NO), and endothelial NO synthase results in the recovery of endothelial tissues in rats. Therefore, this plant might be used in decreasing stress related to cardiovascular diseases (Yang et al., 2014).

KEYWORDS

- ***Salvia sclarea***
- **antimicrobial**
- **essential oil**
- **diterpenoids**
- **linalool**
- **linalyl acetate**

REFERENCES

Abdelkader, M.; Ahcen, B.; Rachid, D.; Hakim, H. Phytochemical Study and Biological Activity of Sage (*Salvia officinalis* L.). *Intern. J. Biol. Biomol. Agric. Food Biotech. Eng.* **2014,** *8* (11), 1231–1235.

Aćimović, M.; Kiprovski, B.; Rat, M.; Sikora, V.; Popović, V.; Koren, A.; Brdar-Jokanović, M. *Salvia sclarea*: Chemical Composition and Biological Activity. *J. Agron.* **2018,** *21*, 18–28.

Angelova, V. R.; Ivanova, R. V.; Todorov, G. M.; Ivanov, K. I. Potential of *Salvia sclarea* L. for Phytoremediation of Soils Contaminated with Heavy Metals. *Intern. J. Agric. Biosyst. Eng.* **2016,** *10* (12), 780–790.

Caniard, A.; Zerbe, P.; Legrand, S.; Cohade, A.; Valot, N.; Magnard, J. L. et al. Discovery and Functional Characterization of Two Diterpene Synthases for Sclareol Biosynthesis in *Salvia sclarea* (L.) and Their Relevance for Perfume Manufacture. *BMC Plant Biol.* **2012,** *12* (1), 1–13.

Coisin, M.; Necula, R.; Grigoras, V.; Gille, E.; Rosenhech, E.; Zamfirache, M. M. Phytochemical Evaluation of Some *Salvia* Species from Romanian flora. *Analele Stiintifice ale Universitatii "Al. I. Cuza" din Iasi* **2012,** *58* (1), 35.

Cui, H.; Zhang, X.; Zhou, H.; Zhao, C.; Lin, L. Antimicrobial Activity and Mechanisms of *Salvia sclarea* Essential Oil. *Botan. Stud.* **2015,** *56* (1), 1–8.

Gross, M.; Nesher, E.; Tokhonov, T.; Raz, O.; Pinhasov, A. Chronic Food Administration of *Salvia sclarea* Oil Reduces Animals' Anxious and Dominant Behavior. *J. Med. Food* **2013,** *16*, 216–222.

Gülçin, I.; Uğuz, M. T.; Oktay, M.; Beydemir, Ş.; Küfrevioğlu, Ö. İ. Evaluation of the Antioxidant and Antimicrobial Activities of Clary Sage (*Salvia sclarea* L.). *Turk. J. Agric. Forest.* **2004,** *28* (1), 25–33.

Jasicka-Misiak, I.; Poliwoda, A.; Petecka, M.; Buslovych, O.; Shlyapnikov, V. A.; Wieczorek, P. P. Antioxidant Phenolic Compounds in *Salvia officinalis* L. and *Salvia sclarea* L. *Ecol. Chem. Eng.* **2018,** *25* (1), 133–142.

Kostić, M.; Kitić, D.; Petrović, M. B.; Jevtović-Stoimenov, T.; Jović, M.; Petrović, A.; Živanović, S. Anti-Inflammatory Effect of the *Salvia sclarea* L. Ethanolic Extract on Lipopolysaccharide-Induced Periodontitis in Rats. *J. Ethnopharmacol.* **2017,** *199*, 52–59.

Küçük, S.; Soyer, P.; Tunalı, Y. Determination of Antimicrobial and Biological Activities of *Salvia sclarea* L. (Lamiaceae) Extracts. *J. Turk. Chem. Soc.* **2019,** *6* (1), 15–20.

Kuźma, Ł.; Skrzypek, Z.; Wysokińska, H. Diterpenoids and Triterpenoids in Hairy Roots of *Salvia sclarea. Plant Cell Tiss. Org. Cult.* **2006,** *84* (2), 171–179.

Peana, A. T.; Moretti, M. D. Pharmacological Activities and Applications of *Salvia sclarea* and *Salvia desoleana* Essential Oils. *Stud. Nat. Prod. Chem.* **2002,** *26*, 391–423.

Raafat, K.; Habib, J. Phytochemical Compositions and Antidiabetic Potentials of *Salvia sclarea* L. Essential Oils. *J. Oleo Sci.* **2018,** ess17187.

Różalski, M.; Kuźma, Ł.; Wysokińska, H.; Krajewska, U. Cytotoxic and Proapoptotic Activity of Diterpenoids from In Vitro Cultivated *Salvia sclarea* Roots. Studies on the Leukemia Cell Lines. *Zeitschrift für Naturforschung C* **2006,** *61* (7–8), 483–488.

Seol, G. H.; Shim, H. S.; Kim, P. J.; Moon, H. K.; Lee, K. H.; Shim, I. et al. Antidepressant-Like Effect of *Salvia sclarea* Is Explained by Modulation of Dopamine Activities in Rats. *J. Ethnopharmacol.* **2010,** *130* (1), 187–190.

Sharopov, F. S.; Setzer, W. N. The Essential Oil of *Salvia sclarea* L. from Tajikistan. *Records Nat. Prod.* **2012,** *6* (1), 75–79.

Sienkiewicz, M.; Głowacka, A.; Poznańska-Kurowska, K.; Kaszuba, A.; Urbaniak, A.; Kowalczyk, E. The Effect of Clary Sage Oil on Staphylococci Responsible for Wound Infections. *Postępy Dermatol. Alergol.* **2015,** *32* (1), 21–26.

Šućur, J., Popović, A., Petrović, M., Anačkov, G., Malenčić, Đ.; Prvulović, D. Allelopathic Effects and Insecticidal Activity of *Salvia sclarea* L. *Studia UBB Chemia* **2015,** *60*, 253–264.

Šućur, J.; Prvulović, D.; Anačkov, G.; Malenčić, Đ. Oxidative Stress in Bromus (*Bromus mollis* L.) Seedlings Treated with Clary Sage (*Salvia sclarea* L.) Aqueous Extract. *Biologica Nyssana* **2016,** *7*, 141–144.

Vaccaro, M. C.; Mariaevelina, A.; Malafronte, N.; De Tommasi, N.; Leone, A. Increasing the Synthesis of Bioactive Abietane Diterpenes in *Salvia sclarea* Hairy Roots by Elicited Transcriptional Reprogramming. *Plant Cell Rep.* **2017,** *36* (2), 375–386.

Yang, H. J.; Kim, K. Y.; Kang, P.; Lee, H. S.; Seol, G. H. Effects of *Salvia sclarea* on Chronic Immobilization Stress Induced Endothelial Dysfunction in Rats. *BMC Compl. Altern. Med.* **2014,** *14*, 396.

Zengin, G.; Senkardes, I.; Mollica, A.; Picot-Allain, C. M. N.; Bulut, G.; Dogan, A.; Mahomoodally, M. F. New Insights into the In Vitro Biological Effects, In Silico Docking and Chemical Profile of Clary Sage–*Salvia sclarea* L. *Comput. Biol. Chem.* **2018,** *75*, 111–119.

CHAPTER 23

BIOACTIVE COMPOUNDS AND PHARMACOLOGICAL EFFICACY OF *THYMUS VULGARIS* L.: A TRADITIONAL FOLK MEDICINAL PLANT

DEEPIKA TRIPATHI[1] and DHARAM CHAND ATTRI[2]

[1]*G. B. Pant National Institute of Himalayan Environment, Kosi-Katarmal, Almora, Uttarakhand, India*

[2]*Department of Botany, Central University of Jammu, Rahya-Suchani (Bagla), Jammu, Jammu & Kashmir, India*

ABSTRACT

Thymus vulgaris L. is an important aromatic medicinal plant belonging to the Lamiaceae family. It is an evergreen, perennial, small shrub with woody base. Traditionally, the leaves of the species are extensively used to provide flavor in meat dishes, soups, salads, and herbal tea. It contains diverse group of chemical constituents categorized as phenolic compounds, terpenoids, flavonoids, steroids, alkaloids, tannins, and saponins. Studies have been reported various pharmacological activities of both plant extracts and essential oil. In the present review, bioactive compounds and pharmacological activities such as expectorant, antibroncholitic, antimicrobial, antispasmodic, anthelmintic, antioxidant, anticancer, anti-inflammatory and neuroprotective, etc. are documented.

Bioactives and Pharmacology of Lamiaceae. T. Pullaiah, PhD (Ed.)

23.1 INTRODUCTION

Thymus vulgaris L. is an important aromatic medicinal plant belonging to the Lamiaceae family. It is native to the Mediterranean region of North Africa, Southern Europe, and Asia (Abd Elbar et al., 2019). *T. vulgaris* is an evergreen, perennial, and small shrub with a woody base, which grows up to 30 cm in height by 40 cm in width in coarse and rough soil with 5–8 pH and full sunny condition (Prasanth Reddy et al., 2014). It is a highly cultivated species in the genus *Thymus* for its ornamental, culinary, and medicinal uses (Nieto, 2020; Dauqan and Abdullah, 2017). *T. vulgaris* leaves are extensively used to provide flavor in meat dishes, soups, and salads. Many studies have reported the pharmacological activities of both plant extracts and essential oil of *T. vulgaris* (Dauqan and Abdullah, 2017; Gedikoğlu et al., 2019). As a folk medicine, local people also prepare herbal tea from leaves to relieve headaches, toothaches, colds, arthritis, and asthma. It is also used for various other pharmacological activities such as expectorant, antibroncholitic, antimicrobial, antispasmodic, anthelmintic, antioxidant, anticancer, anti-inflammatory, and neuroprotective. (Hosseinzadeh et al., 2015; Khazdair et al., 2019; Zeng et al., 2020).

23.2 BIOACTIVE COMPOUNDS

Many studies have been carried out for chemical analysis and structural determination of bioactive compounds from *T. vulgaris* (Zambonelli et al., 2004; Aljabeili et al., 2018; Gedikoğlu et al., 2019). It contains diverse group of chemical constituents categorised as phenolic compounds, terpenoids, flavonoids, steroids, alkaloids, tannins, and saponins. However, the medicinal virtues of this plant are mainly depending on phenolic monoterpene compounds thymol (5-methyl-2-propan-2-yl phenol) and carvacrol (2-methyl-5-propan-2-yl phenol) (Khan et al., 2012; Khazdair et al., 2019). Apart from these compounds, various other bioactive compounds like geraniol, linalool, α-and β-pinene, p-cymene, γ-terpinene, zeaxanthin, apigenin, caffeic acid, ferulic acid, and gentisic acid are also present in *T. vulgaris* (Patil et al., 2021). Most of the compounds are volatile that extracted from plant oil. *T. vulgaris* is highly used in pharmaceutical dosage forms because it contains more essential oil than other species of *Thymus* with a high amount of the main bioactive compound thymol (51.34%). In the category of volatile components, 56.53% oxygenated monoterpenes, 28.69% monoterpene hydrocarbons, 5.04% sesquiterpene hydrocarbons and 1.84% oxygenated sesquiterpenes have been reported (Prasanth Reddy et al.,

2014). In 2018, Aljabeili et al. estimated 1,8-cineole (14.26%), γ-terpinene (12.06%), p-cymene (10.50%), α-terpinene (9.22%), linalool (2.80%), and carvacrol (2.77%) in essential oil of *T. vulgaris* as major compounds. The chemical structures of some important bioactive compounds of *T. vulgaris* are presented in Figure 23.1.

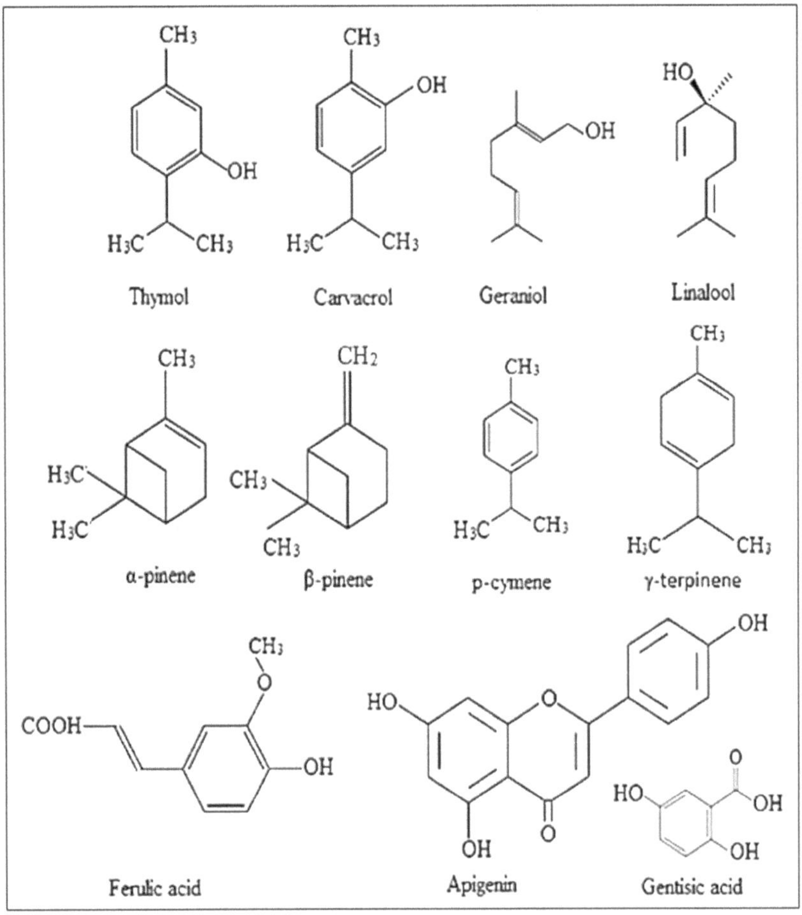

FIGURE 23.1 Chemical structure of important bioactive components of *T. vulgaris* L.

23.3 PHARMACOLOGY

23.3.1 ANTIBACTERIAL ACTIVITY

Many studies have previously reported the antibacterial activity of *T. vulgaris* (Fani and Kohanteb, 2017; Aljabeili et al., 2018). Both plant extract and essential oil of *T. vulgaris* have been tested for their antibacterial activity against food-borne pathogenic bacteria. However, essential oil of *T. vulgaris* showed excellent antibacterial response against *Bacillus cereus*, *Staphylococcus aureus*, *Staphylococcus epidermidis*, *Salmonella enteritidis*, and *Salmonella typhimurium*, while plant extract of *T. vulgaris* was not showing any antibacterial activity (Gedikoğlu et al., 2019). The antimicrobial activity of essential oil is related to the presence of bioactive compounds thymol, carvacrol, and γ-terpinene (Dauqan and Abdullah, 2017). Another study by Aljabeili et al. (2018) showed the high antibacterial potential of essential oil of *T. vulgaris* against pathogenic bacterial strains in comparison to many standard antibiotics. Further, the antibacterial activity of the essential oil of *T. vulgaris* was also observed to suppress the growth of multidrug resistant (MDR) Enterobacteriaceae isolates (Benameur et al., 2019). Therefore, *T. vulgaris* can be used as a promising source of the therapeutic agent in the prevention of bacterial contamination.

23.3.2 ANTIFUNGAL ACTIVITY

The significant antifungal activity of *T. vulgaris* is well-known against pathogenic fungi. In 2018, Diánez et al. estimated the antifungal potential of *T. vulgaris* essential oil against eight pathogenic fungi (*Botrytis cinerea*, *Sclerotinia sclerotiorum*, *Fusarium oxysporum*, *Phytophthora parasitica*, *Pythium aphanidermatum*, *Alternaria brassicae*, *Cladobotryum mycophilum*, and *Trichoderma aggressivum* f.sp. *europaeum*) using the disc–diffusion method. However, the essential oil of *Thymus* only showed antifungal activity for *S. sclerotiorum*, *F. oxysporum*, *P. parasitica*, *C. mycophilum*, and *A. brassicae*. Gucwa et al. (2018) also assessed the antifungal activity of *T. vulgaris* essential oil with a minimum inhibitory concentration (MIC) value 5.731 mg mL^{-1} against clinical isolates of *Candida albicans* and *C. glabrata*. Similarly, Fani and Kohanteb (2017) also observed the antifungal activity of thyme oil against the oral pathogen *C. albicans*. The antifungal and antiaflatoxigenic activity of *T. vulgaris* essential oil against *Aspergillus flavus* has also been evaluated by Tian et al. (2019). They observed the suppression of fungus growth with the downregulation of development-related genes (*brlA*, *abaA*, and *wetA*) and aflatoxin biosynthesis-related genes (*aflR*, *aflD*, and *aflK*) expression.

23.3.3 NEUROPROTECTIVE ACTIVITY

The use of *T. vulgaris* essential oil is suggested for neurodegenerative disorders. Recently, the neuroprotective activity of the bioactive compound thymol (isolated from *T. vulgaris*) has been observed on a chronic unpredictable mild stress (CUMS) model of depression in mice. This study showed that thymol significantly restored the CUMS-induced alterations of monoamine neurotransmitters and negatively regulated the induction of proinflammatory cytokines including interleukin (IL)-1β and IL-6 in CUMS mice (Deng et al., 2015). Previous studies also suggested the neuroprotective effects of thymol can attribute to its potential activity on GABA-mediated inhibition of synaptic transmission (Khazdair et al., 2019). In 2017, Sammi et al. found that thyme oil augments neurotransmission by modulating synaptic acetylcholine levels and nicotinic acetylcholine receptor activity. On this basis, they recommended thyme oil and its constituents, especially para-cymene as a therapeutic agent for the amelioration of cholinergic dysfunction.

23.3.4 ANTIOXIDANT ACTIVITY

Antioxidative activities of *T. vulgaris* have been earlier reported in many studies (Hosseinzadeh et al., 2015; Aljabeili et al., 2018). The antioxidative potential of the aqueous extract of *T. vulgaris* different plant parts (leaf and stem) have been analyzed with 2,2-diphenyl-1-picrylhydrazyl (DPPH) assay. The obtained results showed the dose-dependent free radical scavenging activity of plant extracts (with 92.0 ± 0.5% scavenging activity at a concentration of 1.5 mg mL^{-1}) in accordance with standard antioxidants butylated hydroxyanisole (BHA, 95.7 ± 0.3%) and butylated hydroxytoluene (BHT, 96.6 ± 0.4%) (Banerjee et al., 2019). They also suggested that antioxidative activity may be attributed due to the presence of polysaccharides in *T. vulgaris* leaves such as starch, homogalacturonan, rhamnogalacturonan I (RG-I), and cellulose. In 2019, Gedikoğlu et al. also estimated the antioxidant activity of *T. vulgaris* in different solvent extracts. It was found that the DPPH free radical scavenging activity (IC_{50} value) of the methanol extract (29.22 ± 0.385 μg mL^{-1}) was significantly higher than the ethanol extract (36.77 ± 0.45 μg mL^{-1}). This study also showed the antioxidative potential of essential oil of *T. vulgaris* extracted by two different methods—hydrodistillation and microwave-assisted extraction in which the antioxidant activity was found maximum in microwave-assisted extracted oil than the oil obtained with hydrodistillation method. Abdel–Gabbar et al. (2019) assessed the in vivo activity of antioxidative enzymes such as catalase, glutathione (GSH), glutathione-S-transferase, and superoxide dismutase with the application of an aqueous extract of *T. vulgaris* on the experimental rabbits, where they recorded the increased levels of these enzymes by 14.12, 27.69, 98.75, and 78.29%, respectively.

23.3.5 ANTIVIRAL ACTIVITY

Significant antiviral activity of *T. vulgaris* has been investigated. For example, the inhibitory activity of *T. vulgaris* aqueous extract for herpes simplex virus type 1 (HSV-1) and herpes simplex virus type 2 (HSV-2) has been observed with IC_{50} concentration 0.065 μg mL^{-1} and 0.077 μg mL^{-1}, respectively (Nolkemper et al., 2006). Further, Santoyo et al. (2014) analyzed the antiviral activity of the supercritical CO_2 extract of *T. vulgaris* against HSV-1, where observations indicated a 50% reduction in pre-treated virus infectivity with the *T. vulgaris* extract. In, 2017 Kaewprom et al. investigated the antiviral activity of *T. vulgaris* hydrosol against the Porcine reproductive and respiratory syndrome (PRRS) virus. The effects of *T. vulgaris* essential oil on trans-activator of transcription (Tat)-induced human immunodeficiency virus type-1 (HIV-1) LTR transcription activity has also been evaluated, where *T. vulgaris* [50% Reduction of Transcription (RT_{50}) 0.83 μg mL^{-1}] showed 52% inhibitory potentials on viral transcription with the application of 0.30 μg mL^{-1} concentration (Feriotto et al., 2018).

23.3.6 ANTI-INFLAMMATORY ACTIVITY

Many studies showed the anti-inflammatory activity of *T. vulgaris* plant extracts and essential oil (Ocaña and Reglero, 2012; de Oliveira et al., 2017; Habashy et al., 2018; Kindl et al., 2019). Human macrophages derived from THP-1 monocytes incubated with the *T. vulgaris* fraction oil showed a significant reduction in production and gene expression of the proinflammatory cytokines [tumor necrosis factor (TNF-α), IL-1B, and IL-6] and enhancement in the anti-inflammatory cytokine IL-10 (Ocaña and Reglero, 2012). In, 2017 de Oliveira et al. also reported the anti-inflammatory activity of *T. vulgaris* leaves extract in RAW 264.7 cells exposed to oligosaccharides (LPS, 1 μg mL^{-1}) with an increase in IL-1β levels along with significant inhibition of TNF-α. Methanol extract of *T. vulgaris* also showed anti-inflammatory activity with the potential to inhibit Src tyrosine kinase activity and reduced the production of proinflammatory cytokine interleukin-6 in Balb/C mice splenocytes (Kindl et al., 2019). Fachini-Queiroz et al. (2012) analyzed the in vivo anti-inflammatory activity of *T. vulgaris* essential oil and their bioactive constituent thymol and carvacrol. However, results showed the antagonistic effect of thymol and carvacrol and suggested the key role of carvacrol in the anti-inflammatory activity of *T. vulgaris* essential oil.

23.3.7 HEPATOPROTECTIVE ACTIVITY

The hepatoprotective activity of *T. vulgaris* has earlier been investigated in several studies (Grespan et al., 2014; El-Newary et al., 2017; Soliman et al., 2021). *T. vulgaris* essential oil showed its hepatoprotective activity in acetaminophen-induced hepatic damaged mice by reducing serum marker enzymes aspartate aminotransferase (AST), alanine aminotransferase (ALT), and alkaline phosphatase (ALP) (Grespan et al., 2014). El-Sayed and Ramadan (2017) observed the hepatoprotective activity of plant ethanolic extract against *Toxoplasma gondii* infection by improving the pathological lesions, mitigating the altered liver function, and decreasing the genotoxicity. In another study, Sobhy et al. (2020) demonstrated the elevation in catalase and glutathione activities with low MDA content in experimental model rats (suffering from obesity). They elucidated the role of *T. vulgaris* plant extract and essential oil as a hepatoprotective agent in preventing liver tissue damage. Recently, Soliman et al. (2021) also suggested the hepatoprotective activity of *T. vulgaris* leaf extract in sodium nitrite induced oxidative stressed mice by analyzing the protective activity of plant extract via regulation of antioxidant system enzymes (SOD, GSH, and thioredoxin), serum enzymes (ALT and AST) and pro-inflammatory cytokine (IL-1β, IL-6, and TNF-α).

23.3.8 ANTICANCER ACTIVITY

The anticancer activity of *T. vulgaris* has previously been well documented. Many studies have explained the role of both plant extract and essential oil of *T. vulgaris* as promising herbal agents for chemoprevention and chemotherapy of human cancers (Al-Menhali et al., 2015; Kubatka et al., 2019). Cytotoxicity of essential oil of this plant toward UMSCC1 (head and neck squamous cell carcinoma cell lines) has been recorded with 369 $\mu g\ mL^{-1}$

IC_{50} (Sertel et al., 2011). Kubatka et al. (2019) observed the in vivo anticancer activity of *T. vulgaris* in the treated mammary carcinoma rat model with CD44 and ALDH1A1 downregulation and Bax upregulation. They also investigated the in vitro antiproliferative and proapoptotic effects of *Thymus* essential oils in MCF-7 and MDA-MB-231 cell lines. Zeng et al. (2020) reported the anticancer activity of thymol extracted from *T. vulgaris* and observed apoptosis by activation of the BAX/Bcl-2 signaling pathway and cell cycle arrest in colorectal cancer. They also revealed thymol-mediated antiproliferative and antimetastatic effects in colorectal cancer through inhibition of Wnt/β-catenin pathway activation. In 2018, Heidari et al. synthesized the silver nanoparticles using *T. vulgaris* and analyzed their anticancer activity in comparison to plant ethanolic extract against human breast cancer T47D cell line. This study revealed that *T. vulgaris* biosynthesized nanoparticles had high cytotoxic potential than the plant extract.

23.3.9 INSECTICIDAL ACTIVITY

T. vulgaris is well-reputed plant for its insecticidal activity. In 2018, Szczepanik et al. evaluated the comparative insecticidal activity of major bioactive compounds (thymol and carvacrol) and essential oil of *T. vulgaris* against larvae of *Alphitobius diaperinus* Panzer. This study suggested that thymol and carvacrol compounds having maximum insecticidal activity against *A. diaperinus* larvae than the Thyme essential oil. Similarly, Park et al. (2017) observed the insecticidal activity of *T. vulgaris* essential oil against *Pochazia shantungensis* (hemipterous insect) with 50% lethal concentration (LC_{50}) 75.80 mg L^{-1} after 72 h treatment. In 2018, Pavela and Sedlák also reported the insecticidal activity of essential of Thyme for larvae of *Spodoptera littoralis* (tobacco cutworm) and *Culex quinquefasciatus*.

23.3.10 ANTHELMINTIC POTENTIAL

In 2016, Ferreira et al. observed the anthelmintic activity of *T. vulgaris* essential oil against different life cycle stages of *Haemonchus contortus* (one of the most important parasite species of sheep) and validated that this activity of *T. vulgaris* attributed due to the presence of thymol.

KEYWORDS

- **bioactive compounds**
- **pharmacology**
- **medicinal plants**
- **folk medicine**

REFERENCES

Abd Elbar, O. H.; Farag, R. E.; Shehata, S. A. Effect of Putrescine Application on Some Growth, Biochemical and Anatomical Characteristics of *Thymus vulgaris* L. Under Drought Stress. *Ann. Agric. Sci.* **2019,** *64* (2), 129–137.

Abdel-Gabbar, M.; Ahmed, R. R.; Kandeil, M. A.; Mohamed, A. E. D. H.; Ali, S. M. Administration of Ginger and/or Thyme Has Ameliorative Effects on Liver and Kidney Functions of V-Line Rabbits: Histological and Biochemical Studies. *J. Anim. Physiol. Anim. Nutr.* **2019,** *103* (6), 1758–1767.

Aljabeili, H. S.; Barakat, H.; Abdel-Rahman, H. A. Chemical Composition, Antibacterial and Antioxidant Activities of Thyme Essential Oil (*Thymus vulgaris*). *Food Nutr. Sci.* **2018,** *9* (05), 433.

Al-Menhali, A.; Al-Rumaihi, A.; Al-Mohammed, H.; Al-Mazrooey, H.; Al-Shamlan, M.; AlJassim, M.; Eid, A. H. *Thymus vulgaris* (Thyme) Inhibits Proliferation, Adhesion, Migration, and Invasion of Human Colorectal Cancer Cells. *J. Med. Food.* **2015,** *18* (1), 54–59.

Banerjee, P.; Mukherjee, S.; Bera, K.; Ghosh, K.; Ali, I.; Khawas, S.; Ray, S. Polysaccharides from Thymus vulgaris Leaf: Structural Features, Antioxidant Activity and Interaction with Bovine Serum Albumin. *Intern. J. Biol. Macromol.* **2019,** *125*, 580–587.

Benameur, Q.; Gervasi, T.; Pellizzeri, V.; Pl'uchtová, M.; Tali-Maama, H.; Assaous, F.; Ben-Mahdi, M. H. Antibacterial Activity of *Thymus vulgaris* Essential Oil Alone and in Combination with Cefotaxime Against Bla ESBL Producing Multidrug Resistant Enterobacteriaceae Isolates. *Na. Prod. Res.* **2019,** *33* (18), 2647–2654.

Dauqan, E. M.; Abdullah, A. Medicinal and Functional Values of Thyme (*Thymus vulgaris* L.) Herb. *J. Appl. Biol. Biotechnol..* **2017,** *5* (2), 17–22.

de Oliveira, J. R.; de Jesus Viegas, D.; Martins, A. P. R.; Carvalho, C. A. T.; Soares, C. P.; Camargo, S. E. A.; de Oliveira, L. D. *Thymus vulgaris* L. Extract Has Antimicrobial and Anti-Inflammatory Effects in the Absence of Cytotoxicity and Genotoxicity. *Arch. Oral Biol.* **2017,** *82*, 271–279.

Deng, X. Y.; Li, H. Y.; Chen, J. J.; Li, R. P.; Qu, R.; Fu, Q.; Ma, S. P. Thymol Produces an Antidepressant-Like Effect in a Chronic Unpredictable Mild Stress Model of Depression in Mice. *Behavioural Brain Res.* **2015,** *29* (1), 12–19.

Diánez, F.; Santos, M.; Parra, C.; Navarro, M. J.; Blanco, R.; Gea, F. J. Screening of Antifungal Activity of 12 Essential Oils Against Eight Pathogenic Fungi of Vegetables and Mushroom. *Letters Appl. Microbiol.* **2018,** *67* (4), 400–410.

El-Newary, S. A.; Shaffie, N. M.; Omer, E. A. The Protection of *Thymus vulgaris* Leaves Alcoholic Extract Against Hepatotoxicity of Alcohol in Rats. *Asian Pacific J. Trop. Med.* **2017,** *10* (4), 361–371.

El-Sayed, N. M.; Ramadan, M. E. Hepatoprotective Activity of *Thymus vulgaris* Extract Against *Toxoplasma gondii* Infection. *Asian Pacific J. Trop. Dis.* **2017,** *7* (5), 280–285.

Fachini-Queiroz, F. C.; Kummer, R.; Estevao-Silva, C F.; Carvalho, M. D. D. B., Cunha, J. M.; Grespan, R.; Cuman, R. K. N. Effects of Thymol and Carvacrol, Constituents of *Thymus vulgaris* L. Essential Oil, on the Inflammatory Response. *Evid-Based Complem. Altern. Med.* **2012,** *2012*, 657026. DOI: 10.1155/2012/657026.

Fani, M.; Kohanteb, J. In Vitro Antimicrobial Activity of *Thymus vulgaris* Essential Oil Against Major Oral Pathogens. *J. Evid-Based Complem. Altern. Med.* **2017,** *22* (4), 660–666.

Feriotto, G.; Marchetti, N.; Costa, V.; Beninati, S.; Tagliati, F.; Mischiati, C. Chemical Composition of Essential Oils from *Thymus vulgaris*, *Cymbopogon citratus*, and *Rosmarinus officinalis*, and Their Effects on the HIV-1 Tat Protein Function. *Chem. Biodivers.* **2018,** *15* (2), e1700436.

Ferreira, L. E.; Benincasa, B. I.; Fachin, A. L.; Franca, S. C.; Contini, S. S.; Chagas, A. C.; Beleboni, R. O. *Thymus vulgaris* L. Essential Oil and Its Main Component Thymol: Anthelmintic Effects Against *Haemonchus contortus* from Sheep. *Veter. Parasitol.* **2016,** *228*, 70–76.

Gedikoğlu, A.; Sökmen, M.; Çivit, A. Evaluation of *Thymus vulgaris* and *Thymbra spicata* Essential Oils and Plant Extracts for Chemical Composition, Antioxidant, and Antimicrobial Properties. *Food Sci. Nutr.* **2019,** *7* (5), 1704–1714.

Grespan, R.; Aguiar, R. P.; Giubilei, F N.; Fuso, R. R.; Damião, M. J.; Silva, E. L.; Cuman, R. K. N. Hepatoprotective Effect of Pretreatment with *Thymus vulgaris* Essential Oil in Experimental Model of Acetaminophen-Induced Injury. *Evid.-Based Complem. Altern. Med.* **2014,** *2014*, 954136. DOI: 10.1155/2014/954136.

Gucwa, K.; Milewski, S.; Dymerski, T.; Szweda, P. Investigation of the Antifungal Activity and Mode of Action of *Thymus vulgaris*, *Citrus limonum*, *Pelargonium graveolens*, *Cinnamomum cassia*, *Ocimum basilicum*, and *Eugenia caryophyllus* Essential Oils. *Molecules*. **2018,** *23* (5), 1116.

Habashy, N. H.; Serie, M. M. A.; Attia, W. E.; Abdelgaleil, S. A. Chemical Characterization, Antioxidant and Anti-Inflammatory Properties of Greek *Thymus vulgaris* Extracts and Their Possible Synergism with Egyptian *Chlorella vulgaris*. *J. Funct. Foods*. **2018,** *40*, 317–328.

Heidari, Z.; Salehzadeh, A.; Shandiz, S. A. S.; Tajdoost, S. Anti-Cancer and Anti-Oxidant Properties of Ethanolic Leaf Extract of *Thymus vulgaris* and Its Bio-Functionalized Silver Nanoparticles. *3 Biotech*. **2018,** *8* (3), 1–14.

Hosseinzadeh, S.; Jafarikukhdan, A.; Hosseini, A.; Armand, R. The Application of Medicinal Plants in Traditional and Modern Medicine: A Review of *Thymus vulgaris*. *Intern. J. Clin. Med.* **2015,** *6* (09), 635.

Kaewprom, K.; Chen, Y. H.; Lin, C. F.; Chiou, M. T.; Lin, C. N. Antiviral Activity of *Thymus vulgaris* and *Nepeta cataria* Hydrosols Against Porcine Reproductive and Respiratory Syndrome Virus. *Thai J. Vet. Med.* **2017,** *47* (1), 25.

Khan, R. U.; Naz, S.; Nikousefat, Z.; Tufarelli, V.; Laudadio, V. *Thymus vulgaris*: Alternative to Antibiotics in Poultry Feed. *World's Poultry Sci. J.* **2012,** *68* (3), 401–408.

Khazdair, M. R.; Anaeigoudari, A.; Hashemzehi, M.; Mohebbati, R. Neuroprotective Potency of Some Spice Herbs, a Literature Review. *J. Trad. Complem. Med.* **2019,** *9* (2), 98–105.

Kindl, M.; Bucar, F.; Jelić, D.; Brajša, K.; Blažeković, B.; Vladimir-Knežević, S. Comparative Study of Polyphenolic Composition and Anti-Inflammatory Activity of *Thymus species*. *Eur. Food Res. Technol.* **2019,** *245* (9), 1951–1962.

Kubatka, P.; Uramova, S.; Kello, M.; Kajo, K.; Samec, M.; Jasek, K.; Büsselberg, D. Anticancer Activities of *Thymus vulgaris* L. in Experimental Breast Carcinoma In Vivo and In Vitro. *Intern. J. Mol. Sci.* **2019,** *20* (7), 1749.

Nieto, G. A. Review on Applications and Uses of *Thymus* in the Food Industry. *Plants*. **2020,** *9* (8), 961.

Nolkemper, S.; Reichling, J.; Stintzing, F. C.; Carle, R.; Schnitzler, P. Antiviral Effect of Aqueous Extracts from Species of the Lamiaceae Family Against Herpes Simplex Virus Type 1 and Type 2 In Vitro. *Planta Medica*. **2006,** *72* (15), 1378–1382.

Ocaña, A.; Reglero, G. Effects of Thyme Extract Oils (from *Thymus vulgaris*, *Thymus zygis*, and *Thymus hyemalis*) on Cytokine Production and Gene Expression of oxLDL-Stimulated THP-1-Macrophages. *J. Obes.* **2012,** *012*, 104706. DOI: 10.1155/2012/104706

Park, J. H.; Jeon, Y. J.; Lee, C H.; Chung, N.; Lee, H. S. Insecticidal Toxicities of Carvacrol and Thymol Derived from *Thymus vulgaris* Lin. Against *Pochazia shantungensis* Chou & Lu., Newly Recorded Pest. *Sci. Rep.* **2017,** *7* (1), 1–7.

Patil, S. M.; Ramu, R.; Shirahatti, P. S.; Shivamallu, C.; Amachawadi, R. G. A Systematic Review on Ethnopharmacology, Phytochemistry and Pharmacological Aspects of *Thymus vulgaris* Linn. *Heliyon*. **2021,** *7* (5), e07054.

Pavela, R.; Sedlák, P. Post-Application Temperature as a Factor Influencing the Insecticidal Activity of Essential Oil from *Thymus vulgaris*. *Ind. Crops Prod.* **2018,** *113*, 46–49.

Prasanth Reddy, V.; Ravi Vital, K.; Varsha, P. V.; Satyam, S. Review on *Thymus vulgaris* Traditional Uses and Pharmacological Properties. *Med Aromat Plants*. **2014,** *3* (164), 2167–0412.

Sammi, S. R.; Trivedi, S.; Rath, S. K.; Nagar, A.; Tandon, S.; Kalra, A.; Pandey, R. 1-Methyl-4-propan-2-ylbenzene from *Thymus vulgaris* Attenuates Cholinergic Dysfunction. *Mol. Neurobiol.* **2017,** *54* (7), 5468–5481.

Santoyo, S.; Jaime, L.; García-Risco, M. R.; Lopez-Hazas, M.; Reglero, G. J. I. C. Supercritical Fluid Extraction as an Alternative Process to Obtain Antiviral Agents from Thyme Species. *Indust. Crops Prod.* **2014,** *52*, 475–480.

Sertel, S.; Eichhorn, T.; Plinkert, P. K.; Efferth, T. Cytotoxicity of *Thymus vulgaris* Essential Oil Towards Human Oral Cavity Squamous Cell Carcinoma. *Anticancer Res*. **2011,** *31* (1), 81–87.

Sobhy, H. M.; Hassanen, N. H. M.; Ahmed, M. A. I. Hepatoprotective Activities of Thyme (*Thymus vulgaris* L.) in Rats Suffering from Obesity. *Egyptian J. Chemistry*. **2020,** *63* (12), 5087–5101.

Soliman, M. M.; Aldhahrani, A.; Metwally, M. M. Hepatoprotective effect of *Thymus vulgaris* Extract on Sodium Nitrite-Induced Changes in Oxidative Stress, Antioxidant and Inflammatory Marker Expression. *Sci. Rep*. **2021,** *11* (1), 1–13.

Szczepanik, M.; Zawitowska, B.; Szumny, A. Insecticidal Activities of *Thymus vulgaris* Essential Oil and Its Components (Thymol and Carvacrol) Against Larvae of Lesser Mealworm, *Alphitobius diaperinus* Panzer (Coleoptera: Tenebrionidae). *Allelopathy J.* **2018,** *30* (1), 129–142.

Tian, F.; Lee, S. Y.; Chun, H. S. Comparison of the Antifungal and Antiaflatoxigenic Potential of Liquid and Vapor Phase of *Thymus vulgaris* Essential Oil Against *Aspergillus flavus*. *J. Food Protect*. **2019,** *82* (12), 2044–2048.

Zambonelli, A.; D'Aulerio, A. Z.; Severi, A.; Benvenuti, S.; Maggi, L.; Bianchi, A. (Chemical Composition and Fungicidal Activity of Commercial Essential Oils of *Thymus vulgaris* L. *J. Essent. Oil Res.* **2004,** *16* (1), 69–74.

Zeng, Q.; Che, Y.; Zhang, Y.; Chen, M.; Guo, Q.; Zhang, W. Thymol Isolated from *Thymus vulgaris* L. Inhibits Colorectal Cancer Cell Growth and Metastasis by Suppressing the Wnt/β-Catenin Pathway. *Drug Design Dev. Therap.* **2020,** *14*, 2535.

CHAPTER 24

VITEX NEGUNDO L., AN INDIGENOUS PLANT: A SYSTEMATIC REVIEW ON TRADITIONAL USE, BIOACTIVES, AND PHARMACOLOGICAL ACTIVITIES

SAMRAT GOSWAMI and BISWAJIT ROY

Taxonomy and Biosystematics Laboratory, Post Graduate Department of Botany, Ramakrishna Mission Vivekananda Centenary College, Rahara, Kolkata, West Bengal, India

ABSTRACT

Vitex negundo L. (Lamiaceae) is an aromatic, deciduous, small tree, native to South East Asia, along with Eastern Africa and Madagascar. This plant has been stated medicinally important plant in ancient literatures like Indian Ayurveda, Arabic Unani, and traditional Chinese medicines. Study of *V. negundo* revealed presence of many important bioactive compounds like coumarins, essential oils, flavonoids, lignans, steroids, terpenoids, and organic acids where flavonoids and terpenoids were found predominantly. Among the terpenoids the most abundant were triterpenoids with 18 oleanane, 6 ursane, 2 norursane, 3 lupine, and 9-epi-cucurbitane-type terpenoids. According to ancient literatures, leaf and root can be used in treatment of excessive vaginal discharges, pruritus, helminthiasis, rheumatism, oedema, skin diseases, and puerperal fever. In Unani medicine, seeds can cure spermatorrhoea and promote spermiogenesis. Extracts of various parts of *V. negundo* like leaves, stem bark, fruit, seed and roots were reported to possess analgesic, anti-inflammatory, antimicrobial, anticancer, antioxidant and cytotoxic activities. Moreover, effect on reproductive potential, enzyme

Bioactives and Pharmacology of Lamiaceae. T. Pullaiah, PhD (Ed.)

inhibitory activity, hepatoprotective activity, histomorphological, cytotoxic effects, immuno-stimulant activity, insecticidal activity, snake venom antagonistic activity and wound healing activity were also noticed.

24.1 INTRODUCTION

Vitex negundo L., belonging to the family Lamiaceae, is a medicinally important plant which prefers to grow in humid places and has been native to South East Asia, parts of Central Asia, West Asia along with Eastern Africa and Madagascar (POWO, 2019). It was introduced in parts of Europe, North America, and West Indies.

There were reports of more than 238 vernacular names in more than 9 languages (NMPB; FRLHT, *2010*). It is commonly known as Five-Leafed Chaste Tree, Chasteberry, Chaste tree, Monk's pepper, and Cloister pepper (Karunamoorthi et al., 2008). It is known by in different states of India, by various vernacular names, Tella vavili in Andhra Pradesh (Basavaraju et al., 2009), Pachatia (Purkayastha et al., 2005), Pasatiya (Saikia et al., 2006) and Pochatia (Kotoky and Das, 2008) in Assam, Sambhaalu (Kirtikar and Basu, 1984) and Bana (Sharma et al., 2004) in Himachal Pradesh, Nekki (Bhandary et al., 1995), Lakki, and Karilakki (Hebbar et al., 2004) in Karnataka, Lingur (Kirtikar and Basu, 1984), Samalu (Jagtap et al., 2006) and Nirgundi (Kosalge and Fursule, 2009) in Maharashtra, Languni in Orissa (Pattanaik et al.,2008), Notchi (Ignacimuthu et al., 2006) and Nochi (Samy et al., 2008) in Tamil Nadu, Somi (Jain and Puri, 1984) and Mala (Ali, 1999) in Uttar Pradesh, Nisinda and Nirgundi in West Bengal (Ahuja et al., 2015). Some trade names for leaves are Karinochiyila pacha, Lakki soppu, Nigodpan (Nirgundipan), Nirgudi paan, and Nochi. Roots and seeds have trade names Karinochi and Nirgudi beej, respectively (*FRLHT, 2016). Agnus-castus negundo* (L.) Carrière and *Vitex paniculata* Lam. are considered as widely accepted synonyms for this plant species (*FRLHT, 2016;* POWO, 2019).

V. negundo is an aromatic, deciduous, shrub or small tree with dense silvery tomentose, obtuse quadrangular branchlets. Leaves decussate, palmately compound with three to five leaflets; leaflets narrowly oblong or elliptic to lanceolate, cuneate to short attenuate at base, apex accuminate, glabrous, almost entire to sinuate or with a few teeth, terminal leaflets are longer than the lateral ones; petioles long, hairy. Inflorescence is usually an axillary or terminal panicle with opposite or subverticillate cymose branches. Flowers are small, scented, hermaphrodite, zygomorphic, and bluish–purple in color. Fruit is drupe, globose, four-seeded, and black when ripe. Seeds are provided with an enlarged calyx at the base. Flowering and fruiting time throughout the year (Paria and Chattopadhyay, 2005; Karunamoorthi et al., 2008; Meena et al., 2011; POWO, 2019).

The medicinal importance of *V. negundo* has been mentioned in ancient literature as Indian Ayurveda, Arabic Unani, and traditional Chinese medicine. The various folk medicine of *V. negundo* included the uses of leaves, flowers, roots, and stem bark. The mixture of *Piper nigrum* and *V. negundo* leaf decoction was effective against catarrhal fever with the heaviness of the head and dull hearing. Headaches, neck gland sores, tubercular neck swellings, and sinusitis can be cured by applying a poultice of the crushed leaf. Venereal diseases and other syphilitic skin disorders were treated with leaf essential oils. Root and stem bark tincture relieves rheumatism and bladder irritability (Tirtha, 1998). In Ayurvedic medicine, it acts as a rejuvenating tonic for retarding old age and retains and promotes virility. According to The Ayurvedic Pharmacopoeia of India, juice of leaf and root can be used in the treatment of excessive vaginal discharges, pruritus, helminthiasis, rheumatism, edema, skin diseases, and puerperal fever. In Unani medicine, seeds can cure spermatorrhea and promote spermiogenesis (Khare, 2004). Leaf infusions were used to treat jaundice (Purkayastha et al., 2005; Pattanaik et al., 2008). Crushed leaves were applied on the skin and juice was orally taken in treatment of urticaria, however, crushed leaves and root paste were applied on the infected place and on boils in case of cellulitis and abscesses, respectively. Juice of crushed leaves and roots was mixed with *Sesamum indicum* oil, which was applied on carbuncles and its paste on eczema (Saikia et al., 2006). Fresh leaf juice with sugar is taken orally 2–3 times/day for 7–15 days to treat liver disorders (Kotoky and Das, 2008). The use of *V. negundo* against diseases such as kwashiorkor and rheumatism was reported. It can also be used as an expectorant (Kirtikar and Basu, 1984; Vishwanathan and Basavaraju, 2010). Leaves boiled in cow urine were made into a paste and applied on wounds and body aches (Jain and Puri, 1984; Sharma et al., 2004). Leaves are crushed with common salt and placed on the painful teeth to get relief from toothache (Hebbar et al., 2004). Leaf juice with rhizome of *Zingiber officinale* was applied to the nasal cavity to treat migraine (Bhandary et al., 1995). Encephalitis can be cured when legs were warmed in water-boiled extract of *V. negundo* and *Azadirachta indica* leaves (Jagtap et al., 2006). Joint pain can be treated with the juice of fresh leaves (Kosalge and Fursule, 2009). Leaf paste when applied externally acts as a snakebite antidote (Samy et al., 2008). Fresh leaves boiled in water and the vapor was inhaled to cure cough, cold, fever, and headache (Ignacimuthu et al., 2006). In the case of eye pain leaf juice was applied (Jain and Puri, 1984). Fresh leaves with fodder are used as a refrigerant for cattle (Ali, 1999). These applications prioritize the ethnobotanical diversity of this plant (Vishwanathan and Basavaraju, 2010).

This review provides a comprehensive study of the medicinal aspects of *V. negundo* and thus consists latest information on traditional uses, bioactive compounds, and pharmacological activities of the plant.

24.2 BIOACTIVES

Phytochemical screening indicated the presence of coumarins, essential oils, flavonoids, lignans, steroids, terpenoids, and organic acids (Fig. 24.1). The most dominant bioactive substances found in *V. negundo* are flavonoids and terpenoids. Tris (2,4-ditert-butylphenyl) phosphate a coumarin, was extracted from leaves of *V. negundo* (Vinuchakkaravarthy et al., 2011; Zheng et al., 2015). However, *iso*-fraxidin, xanthotoxin, and 5,8-d imethoxypsoralen were isolated from seeds (Zhao et al., 2012; Zheng et al., 2015).

Essential oils isolated from leaves of *V. negundo* contain *p*-cymene, *cis*-ocimene, citronellal, β-caryophyllene, α-guaiene, δ-guaiene, valencene, caryophyllene epoxide, β-curcumene, ethyl-9–hexadecenoate, guaia-3,7-diene, palmitic acid, (E)-nerolidol, humulene epoxide 1, humulene epoxide 2, globulol, α-cadinol, *epi*-α-cadinol, and α-murolol. While *p*-cymene, β-caryophyllene, valencene, α-selinene, germacren-4-ol, caryophyllene epoxide, β-selinene, formic acid, n-heptane, trans-α-bergamotene, (E)-nerolidol and *P*-(1,1-dimethylethyl) toluene were isolated from flowers. Moreover, α-copaene, α-cedrene, α-guaiene, guaia-3,7-diene, α-humulene, aristolene, germacrene D, β-selinene, β-caryophyllene, caryophyllene oxide, palmitolic acid, n-hexadecanoic acid, and acetyl lactylglycerate were obtained from the fruits (Khokra et al., 2008).

Many flavonoids were isolated from leaves of *V. negundo* which include acerosin-5-O-glucoside, 5,3′-hydroxy-6,7,4′-trimethoxy flavones, 5,7,3′-Trihydroxy-6,8,4′-trimethoxy flavones, 5,6,7,8,3′,4′,5′-heptamethoxy flavones, corymbosin, 5-hydroxy-6,7,8,3′,4′-pentamethoxyflavone, 5-hydroxy-6,7,8,4′-tetramethoxy flavones, 3,4,5,7,3′,4′,5′-hexahydroxy-6,8-dimethoxyflavone, and 4,5-dihydroxy-3′,4′-dimethoxyflavone-6-O-rhamnoglucoside (Subramanian and Misra, 1979; Ferdous et al., 1984; Banerji et al., 1988). Vitexicarpin, 3,5,6,7,3′,4′-hexamethoxyflavone, 5,3′-diacetoxy-3,6,7,4′-tetramethoxyflavone, and six acylated derivatives of vitexicarpin, namely, 3′-benzoyloxy-5-hydroxy-3,6,7,4′-tetramethoxyflavone, 5,3′-dibenzoyloxy-3,6,7,4′-tetramethoxyflavone, 5,3′-dipropanoyloxy-3,6,7,4′-tetramethoxyflavone, 5,3′-dibutanoyloxy-3,6,7,4′-tetramethoxyflavone, 5,3′-dipent-4-enoyloxy-3,6,7,4′-tetramethoxyflavone, and 5,3′-dihexanoyloxy-3,6,7,4′-tetramethoxyflavone were also extracted (Díaz et al., 2003). However, luteolin, vitegnoside, iso-orientin, 5-hydroxy-3,6,7,3′,4′-pentamethoxy flavones, 5-hydroxy-7,4′-dimethoxy flavones, 5,7-dihydroxy-6,4′-dimethoxy flavones, 7,8-dimethyl herbacetin-3-rhamnoside, 5-hydroxy-3,6,7,3′,4′-pentamethoxy flavones, vitegnoside, luteolin-7-O-β-D-glucoside, catechin, epicatechin, quercetin, myricetin, kaempferol, and naringenin were isolated from the leaves (Sathiamoorthy et al., 2007; Gautam et al., 2008; Sharma et al., 2009; Zargar et al., 2011a; Zheng et al., 2015). 5, 3′-hydroxy-7,8,4′-trimethoxyflavone, 3,6,7,3′,4′-pentamethoxyflavone-5-O-glucopyransylrhamnoside, vitex-incafeate, 3,6,7,3′,4′-pentamethoxy-5-Oglucopyranosyl-rhamnoside, and 4′-O-methylmyricetin-3-O-[4″-O-β-Dgalactosyl]-β-D-galactopyranoside were obtained from stem bark (Misra and Subramanian, 1980; Gautam et al., 2008; Zheng et al., 2015). Five flavonoids, chrysoplenetin, chrysosplenol d, 4, 5-dihydroxy-3, 6, 7-trimethoxyflavone, casticin, and 5, 7-dihydroxychromone were extracted from the seeds (Awale et al., 2011; Zhao et al., 2012; Zheng et al., 2015).

Phytochemical analysis revealed the presence of phenylnaphthalene type lignans as vitrofolal E, vitrofolal F, negundin A, negundin B, (+)-lyoniresinol, (+)-lyoniresinol-3α-O-β-D-glucoside, and (+)-diasyringaresinol, a furanofuran lignin in the roots (Haq et al., 2004, 2006; Zheng et al., 2015). Sixteen phenylnaphthalene-type lignans were isolated from seeds which include vitrofolal E, vitrofolal F, vitedoamine A, vitedoamine B, vitedoin A, vitexin B, vitexdoin A, vitexdoin C, vitexdoin D, vitexdoin G, vitexdoin H, vitexdoin I, detetrahydroconidendrin, 6-hydroxy-4-(4-hydroxy-3-methoxyphenyl)-3-hydroxymethyl-7-methoxy-3,4-dihydro-2-naphthaldehyde, secoisolariciresinoldiglucoside, and matairesinol. Moreover, phenylpropanoids such as coniferyl aldehyde, trans-3,5-dimethoxy-4-hydroxycinnamic aldehyde, 2-methoxy-4-(3-methoxy-1-propenyl)-phenol; phenylindenes such as 2α,3β-7-O-methylcedrusin, vitexdoin F, and furanofuranlignans sich as vitelignin A, 4-oxosesamin, (+)-sesamin, (+)-paulownin, 4-hydroxysesamin, 4,8-dihydroxysesamin, 4-oxopaulownin, (+)-2-(3-methoxy-4-hydroxyphenyl)-6-(3,4-methylenedioxy)phenyl-3,7- dioxabicyclo[3.3.0] octane, and (+)-pinoresinol were also obtained from seeds (Ono et al., 2004; Zhou et al., 2009; Zheng et al., 2009, 2011, 2014, 2015; Zhao et al., 2012).

Leaves of *V. negundo* contained steroids, namely, *β*-sitosterol, stigmasterone, and 22,23-dihydro-*α*-spinasterol-*β*-D-glucoside (Chandramu et al., 2003; Chowdhury et al., 2011). 24ζ -methyl-5α-lanosta-25-one, 7-oxositosterol, stigmast-4-en-6*β*-ol-3-one, ergosterol peroxide, and lanostan-8, 25-dien-3*β*-ol were isolated from the seeds (Chawla et al., 1992; Zhao et al., 2012; Zheng et al., 2015).

A wide range of monoterpenes, sesquiterpenes, diterpenes, and triterpenes have been isolated from different parts of the plant. Monoterpenoids were isolated which were mostly iridoids and their glycosides (Yao et al., 2016). *V. negundo* leaves contain nishindaside, 1,4a,5,7a-tetrahydro-1-β-D-glucosyl-7-(30,40- dihydroxybenzoyloxymethyl)-5-ketocyclopenta[c] pyran-4-carboxylic acid, and negundoside (Sharma et al., 2009). Negundoside was most abundant monoterpenoid in *V. negundo* (Tasduq et al., 2008; Gautam et al., 2008; Sundaram et al., 2012a, b; Yao et al., 2016). Agnuside was detected in the leaves of *V. negundo* (Gautam et al., 2008; Pandey et al., 2012; Arai et al., 2013; Yao et al., 2016). 1, 2-di-substituted idopyranose was isolated from the leaves (Manikandan et al., 2011). Whole plant body extraction confirmed the presence of 6′-O-p-hydroxybenzoylmussaenosidic acid, 2′-O-trans-p-coumaroylloganic acid and 6′-O-p-hydroxybenzoyl-gardoside (Huang et al. 2013a; Yao et al., 2016).

Four sesquiterpenoids named 1,6-dioxo-2(3), 9(10)-dehydrofuranoeremophilane, 4,6-dimethyl-11-formyl-1-oxo-4h, 2,3-dihydronaphthofuran, and 4,6-dimethyl-11-dimethoxymethyl-1-oxo-4h, 2,3-dihydronaphthofuran isolated from the stem (Tiwari et al., 2013; Yao et al., 2016). Negunfurol and 4,6-dimethyl-11-formyl-1-oxo-4H, 2,3-dihydronaphthofuran were also obtained from the seeds (Zheng et al., 2012b; Yao et al., 2016).

A total of 18 diterpenoids were reported which included 8 labdane, 5 norlabdane, and 3 abietane compounds as well as halimane and isopimarane compounds. The two labdanes, namely, nishindanol, 8-*epi*-sclareol and one halimane vitetrifolin D, respectively, were extracted from leaves and stem (Arai et al., 2013; Yao et al., 2016). Labdane-type polyalthic acid was isolated from the fruits (Huang et al., 2013b; Yao et al., 2016). The majority of labdanes type diterpenoids were isolated from seeds which include negundol, negundoin D, negundoin E, vitexilactone B, viteagnusin C, 8-*epi*-sclareol and norlabdanes, namely, negundoin A, negundoin B, negundoin C, negundoal, and vitedoin B. Moreover, abietanes such as 3β-hydroxyabieta-8,11,13-trien-7-one, negundoin F; isopimarane as negundoin G; halimane such as vitetrifolin D was also found from the seeds (Ono et al., 2004; Zheng et al., 2010a, 2012a, b; Yao et al., 2016).

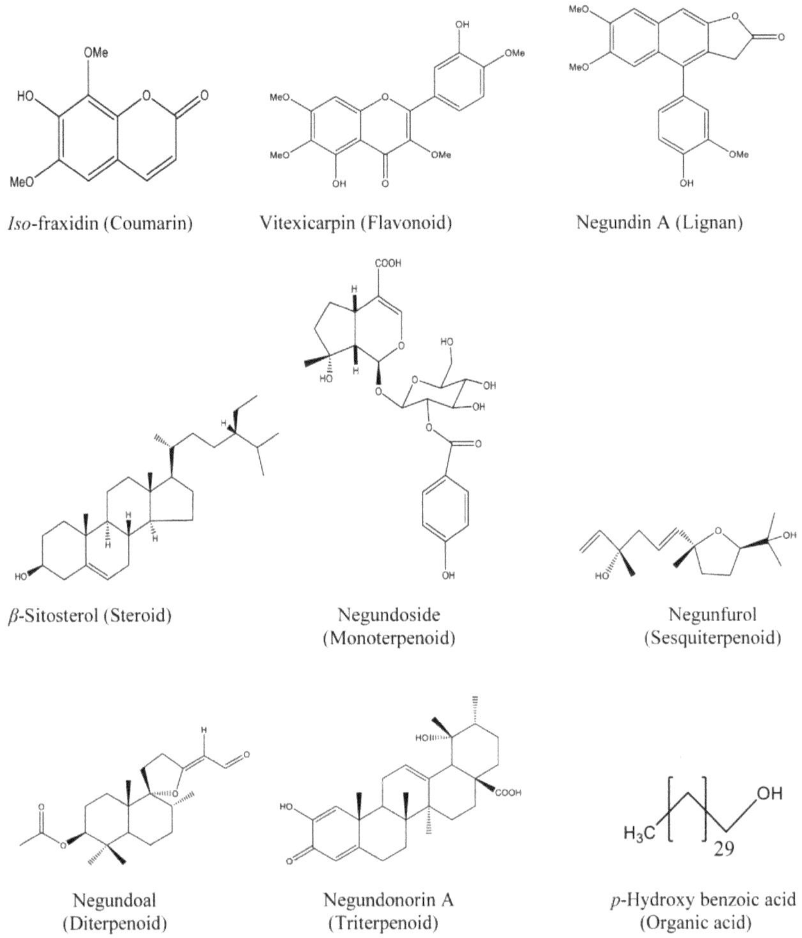

FIGURE 24.1 Chemical structure of different bioactive compounds.

The most abundant terpenes were triterpenoids, which included, 18 oleanane, 6 ursane, 2 norursane, 3 lupine, and a 9-*epi*-cucurbitane-type terpenoids. The oleanane type triterpenoids as3β-acetoxyolean-12-en-27-oic acid, 2α, 3α-dihydroxyoleana-5, 12-dien-28-oic acid, 2β, 3α-diacetoxyoleana-5, 12-dien-28-oic acid, and 2α, 3β-diacetoxy-18-hydroxyoleana-5,12-dien-28-oic acid were isolated from the leaves (Chawla et al., 1992; Yao et al., 2016). Ursolic acid and betulinic acid were obtained from leaves (Chandramu et al., 2003; Yao et al., 2016). Moreover, oleanane-type vulgarsaponin A; ursane-type corosolic acid, 2α,3α,24-trihydroxyurs-12-en-28-oic acid-28-O-β-D-glucopyranosyl ester, and 2α,3α,24-trihydroxyurs-12,20(30)-dien-28-oic acid-28-O-β-D-glucopyranosyl ester were extracted from stem (Chen et al., 2014; Yao et al., 2016). Oleanane triterpenes which include 2α,3α-dihydroxyolean-12-en-28-oic-acid, 2α,3α,23-trihydroxyolean-12-en-28-oic acid, oleanolic acid, 2α,3α,23-trihydroxyolean-12-en-28-oic acid methyl ester, 3-acetyloxy-11-en-28-oic acid prophyl ester, 3-acetyloxy-11-oxo-olean-12-en-28-oic acid butyl ester, and β-amyrin obtained from callus culture of leaves (Noel and Dayrit, 2005). Oleanane-type vulgarsaponin A as well as ursane-type corosolic acid, 2α,3α,24-trihydroxyurs-12-en-28-oic acid-28-O-β-D-glucopyranosyl ester, and 2α,3α,24-trihydroxyurs-12,20(30)-dien-28-oic acid-28-O-β-D-glucopyranosyl ester were isolated from the stem (Chen et al., 2014; Yao et al., 2016). 3β-Acetoxyolean-12-en-27-oic acid and 3β-hydroxyolean-5, 12-dien-28-oic acid, the oleanane-type triterpene were extracted from the barks (Verma et al., 2011). Another oleanane-type triterpenoid oleanolic acid acetate was isolated from the roots of *V. negundo* (Mishra et al., 2014). Fruits of *V. negundo* contained (24R/S)-24-hydroxy-3α, 10α-epoxy-9-*epi*-cucurbita-25-ene, and a 9-*epi*-cucurbitane (Huang et al. 2013b; Yao et al., 2016). Five oleanane-type triterpenoids, namely, oleanolic acid, maslinic acid, 2α,3α,24-trihydroxyolean-12-en-28-oic acid, 2α,3β,19α,23-tetrahydroxyolean-12-en-28-oic acid, and 2α,3β,23-trihydroxyolean-12-en-28-oic acid; ursane-type as ursolic acid, 3-epicorosolic acid, and tormentic acid; norursane type, namely, negundonorin A, negundonorin B and betulinic acid, lup-20(29)-en-3β, and 30-diol as lupine type were isolated from the seeds (Zheng et al., 2010b, 2012a; Wang et al., 2012; Yao et al., 2016).

Organic acids such as salicylic acid, *p*-hydroxy benzoic acid, and *n*-hentriacontanol were obtained from the leaves of *V. negundo* (Chandramu et al., 2003; Chowdhury et al., 2011; Zheng et al., 2015).

24.3 PHARMACOLOGY

24.3.1 ANALGESIC ACTIVITY

Hydroalcoholic leaf extract (500 and 1000 mg/kg) significantly increased the reaction time in the writhing test and tail immersion test whereas it decreased the writhing movements in mice in acetic acid-induced writhing test (Telang et al., 1999). Aqueous leaf extract with different doses (1.25, 2.5, and 5 g/kg) possesses analgesic activity on Wister rats using a hot plate and tail flick test along with formalin test (Dharmasiri et al., 2003).

24.3.2 ANTIAMNESIC ACTIVITY

Wistar rats were treated with aqueous extract of whole plant body with three different doses (100, 300, 900 mg/kg) to decrease in amnesic activity and to increased learning and memory (Kanwal et al., 2010).

24.3.3 ANTI-INFLAMMATORY ACTIVITY

Chloroform extracts of seeds exhibited anti-inflammatory activity in Sprague–Dawley rats, subjected to carrageenan-induced rat paw edema (Chawla et al., 1992). Hydroalcoholic extract (500 and 1000 mg/kg) significantly decreased the rat paw edema and formation of granuloma pouch in rats (Telang et al., 1999). Moreover, doses of aqueous extract of leaves (1.25, 2.5, and 5 g/kg) suppressed carrageenan-induced rat paw edema in Wistar rats in an inverse dose-independent manner (Dharmasiri et al., 2003). Ethanolic extract of leaves inhibited carrageenan-induced rat paw edema and cotton pellet granuloma models (Tandon and Gupta, 2006).

24.3.4 ANTIBACTERIAL ACTIVITY

Essential oils from leaves showed high effectiveness against *Staphylococcus aureus*, whereas, essential oils from fruits were also effective against *Escherichia coli* and *Bacillus subtilis.* Moreover, ethyl acetate extract found to be more impressive toward antibacterial property (Khokra et al., 2008). Methanolic extract of leaves inhibited the growth of *Bacillus cereus* and *Pseudomonas aeruginosa. Bacillus megaterium*, *B. subtilis* and *Vibrio mimicus* were neutralized by pet-ether and carbon tetrachloride extracts. However, all the extracts proved potent against *Salmonella typhi* (Chowdhury et al., 2009). Methanolic and ethanolic extracts of leaves as well as petroleum ether and chloroform extracts of bark successfully resisted the growth of *Staphylococcus epidermidis*, *B. subtilis*, *E. coli*, *S. aureus*, *Klebsiella pneumoniae*, *Clostridium* and *Streptococcus pneumoniae* (Panda et al., 2009; Arumanayagam and Arunmani, 2018; Prasad and Das, 2018; Almuslih and AL-Assie, 2020). Phytopathogenic bacteria such as *Pseudomonas solanacearum* and *Xanthomonas axonopodis* pv. *citri* were highly resisted by ethyl acetate extracts of flower (Jeyaseelan et al., 2010). Silver nanoparticles were biosynthesized using silver nitrate as a silver precursor and methanolic extract of leaves as a reducing agent and stabilizer. This

Ag/*V. negundo* emulsion (20 μl) displayed antibacterial property against *E. coli* and *S. aureus* (Zargar et al., 2011b).

24.3.5 ANTICANCER, ANTINEOPLASTIC, AND CYTOTOXIC ACTIVITIES

Ethanolic and aqueous extract of leaves acted against Dalton's (DAL) and Ehrlich's ascetic carcinoma (EAC) in Swiss albino mice. The treatment reduces tumor weight, increased life duration and revived hematological parameters to normal levels (Chitra et al., 2009; Dewade et al., 2010; Kannikaparameswari and Indhumathi, 2013). Silver nanoparticles biosynthesis from leaf extract inhibited the proliferation of human colon cancer cell line HCT15 with an IC50 of 20 g/mL (Prabhu et al., 2013). Chrysoplenetin and chrysosplenol D (PC50 value of 3.4 and 4.6 μg/mL respectively) isolated from ethanolic extract, showed cytotoxicity toward PANC-1 Human pancreatic cancer cells (Awale et al., 2011). Chloroform extract from leaves displayed cytotoxicity in the human cancer cell line panel (Díaz et al., 2003). The free and bound flavonoids, contained in ethanolic extract of leaves and seeds, proved non-toxic on human erythrocytes (Sharma et al., 2011). Moreover, ethanolic extract of leaves did not show any cytotoxic effects on mammary and genitor-urinary cell lines in mice (Yunos et al., 2005).

24.3.6 ANTICONVULSANT AND CNS DEPRESSANT ACTIVITY

Methanolic extract of leaves (0.15, 0.20, and 0.30 g/kg) increased the GABA levels in brain of albino mice to protect against strychnine and leptazole-induced convulsions. Moreover, petroleum and butanol extracts of leaves proved noteworthy against maximal electroshock shock (MES) seizures and leptazole-induced convulsions (Ravishankar et al., 1985, 1986; Gupta et al., 1999). The ethanolic leaf extract when administered orally in graded doses (250, 500, and 1000 mg/kg) showed significant post-ictal depression at 1000 mg/kg dose (Tandon and Gupta, 2005).

24.3.7 ANTIFUNGAL ACTIVITY

Growth of *Alternaria alternata* and *Curvularia lunata* were successfully resisted by dichloromethane extract of leaves (Guleria and Kumar, 2006).

Aqueous and ethanolic extracts of leaves inhibited the growth of *Pyricularia oryzae*, *Cryptococcus neoformans*, and *Trichophyton mentagrophytes* (Sathiamoorthy et al., 2007; Gautam et al., 2008), however, crude ethanolic extract of fruits neutralized the growth of *Fusarium solani* and *Microsporum canis* (Mahmud et al., 2009). Ethanolic extract of leaves showed antifungal activity against *Candida albicans*, *Aspergillus niger*, *C. neoformans*, *Penicillium* sp., and *Mucor* sp., whereas chloroform and methanolic extract highly suppressed the growth of *Penicillium* sp. (Anbalagan et al., 2017).

24.3.8 ANTIHELMINTIC ACTIVITY

Ethanolic extract of roots with different doses (25, 50, and 100 mg/mL) were used to express antihelminthic activity in terms of paralysis and death of *Pheritima posthuma* (Rastogi et al., 2009).

24.3.9 ANTIHISTAMINE ACTIVITY

Different doses of aqueous extract of mature leaves (1.25, 2.5, and 5 g/kg), administered as histamine injection in Wistar rats, indicated antihistamine activity (Dharmasiri et al., 2003).

24.3.10 ANTIHYPERPIGMENTATION ACTIVITY

The methanolic extract of roots contained lignans which inhibited tyrosinase activity and other lignans (+) -lyoniresinol was potent to treat hyperpigmentation associated with high production of melanocytes (Haq et al., 2006).

24.3.11 ANTINOCICEPTIVE ACTIVITY

The ethanolic leaf extract with different doses (100, 250, and 500 mg/kg) exhibited antinociceptive activity in both tail-flick test and writhing test (Gupta and Tandon, 2005). Petroleum ether fractions from the aqueous ethanol extract of seeds when administered orally, at different doses (12, 24, and 48 mg/kg) showed dose-related inhibitions in the hot-plate test (Zheng et al., 2010c).

24.3.12 ANTIOXIDANT ACTIVITY

V. negundo leaves contained polar compounds such as vitexin (Tandon and Gupta, 2005), catechin, epicatechin, quercetin, and myricetin (Zargar et al., 2011a) which possess high antioxidant properties. Ethyl acetate extracts of leaves (50 mg/kg) showed promising results in decreasing the levels of superoxide dismutase, catalase, glutathione peroxidase, and glucose-6 phosphate dehydrogenase in Freund's adjuvant induced rats (Devi et al., 2007). Ethanolic leaves extract showed antioxidant activity (23.21 mg/100 g) of ascorbic acid equivalent antioxidant capacity which was evaluated by the DPPH method (Kumar et al., 2010) and (69.82 ±2.99%) at a concentration of 500 µg/mL of plant extract expressed in ascorbic acid equivalents (Lakshmanashetty et al., 2010).

24.3.13 ANTIPYRETIC ACTIVITY

Methanolic extract of leaves (250 mg/kg and 300 mg/kg), showed significant antipyretic effects by controlling the elevation of body temperature in rabbits and rats (Murthy et al., 2010; Miskin et al., 2012).

24.3.14 ANTITUSSIVE ACTIVITY

Sulfur dioxide-induced cough in mice encountered with antitussive actions of n-butanol fractions in a dose (250–1000 mg/kg) dependent manner. Moreover, it also provided efficacy in respiratory conditions like asthma, bronchitis, tuberculosis, and other respiratory tract infections (Haq et al., 2011).

24.3.15 ANTIVIRAL ACTIVITY

Aqueous and ethanolic extract of aerial parts inhibited HIV type 1 reverse transcriptase at a ratio higher than 90% at 200 µg/mL and 92.8% at 200 µg/mL concentration (Woradulayapinij et al., 2005; Kannan et al., 2012).

24.3.16 ANXIOLYTIC ACTIVITY

Ethanolic roots extract (100 and 200 mg/kg) exhibited an anxiolytic effect with less locomotor activity in Swiss albino mice. This extract also increased the concentration of GABA, an inhibitory neurotransmitter in CNS (Adnaik et al., 2009).

24.3.17 EFFECT ON REPRODUCTIVE POTENTIAL

Two flavonoids 5,7,3′-trihydroxy flavones and 6,8,4′-trihydroxy flavones obtained from seeds showed both estrogenic and anti-implantation properties (Bhargava, 1986). Seed extract (10 mg/kg) disrupted spermatogenesis in dogs and interfered with functions of male reproductive system in rats (Bhargava, 1989; Das et al., 2004). Ethanolic extracts of fruits showed estrogen-like activity (Hu et al., 2007).

24.3.18 ENZYME INHIBITORY ACTIVITY

The root extracts have suppressed lipoxygenase and butyrylcholinesterase, due to the presence of two lignans negundin B and vitrofolal F (Haq et al., 2004). Methanolic extract of roots showed antityrosinase potential (Haq et al., 2006). Activity of xanthine oxidase was inhibited by methanolic extracts (100 μg/mL) of roots (Umamaheswari et al., 2007). The presence of lignans in methanolic extract showed inhibitory activity against α-chymotrypsin (Lodhi et al., 2008). Methanol, acetone, and aqueous leaf extracts showed ≥50% α-amylase and urease inhibition at 1 mg/mL (Prakash et al., 2017).

24.3.19 HEPATOPROTECTIVE ACTIVITY

Ethanolic leaves extract (100, 250, and 500 mg/kg) was responsive in antitubercular treatment (Tandon et al., 2008). Negundoside and agnuside showed hepatoprotective activity and were used in treatment of liver diseases (Sharma et al., 2009). However, ethanolic extracts of leaves (100 and 300 mg) inhibited acute hepatotoxicity which was measured through glucose, protein, triglycerides, bilirubin, urea, creatinine, ALP, ACP, SGPT, SGOT and histopathological parameters of rats (Mahalakshmi et al., 2010).

24.3.20 HISTOMORPHOLOGICAL AND CYTOTOXIC EFFECTS

Doses of ethanolic extract of leaves (1–10 g/kg) exhibit histomorphological effects in the heart, lungs, and liver tissue of rats (Tandon and Gupta, 2004).

24.3.21 IMMUNO-STIMULANT ACTIVITY

The immunostimulatory effect of methanolic extracts of seeds on human polymorph nuclear cells was evaluated through oxyburst phagocytosis assay (Singh et al., 2005).

24.3.22 INSECTICIDAL AND LARVICIDAL ACTIVITY

Essential oils of leaf exhibited resistance against pests of the stored products (Hebbalkar et al., 1992). Aqueous extract of leaves showed 94% repellency up to 4 h against mosquitoes (Amancharla et al. 1999). Hexane extracts from leaves were a potent eco-friendly mosquito larvicide against the larvae of *Anopheles subpictus* and *Culex tritaeniorhynchus* (Kamaraj et al., 2008).

24.3.23 LAXATIVE STUDY

Doses of aqueous extracts of leaves (100 and 200 mg/kg) increased the fecal output in albino rats (Adnaik et al., 2008).

24.3.24 SNAKE VENOM ANTAGONISTIC ACTIVITY

The toxin of *Vipera russellii* and *Naja kaouthia* induced hemorrhage, coagulant, defibrinogenating and inflammatory effects, which were efficiently eliminated by the methanolic extracts of root (Alam and Gomes, 2003).

24.3.25 WOUND HEALING ACTIVITY

Methanolic extract at dose levels of (200 and 400 mg/kg) support wound healing by increasing superoxide dismutase and catalase in granuloma tissue in Wistar albino rats (Roosewelt et al., 2011). Aqueous and ethanolic extract of leaves (200 mg/kg) proved useful in healing wounds in Wister rats (Talekar et al., 2012).

KEYWORDS

- **anticancer**
- **antiviral**
- **negundoside**
- **vitrofolal F**
- ***Vitex negundo***

REFERENCES

Adnaik, R. S.; Pai, P. T.; Mule, S. N.; Naikwade, N. S.; Magdum, C. S. Laxative Activity of *Vitex negundo* Linn. Leaves. *Asian J. Exp. Sci.* **2008,** *22* (1), 159–160.

Adnaik, R. S.; Pai, P. T.; Sapakal, V. D.; Naikwade, N. S.; Magdum, C. S. Anxiolytic Activity of *Vitex negundo* Linn. in Experimental Models of Anxiety in Mice. *Intern. J. Green Pharmacy.* **2009,** *3* (3), 243–247.

Ahuja, S. C.; Ahuja, S.; Ahuja, U. Nirgundi (*Vitex negundo*)– Nature's Gift to Mankind. *Asian Agri-History.* **2015,** *19* (1), 5–32.

Alam, M. I.; Gomes, A. Snake Venom Neutralization by Indian Medicinal Plants (*Vitex negundo* and *Emblica officinalis*) Root Extracts. *J. Ethnopharmacol.* **2003,** *86* (1), 75–80.

Ali, Z. A. Folk Veterinary Medicine in Moradabad District (Uttar Pradesh), India. *Fitoterapia.* **1999,** *70* (4), 340–347.

Almuslih, S. Q.; AL-Assie, A. H. Study of Antibacterial Phenolic Compounds of *Vitex* Genus Leaves Extractions by HPLC and SNPS of the First Structural Biosynthesis Pathway Gene (PAL). *Plant Arch.* **2020,** *20* (1), 93–98.

Amancharla, P. K.; Muthuraj, P. S.; Rao, G. V.; Singh, O. V. Isolation of a Potent Mosquito Repellent from *Vitex negundo* L.: An Alternative Source of Rotundial. *Nat. Prod. Sci.* **1999,** *5* (2), 104–106.

Anbalagan, S.; Sankareswaran, M.; Moorthy, M.; Elakkia, B.; Fahamitha, E. Phytochemical Analysis and Antifungal Activity of *Vitex negundo* Leaf Extracts Against Clinically Isolated Fungal Pathogens. *Indian J. Appl. Microbiol.* **2017,** *20* (2), 119–125.

Arai, M. A.; Fujimatsu, T.; Uchida, K.; Sadhu, S. K.; Ahmed, F.; Ishibashi, M. Hh Signaling Inhibitors from *Vitex negundo*; Naturally Occurring Inhibitors of the GLI1–DNA Complex. *Mol. BioSyst.* **2013,** *9* (5), 1012–1018.

Arumanayagam, S.; Arunmani, M. Antibacterial Activity of *Vitex negundo* Linn. *J. Microbiol. Biotech. Food Sci.* **2018,** *8* (2), 829–834.

Awale, S.; Linn, T. Z.; Li, F.; Tezuka, Y.; Myint, A.; Tomida, A.; Yamori, T.; Esumi, H.; Kadota, S. Identification of Chrysoplenetin from *Vitex negundo* as a Potential Cytotoxic Agent Against PANC-1 and a Panel of 39 Human Cancer Cell Lines (JFCR-39). *Phytother. Res.* **2011,** *25* (12), 1770–1775.

Banerji, J.; Das, B.; Chakrabarty, R. Isolation of 4, 4'-Dimethoxy-Trans-Stilbene and flavonoids from Leaves and Twigs of *Vitex negundo* Linn. *Indian J. Chem. Sect. B-Org. Chem. Incl. Med. Chem.* **1988,** *27* (6), 597–599.

Basavaraju, R.; Vennel Raj, J.; Bhiravamurthy, P. V. Medicinal Plant Resources of Puttaparthi Mandal: Taxonomic Overview and Need for Conservation. *Ethnobot. Leaflets* **2009,** *2009* (11), 1382–1400.

Bhandary, M. J.; Chandrashekar, K. R.; Kaveriappa, K. M. Medical Ethnobotany of the Siddis of Uttara Kannada District, Karnataka, India. *J. Ethnopharmacol.* **1995,** *47* (3), 149–158.

Bhargava, S. K. Antiandrogenic Effects of a Flavonoid-Rich Fraction of *Vitex negundo* Seeds: A Histological and Biochemical Study in Dogs. *J. Ethnopharmacol.* **1989,** *27* (3), 327–339.

Bhargava, S. K. Antifertility Effects of the Flavonoids (VI–VII) of *Vitex negundo* L. Seeds in Dogs. *Plant Med. Phytother.* **1986,** *20* (2), 188–198.

Chandramu, C.; Manohar, R. D.; Krupadanam, D. G.; Dashavantha, R. V. Isolation, Characterization and Biological Activity of Betulinic Acid and Ursolic Acid from *Vitex negundo* L. *Phytother. Res.* **2003,** *17* (2), 129–134.

Chawla, A. S.; Sharma, A. K.; Handa, S. S.; Dhar, K. L. Chemical Investigation and Anti-Inflammatory Activity of *Vitex negundo* Seeds. *J. Nat. Prod.* **1992,** *55* (2), 163–167.

Chen, J.; Fan, C. L.; Wang, Y.; Ye, W. C. A New Triterpenoid Glycoside from *Vitex negundo*. *Chin. J. Nat. Med.* **2014,** *12*, 0218–0221.

Chitra, V.; Sharma, S.; Kayande, N. Evaluation of Anticancer Activity of *Vitex negundo* in Experimental Animals: An In Vitro and In Vivo Study. *Int. J. Pharm. Tech. Res.* **2009,** *1*, 1485–1489.

Chowdhury, J. A.; Islam, M. S.; Asifuzzaman, S. K.; Islam, M. K. Antibacterial and Cytotoxic Activity Screening of Leaf Extracts of *Vitex negundo* (Fam: Verbenaceae). *J. Pharmaceut. Sci. Res.* **2009,** *1* (4), 103–108

Chowdhury, N. Y.; Islam, W.; Khalequzzaman, M. Insecticidal Activity of Compounds from the Leaves of *Vitex negundo* (Verbenaceae) Against *Tribolium castaneum* (Coleoptera: Tenebrionidae). *Intern. J. Trop. Insect Sci.* **2011,** *31* (3), 174–181.

Das, S.; Parveen, S.; Kundra, C. P.; Pereira, B. M. Reproduction in Male Rats Is Vulnerable to Treatment with the Flavonoid-Rich Seed Extracts of *Vitex negundo*. *Phytother. Res.* **2004,** *18* (1), 8–13.

Devi, P. R.; Kumari, S. K.; Kokilavani, C. Effect of *Vitex negundo* Leaf Extract on the Free Radicals Scavengers in Complete Freund's Adjuvant Induced Arthritic Rats. *Indian J. Clin. Biochem.* **2007,** *22* (1), 143–147.

Dewade, D. R.; Christina, A. J. M.; Chidambaranathan, N.; Bhajipale, N. S.; Tekade, N. P. Antitumor Activity of *Vitex negundo* Linn. Against Dalton's Ascitic Lymphoma. *Intern. J. PharmTech Res.* **2010,** *2* (2), 1101–1104.

Dharmasiri, M. G.; Jayakody, J. R. A. C.; Galhena, G.; Liyanage, S. S. P.; Ratnasooriya, W. D. Anti-Inflammatory and Analgesic Activities of Mature Fresh Leaves of *Vitex negundo*. *J. Ethnopharmacol.* **2003,** *87* (2–3), 199–206.

Díaz, F.; Chávez, D.; Lee, D.; Mi, Q.; Chai, H. B.; Tan, G. T.; Kardono, L. B.; Riswan, S.; Fairchild, C. R.; Wild, R.; Farnsworth, N. R.; Cordell, G. A.; Pezzuto, J. M.; Kinghorn, A. D. Cytotoxic Flavone Analogues of Vitexicarpin, a Constituent of the Leaves of *Vitex negundo*. *J. Nat. Prod.* **2003,** *66* (6), 865–867.

Ferdous, A. J.; Jabbar, A.; Hasan, C. M. Flavonoids from *Vitex negundo*. *Bangladesh Acad. Sci.* **1984,** *8* (2), 23–27.

FRLHT. FRLHT's ENVIS Centre on Medicinal Plants, Bengaluru. 2016. *envis.frlht.org/frlhtenvis.nic.in/* **(accessed on 23 Mar 2021).**

Gautam, L. N.; Shrestha, S. L.; Wagle, P.; Tamrakar, B. M. Chemical Constituents from *Vitex negundo* Linn. of Nepalese Origin. *Sci. World* **2008,** *6* (6), 27–32.

Guleria, S.; Kumar, A. Antifungal activity of some Himalayan Medicinal Plants Using Direct Bioautography. *J. Cell Mol. Biol.* **2006,** *5*, 95–98.

Gupta, M.; Mazumder, U. K.; Bhawal, S. R. CNS Activity of *Vitex negundo* Linn. in Mice. *Indian J. Exp. Biol.* **1999,** *37* (2), 143–146.

Gupta, R. K.; Tandon, V. R. Antinociceptive Activity of *Vitex-negundo* Linn. Leaf extract. *Indian J. Physiol. Pharmacol.* **2005,** *49* (2), 163–170.

Haq, A. U.; Malik, A.; Anis, I.; Khan, S. B.; Ahmed, E.; Ahmed, Z.; Nawaz, S. A.; Choudhary, M. I. Enzyme Inhibiting Lignans from *Vitex negundo. Chem Pharm Bull* (Tokyo). **2004,** *52*, 1269–1272.

Haq, A. U.; Malik, A.; Khan, M. T. H.; Haq, A. U.; Khan, S. B.; Ahmad, A.; Choudhary, M. I. Tyrosinase Inhibitory Lignans from the Methanol Extract of the Roots of *Vitex negundo* Linn. and Their Structure–Activity Relationship. *Phytomedicine* **2006,** *13* (4), 255–260.

Haq, R. U.; Shah, A. U. H. A.; Khan, A. U.; Ullah, Z.; Khan, H. U.; Khan, R. A.; Malik, A. Antitussive and Toxicological Evaluation of *Vitex negundo. Nat. Prod. Res.* **2011,** *26* (5), 484–488.

Hebbalkar, D. S.; Hebbalkar, G. D.; Sharma, R. N.; Joshi, V. S.; Bhat, V. S. Mosquito Repellent Activity of Oils from *Vitex negundo* Linn. Leaves. *Indian J. Med. Res.* **1992,** *95*, 200–203.

Hebbar, S. S.; Harsha, V. H.; Shripathi, V.; Hegde, G. R. Ethnomedicine of Dharwad District in Karnataka, India—Plants Used in Oral Health Care. *J. Ethnopharmacol.* **2004,** *94* (2–3), 261–266.

Hu, Y.; Zhang, Q. Y.; Hou, T. T.; Xin, H. L.; Zheng, H. C.; Rahman, K.; Qin, L. P. Estrogen-Like Activities in *Vitex* Species from China Determined by a Cell Based Proliferation Assay. *Die Pharmazie.* **2007,** *62* (11), 872–875.

Huang, J.; Wang, G. C.; Wang, C. H.; Huang, X. J.; Ye, W. C. Two New Glycosides from *Vitex negundo. Nat. Prod. Res.* **2013a,** *27* (20), 1837–1841.

Huang, D.; Qing, S.; Zeng, G.; Wang, Y.; Guo, H.; Tan, J.; Zhou, Y. Lipophilic Components from *Fructus Viticis Negundo* and Their Anti-Tumor Activities. *Fitoterapia.* **2013b,** *86*, 144–148.

Ignacimuthu, S.; Ayyanar, M.; Sivaraman, K. Ethnobotanical Investigations Among Tribes in Madurai District of Tamil Nadu (India). *J. Ethnobiol. Ethnomedicine.* **2006,** *2* (1), *25*, https://doi.org/10.1186/1746–4269–2-25.

Jagtap, S. D.; Deokule, S. S.; Bhosle, S. V. Some Unique Ethnomedicinal Uses of Plants Used by the Korku Tribe of Amravati District of Maharashtra, India. *J. Ethnopharmacol.* **2006,** *107* (3), 463–469.

Jain, S. P.; Puri, H. S. Ethnomedicinal Plants of Jaunsar-Bawar Hills, Uttar Pradesh, India. *J. Ethnopharmacol.* **1984,** *12* (2), 213–222.

Jeyaseelan, E. C.; Pathmanathan, M. K.; Jeyadevan, J. P. Inhibitory Effect of Different Solvent Extracts of *Vitex negundo* L. and *Allium sativum* L. on Phytopathogenic Bacteria. *Arch. Appl. Sci. Res.* **2010,** *2* (6), 325–331.

Kamaraj, C.; Bagavan, A.; Rahuman, A. A.; Zahir, A. A.; Elango, G.; Pandiyan, G. Larvicidal Potential of Medicinal Plant Extracts Against *Anopheles subpictus* Grassi and *Culex tritaeniorhynchus* Giles (Diptera: Culicidae). *Parasitol. Res.* **2008,** *104* (5), 1163–1171.

Kannan, M.; Rajendran, P.; Vedha, V.; Ashok, G.; Anushka, S.; Chandran, P.; Nair, R. HIV-1 Reverse Transcriptase Inhibition by *Vitex negundo* L. Leaf Extract and Quantification of Flavonoids in Relation to Anti-HIV Activity. *J. Cell. Mol. Biol.* **2012,** *10* (2), 53–59.

Kannikaparameswari, N.; Indhumathi, T. Haematological and Cytotoxic Effect of the Ethanolic Extract of *Vitex negundo. Int. J. Life Sci. Biotec. Pharm. Res.* **2013,** *2* (1), 247–253.

Kanwal, A.; Mehla, J.; Kuncha, M.; Naidu, V. G. M.; Gupta, Y. K.; Sistla, R. Anti-Amnesic Activity of *Vitex negundo* in Scopolamine Induced Amnesia in Rats. *Pharmacol. Pharm.* **2010,** *1* (1), 1–8.

Karunamoorthi, K.; Ramanujam, S.; Rathinasamy, R. Evaluation of Leaf Extracts of *Vitex negundo* L. (Family: Verbenaceae) Against Larvae of *Culex tritaeniorhynchus* and Repellent Activity on Adult Vector Mosquitoes. *Parasitol. Res.* **2008,** *103* (3), 545–550.

Khare, C. P. *Encyclopedia of Indian Medicinal Plants*; Springer-Verlange: New York, 2004; pp 710–711.

Khokra, S. L.; Prakash, O.; Jain, S.; Aneja, K. R.; Dhingra, Y. Essential Oil Composition and Antibacterial Studies of *Vitex negundo* Linn. Extracts. *Indian J. Pharma. Sci.* **2008,** *70* (4), 522–526.

Kirtikar, K. R.; Basu, B. D. *Indian Medicinal Plants*, Vol. 3; Bishen Singh Mahendra Pal Singh: Dehradun, 1984.

Kosalge, S. B.; Fursule, R. A. Investigation of Ethnomedicinal Claims of Some Plants Used by Tribals of Satpuda Hills in India. *J.Ethnopharmacol.* **2009,** *121* (3), 456–461.

Kotoky, J.; Das, P. Medicinal Plants Used for Liver Diseases in Some Parts of Kamrup District of Assam, a North Eastern State of India. *Fitoterapia*. **2008,** *79* (5), 384–387.

Kumar, P. P.; Kumaravel, S.; Lalitha, C. Screening of Antioxidant Activity, Total Phenolics and GC-MS Study of *Vitex negundo*. *Afr. J. Biochem. Res.* **2010,** *4* (7), 191–195.

Lakshmanashetty, R. H.; Nagaraj, V. B.; Hiremath, M. G.; Kumar, V. In Vitro Antioxidant Activity of *Vitex negundo* L. Leaf Extracts. *Chiang Mai J. Sci.* **2010,** *37* (3), 489–497.

Lodhi, M. A.; Haq, A. U.; Choudhary, M. I.; Malik, A.; Ahmad, S. α-Chymotrypsin Inhibition Studies on the Lignans from *Vitex negundo* Linn. *J. Enzyme Inhib. Med. Chem.* **2008,** *23* (3), 400–405.

Mahalakshmi, R.; Rajesh, P.; Ramesh, N.; Balasubramanian, V.; Kannan, V. R. Hepatoprotective Activity on *Vitex negundo* Linn. (Verbenaceae) by Using Wistar Albino Rats in Ibuprofen Induced Model. *Intern. J. Pharmacol.* **2010,** *6* (5), 658–663.

Mahmud, S.; Shareef, H.; Farrukh, U.; Kamil, A.; Rizwani, G. H. Antifungal Activities of *Vitex negundo* Linn. *Pak. J. Bot.* **2009,** *41* (4), 1941–1943.

Manikandan, R.; Thiagarajan, R.; Beulaja, S.; Sivakumar, M. R.; Meiyalagan, V.; Sundaram, R.; Arumugam, M. 1, *2* Di-Substituted Idopyranose from *Vitex negundo* L. Protects Against Streptozotocin-Induced Diabetes by Inhibiting Nuclear Factor-Kappa B and Inducible Nitric Oxide Synthase Expression. *Microsc. Res. Techniq.* **2011,** *74* (4), 301–307.

Meena, A. K.; Singh, U.; Singh, A.; Mishra, S.; Rao, M. M.; Padhi, M. M.; Babu, R.; Gaurav, A. A Comparative Study of the Quality Assessment of *Vitex negundo* Linn Leaves Collected from Three Different Geographical Locations. *Asian J. Res. Chem.* **2011,** *4* (6), 986–989.

Mishra, S.; Pani, S. R.; Rout, K. K.; Nayak, S. K.; Sahoo, S. Bioassay Guided Fractionation and Hepatoprotective Activity of Oleanolic Acid Acetate Isolated from *Vitex negundo* Linn. *J. Biol. Active Prod. Nat.* **2014,** *4* (2), 89–100.

Miskin, N.; Manjunath, K. P.; Bhandarkar, A.; Bolakatti, G.; Katagi, M. S. Antipyretic Activity of *Vitex negundo* Linn Leaves Extracts. *RGUHS J. Pharm. Sci.* **2012,** *2* (2), 78–81.

Misra, G. S.; Subramanian, P. M. Three New Flavone Glycosides from *Vitex negundo*. *Planta Medica.* **1980,** *38* (02), 155–160.

Murthy, J. R.; Venkataraman, S.; Meera, R.; Desmukh, K. S.; Chidambaranathan, N.; Devi, P. Phytochemical Investigation and Antipyretic Activity of Leaf Extract of *Vitex negundo* Linn. *Intern. J. Pharmtech Res.* **2010,** *2* (2), 1068–1073.

NMPB and FRLHT, 2010. http://www.medicinalplants.in/ (accessed on 23 Mar 2021).

Noel, M. G.; Dayrit, F. M. Triperpenes in the Callus Culture of *Vitex negundo* L. *Philippine J. Sci.* **2005,** *134* (1), 5–19.

Ono, M.; Nishida, Y.; Masuoka, C.; Li, J. C.; Okawa, M.; Ikeda, T.; Nohara, T. Lignan Derivatives and a Norditerpene from the Seeds of *Vitex negundo*. *J. Nat. Prod.* **2004,** *67* (12), 2073–2075.

Panda, S. K.; Thatoi, H. N.; Dutta, S. K. Antibacterial Activity and Phytochemical Screening of Leaf and Bark Extracts of *Vitex negundo* L. from Similipal Biosphere Reserve, Orissa. *J. Med. Plants Res.* **2009,** *3* (4), 294–300.

Pandey, A.; Bani, S.; Satti, N. K.; Gupta, B. D.; Suri, K. A. Anti-arthritic Activity of Agnuside Mediated Through the Down-Regulation of Inflammatory Mediators and Cytokines. *Inflamm. Res.* **2012,** *61* (4), 293–304.

Paria, N. D.; Chattopadhyay, S. P. Flora of Hazaribagh District, Bihar. *Botanical Survey India* **2005,** *2*, 682–683.

Pattanaik, C.; Reddy, C. S.; Murthy, M. S. R. An Ethnobotanical Survey of Medicinal Plants Used by the Didayi Tribe of Malkangiri District of Orissa, India. *Fitoterapia* **2008,** *79* (1), 67–71.

POWO. Plants of the World Online. Facilitated by the Royal Botanic Gardens, Kew. Published on the Internet, 2019. http://www.plantsoftheworldonline.org/ (accessed on 15 Mar 2021).

Prabhu, D.; Arulvasu, C.; Babu, G.; Manikandan, R.; Srinivasan, P. Biologically Synthesized Green Silver Nanoparticles from Leaf Extract of *Vitex negundo* L. Induce Growth-Inhibitory Effect on Human Colon Cancer Cell Line HCT15. *Process Biochem.* **2013,** *48* (2), 317–324.

Prakash, V.; Rana, S.; Sagar, A. Studies on Analysis of Antioxidant and Enzyme Inhibitory Activity of *Vitex negundo* Linn. *Int. J. Pharmacogn. Pythochem. Res.* **2017,** *9* (6), 833–839.

Prasad, T. G.; Das, V. Evaluation of Phytochemical and Antimicrobial Activities of *Vitex negundo* L. Leaf Extracts. *Intern. J. Res. Analyt. Rev.* **2018,** *5* (3), 2348–1269.

Purkayastha, J.; Nath, S.; Islam, M. Ethnobotany of Medicinal Plants from Dibru-Saikhowa Biosphere Reserve of North-East India. *Fitoterapia*. **2005,** *76* (1), 121–127.

Rastogi, T.; Bhutda, V.; Moon, K.; Aswar, P. B.; Khadabadi, S. S. Comparative Studies on Anthelmintic Activity of *Moringa oleifera* and *Vitex negundo*. *Asian J. Res. Chem.* **2009,** *2* (2), 181–182.

Ravishankar, B.; Bhaskaran, N. R.; Sasikala, C. K. Pharmacological Evaluation of *Vitex negundo* Leaves, *Bull. Medico. Ethno-Bot. Res.* **1985,** *6* (1), 72–92.

Ravishankar, B.; Bhaskaran, N. R.; Sasikala, C. K. Pharmacological Evaluation of *Vitex negundo* Roots. *J. Res. Ayurv. Siddha* **1986,** *7* (1–2), 62–77.

Roosewelt, C.; Vincent, S.; Sujith, K.; Darwin, C. R. Wound Healing Activity of Methanolic Extract of *Vitex negundo* Leaves in Albino Wistar Rats. *J. Pharm. Res.* **2011,** *4* (8), 2553–2555.

Saikia, A.; Ryakala, V.; Sharma, P.; Goswami, P.; Bora, U. Ethnobotany of Medicinal Plants Used by Assamese People for Various Skin Ailments and Cosmetics. *J. Ethnopharmacol.* **2006,** *106* (2), 149–157.

Samy, R. P.; Thwin, M. M.; Gopalakrishnakone, P.; Ignacimuthu, S. Ethnobotanical Survey of Folk Plants for the Treatment of Snakebites in Southern Part of Tamilnadu, India. *J. Ethnopharmacol.* **2008,** *115* (2), 302–312.

Sathiamoorthy, B.; Gupta, P.; Kumar, M.; Chaturvedi, A. K.; Shukla, P. K.; Maurya, R. New Antifungal Flavonoid Glycoside from *Vitex negundo*. *Bioorg. Med. Chem. Lett.* **2007,** *17* (1), 239–242.

Sharma, A.; Tyagi, S.; Nag, R.; Chaturvedi, A.; Nag, T. N. Antimicrobial Activity and Cellular Toxicity of Flavonoid Extracts from *Pongamia pinnata* and *Vitex negundo*. *Roman. Biotech. Lett.* **2011,** *16* (4), 6396–6400.

Sharma, P.; Chauhan, N.; Lal, B. Observations on the Traditional Phytotherapy Among the Inhabitants of Parvati Valley in Western Himalaya, India. *J. Ethnopharmacol.* **2004,** *92* (2–3), 167–176.

Sharma, R. L.; Prabhakar, A.; Dhar, K. L.; Sachar, A. A New Iridoid Glycoside from *Vitex negundo* Linn (Verbenaceae). *Nat. Prod. Res.* **2009,** *23* (13), 1201–1209.

Singh, D. D.; Chitra, G.; Singh, I. P.; Bhutani, K. K. Immunostimulatory Compounds from *Vitex negundo*. *Indian J. Chem.* **2005,** *44B* (6), 1288–1290.

Subramanian, P. M.; Misra, G. S. Flavonoids of *Vitex negundo*. *J. Nat. Prod.* **1979,** *42* (5), 540–542.

Sundaram, R.; Naresh, R.; Ranadevan, R.; Shanthi, P.; Sachdanandam, P. Effect of Iridoid Glucoside on Streptozotocin Induced Diabetic Rats and Its Role in Regulating Carbohydrate Metabolic Enzymes. *Eur. J. Pharmacol.* **2012a,** *674* (2–3), 460–467.

Sundaram, R.; Naresh, R.; Shanthi, P.; Sachdanandam, P. Antihyperglycemic Effect of Iridoid Glucoside, Isolated from the Leaves of *Vitex negundo* in Streptozotocin-Induced Diabetic Rats with Special Reference to Glycoprotein Components. *Phytomedicine* **2012b,** *19* (3–4), 211–216.

Talekar, Y. P.; Das, B.; Paul, T.; Talekar, D. Y.; Apte, K. G.; Parab, P. B. Wound Healing Potential of *Vitex negundo* Linn in Experimental Animals. *Int. J. Pharm. Sci.* **2012,** *4* (4), 543–546.

Tandon, V.; Gupta, R. K. Histomorphological Changes Induced by *Vitex negundo* in Albino Rats. *Indian J. Pharmacol.* **2004,** *36* (3), 176–177.

Tandon, V. R.; Gupta, R. K. An Experimental Evaluation of Anticonvulsant Activity of *Vitex-negundo*. *Indian J. Physiol. Pharmacol.* **2005,** *49* (2), 199–205.

Tandon, V. R.; Gupta, R. K. Anti-Inflammatory Activity and Mechanism of Action of *Vitex negundo* Linn. *Int. J. Pharmacol.* **2006,** *2* (3), 303–308.

Tandon, V. R.; Khajuria, V.; Kapoor, B.; Kour, D.; Gupta, S. Hepatoprotective Activity of *Vitex negundo* Leaf Extract Against Anti-Tubercular Drugs Induced Hepatotoxicity. *Fitoterapia.* **2008,** *79* (7–8), 533–538.

Tasduq, S. A.; Kaiser, P. J.; Gupta, B. D.; Gupta, V. K.; Johri, R. K. Negundoside, an Iridiod Glycoside from Leaves of *Vitex negundo*, Protects Human Liver Cells Against Calcium-Mediated Toxicity Induced by Carbon Tetrachloride. *World J. Gastroenterol.* **2008,** *14* (23), 3693–3709.

Telang, R. S.; Chatterjee, S.; Varshneya, C. Study on analgesic and anti-inflammatory activities of *Vitex negundo* Linn. *Indian J. Pharmacol.* **1999,** *31* (5), 363–366.

Tirtha, S. *The Ayurveda Encyclopedia*; Ayurveda Holistic Center Press, Bayville: New York, 1998.

Tiwari, N.; Yadav, A. K.; Vasudev, P. G.; Gupta, M. M. Isolation and Structure Determination of Furanoeremophilanes from *Vitex negundo*. *Tetrahedron Lett.* **2013,** *54* (19), 2428–2430.

Umamaheswari, M.; AsokKumar, K.; Somasundaram, A.; Sivashanmugam, T.; Subhadradevi, V.; Ravi, T. K. Xanthine Oxidase Inhibitory Activity of SOME INDIAN MEDICAL PLANTS. *J. Ethnopharmacol.* **2007,** *109* (3), 547–551.

Verma, V. K.; Siddiqui, N. U.; Aslam, M. Phytochemical Constituents from the Bark of *Vitex negundo* Linn. *Int. J. Pharm. Sci. Rev. Res.* **2011,** *7*, 93–95.

Vinuchakkaravarthy, T.; Kumaravel, K. P.; Ravichandran, S.; Velmurugan, D. Active Compound from the Leaves of *Vitex negundo* L. Shows Anti-Inflammatory Activity with Evidence of Inhibition for Secretory Phospholipase A2 Through Molecular Docking. *Bioinformation* **2011,** *7* (4), 199–206.

Vishwanathan, A. S.; Basavaraju, R. A Review on *Vitex negundo* L.: A Medicinally Important Plant. *Eur. J. Biol. Sci.* **2010,** *3* (1), 30–42.

Wang, Y. J.; He, X.; Zeng, G. Y.; Tan, J. B.; Li, X.; Zhou, Y. J. Constituents of Triterpenes in the Seeds of *Vitex negundo* L. *Cent. South Pharm.* **2012,** *10*, 409–412.

Woradulayapinij, W.; Soonthornchareonnon, N.; Wiwat, C. In Vitro HIV Type 1 Reverse Transcriptase Inhibitory Activities of Thai Medicinal Plants and *Canna indica* L. Rhizomes. *J. Ethnopharmacol.* **2005,** *101* (1–3), 84–89.

Yao, J. L.; Fang, S. M.; Liu, R.; Oppong, M. B.; Liu, E. W.; Fan, G. W.; Zhang, H. A Review on the Terpenes from Genus *Vitex*. *Molecules* **2016,** *21* (9), 1179. DOI: 10.3390/molecules21091179.

Yunos, N. M.; Ali, R. M.; Kean, O. B.; Abas, R. Cytotoxicity Evaluations on *Vitex negundo* Anti-Inflammatory Extracts. *Malaysian J. Sci.* **2005,** *24* (1), 213–217.

Zargar, M.; Azizah, A. H.; Roheeyati, A. M.; Fatimah, A. B.; Jahanshiri, F.; Pak-Dek, M. S. Bioactive Compounds and Antioxidant Activity of Different Extracts from *Vitex negundo* Leaf. *J. Med. Plants Res.* **2011a,** *5* (12), 2525–2532.

Zargar, M.; Hamid, A. A.; Bakar, F. A.; Shamsudin, M. N.; Shameli, K.; Jahanshiri, F.; Farahani, F. Green Synthesis and Antibacterial Effect of Silver Nanoparticles Using *Vitex negundo* L. *Molecules.* **2011b,** *16* (8), 6667–6676.

Zhao, X. X.; Zheng, C. J.; Qin, L. P. Chemical Constituents from Fruits of *Vitex negundo*. *Chinese Trad. Herbal Drugs* **2012,** *43*, 2346–2350.

Zheng, C. J.; Huang, B. K.; Han, T.; Zhang, Q. Y.; Zhang, H.; Rahman, K.; Qin, L. P. Nitric Oxide Scavenging Lignans from *Vitex negundo* Seeds. *J. Nat. Prod.* **2009,** *72* (9), 1627–1630.

Zheng, C. J.; Huang, B. K.; Wang, Y.; Ye, Q.; Han, T.; Zhang, Q. Y.; Zhang, H.; Qin, L. P. Anti-Inflammatory Diterpenes from the Seeds of *Vitex negundo*. *Bioorg. Med. Chem.* **2010a,** *18* (1), 175–181.

Zheng, C. J.; Huang, B. K.; Wu, Y. B.; Han, T.; Zhang, Q. Y.; Zhang, H.; Qin, L. P. Terpenoids from *Vitex negundo* Seeds. *Biochem. Syst. Ecol.* **2010b,** *38* (2), 247–249.

Zheng, C. J.; Huang, B. K.; Han, T.; Zhang, Q. Y.; Zhang, H.; Rahman, K.; Qin, L. P. Antinociceptive Activities of the Liposoluble Fraction from *Vitex negundo* Seeds. *Pharma. Biol.* **2010c,** *48* (6), 651–658.

Zheng, C. J.; Lan, X. P.; Cheng, R. B.; Huang, B. K.; Han, T.; Zhang, Q. Y.; Zhang, H.; Rahman, K.; Qin, L. P. Furanofuranlignans from *Vitex negundo* Seeds. *Phytochem, Lett.* **2011,** *4* (3), 298–300.

Zheng, C. J.; Pu, J.; Zhang, H.; Han, T.; Rahman, K.; Qin, L. P. Sesquiterpenoids and norterpenoids from *Vitex negundo*. *Fitoterapia.* **2012a,** *83* (1), 49–54.

Zheng, C. J.; Lan, X. P.; Wang, Y.; Huang, B. K.; Han, T.; Zhang, Q. Y.; Qin, L. P. A new labdane diterpene from *Vitex negundo*. *Pharma. Biol.* **2012b,** *50* (6), 687–690.

Zheng, C. J.; Li, H. Q.; Ren, S. C.; Xu, C. L.; Rahman, K.; Qin, L. P.; Sun, Y. H. Phytochemical and Pharmacological Profile of *Vitex negundo*. *Phytother. Res.* **2015,** *29* (5), 633–647.

Zheng, C. J.; Zhang, X. W.; Han, T.; Jiang, Y. P.; Tang, J. Y.; Brömme, D.; Qin, L. P. Anti-Inflammatory and Anti-Osteoporotic Lignans from *Vitex negundo* Seeds. *Fitoterapia* **2014,** *93*, 31–38.

Zhou, Y.; Liu, Y. E.; Cao, J.; Zeng, G.; Shen, C.; Li, Y.; Zhou, M.; Chen, Y.; Pu, W.; Potters, L.; Shi, Y. E. Vitexins, Nature-Derived Lignan Compounds, Induce Apoptosis and Suppress Tumor Growth. *Clin. Cancer Res.* **2009,** *15* (16), 5161–5169.

CHAPTER 25

PHYTOCHEMICAL AND PHARMACOLOGICAL ACTIVITIES OF *MELISSA OFFICINALIS* (LEMON BALM)

REEMA GABRANI, MARIA ISHAQUE, and KANISHKA JAIN

Department of Biotechnology, Jaypee Institute of Information Technology, Noida, Uttar Pradesh, India

ABSTRACT

Melissa officinalis, also known as lemon balm, consists of many active compounds, which are beneficial due to their medicinal activities. Traditionally, this herb has been used to improve memory, to treat depression, anxiety, and other heart injuries. The traditional uses of the *Melissa officinalis* suggest the therapeutic importance of this plant. Some of the bioactive compositions and their structures of *M. officinalis* listed in this chapter are flavonoids, terpenes, hesperidin, and essential oils. The plant extract of *M. officinalis* can help in curing cardiac injuries and can be used as neuroprotective agent. The essential oil of lemon balm can work as anti-infectious agent against bacteria and viruses, and have elicited promising high efficacy against resistant strains. Some other therapeutic activities of *M. officinalis* are also discussed in detail in this chapter.

25.1 INTRODUCTION

Melissa officinalis commonly known as lemon balm belongs to the mint family Lamiaceae. The plant is traditionally used as a medicinal herb

Bioactives and Pharmacology of Lamiaceae. T. Pullaiah, PhD (Ed.)

because of its astonishing phytochemical properties. Earlier, it was found to be largely cultivated in Iran (Miraj et al., 2017). Since ages, *M. officinalis* has been used for therapeutic purposes in many countries like European countries, Mediterranean countries, and Middle East Countries. European traditional Medicine and Iranian Traditional Medicine have been using this perennial herb to treat numerous diseases like high blood, nervous system disorders, baldness, and morning sickness. *M. officinalis* tea consumption helps in improving brain capacity and memory (Shakeri et al., 2016). Apart from the pharmaceutical benefits, they have been used in food industry and aromatherapy as well.

25.2 BIOACTIVES

The main bioactive constituents of *M. officinalis* include polyphenolic compounds, flavonoids, tannins, terpenes, and essential oils. The flavonoid concentration can vary in different parts of the plant and is also dependent on seasons and environmental conditions (Moacă et al., 2018). The total phenolic content has been reported to be more in leaves than the stem. The polyphenolic compounds are caffeic acid derivatives, sulfated rosmarinic acid, rosmarinic acid, salvianolic acid A, B and C subtypes, hesperidin, sangerinic acid, lithospermic acid A, and protocatechuic acid (Barros et al., 2013). Among the phenolic compounds, rosmarinic and lithospermic acid A are present in abundance (Carocho et al., 2015). The flavonoids (quercitrin, rhamoatrin, and luteolin-o-glucuronide) add some more beneficial properties to the herb (Caleja et al., 2019; Barros et al., 2013). Chemical structures of important phenolic constituents of *M. officinalis* are depicted in Figure 25.1.

Also, some quantity of tannins, terpenes like monoterpenoid aldehydes (citronellal, neral, and geranial), sesquiterpenes (β-caryophyllene) (Chizzola et al., 2018), triterpenoids, monoterpene glycosides, triterpenes (ursolic and oleanolic acids), and essential oils like citral are found to improve the therapeutic properties of *M. officinalis* (Fig. 25.2) (Allahverdiyev et al., 2013).

25.3 PHARMACOLOGY

25.3.1 ANTIVIRAL ACTIVITY

Traditionally, *M. officinalis* has been used against the viral infections like herpes, rabies, hepatitis C and acquired immunodeficiency syndrome. The

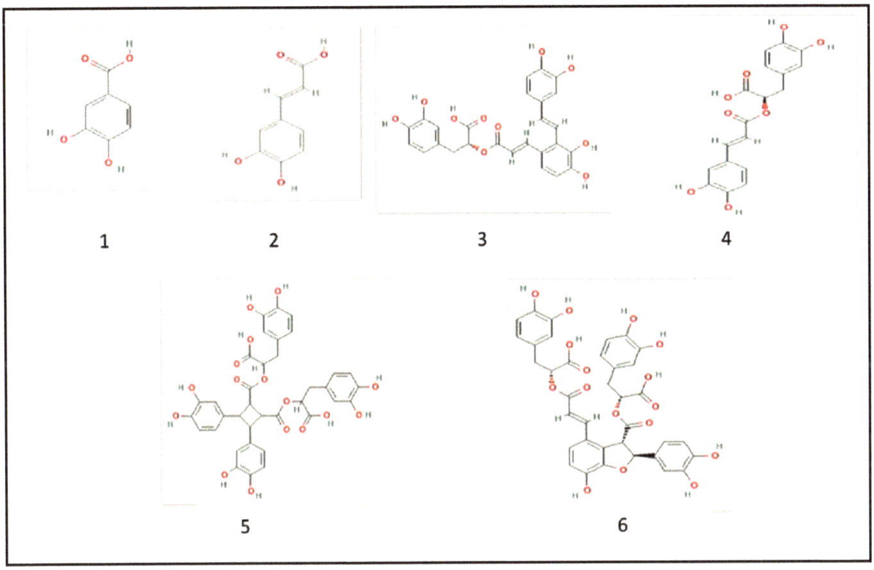

FIGURE 25.1 Chemical structures of phenolic compound in *Melissa officinalis*. (**1**) Protocatechuic acid, (**2**) Caffeic acid, (**3**) Salvianolic, (**4**) Rosmarinic acid, (**5**) Sangerinic acid, and (**6**) Salvianolic acid B. (Miron et al., 2013).

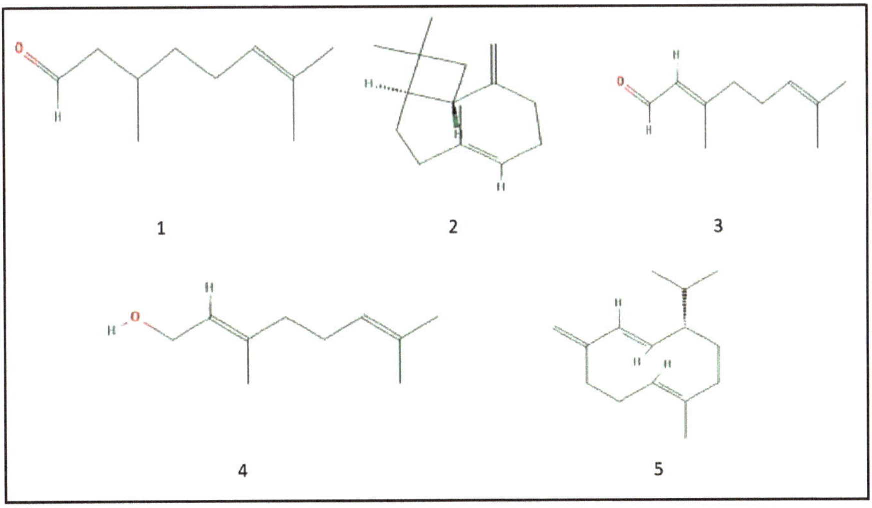

FIGURE 25.2 Chemical structure of essential oil constituents in Melissa officinalis. (**1**) Citronellal, (**2**) b-caryophyllene, (**3**) E-citral, (**4**) Geraniol, and (**5**) Germacrene-D (Rehman et al., 2017).

bioactives like rosmarinic acid, citral and citronellal are responsible for the therapeutic property of *M. officinalis* extract (Shakeri et al., 2016). Rosmarinic acid is found to play a major role in antiviral activity (Noguchi-Shinohara et al., 2015). The rosmarinic acid showed the higher virucidal activity compared to caffeic acid and p-coumaric acid by interrupting the interaction of Herpes simplex virus type 1 (HSV-1) with the host cell (Astani et al., 2012). Later, Astani et al. (2014) reported that *M. officinalis* extract demonstrated anti-HSV-1 activity in both acyclovir-sensitive and acyclovir-resistant cells (Astani et al., 2014). The aqueous extract of *M. officinalis* could restrict human immunodeficiency virus 1 (HIV-1) replication cycle at the early stage and inhibit the infection (Geuenich et al., 2008).

Due to developed resistance to the current antiviral drugs, the bioactive constituents like monoterpene aldehydes, citral a, citral b have contributed to the antiviral activity against Influenza A virus subtype H9N2 (A/H9N2) (Pourghanbari et al., 2016). It has been found to act in different phases of replication cycle and inhibit the replication of avian influenza virus. The rosmarinic acid has shown anti-Enterovirus 71 (EV71) activity to interrupt its life cycle. The mechanism was attributed to hindrance in the attachment of the virus to the host cell and cleavage of translation initiation factor (Chen et al., 2017). Rosmarinic acid was found to be effective in curtailing the early stage of EV71 viral infection in the mouse model and thus can be developed as preventive and therapeutic medicine (Lin et al., 2019).

25.3.2 ANTIBACTERIAL ACTIVITY

The antibacterial activity of *M. officinalis* has been mostly found against food-borne bacteria. The essential oils of *M. officinalis* have constituents like citrals, citronellal, thymol, and β-caryophyllene that are helpful in the treatment of food-borne diseases (Caleja et al., 2019; Mimica-Dukic et al., 2004). The chloroform, ethyl acetate, and n-butanol *M. officinalis* extract having rosmarinic acid caffeic acid, and p-coumaric acid, respectively, have prominent bactericidal activity against *Staphylococcus aureus, Sarcina lutea,* and *Bacillus cereus* (Canadanović-Brunet et al., 2008). Lithospermic acid A and rosmarinic acid were effective against *Salmonella typhimurium* and *Pseudomonas aeruginosa* (Carocho et al., 2015). The lemon balm essential oil was more effective against Gram-positive bacteria *S. aureus,* compared to Gram-negative bacteria *E. coli*. The higher resistance of *E. coli* is due to the hydrophilic outer membrane of Gram-negative bacteria that interrupts the hydrophobic contact of essential oil to the cell membrane of the bacteria (Ehsani et al., 2017). The efficacy of *M. officinalis* extract was studied against antibiotic-resistant strains. *M. officinalis* extract could potentiate the efficacy of streptomycin against *Streptococcus mutans* (Araújo et al., 2015). In another study, it was found that the multi-resistant strain of *Shigella sonnei* was highly sensitive to the extract (Mimica-Dukic et al., 2004).

25.3.3 ANTIFUNGAL ACTIVITY

The antifungal activity of *M. officnalis* has been reported against many fungal strains like *Penicillium funiculosum* (Carocho et al., 2015), *Candida albicans, Candida glabrata, C. krusei,* and *Aspergillus fumigatus* (Tantry et al., 2014). The fungal strain *C. albicans* responsible for candidiasis has been reported to be resistant to some of the currently available drugs. In the latest study, a new technique, methylcellulose hydrogel containing *M. officinalis* essential oil was used to observe its antibacterial potential (Serra et al., 2020). It showed higher activity than the ketoconazole and bifonazole against *Aspergillus ochraceus, A. fumigatus, P. funiculosum*, and *P. ochrochloron* (Caleja et al., 2018). The essential oil showed antifungal effects against *A. castellani* by killing the trophozoites and cysts (Ergüden et al., 2016). The constituents like monoterpene aldehydes, isomenthone, menthone, and E-caryophyllene were successful in antifungal action against *Trichophyton* species (Mimica-Dukic et al., 2004).

25.3.4 ANTIOXIDANT ACTIVITY

Oxidation causes degenerative diseases through DNA mutation, protein oxidation, and lipid peroxidation. The antioxidant activity of *M. officinalis* essential oil was attributed to geraniol, thymol, geranial, citral, and citronellal as determined by 2,2-diphenyl-1-picrylhydrazyl (DPPH), and other antioxidative tests (Ehsani et al., 2017). In another study, it was found that the lemon balm extract had the highest ability to scavenge natural and synthetic-free radicals. It could scavenge DPPH in a concentration-dependent manner (Miraj et al., 2017). The antioxidant activity against lipid peroxidation both by auto-oxidation and EDTA-mediated oxidation was investigated in one of the studies. The main aim was to protect the oxidation of linoleic acid, the main fatty acid for the human body. The *M. officinalis* extract could protect by 80–90% against auto-oxidation and to a higher degree in the case of EDTA-mediated oxidation (Marongiu et al., 2004). The extract has been envisaged for the treatment of various neurological disorders caused by oxidative damage. The extract has been shown to increase the cell viability and prevent the cardiovascular diseases due to oxidative stress (Safaeian et al., 2016). The antioxidant activity of the same can also be used to protect human retinal pigment epithelial cells from degradation due to oxidative damage (Jeung et al., 2016).

25.3.5 ANTI-INFLAMMATORY ACTIVITY

The rosmarinic acid in *M. officinalis* can reduce inflammation. In one of the studies, it was reported to have anti-inflammatory activity against rat liver ischemia—reperfusion (Javid et al., 2018). In another study, rosmarinic acid exhibited the anti-inflammatory activity against edema induced by carrageenan in mice (Bounihi et al., 2013).

25.3.6 HEPATOPROTECTIVE ACTIVITY

Liver plays a critical role in the metabolism of drugs. The expression of the liver marker enzymes increases upon oxidative damage. The extract from *M. officinalis* has been reported to reduce the liver marker enzymes in hypercholesterolemic rat model and the reduction was comparable to the commercial drug, Atorvastatin (Zarei et al., 2014). Another study proved the hypolipidemic effect and protective effect of the lemon balm extract on the liver of hyper lipidemic mice (Bolkent et al., 2005).

25.3.7 ANTI-TUMOROGENIC ACTIVITY

The anti-cancerous effect of the *M. officinalis* was found due to the presence of rosmarinic and ursolic acid. The ethanolic extract of lemon balm showed the anti-tumor effect against breast cancer without any cytotoxic effect on the nearby healthy cells. Their main mechanism was due to the inhibition of angiogenesis (Ghiulai et al., 2020). In another study, different extracts like ethanolic, hydroalcoholic, and aqueous extract of lemon balm were tested to stop the cancerous growth. The ethanolic extract of lemon balm, in a dose-dependent manner, induced apoptosis in lung cancer cell line by reducing pro-caspase 3 and increasing p53 levels (Magalhães et al., 2018). The comparative study revealed that the stem extract of *M. officinalis* had higher antitumor activity than the leaves extract (Moacă et al., 2018). In one of the studies, an antiproliferative activity of hydroalcoholic extract of lemon balm has shown potency in inhibiting the expression of oncogenes responsible for prostate cancer (Jahanban-Esfahlan et al., 2017). The lemon balm could potentiate the effect of doxorubicin drug against human breast cancer cell line (MCF-7)/animal model anti-cancer effects (Hamza et al., 2016). The lemon extract and its constituent rosmarinic acid have been reported to protect keratinocytes from UV-exposure and DNA damage (Pérez-Sánchez et al., 2016). The citral found in essential oil-induced apoptosis in hematopoietic cancer cell lines and showed caspase-3 catalytic activity (Dudai et al., 2005).

25.3.8 ANTI-ULCER ACTIVITY

The methanolic extract of lemon balm, due to the presence of flavonoids, exhibited gastroprotective effect in rats having gastric ulcers (Saberi et al., 2016). In another study, the extract showed the dose-dependent anti-ulcerogenic activity in severe gastric ulcer induced by indomethacin (Khayyal et al., 2001).

25.3.9 ANTI-HYPERGLYCEMIC AND ANTI-HYPERLIPIDEMIC ACTIVITY

As we know with the changes in lifestyle, diabetes mellitus patients are on the rise and there is a need to explore other treatment options. The ethanolic extract of *M. officinalis* was shown to reduce plasma glucose levels and helped in improving painful diabetic neuropathy (Hasanein and Riahi, 2015). The extract in another study was found to activate peroxisome proliferator-activated receptor that helps in maintaining body glucose and lipid metabolism (Weidner et al., 2014). The extract also reduced the triglyceride levels and facilitated the treatment of dyslipidemic diabetes patients (Nayebi et al., 2019). The random, double-blind clinical trial on type-II diabetic patients revealed that hydroalcoholic extract of *M. officinalis* could significantly alter the quantitative insulin sensitivity check index (QUICKI), HDL-c and reduce inflammation (Asadi et al., 2019). *M. officinalis* citral and some other components could improve the glucose uptake and inhibit the accumulation of lipids in 3T3-L1 adipocytes (Yen et al., 2015).

25.3.10 CARDIOPROTECTIVE ACTIVITY

The bioactive components of lemon balm can treat the cardiac injuries. An anthracycline antibiotic doxorubicin used as anticancer drug is reported to be cardiotoxic. Hamza et al. (2016) have shown the protective effect of *M. officinalis* against doxorubicin. The aqueous extract of *M. officinalis* could significantly decrease the heart rate and blood pressure in the Wistar rat model, which was induced for heart injury with isoproterenol (Joukar et al., 2016). The ethanolic extract was found to react against many cardiovascular diseases. The rosmarinic acid in the extract was found to reduce cardiac hypertrophy and cardio-protection and showed hypotensive activities (Sobhani et al., 2017).

25.3.11 NEUROPROTECTIVE ACTIVITY

M. officinalis has been traditionally used as a neuroprotective agent due to its antioxidant activity. The extract can reduce intracellular reactive oxygen species (ROS) formation (López et al., 2009). The main phenolic compounds like rosmarinic acid, quercetin, gallic acid, caffeic acid, and chlorogenic acid are responsible for the neuroprotective effects. It can be used to treat dementia, epilepsy, stroke, and paralysis (Shakeri et al., 2016). The essential oil was found to inhibit hypoxia-inducible factor-1α and apoptosis and thus showed protective effect against CNS diseases related to ischemic brain injury (Bayat et al., 2012). The various extracts of lemon balm at different concentrations showed their apoptotic and necrotic activity on glioblastoma C6 cells. The methanol extract of *M. officinalis* was found to protect PC12 cells from H_2O_2 toxicity. The bioactive constituent polyphenols, flavonoids, and terpenoids of the acidic fraction of lemon balm ethanolic extract exerted a significant neuroprotective effect on the amyloid beta-induced cytotoxicity and oxidative damage in PC12 cells (Sepand et al., 2013). The acidic fraction was also found to protect cerebellar granule neurons due to its antioxidant and antiapoptotic properties. The fraction attenuated the acetylcholinesterase, caspase-3 activity, and lipid peroxidation (Soodi et al., 2017).

KEYWORDS

- **antiviral**
- **bioactive compounds**
- **cardioprotective**
- **essential oil**
- **neuroprotective**

REFERENCES

Allahverdiyev, A.; Bagirova, M.; Yaman, S.; Koc, R.; Abamor, E.; Ates, S.; Baydar, S.; Elcicek, S.; Oztel, O. Chapter 17—Development of New Antiherpetic Drugs Based on Plant Compounds. In *Fighting Multidrug Resistance with Herbal Extracts, Essential Oils and Their Components*; Rai, M.; Kon, K., Eds.; Academic Press, 2013; pp. 245–259.

Araújo, S. G.; Alves, L. F.; Pinto, M. E.; Oliveira, G. T.; Siqueira, E. P.; Ribeiro, R. I.; Ferreira, J. M.; Lima, L. A. Volatile Compounds of Lamiaceae Exhibit a Synergistic Antibacterial Activity with Streptomycin. *Braz. J. Microbiol.* **2015,** *45* (4), 1341–1347.

Asadi, A.; Shidfar, F.; Safari, M.; Hosseini, A. F.; Fallah Huseini, H.; Heidari, I.; Rajab, A. Efficacy of *Melissa officinalis* L. (Lemon Balm) Extract on Glycemic Control and Cardiovascular Risk Factors in Individuals with Type 2 Diabetes: A Randomized, Double-Blind, Clinical Trial. *Phytother Res.* **2019,** *33* (3), 651–659.

Astani, A.; Navid, M. H.; Schnitzler, P. Attachment and Penetration of Acyclovir-Resistant Herpes Simplex Virus Are Inhibited by *Melissa officinalis* Extract. *Phytother Res.* **2014,** *28* (10), 1547–1552.

Astani, A.; Reichling, J.; Schnitzler, P. *Melissa officinalis* Extract Inhibits Attachment of Herpes Simplex Virus In Vitro. *Chemotherapy* **2012,** *58* (1), 70–77.

Barros, L.; Dueñas, M.; Dias, M. I.; Sousa, M. J.; Santos-Buelga, C.; Ferreira, I. C. Phenolic Profiles of Cultivated, In Vitro Cultured and Commercial Samples of *Melissa officinalis* L. Infusions. *Food Chem.* **2013,** *136* (1), 1–8.

Bayat, M.; Azami Tameh, A.; Hossein Ghahremani, M.; Akbari, M.; Mehr, S. E.; Khanavi, M.; Hassanzadeh, G. Neuroprotective Properties of *Melissa officinalis* After Hypoxic-Ischemic Injury Both In Vitro and In Vivo. *Daru* **2012,** *20* (1), 42.

Bolkent, S.; Yanardag, R.; Karabulut-Bulan, O.; Yesilyaprak, B. Protective Role of *Melissa officinalis* L. Extract on Liver of Hyperlipidemic Rats: A Morphological and Biochemical Study. *J. Ethnopharmacol.* **2005,** *99* (3), 391–398.

Bounihi, A.; Hajjaj, G.; Alnamer, R.; Cherrah, Y.; Zellou, A. In Vivo Potential Anti-Inflammatory Activity of *Melissa officinalis* L. Essential Oil. *Adv Pharmacol Sci.* **2013,** 101759.

Caleja, C.; Barros, L.; Barreira, J.; Ciric, A.; Sokovic, M.; Calhelha, R.; Beatriz, M.; Oliveira, P. P.; Ferreira, I. Suitability of lemon balm (*Melissa officinalis* L.) Extract Rich in Rosmarinic Acid as a Potential Enhancer of Functional Properties in Cupcakes. *Food Chem.* **2018,** *250*, 67–74.

Caleja, C.; Finimundy, T.; Pereira, C.; Barros, L.; Calhelha, R.; Sokovic, M.; Ivanov, M.; Carvalho, A.; Rosa, E.; Ferreira, I. Challenges of Traditional Herbal Teas: Plant Infusions and Their Mixtures with Bioactive Properties. *Food Funct.* **2019,** *10* (9), 5939–5951.

Canadanović-Brunet, J., Cetković, G., Djilas, S., Tumbas, V., Bogdanović, G., Mandić, A., Markov, S., Cvetković, D., & Canadanović, V. Radical Scavenging, Antibacterial, and Antiproliferative Activities of *Melissa officinalis* L. Extracts. *J. Med. Food.* **2008,** *11* (1), 133–143.

Carocho, M.; Barros, L.; Calhelha, R. C.; Ćirić, A.; Soković, M.; Santos-Buelga, C.; Morales, P.; Ferreira, I. C. *Melissa officinalis* L. Decoctions as Functional Beverages: A Bioactive Approach and Chemical Characterization. *Food Funct.* **2015,** *6* (7), 2240–2248.

Chen, S. G.; Leu, Y. L.; Cheng, M. L.; Ting, S. C.; Liu, C. C.; Wang, S. D.; Yang, C. H.; Hung, C. Y.; Sakurai, H.; Chen, K. H.; Ho, H. Y. Anti-Enterovirus 71 Activities of *Melissa officinalis* Extract and Its Biologically Active Constituent Rosmarinic Acid. *Sci. Rep.* **2017,** *7* (1), 12264.

Chizzola, R.; Lohwasser, U.; Franz, C. Biodiversity within *Melissa officinalis*: Variability of Bioactive Compounds in a Cultivated Collection. *Molecules* **2018,** *23* (2), 294.

Dudai, N.; Weinstein, Y.; Krup, M.; Rabinski, T.; Ofir, R. Citral Is a New Inducer of Caspase-3 in Tumor Cell Lines. *Planta Med.* **2005,** *71* (5), 484–488.

Ehsani, A.; Alizadeh, O.; Hashemi, M.; Afshari, A.; Aminzare, M. Phytochemical, antioxidant and antibacterial properties of *Melissa officinalis* and *Dracocephalum moldavica* essential oils. *Vet. Res. Forum.* **2017,** *8* (3), 223–229.

Ergüden, C.; Özkoç, S.; Öztürk, B.; Delibaş, S. Investigation of the In Vitro Effects of *Melissa officinalis* L., *Mentha x piperita* L. and *Ocimum basilicum* L. (Lamiaceae) Essential Oils on the Cysts and Trophozoites of *Acanthamoeba castellani. Mikrobiyol. Bull.* **2016,** *50* (4), 569–579.

Geuenich, S.; Goffinet, C.; Venzke, S.; Nolkemper, S.; Baumann, I.; Plinkert, P.; Reichling, J.; Keppler, O. T. Aqueous Extracts from Peppermint, Sage and Lemon Balm Leaves Display Potent Anti-HIV-1 Activity by Increasing the Virion Density. *Retrovirology* **2008,** *5*, 27.

Ghiulai, R.; Avram, S.; Stoian, D.; Pavel, I. Z.; Coricovac, D.; Oprean, C.; Vlase, L.; Farcas, C.; Mioc, M.; Minda, D.; Motoc, A.; Szuhanek, C.; Danciu, C.; Soica, C.; Sima, L. Lemon Balm Extracts Prevent Breast Cancer Progression In Vitro and In Ovo on Chorioallantoic Membrane Assay. *Evid. Based Complem. Altern. Med.* **2020,** 6489159.

Hamza, A. A.; Ahmed, M. M.; Elwey, H. M.; Amin, A. *Melissa officinalis* Protects Against Doxorubicin-Induced Cardiotoxicity in Rats and Potentiates Its Anticancer Activity on MCF-7 Cells. *PLoS One* **2016,** *11* (11), 0167049.

Hasanein, P., Riahi, H. Antinociceptive and Antihyperglycemic Effects of *Melissa officinalis* Essential Oil in an Experimental Model of Diabetes. *Med. Princ. Pract.* **2015,** *24* (1), 47–52.

Jahanban-Esfahlan, R.; Seidi, K.; Monfaredan, A.; Shafie-Irannejad, V.; Abbasi, M. M.; Karimian, A.; Yousefi, B. The Herbal Medicine *Melissa officinalis* Extract Effects on Gene

Expression of p53, Bcl-2, Her2, VEGF-A and hTERT in Human Lung, Breast and Prostate Cancer Cell Lines. *Gene* **2017,** *613*, 14–19.

Javid, A. Z.; Haybar, H.; Dehghan, P.; Haghighizadeh, M. H.; Mohaghegh, S. M.; Ravanbakhsh, M.; Mohammadzadeh, A. The Effects of *Melissa officinalis* (Lemon Balm) in Chronic Stable Angina on Serum Biomarkers of Oxidative Stress, Inflammation and Lipid Profile. *Asia Pac. J. Clin. Nutr.* **2018,** *27* (4), 785–791.

Jeung, I. C.; Jee, D.; Rho, C. R.; Kang, S. *Melissa officinalis* L. Extracts Protect Human Retinal Pigment Epithelial Cells Against Oxidative Stress-Induced Apoptosis. *Int. J. Med. Sci.* **2016,** *13* (2), 139–146.

Joukar, S.; Asadipour, H.; Sheibani, M.; Najafipour, H.; Dabiri, S. The Effects of *Melissa officinalis* (Lemon Balm) Pretreatment on the Resistance of the Heart to Myocardial Injury. *Pharm. Biol.* **2016,** *54* (6), 1005–1013.

Khayyal, M. T.; El-Ghazaly, M. A.; Kenawy, S. A.; Seif-el-Nasr, M.; Mahran, L. G.; Kafafi, Y. A.; Okpanyi, S. N. Antiulcerogenic Effect of Some Gastrointestinally Acting Plant Extracts and Their Combination. *Arzneimittelforschung* **2001,** *51* (7), 545–553.

Lin, W. Y.; Yu, Y. J.; Jinn, T. R.; Evaluation of the Virucidal Effects of Rosmarinic Acid Against Enterovirus 71 Infection Via In Vitro and In Vivo Study. *Virol. J.* **2019,** *16* (1), 94.

López, V.; Martín, S.; Gómez-Serranillos, M. P.; Carretero, M. E.; Jäger, A. K.; Calvo, M. I. Neuroprotective and Neurological Properties of *Melissa officinalis. Neurochem Res.* **2009,** *34* (11), 1955–1961.

Magalhães, D. B.; Castro, I.; Lopes-Rodrigues, V.; Pereira, J. M.; Barros, L.; Ferreira, I.; Xavier, C.; Vasconcelos, M. H. *Melissa officinalis* L. Ethanolic Extract Inhibits the Growth of a Lung Cancer Cell Line by Interfering with the Cell Cycle and Inducing Apoptosis. *Food Funct.* **2018,** *9* (6), 3134–3142.

Marongiu, B.; Porcedda, S.; Piras, A.; Rosa, A.; Deiana, M.; Dessì, M. A. Antioxidant Activity of Supercritical Extract of *Melissa officinalis* Subsp. *officinalis* and *Melissa officinalis* Subsp. *inodora. Phytother Res.* **2004,** *18* (10), 789–792.

Mimica-Dukic, N.; Bozin, B.; Sokovic, M.; Simin, N. Antimicrobial and Antioxidant Activities of *Melissa officinalis* L. (Lamiaceae) Essential Oil. *J. Agric. Food Chem.* **2004,** *52* (9), 2485–2489.

Miraj, S.; Rafieian-Kopaei; Kiani, S. *Melissa officinalis* L: A Review Study with an Antioxidant Prospective. *J. Evid. Based Complem. Altern. Med.* **2017,** *22* (3), 385–394.

Miron, T.; Herrero, M.; Ibáñez, E. Enrichment of Antioxidant Compounds from Lemon Balm (*Melissa officinalis*) by Pressurized Liquid Extraction and Enzyme-Assisted Extraction. *J. Chromatogr. A* **2013,** *1288*, 1–9.

Moacă, E. A.; Farcaş, C.; Ghiţu, A.; Coricovac, D.; Popovici, R.; Cărăba-Meiţă, N. L.; Ardelean, F.; Antal, D. S.; Dehelean, C.; Avram, S. A Comparative Study of *Melissa officinalis* Leaves and Stems Ethanolic Extracts in Terms of Antioxidant, Cytotoxic, and Antiproliferative Potential. *Evid. Based Complem. Alternat. Med.* **2018,** 7860456.

Nayebi, N.; Esteghamati, A.; Meysamie, A.; Khalili, N.; Kamalinejad, M.; Emtiazy, M.; Hashempur, M. H. The Effects of a *Melissa officinalis* L. Based Product on Metabolic Parameters in Patients with type 2 Diabetes Mellitus: A Randomized Double-Blinded Controlled Clinical Trial. *J. Complem. Integr. Med.* **2019,** *16* (3), 20180088.

Noguchi-Shinohara, M.; Ono, K.; Hamaguchi, T.; Iwasa, K.; Nagai, T.; Kobayashi, S; Nakamura, H.; Yamada, M. Pharmacokinetics, Safety and Tolerability of *Melissa officinalis* Extract Which Contained Rosmarinic Acid in Healthy Individuals: A Randomized Controlled Trial. *PloS One* **2015,** *10* (5), e0126422.

Pérez-Sánchez, A.; Barrajón-Catalán, E.; Herranz-López, M.; Castillo, J.; Micol, V. Lemon Balm Extract (*Melissa officinalis* L.) Promotes Melanogenesis and Prevents UVB-Induced Oxidative Stress and DNA Damage in a Skin Cell Model. *J. Dermatol. Sci.* **2016,** *84* (2), 169–177.

Pourghanbari, G.; Nili, H.; Moattari, A.; Mohammadi, A.; Iraji, A. Antiviral Activity of the Oseltamivir and *Melissa officinalis* L. Essential Oil Against Avian Influenza A Virus (H9N2). *Virus Dis.* **2016,** *27* (2), 170–178.

Rehman, S.; Latief, R.; Bhat, K.; Khuroo, M.; Shawl, A.; Chandra, S. Comparative Analysis of the Aroma Chemicals of *Melissa officinalis* Using Hydrodistillation and HS-SPME Techniques. *Arab. J. Chem.* **2017,** *10*, S2485–S2490.

Saberi, A.; Abbasloo, E.; Sepehri, G.; Yazdanpanah, M.; Mirkamandari, E.; Sheibani, V.; Safi, Z. The Effects of Methanolic Extract of *Melissa officinalis* on Experimental Gastric Ulcers in Rats. *Iran Red Crescent Med. J.* **2016,** *18* (7), e24271.

Safaeian, L.; Sajjadi, S. E.; Javanmard, S. H.; Montazeri, H.; Samani, F. Protective Effect of *Melissa officinalis* Extract Against H_2O_2-Induced Oxidative Stress in Human Vascular Endothelial Cells. *Res. Pharm. Sci.* **2016,** *11* (5), 383–389.

Sepand, M. R.; Soodi, M.; Hajimehdipoor, H.; Soleimani, M.; Sahraei, E. Comparison of Neuroprotective Effects of *Melissa offiçinalis* Total Extract and Its Acidic and Non-Acidic Fractions Against a β-Induced Toxicity. *Iran J. Pharm. Res.* **2013,** *12* (2), 415–423.

Serra, E.; Saubade, F.; Ligorio, C.; Whitehead, K.; Sloan, A.; Williams, D. W.; Hidalgo-Bastida, A.; Verran, J.; Malic, S. Methylcellulose Hydrogel with *Melissa officinalis* Essential Oil as a Potential Treatment for Oral Candidiasis. *Microorganisms.* **2020,** *8* (2), 215.

Shakeri, A.; Sahebkar, A.; Javadi, B. *Melissa officinalis* L.—A Review of Its Traditional Uses, Phytochemistry and Pharmacology. *J. Ethnopharmacol.* **2016,** *188*, 204–228.

Sobhani, Z.; Nami, S. R.; Emami, S. A.; Sahebkar, A.; Javadi, B. Medicinal Plants Targeting Cardiovascular Diseases in View of Avicenna. *Curr. Pharm. Des.* **2017,** *23* (17), 2428–2443.

Soodi, M.; Dashti, A.; Hajimehdipoor, H.; Akbari, S.; Ataei, N. *Melissa officinalis* Acidic Fraction Protects Cultured Cerebellar Granule Neurons Against Beta Amyloid-Induced Apoptosis and Oxidative Stress. *Cell J.* **2017,** *18* (4), 556–564.

Tantry, M.; Bhat, G.; Idris, A.; Dar, J.; Yousef Al Omar, S.; Masoodi, K.; Ganai, B.; Kamili, A.; Shawl, A. Sulfated Triterpenes from Lemon Balm. *Helv. Chim. Acta.* **2014,** *97* (11), 1497–1506.

Weidner, C.; Wowro, S. J.; Freiwald, A.; Kodelja, V.; Abdel-Aziz, H.; Kelber, O.; Sauer, S. Lemon Balm Extract Causes Potent Antihyperglycemic and Antihyperlipidemic Effects in Insulin-Resistant Obese Mice. *Mol. Nutr. Food Res.* **2014,** *58* (4), 903–907.

Yen, H. F.; Hsieh, C. T.; Hsieh, T. J.; Chang, F. R.; Wang, C. K. In Vitro Anti-Diabetic Effect and Chemical Component Analysis of 29 Essential Oils Products. *J. Food Drug Anal.* **2015,** *23* (1), 124–129.

Zarei, A.; Changizi-Ashtiyani, S.; Taheri, S.; Rasekh, F. Comparison Between Effects of Different Doses of *Melissa officinalis* and Atorvastatin on the Activity of Liver Enzymes in Hypercholesterolemia Rats. *Avicenna J. Phytomed.* **2014,** *4* (1), 15–23.

CHAPTER 26

ETHNOPHARMACOLOGICAL APPLICATIONS AND PHYTOCHEMICAL COMPOUNDS OF THE GENUS *CALLICARPA* L.

F. NGHAKLIANA, MARY ZOSANGZUALI, MARINA LALREMRUATI, C. LALMUANSANGI, and ZOTHANSIAMA

Department of Zoology, Mizoram University (A Central University), Aizawl, Mizoram, India

ABSTRACT

Callicarpa L. is a large genus of herbaceous plants in the Lamiaceae family that is widely distributed in the East and Southeast Asia, Australia, North America, and Central America. It contains about 160 species. Many species of this genus have been extensively used in various traditional medicines. Ethnomedical reports include the use of the *Callicarpa* genus for the treatment of inflammation, rheumatism, haematuria, hematemesis, fractures, women amenorrhea, gastrointestinal bleeding, scrofula, abdominal colic and for the treatment of stomach cancer and haemostatic agent. The *Callicarpa* species contain a variety of secondary metabolites such as terpenoids, flavonoids, phenylethanoids, and phenylpropanoids among which calliterpenone, ursolic acid, oleanolic acid, quercetin, etc are common. Owing to this, many species of this genus have attracted the interest of many researchers and have been studied for their anticancer, anti-inflammatory, immunogenic and wound healing activity, antibacterial activity, antifungal activity, antidiabetic, antioxidant and antiparasitic activities.

Bioactives and Pharmacology of Lamiaceae. T. Pullaiah, PhD (Ed.)

26.1 INTRODUCTION

Callicarpa L. is a large genus that belongs to the family Lamiaceae and consists of about 160 species of herbaceous plants (Bramley, 2013). They are widely distributed in East and Southeast Asia, Australia, North America and Central America, and contribute largely to traditional medicinal practices in different parts of the world including Chinese and Indian traditional medicine (Jones and Kinghorn, 2008: Tu et al., 2013; Shihan et al., 2015). Plants of this genus have been widely used in indigenous medicine for thousands of years in the treatment of inflammation, rheumatism, hematuria, hematemesis, fractures, women amenorrhea, gastrointestinal bleeding, scrofula, etc. (Jones and Kinghorn, 2008).

In India, many species of this genus such as *C. arborea* and *C. macrophylla* constitute an important part of the *Ayurveda* (Shankar et al., 2014). Different plants of these species have been extensively used in indigenous Chinese medicine for thousands of years for the treatment of various ailments including inflammation (Mi et al., 1984), intestinal disorders (Chen et al., 1986), etc. The local people of Mizoram in India use the leaves and bark of *C. arborea* as hemostatic agent, abdominal colic (Sharma et al., 2016) and for the treatment of stomach cancer (Lalfakzuala et al., 2007).

26.2 BIOACTIVES

The species of *Callicarpa* exhibit high variance in their phytochemical constituent and are found to contain a number of secondary metabolites such as terpenoids, flavonoids, phenylethanoid, and phenypropanoids. Terpenoids especially the diterpenoids and triterpenoids are the most abundant constituents in this genus. Among the terpenoids, diterpenoids, and triterpenoids have been observed to be most abundant in this genus (Tu et al., 2013). Preliminary screening also revealed the presence of alkaloids, fatty acids, flavones, lignans, monoterpenes, phenylpropanoids, and phytosterols in the genus. The details of compounds identified from *Callicarpa* are summarized in Tables 26.1–26.4.

TABLE 26.1 Summary of Phytochemical Constituents of the Genus *Callicarpa L.*

Species	Parts screened	Phytochemical present	References
C. angustifolia	Leaves and twigs	Alkaloids	Cannon et al. (1980)
C. arborea	Leaves	Saponins	Lalawmpuii et al. (2015)

TABLE 26.1 *(Continued)*

Species	Parts screened	Phytochemical present	References
		Steroids	
		Glycosides	
		Flavonoids	
		Tannins	
C. bodinieri	Leaves	Flavonoids	Taoubi et al. (1992)
C. candicans	Not stated	Essential oil	Goh et al. (1984)
		Flavonoids	
		Terpenoids	
C. macrophylla	Aerial parts	Tannins	Atal et al. (1978)
C. tomentosaa	Leaves	Flavonoids	Taoubi et al. (1992)
C. reevesii	Not stated	Tannins	Siang (1983)

TABLE 26.2 List of Terpenoids Isolated from the Genus *Callicarpa L.*

Sl.	Compound isolated	Name of species	References
Monoterpenoids			
1.	4-Terpineol	*C. kochiana*	Lin et al. (2010)
2.	α-Terpineol	*C. kochiana*	Lin et al. (2010)
3.	α-Pinene	*C. japonica*	Kobaisy et al. (2002)
4.	β-Pinene	*C. japonica*	Kobaisy et al. (2002)
5.	Martynoside	*C. arborea*	Wang et al. (2011a)
6.	Isomartynoside	*C. arborea*	Wang et al. (2011a)
7.	6-O-benzoylphlorigidoside B	*C. formosana*	Wang et al. (2011b)
8.	6-O-trans-cinnamoylphlorigidoside B	*C. formosana*	Wang et al. (2011b)
9.	6-O-trans-p-coumaroylshanzhiside methyl ester	*C. formosana*	Wang et al. (2011b)
10.	4′-O-trans-p-coumaroylmussaenoside	*C. formosana*	Wang et al. (2011b)
11.	6β-Hydroxyipolamiide	*C. formosana*	Wang et al. (2011b)
12.	Phlorigidosides	*C. formosana*	Wang et al. (2011b)
13.	2-Methoxy-9-methyl-3-oxabicyclo [4.3.0] nonane-7,9-diol	*C. formosana*	Wang et al. (2011b)
14.	Nudifloside	*C. nudiflora*	Mei et al. (2010)
15.	Linearoside	*C. nudiflora*	Mei et al. (2010)
16.	β-Caryophyllene	*C. kochiana*	Lin et al. (2010)
17.	γ-Caryophyllene	*C. japonica*	Kim and Shin (2004)
18.	Germacrene B	*C. japonica*	Kim and Shin (2004)
19.	Germacrene D	*C. japonica*	Kobaisy et al. (2002)

TABLE 26.2 *(Continued)*

Sl.	Compound isolated	Name of species	References
20.	AromadendreneII	*C. japonica*	Kim and Shin (2004)
21.	Epiglobulol	*C. japonica*	Kim and Shin (2004)
22.	Viridiflorol	*C. japonica*	Kobaisy et al. (2002)
23.	α-Guaiene	*C. japonica*	Kobaisy et al. (2002)
24.	γ–Elemene	*C. japonica*	Kobaisy et al. (2002)
25.	(-)-Spathulenol	*C. integerrima*	Chai et al. (2010)
		C. japonica	Kobaisy et al. (2002)
		C. formosana	Lin et al. (2009)
26.	Elixene	*C. integerrima*	Chai et al. (2010)
27.	τ-Cadinene	*C. integerrima*	Chai et al. (2010)
28.	7-Epi-α-selinene	*C. japonica*	Kobaisy et al. (2002)
29.	Dendrolasin	*C. macrophylla*	Singh et al. (2010)
30.	(1β,6α)-Euesm-4(15)-ene-1,6-diol	*C. pedunculata*	Liu et al. (2006)
31.	(-)-Clovane-2 β,9α -diol	*C. pedunculata*	Liu et al. (2006)
32.	(9 β)-Caryolane-1,9-diol	*C. pedunculata*	Liu et al. (2006)
33.	Cendrolasin	*C. macrophylla*	Singh et al. (2010)
Diterpenoids			
1.	Pedunculatic acid B	*C. kochiana*	Lin et al. (2012)
2.	7α,15-Dihydroxydehydroabietic acid	*C. kochiana*	Lin et al. (2012)
3.	Kochianic acid A	*C. pedunculata*	Liu et al. (2006)
4.	Pedunculatic acid A	*C. formosana*	Liu et al. (2006)
5.	Dehydroabietic acid	*C. pedunculata*	Lu and Hua (2011)
6.	7β-Hydroxydehydroabietic	*C. pedunculata*	Lu and Hua (2011)
7.	7α-Hydroxydehydroabietic acid	*C. pedunculata*	Lu and Hua (2011)
8.	7-Oxodehydroabietic acid	*C. pedunculata*	Lu and Hua (2011)
9.	6,8,11,13-Abietatrien-18-oic acid	*C. pedunculata*	Lu and Hua (2011)
10.	3-Oxo-abieta-8,11,13-triene	*C. pilosissima*	Chen et al. (2009)
11.	7-Oxo-abieta-8,11,13-triene	*C. pilosissima*	Chen et al. (2009)
12.	2α,3α -Dihydroxyus-12-en-28-oic acid	*C. japonica*	Ono et al. (2009)
13.	12-Deoxy-seco-hinokiol methyl ester	*C. pilosissima*	Chen et al. (2009)
14.	Callicarpic acid A	*C. pilosissima*	Chen et al. (2009)
15.	Callicarpic acid B	*C. pilosissima*	Chen et al. (2009)
16.	12-Deoxy-11,12-dihydro-seco-hinokiol methyl ester	*C. pilosissima*	Chen et al. (2009)
17.	9α-Hydroxycallicarpic acid A	*C. pilosissima*	Chen et al. (2009)
18.	Callilongisin A	*C. longissima*	Liu et al. (2012)

TABLE 26.2 *(Continued)*

Sl.	Compound isolated	Name of species	References
19.	Callilongisin B	*C. longissima*	Liu et al. (2012)
20.	Callilongisin C	*C. longissima*	Liu et al. (2012)
21.	3β-Hydroxy-8,11,13(14),15-abietatetraen-19-oic acid	*C. japonica*	Ono et al. (2009)
22.	Hardwickiic acid	*C. formosana*	Wang et al. (2011b)
23.	Monomethyl kolavate	*C. formosana*	Wang et al. (2011b)
24.	Echinophyllin C	*C. formosana*	Wang et al. (2011b)
25.	Clerodermic acid methyl ester	*C. formosana*	Wang et al. (2011b)
26.	15,16-Dihydro-15-methoxy-16-oxo-hardwickiic acid	*C. formosana*	Wang et al. (2011b)
27.	Pentandralactone	*C. pentandra*	Xu et al. (2000)
28.	Pentandranoic acid A	*C. pentandra*	Xu et al. (2000)
29.	Pentandranoic acid B	*C. pentandra*	Xu et al. (2000)
30.	Pentandranoic acid C	*C. pentandra*	Xu et al. (2000)
31.	16,17-Dihydroxy-3-oxophylloladane	*C. nudiflora.*	Zhang et al. (2010)
32.	17-Norphyllocladane-3,16-dione	*C. macrophylla*	Chung et al. (2006)
33.	Phylloclad-15-en-3,17-dione	*C. furfuracea*	Shao et al. (2006)
34.	3β,16β -Phyllocladane-3,16,17-triol	*C. furfuracea*	Shao et al. (2006)
35.	3β,16β-Phyllocladane-3,16,17-triol	*C. furfuracea*	Shao et al. (2006)
36.	16α,17,19-Trihydroxyphyllocladan-3-one	*C. furfuracea*	Shao et al. (2006)
37.	4,16α,17-Trihydroxy-3,4-secophyllocladan-3-oic acid	*C. furfuracea*	Shao et al. (2006)
38.	5β,16α-4,16,17-Trihydroxy-3,4-secophyllocladan-3-oic acid	*C. furfuracea*	Shao et al. (2006)
39.	16-Hydroxy-17-acetoxy-3-oxophyllocladane	*C. macrophylla*	Chung et al. (2006)
40.	Calliterpenone monoacetate	*C. furfuracea*	Shao et al. (2006)
41.	Calliterpenone	*C. macrophylla*	Goel et al. (2007)
		C. *kwangtungensis*	Zhou et al. (2005)
		C. nudiflora	Wang et al. (2007)
		C. formosana	Hu et al. (2001)
		C. longifolia	Subramanian et al. (1974)
		C. furfuracea	Shao et al. (2006)
42.	Calliterpenone-17-acetate	*C. furfuracea*	Shao et al. (2006)
		C. longifolia	Subramanian et al. (1974)

TABLE 26.2 *(Continued)*

Sl.	Compound isolated	Name of species	References
		C. macrophylla	Singh and Agrawal (1994)
43.	Callilongisin D	*C. longissima*	Liu et al. (2012)
44.	7α-Hydroxy sandaracopimaric acid	*C. nudiflora*	Zhang et al. (2010)
45.	Calliphyllin	*C. pedunculata*	Hu et al. (2001)
46.	14α-Hydroxy-7,15-isopimaradien-18-oic acid	*C. pedunculata*	Hu et al. (2001)
47.	Isopimaric acid	*C. pedunculata*	Hu et al. (2001)
Triterpenoids			
1.	α-Amyrin	*C. kochiana*	Lin et al. (2010)
		C. bodinieri	Ren et al. (2001)
2.	2α-Hydroxyursolic acid	*C. nudiflora*	Wang et al. (2007)
3.	2α,3α-Dihydroxyurs-12-en-28-oic acid	*C. formosana*	Ren et al. (2001)
		C. bodinieri	Chen et al. (1986)
4.	2α,3α,19α -Trihydroxyurs-12-en-28-oic acid	*C. macrophylla*	Pan and Sun (2006)
		C. nudiflora	Wang et al. (2007)
5.	2α,3α,19α,24-Tetrahydroxyurs-12-en-28-oic acid-28-O-β-D-glucose ester	*C. bodinieri*	Ren et al. (2004)
6.	24-Ethylcholesta-7,22-dien-3β-ol	*C. bodinieri*	Ren et al. (2004)
7.	Pomolic acid	*C. kochiana*	Lin et al. (2010)
		C. integerrima	Zhu et al. (2012)
8.	Myrianthic acid	*C. kochiana*	Lin et al. (2010)
9.	2β,3β,19α-Trihydroxy-12-en-28-ursolic acid	*C. kochiana*	Lin et al. (2010)
10.	α-Amyrin-3-O-β-D-glucopyranoside	*C. kochiana*	Lin et al. (2010)
11.	Pruvuloside A	*C. kochiana*	He et al. (2011)
		C. integerrima	Zhu et al. (2012)
12.	Ursolic acid	*C. longipes*	Gao et al. (2011)
		C. macrophylla	Pan and Sun (2006)
		C. kwangtungensis	Chen et al. (2008)
		C. cathayana	Zhou et al. (2005)
		C. nudiflora	Dong et al. (2009)
		C. arborea	Wang et al. (2011a)
		C. integerrima	Zhu et al. (2012)
		C. longissima	Liu et al. (2012)
		C. japonica	Ono et al. (2009)

TABLE 26.2 *(Continued)*

Sl.	Compound isolated	Name of species	References
13.	2α,3α,19,24-Tetrahydroxyurs-12-en-oic acid-28-O-β-D-glucopyranoside	*C. bodinieri*	Ren et al. (2004)
14.	Pomonic acid 3-Hydroxy-12-ursen-28-oic acid	*C. pedunculata*	Zhou (2011)
15.	3-Oxo-12-ursen-28-oic acid	*C. pedunculata*	Zhou (2011)
16.	3β,19α-Dihydroxy-12-ursen-28-oic acid	*C. pedunculata*	Zhou (2011)
17.	2β, 3β, 19α-Trihydroxy-12-ursen-28-oic acid	*C. pedunculata*	Zhou (2011)
18.	Urs-12-en-3β-ol	*C. nudiflora*	Dong et al. (2009)
19.	Arjunglucoside I	*C. nudiflora*	Gao et al. (2010)
20.	Euscaphic acid	*C. nudiflora*	Wang et al. (2007)
21.	2α-Hydroxyursolic acid	*C. nudiflora*	Wang et al. (2007)
		C. integerrima	Zhu et al. (2012)
22.	3β-O-Acetyl ursolic acid	*C. integerrima*	Zhu et al. (2012)
23.	2α, 3β, 19α, 23-Tetrahydroxy-urs-12-en-28-oic acid.	*C. integerrima*	Zhu et al. (2012)
24.	2α,19α -Dihydroxyursolic acid	*C. integerrima*	Zhu et al. (2012)
25.	2α, 3α,24-Trihydroxy-ursen-12-en-28-oic acid	*C. nudiflora*	Zhang et al. (2010)
26.	2α,3α,19α -Trihydroxyurs-12-en-28-oic acid-28-O-β-D-glucopyranoside	*C. nudiflora*	Zhang et al. (2010)
27.	2α, 3α, 19α, 23-Tetrahydroxyurs-12-en-28-oic acid-28-O-β-D-glucopyranoside	*C. nudiflora*	Zhang et al. (2010)
28.	β-Amyrin	*C. pedunculata*	Zhou (2011)
		C. japonica	Ono et al. (2009)
		C. macrophylla	Chung et al. (2006)
		C. formosana	Hu et al. (2001)
29.	Oleanolic acid	*C. kochiana*	Lin et al. (2010)
		C. pedunculata	Zhou (2011)
		C. kwangtungensis	Chen et al. (2008)
		C. nudiflora	Wang et al. (2007)
		C. integerrima	Zhu et al. (2012)
		C. japonica	Ono et al. (2009)
30.	Erythrodiol	*C. pedunculata*	Zhou (2011)
31.	3-Acetyl oleanolic acid	*C. integerrima*	Zhu et al. (2012)
32.	2α,3β,23-Trihydroxyoleana-12-en-28-oic-O-β–D-glucopyranoside	*C. kochiana*	Lin et al. (2010)

TABLE 26.2 *(Continued)*

Sl.	Compound isolated	Name of species	References
33	2α,3α,19α,23-Tetrahydroxy-olean-12-en-28-oic acid-28-O-β-D-glucopyranosyl ester	*C. kochiana*	Lin et al. (2010)
34.	2α, 3β, 19α, 23-Tetrahydroxy-olean-12- en-28-oic acid-28-O-β-D-glucopyranosyl-(1→4)- β-D glucopyranoside	*C. integerrima*	Zhu et al. (2012)
35.	2α,3α,19α, 23-Tetrahydroxy-olean-12-en-28-oic acid	*C. bodinieri*	Ren et al. (2003)
36.	Sericoside	*C. japonica*	Ono et al. (2009)
37.	2α,3α,19α,24-Tetrahydroxyolea-12-en-28-oic acid-β-D-glucopyranosyl ester	*C. japonica*	Ono et al. (2009)
38.	2α,3β,19α,23-Tetrahydroxy-olean-12-en-28-oic acid	*C. integerrima*	Zhu et al. (2012)
39.	2α -hydroxy-oleanolic acid	*C. integerrima*	Zhu et al. (2012)
40.	Hederagenin	*C. integerrima*	Zhu et al. (2012)
41.	Betulinic acid	*C. longipes*	Gao et al. (2011)
		C. bodinieri	Ren et al. (2001)
		C. kwangtungensis	Chen et al. (2008)
		C. integerrima	Zhu et al. (2012)
		C. kochiana	Lin et al. (2010)
42.	Squalene	*C. pilosissima*	Chen et al. (2009)
43.	2,6,14,18,22-Pentamethyl-n-tetracos-9-en-17α-ol-27-oic acid	*C. macrophylla*	Chung et al. (2005)

TABLE 26.3 List of Flavonoid Compounds Isolated from the Genus *Callicarpa L.*

Sl.	Compound isolated	Name of species	References
Flavones			
1.	Apigenin-6,8-di-C-β-D-glucopyranoside	*C. kochiana*	He et al. (2011)
2.	Kumatakenin	*C. longipes*	Gao et al. (2011)
		C. kwangtungensis	Chen et al. (2008)
		C. macrophylla	Singh and Agrawal (1994)
3.	Apigenin	*C. longipes*	Gao et al. (2011)
		C. nudiflora	Liang et al. (2011)
4.	Ayanin	*C. longipes*	Gao et al. (2011)
		C. nudiflora	Liang et al. (2011)

TABLE 26.3 *(Continued)*

Sl.	Compound isolated	Name of species	References
5.	5,4′-Dihydroxy-3, 6,7-trimethoxyflavone	*C. longipes*	Gao et al. (2011)
6.	5, 7, 3′,4′-Tetrahydroxy-3-methoxyflavone	*C. longipes*	Gao et al. (2011)
7.	Rhamnatin	*C. kwangtungensis*	Chen et al. (2008)
8.	Velutin	*C. kwangtungensis*	Chen et al. (2008)
9.	5, 4′-Dihydroxy-3,7,3′ trimethoxyflavone	*C. nudiflora*	Dong et al. (2009)
		C. macrophylla	Singh and Agrawal (1994)
		C. kwangtungensis	Chen et al. (2008)
10.	5-Hydroxy-3,7,3′4′-tetramethoxyflavone	*C. nudiflora*	Dong et al. (2009)
11.	Luteolin-7-O-β-D-glucopyranoside	*C. nudiflora*	Dong et al. (2009)
12.	Luteolin	*C. nudiflora*	Dong et al. (2009)
		C. kwangtungensis	Liu et al. (2006)
13.	Juncein	*C. nudiflora*	Wang et al. (2007)
14.	Dracocephaloside	*C. nudiflora*	Wang et al. (2007)
15.	Luteolin-7-O-(6″-trans-caffeoyl)-ß-D-glucopy ranoside	*C. nudiflora*	Gao et al. (2010)
16.	Luteolin-7-O-(6″-trans-cinnamoyl)-ß-D-glucopy ranoside	*C. nudiflora*	Gao et al. (2010)
17.	Luteolin-7- O-(6″-p-coumaryl)-ß-D-glucopyranoside	*C. nudiflora*	Gao et al. (2010)
18.	5,7,4′-Trihydroxy-3′-methoxyflavone	*C. nudiflora*	Gao et al. (2010)
19.	Luteolin-3′–O-β-D-glucopyranoside	*C. nudiflora*	Gao et al. (2010)
20.	Apigenin-7-O-β-D-glucopyranoside	*C. nudiflora*	Gao et al. (2010)
21.	5,7,4′-Trihydroxyflavone	*C. nudiflora*	Gao et al. (2010)
22.	Ermanine	*C. nudiflora*	Zhang et al. (2010)
23.	5-Hydroxy-3,7,4′-trimethoxyflavone	*C. nudiflora*	Zhang et al. (2010)
24.	Luteolin-7-O-glucoside	*C. japonica*	Ono et al. (2009)
		C. nudiflora	Zhang et al. (2010)
25.	3′,5,7-Trihydroxyflavone-4′-O-β-D-glucoside	*C. bodinieri*	Ren et al. (2003)
26.	5,7-Dihydroxy-3′-methoxyflavone-4′-O-β-D-glucoside	*C. bodinieri*	Ren et al. (2003)
27.	Penduletin	*C. pilosissima*	Chen et al. (2009)
28.	Artemetin	*C. pilosissima*	Chen et al. (2009)
29.	3,5,6,7,3′,4′-Hexamethoxyflavone	*C. pilosissima*	Chen et al. (2009)
30.	Casticin	*C. pilosissima*	Chen et al. (2009)
31.	Acacetin	*C. pilosissima*	Chen et al. (2009)

TABLE 26.3 *(Continued)*

Sl.	Compound isolated	Name of species	References
32.	5,7,3′,4′-Tetrahydroxy-8-C-β-D-glucopyranosyl flavone	*C. kwangtungensis*	Jia et al. (2012)
33.	5,7,3′,4′-Tetrahydroxy-6-C-[a-L-rhamnopyranosyl-(1-2)]-β-D-glucopyranosyl flavone	*C. kwangtungensis*	Jia et al. (2012)
34.	5,7,3′,4′-Tetrahydroxy-8-methoxy-6-C-β-D-glucopyranosyl flavone	*C. kwangtungensis*	Jia et al. (2012)
35.	5,4′-Dihydroxy-6,7-dimethoxy-8-C-[β-D-xylcopyranosyl -(1-2)]-β-D-glucopyranosyl flavone	*C. kwangtungensis*	Jia et al. (2012)
36.	5,4′-Dihydroxy-6,7-dimethoxy-8-C-β-D-glucopyranosyl flavone	*C. kwangtungensis*	Jia et al. (2012)
37.	Genkwanin	*C. americana*	Porras et al. (2019)
Flavonols			
1.	Quercetin-7-O-α-L-rhamnopy anoside	*C. cathayana*	Zhou et al. (2005)
2.	Kaempferol-7-O-β-D-glucopyranoside	*C. kochiana*	He et al. (2011)
3.	Quercetin-3, 3′-dimethyl ether	*C. longipes*	Gao et al. (2011)
4.	Quercetin-3, 3′, 4′-trimethyl ether	*C. longipes*	Gao et al. (2011)
5.	Quercetin-7-O-β-D-glucoside	*C. pedunculata*	Zhou (2011)
6.	Quercetin	*C. pedunculata*	Zhou (2011)
		C. kwangtungensis	Zhou et al. (2005)
		C. cathayana	Zhou et al. (2005)
		C. nudiflora	Liang et al. (2011)
7.	Quercetin-3-O-β-D-glucopyranoside	*C. kwangtungensis*	Zhou et al. (2005)
	Phenylethanoids & Phenylpropanoids		
1.	(+)-Sesamin	*C. bodinieri*	Ren et al. (2004)
		C. furfuracea	Shao et al. (2006)
2.	Sesamin-2-ol	*C. furfuracea*	Shao et al. (2006)
3.	Sesamin-2, 2′-diol	*C. furfuracea*	Shao et al. (2006)
4.	(9α)-9-Methoxysesamin-2, 2′-diol	*C. furfuracea*	Shao et al. (2006)
5.	Acteoside	*C. bodinieri*	Ren et al. (2004)
		C. peii	Wu et al. (2012)
		C. pedunculata	Zhou (2011)
		C. nudiflora	Mei et al. (2010)
6.	Isomartynoside	*C. nudiflora*	Gao et al. (2010)
7.	Deacylisomartynoside B	*C. nudiflora*	Gao et al. (2010)
8.	Forsythoside B	*C. pedunculata*	Lu and Shen (2008)

TABLE 26.3 *(Continued)*

Sl.	Compound isolated	Name of species	References
		C. peii	Wu et al. (2012)
		C. kwangtungensis	Zhou et al. (2006)
		C. kochiana	Lin et al. (2010)
		C. nudiflora	Gao et al. (2010)
		C. dichotoma	Koo et al. (2005)
9.	Arenarioside	*C. pedunculata*	Lu and Shen (2008)
10.	Isoacteoside	*C. pedunculata*	Lu and Shen (2008)
		C. kochiana	Lin et al. (2010)
		C. nudiflora	Dong et al. (2009)
		C. cathayana	Zhou et al. (2005)
		C. dichotoma	Koo et al. (2005)
11.	4-Hydroxyphenethyl tetradecanoate	*C. pilosissima*	Chen et al. (2009)
12.	Lamiophlomioside A	*C. kochiana*	Lin et al. (2010)
13.	Cistanoside D	*C. kochiana*	Lin et al. (2010)
14.	Cistanoside C	*C. kochiana*	Lin et al. (2010)
15.	Cistanoside H	*C. dichotoma*	Koo et al. (2005)
16.	Cistanoside F	*C. peii*	Wu et al. (2012)
17.	Callicarposide A	*C. kwangtungensis*	Xie et al. (2009)
18.	Poliumoside	*C. kwangtungensis*	Zhou et al. (2006)
		C. nudiflora	Mei et al. (2010)
		C. kochiana	Lin et al. (2010)
19.	Antiarol rutinoside	*C. arborea*	Wang et al. (2011a)
20.	Kelampayosides A	*C. arborea*	Wang et al. (2011a)
21.	Samioside	*C. nudiflora*	Zhang et al. (2010)
22.	Isovanillic acid	*C. kwangtungensis*	Chen et al. (2008)
23.	Salicylic acid	*C. kwangtungensis*	Chen et al. (2008)
24.	Syringic acid	*C. kwangtungensis*	Chen et al. (2008)
25.	Succinic acid	*C. kwangtungensis*	Chen et al. (2008)
26.	Caffeic acid	*C. nudiflora*	Dong et al. (2010)
27.	Vanilli acid	*C. nudiflora*	Dong et al. (2010)
28.	4-Hydroxycinnamic acid	*C. nudiflora*	Dong et al. (2010)
29.	Ferulic acid	*C. nudiflora*	Dong et al. (2010)
30.	Protocatechuic acid	*C. nudiflora*	Dong et al. (2010)
31.	2′-Acetylacteoside	*C. dichotoma*	Koo et al. (2005)
32.	Poliumoside	*C. dichotoma*	Koo et al. (2005)
33.	Brandioside	*C. dichotoma*	Koo et al. (2005)

TABLE 26.3 *(Continued)*

Sl.	Compound isolated	Name of species	References
34.	Echinacoside	*C. dichotoma*	Koo et al. (2005)
35.	E-tubuloside E	*C. dichotoma*	Koo et al. (2005)
36.	Z-tubuloside E	*C. dichotoma*	Koo et al. (2005)

TABLE 26.4 List of Other Compounds Isolated from the Genus *Callicarpa L.*

Sl.	Compound isolated	Name of species	References
1.	O-hydroxybenzoic acid	*C. kochiana*	Lin et al. (2010)
2.	β-Sitosterol	*C. kochiana*	Lin et al. (2010)
		C. macrophylla	Pan and Sun (2006)
3.	Dauricine	*C. kochiana*	Lin et al. (2010)
4.	α-Tocopherol trimer B	*C. pilosissima*	Chen et al. (2009)
5.	Antiaiol	*C. arborea*	Wang et al. (2011a)
6.	Lirioresinol B	*C. arborea*	Wang et al. (2011a)
7.	Lutein	*C. arborea*	Wang et al. (2011a)
8.	Aesculetin	*C. nudiflora*	Dong et al. (2009)
9.	Acanthoside B	*C. nudiflora*	Dong et al. (2009)
10.	1,1-Diphenyl-2-picrylhydrazyl	*C. japonica*	Ono et al. (2009)
11.	(+)-Lyoniresinol3α–O-(6″-3,5-dimethoxy–hydroxybenzoyl)-β-D-glucopyranoside	*C. japonica*	Ono et al. (2009)
12.	β-Sitosterol-3-O-β-D-glucoside	*C. macrophylla*	Chung et al. (2006)
13.	(+)-Glaberide	*C. pilosissima*	Chen et al. (2009)
14.	Aurantiamide acetate	*C. nudiflora*	Zhang et al. (2010)
15.	Protocatechuic aldehyde	*C. nudiflora*	Dong et al. (2009)
16.	Lariciresinol	*C. furfuracea*	Shao et al. (2006)

26.3 PHARMACOLOGY

As conspicuous from the lavish content of different bioactive compounds, the genus *Callicarpa* exhibits a diverse range of different pharmacological activities which are summarized as follows:

26.3.1 ANTICANCER ACTIVITY

A number of species from the Genus *Callicarpa* have been screened for their anti-cancer activity and showed promising results. *C. americana* was fractionized and screened by Jones et al. (2007) against LNCaP, Lu1, and MCF-7 cell lines and among the different compounds isolated, three unknown compounds as well as known compounds such as genkwanin, 16ξ-hydroxycleroda-3,13-dien-15,16-olide, and 2-formyl-16ξ-hydroxy-3-A-norcleroda-2,13-dien-15,16-olide were observed to possess anticancer activity. The chloroform extract of leaves of *C. arborea* was found to be effective in a dose-dependent manner against Type-II Human Lung Adenocarcinoma cell line (A549) via disruption of redox balance of the cells. The chloroform extract was further tested for its ability to inhibit cellular growth and clonogenicity and was observed to significantly reduce the reproductive integrity of the A549 cells as compared to the untreated control as determined by the reduction of number of colonies in the treated groups (Nghakliana et al., 2020). In a screening done by Suffness et al. (1988), *C. pilosissima* extract was found to be weakly effective against P388 tumor but not against colon 38 tumor and melanoma B-16 tumor. Feng et al. (2017) demonstrated the cytotoxicity of *C. nudiflora* against HeLa and Hey A8 cells using the MTT Assay. The ethyl acetate extract of *C. nudiflora* was also able interfere with the adhesion between tumor cells and extracellular matrix protein, inhibit p-snail, and increase the expression of e-cadherin, thereby reducing metastatic progression (Bin et al., 2015). Supportively, cytotoxic bioactive compounds nudifloside and linearoside were isolated from the ethanolic extract of the aerial parts of *C. nudiflora* (Mei et al., 2010).

26.3.2 ANTI-INFLAMMATORY, IMMUNOGENIC AND WOUND HEALING ACTIVITY

Liang et al. (2009) demonstrated that the aqueous extract of *C. nudiflora* has the ability to reduce swelling which is a clear indication of anti-inflammatory activity. *C. kwangtungensis* was found to contain C-glycosylflavones which is a potent COX-2 inhibitory, thereby reducing pain and inflammation and subsequently promoting wound healing (Jia et al., 2012). Liu et al. (2012) performed an *in vitro* assay that verifies the anti-inflammatory activity of bioactive diterpenes isolated from *C. longissima.* The methanolic extract of *C. arborea* leaves showed significant wound-healing activity in rats (Kumar et al., 2009). Using the excision wound model, six groups of male albino rats were excised at the depilated dorsal thoracic region. Animals treated with *C. arborea* extract showed complete healing within 14 days which was found to be equally effective compared to the standard wound healing drug

Framycetin. Pakyari et al. (2013) demonstrated wound healing activity of *C. nudiflora* by inducing second-degree scald wounds on the dorsal surface of Sprague–Dawley using hot water and the petroleum ether extract was able to treat scald and reduce scar formation by increasing VEGF expression.

26.3.3 ANTIBACTERIAL ACTIVITY

The petroleum ether, chloroform, ethanol, and methanolic extracts of stem bark of *C. arborea* extracts have been shown to possess excellent antibacterial activity against large varieties of both Gram-positive and Gram-negative bacteria (Shihan et al., 2015; Roy et al., 2020) using disc-diffusion method. Volatile substances found in *C. japonica* such as examined for their antibacterial activities against six foodborne microorganisms and out of the 32 compounds identified, 2-hexenal, 2,4-hexadienal, 1-octen-3-ol, 2,4-heptadienal, and epiglobulol strongly inhibited microorganism growth (Kim and Shin, 2004). Zhi-ying et al. (2019) observed that phenylpropanoid compounds isolated from *C. kwangtungensis* possess antibacterial activity. Apart from these studies, various other species have been found to inhibit bacterial growth and are summarized in Table 26.5.

TABLE 26.5 List of Bacterial Species Inhibited by Extracts of Different Species of *Callicarpa L.*

Gram-positive bacteria		
Bacterial species	***Callicarpa* species**	**References**
Bacillus cereus	*C. arborea*	Shihan et al. (2015)
	C. japonica	Kim and Shin (2004)
Bacillus subtilis	*C. macrophylla*	Yadav et al. (2012)
	C. arborea	Shihan et al. (2015), Roy et al. (2020)
Bacillus megaterium	*C. arborea*	Shihan et al. (2015)
Clostridium sporogenes	*C. macrophylla*	Yadav et al. (2012)
Listeria monocytogenes	*C. japonica*	Kim and Shin (2004)
Micrococcus luteus	*C. arborea*	Roy et al. (2020)
	C. macrophylla	Yadav et al. (2012)
Methicillin Resistant *Staphylococcus aureus*	*C. farinosa*	Chung et al. (2004)
Mycobacterium tuberculosis	*C. nudiflora*	Chen et al. (2014)

TABLE 26.5 *(Continued)*

Bacterial species	*Callicarpa* species	References
Sarcina lutea	*C. arborea*	Shihan et al. (2015)
Staphylococcus aureus	*C. arborea*	Shihan et al. (2015)
	C. japonica	Kim and Shin (2004)
	C. kwangtungensis	Zhi-ying et al. (2019)
	C. macrophylla	Chung et al. (2004)
	C. farinosa	Chung et al. (2004)
Staphylococcus epidermidis	*C. macrophylla*	Yadav et al. (2012)
	C. longifolia	Novaryatiin et al. (2018)
Streptococcus faecalis	*C. macrophylla*	Yadav et al. (2012)
Gram-negative Bacteria		
Agrobacterium tumefaciens	*C. macrophylla*	Yadav et al. (2012)
Escherichia coli	*C. arborea*	Shihan et al. (2015), Roy et al. (2020)
	C. kwangtungensis	Zhi-ying et al. (2019)
	C. macrophylla	Yadav et al. (2012)
Klebsiella aerogenes	*C. macrophylla*	Yadav et al. (2012)
Klebsiella pneumoniae	*C. arborea*	Roy et al. (2020)
	C. macrophylla	Yadav et al. (2012)
Pseudomonas aeruginosa	*C. arborea*	Roy et al. (2020) Shihan et al. (2015)
	C. macrophylla	Yadav et al. (2012)
	C. kwangtungensis	Zhi-ying et al. (2019)
Salmonella paratyphi	*C. arborea*	Shihan et al. (2015)
Salmonella typhi	*C. arborea*	Shihan et al. (2015)
Salmonella typhimurium	*C. arborea*	Kim and Shin (2004)
	C. japonica	Yadav et al. (2012)
	C. macrophylla	Roy et al. (2020)
Serratia marcescens	*C. macrophylla*	Yadav et al. (2012)
Shigella boydii	*C. arborea*	Shihan et al. (2015)
Shigella dysenteriae	*C. arborea*	Shihan et al. (2015)
Vibrio mimicus	*C. arborea*	Shihan et al. (2015)
Vibrio parahemolyticus	*C. arborea*	Shihan et al. (2015)
	C. japonica	Kim and Shin (2004)

26.3.4 ANTIFUNGAL ACTIVITY

Ethanolic extract of *C. arborea* has been reported to exhibit antifungal against three fungal species, that is, *Aspergillus niger, Candida albicans,* and *Saccharomyces cerevisiae* (Shihan et al., 2015). Yadav et al. (2012) also reported the antifungal activity of *C. macrophylla* against several fungal strains including *G. fujikoroi, C. neoformans, C. albicans, M. verrucaria,* and *A. niger.* The aqueous extract of leaves of *C. macrophllya* was also tested for its antifungal activity by Tripathi et al. (2017) and was found to be effective against *A. alternata, A. flavus, A. niger, C. cladosporidies , D. halodes,* and *F. moniliforme.*

26.3.5 ANTI-DIABETIC ACTIVITY

The hydroalcoholic extract of stem bark of *C. arborea* was screened for its anti-diabetic activity against streptozotocin (STZ)-induced diabetic rats. Results of the anti-diabetic activity study revealed that *C. arborea* stem bark has the ability to lower the glucose level of hyperglycemic rats as compared to the untreated control after 21 days of treatment with 250 and 500 mg/kg body weight of *C. arborea* extract. Histological observations of the treated pancreas and liver tissues confirmed the anti-diabetic efficacy of *C. arborea.* The levels of triglycerides (TG), total cholesterol (TC), and low-density lipoprotein (LDL) were also reduced significantly in the treated experimental animals as compared to the control groups (Junejo et al., 2017). WenYu et al. (2019) also explored the anti-diabetic activity of *C. nudiflora* in STZ-induced diabetic rat. They observed improvement in oral glucose tolerance, lipid metabolism, and insulin resistance, and reversed the diabetes-related damage in liver and pancreas by activating the AMPK–ACC pathway. Similar anti-diabetic activity of other species such as *M. macrophylla* (Jawaid et al., 2016) and *C. longifolia* (Idar and Aritonang, 2019) has also been reported.

26.3.6 ANTIOXIDANT ACTIVITY

The chloroform and methanolic extract of *C. arborea* leaves have been reported to exhibit significant reducing activity, chelating, and free radical scavenging activities with an IC_{50} of 2.393 μg/mL based on total antioxidant assay (Lalawmpuii et al., 2015). Natural antioxidant compounds such as forsythiaside B, poliumoside, acteoside, alyssonoside, brandioside and their derivatives and one flavone (rhamnazin) were isolated from *C. kwangtungensis* (Cai et al., 2014). Three new lignans, (1R,2R,5R,6R)-1-hydroxy-2-(3,4-methylenedioxyphenyl)-6-(3,4-dihydroxyphenyl)-3,7-dioxabicyclo[3.3.0] octane, (1R,2S,5R,6R)-1-hydroxy-2-(3,4-methylenedioxyphenyl)-6-(3,4-dihydroxyphenyl)-3,7-dioxabicyclo[3.3.0]octane, and (1R,2S,5R,6R)-1-hydroxy-2-(3,4-dihydroxyphenyl)-6-(3,4-methylenedioxyphenyl)-3,7 dioxabicyclo[3.3.0]octane, were isolated from the leaves of *C. nudiflora* which were observed to be potent antioxidant compounds (Luo et al., 2014). Other species showing antioxidant activity include *C. linata* (Qadrie et al., 2014), *C. macrophylla* (Sharma et al., 2021), *C. tomentosa* (Shrilakshmi et al., 2011), and *C. maingayi* (Ado et al., 2016).

26.3.7 ANTI-PARASITIC ACTIVITIES

The anti-parasitic potential of methanolic extract of *C. arborea* stem bark was investigated using the intestinal tapeworm, *Raillietinae chinobothrida*. The methanolic extract of *C. arborea* bark exhibits a dose-dependent anti-parasitic activity which was similar to that of the standard anti-helminth drug albendazole. Scanning electron microscopy revealed damaging effects of *C. arborea* on all over the body of the tapeworm including shrinkage of the tegument on the scolex, neck region, and strobili. Deformity in the suckers and rostellum on the scolex was also observed (Lalthanpuii and Lalchhandama, 2020).

26.3.8 OTHER PHARMACOLOGICAL ACTIVITIES

An experiment conducted by Shihan et al. (2015) using in vitro clot lysis method in human blood revealed the thrombolytic activity of ethanolic extract of *C. arborea* stem bark. *C. arborea* extract was reported to lyse blood clot at the dose of 10 mg/mL only. *C. nudiflora* possesses antiviral activity against RSV, EV71, COXB5, and HSV-1 (Yang et al., 2021). Several triterpene compounds isolated from *C. nudiflora* displayed significant hepatoprotective effects against D-galactosamine-induced toxicity in WB-F344 rat hepatic epithelial stem-like cells (Bo et al., 2014).

Flavonoids from *C. kwangtungensis* and *C. bodinieri* have the ability to reduce writhing in acetic acid-induced writhing, thereby displaying their analgesic activity (Ren et al., 2003; Hong, 2004). Acteoside, a neuroprotective and anti-amnesic bioactive compound has been isolated from *C.*

dichotoma which has the ability to reduce glutamate-induce glutamate as well as scopolamine-induced amnesia in mice (Koo et al., 2006).

Among the different compounds isolated from *Callicarpa* genus, the following listed in Table 26.6 has been reported to be pharmacologically active.

TABLE 26.6 Pharmacologically Active Compounds from *Callicarpa* Genus.

Pharmacological activity	Compounds	References
Cytotoxic activity	Nudifloside	Mei et al. (2010)
	Linearoside	
Antibacterial activity	2-Hexenal	Kim and Shin (2004)
	2,4-Hexadienal	
	1-Octen-3-ol	
	2,4-Heptadienal	
Anti-inflammatory activity	Callilongisins A	Liu et al. (2012)
	Callilongisins D	
	Ursolic	
	5,7,3′,4′ Tetrahydroxy-8C-β-Dglucopyranosyl flavone	Jia et al. (2012)
	5,7,30,40- Tetrahydroxy-6- C-[α-Lrhamnopyranosyl-	
	(1-2)]- β-Dglucopyranosyl flavone	
	5,7,30,40-Tetrahydroxy-8- methoxy-6-C- β-	
	D-glucopyranosyl flavone	
Antitubercular activity	12-Deoxy-secohinokiol methyl ester	Chen et al. (2009)
	12-Deoxy-11,12- dihydro-secohinokiol methyl ester	
	Callicarpic acid A	
	Callicarpic acid B	
	9α Hydroxycallicarpic acid A	
	3-Oxo-abieta-8,11, 13-triene	
	α-Tocopherol trimer B	
Hemostatic activity	2α,3α,19α,23- Tetrahydroxyursa-12-en-28- oic acid-28-O- β D-glucopyranosyl ester	Zhang et al. (2010)
	Rhamnazin	

TABLE 26.6 *(Continued)*

Pharmacological activity	Compounds	References
	2α,3α,24- Trihydroxy-ursa-12-en-28-oic acid	Ren et al. (2003)
Analgesic activity	3′,5,7- Trihydroxyflavone-4′-O- β-D glucoside	Ren et al. (2003)
	5,7-Dihydroxy-3′- methoxyflavone-4′-O-βD-glucoside	
Neuroprotective activity	Acteoside	Koo et al. (2006)
	2′-acetylacteoside	
	Forsythoside B	
	Poliumoside	
	Brandioside	
	Echinacoside	
	Isoacteoside	
	Cistanoside H	
	E-tubuloside E	
	Z-tubuloside E	

KEYWORDS

- ***Callicarpa***
- **Lamiaceae**
- **anticancer**
- **bioactive**
- **antibacterial**
- **ethnobotany**
- **calliterpenone**

REFERENCES

Ado, M. A.; Abas, M.; Leong, S. W.; Shaari, K.; Ismail, I. S.; Ghazali, H. G.; Lajis, N. H. Chemical Constituents and Biological Activities of *Callicarpa maingayi* Leaves. *S. Afr. J. Bot.* **2016,** *104*, 98–104.

Atal, C. K.; Srivastava, J. B.; Wali, B. K.; Chakravarty, R. B.; Dhawan, B. N.; Rastogi, R. P. Screening of Indian Plants for Biological Activity- Part VIII. *Indian J. Exp. Biol.* **1978,** *16*, 330–349.

Bin, C.; Yue-Hua, L.; Xiao-Qin, K.; Jian-Jiang, F.; Shan, W. Anti-Metastatic Effects of Extracts from *Callicarpa nudiflora* via Downregulating Snail Activation on Breast Cancer Cell. *Chin. J. Exp. Trad. Med. Form.* **2015,** *21* (18), 94–98.

Bo, H.; Hui-Zheng, F.; Wei-Kang, C.; Yue-Hua, L. Hepatoprotective Triterpenoid Saponins from *Callicarpa nudiflora. Chem. Pharm. Bull.* **2014,** *62* (7), 695–699.

Bramley, G. L. C. The Genus *Callicarpa* (Lamiaceae) in the Philippines. *Kew Bull.* **2013,** 68, 369–418.

Cai, H.; Xie, Z.; Liu, G.; Sun, X.; Peng, G.; Lin, B.; Liao, Q. Isolation, Identification and Activities of Natural Antioxidants from *Callicarpa kwangtungensis* Chun. *PLoS One* **2014,** *9* (3), e93000.

Cannon, J. R.; Dampawan, P.; Lojanapiwatna, V.; Phuriyakorn, B.; Sinchai, W.; Sriirugsa, P.; Suvatabhandhu, K.; Wiriyachitra, P. A Contribution to the Thai Phytochemical Survey. *J. Sci. Soc. Thailand* **1980,** *6*, 46–53.

Chai, L.; Lin, C. Z.; Zhu, C. C.; Zhao, Z. X.; Zhang, C. X. Analysis of the Chemical Constituents of Essential Oil from the Leaves of *Callicarpa integerrima. J. Chin. Med. Mater.***2010,** *33*, 382–385

Chen, C. H.; Song, T. Y.; Liang, Y. C. Acteoside and 6-O-Acetylacteoside Downregulate Cell Adhesion Molecules Induced by IL-1 Beta Through Inhibition of ERK and JNK in Human Vascular Endothelial Cells. *J. Agric. Food Chem.* **2009,** *57*, 8852–8859.

Chen, R. S.; Lai, J. S.; Wu, T. S. Studies on the Constituents of *Callicarpa formosana. J. Chin. Chem. Soc.* **1986,** *33*, 329–334.

Chen, Y. H.; Feng, F.; Ren, D. C.; Xie, N. Chemical Constituents from the Aerial Part of *Callicarpa kwangtungensis* Chun. *Chin. J. Nat. Med.* **2008,** *6*, 120–122.

Chen, Y.; Zhi-Xin, G.; Yun-hong, T.; Sheng, H. Antibacterial Effect of *Callicarpa nudiflora* Against Clinical Isolated *Mycobacterium tuberculosis* In Vitro Antibacterial. *Chin. J. Trad. Chin. Med. Pharm.* **2014,** *29* (11), 3630–3632.

Chung, I. M.; Ali, M.; Upadhayay, K.; Ahmad, A. Isolation and Cytotoxic Activity of Acyclic Triterpene Callicarpenol from *Callicarpa macrophylla. Asian J. Chem.* **2005,** *17*, 1907–1914.

Chung, I. M.; Ali, M.; Upadhayay, K.; Ahmad, A. Isolation of Fatty Acids and Other Constituents from *Callicarpa macrophylla* Fruits. *Asian J. Chem.* **2006,** *8*, 1751–1758.

Chung, P. Y.; Chung, L. Y.; Ngeow, Y. F.; Goh, S. H.; Imiyabir, Z. Antimicrobial Activities of Malaysian Plant Species. *Pharm. Biol.* **2004,** *42*, 292–300.

Dong, L.; Liu, M. S.; Wang, J. H. Lipophilic Compounds of *Callicarpa nudiflora* Hook. et Arn. *Chin. J. Med. Chem.* **2009,** *19*, 371–374.

Dong, L.; Wang, J. H.; Liu, M. S. Phenolic Acid Constituents of Leaves from *Callicarpa nudiflora* Hook. et Arn. *Shenyang Yaoke Daxue Xuebao* **2010,** *27*, 290–291.

Feng, S. X.; Yi, B.; Zhang, M.; Xu, J.; Lin, H.; Xu, W. T. Iridoid Glycosides from *Callicarpa nudiflora* Hook. *Nat. Prod. Res.* **2017,** *31* (2), 181–189.

Gao, F. P.; Wang, H.; Ye, W. C.; Zhao, S. X. Chemical Constituents from the Leaves of *Callicarpa nudiflora. Zhongguo Yaok eDaxuexuebao* **2010,** *41*, 120–123.

Gao, L.; Lin, C. Z.; Zhu, C. C. Chemical Constituents of *Callicarpa longipes. Chin. Trad. Herb. Drugs.* **2011,** *42*, 1289–1292.

Goel, M. K.; Kukreja, A. K.; Singh, A. K. In Vitro Plant Growth Promoting Activity of Phyllocladane Diterpenoids Isolated from *Callicarpa macrophylla* Vahl in Shoot Cultures of *Rauwolfia serpentina. Nat. Prod. Commun.* **2007,** *2*, 799–802.

Goh, S. H.; Soepadmo, E.; Chang, P.; Barnerjee, U.; Chan, K. C.; Deverre, J. R.; Hadi, H.; Loke, S. E.; Nasrulhawq, A.; Oo, S. L.; Taylor, C. E.; Wong, W. H.; Zakaria, M. Studies on Malaysian Medicinal Plants: Preliminary Results. In *Asian Symposium on Medicinal Plants and Spices*.Seoul, Korea, 1984.

He, Z. M.; Lin, Z. Z.; Wang, Z. Y.; Zhu, C. C. Chemical Constituents of *Callicarpa konchiana. Pharm. Today* **2011,** *21*, 328–344.

Hong, Y. L. The Research About Preparation and Quality Control of the Fuyankang Tablet as Well as the Compounds from *Callicarpa kwangtungersis* Chun. *Beijing University of Chinese Medicine*, Master Thesis, 2004.

Hu, Y. M.; Shen, Y. M.; Gu, Q. X.; Zuo, G. Y.; Hao, X. J. Studies on Chemical Constituents of *Callicarpa pedunculata. Zhong Cao Yao* **2001,** *32*, 1063–1065.

Idar, I.; Aritonang, M. P. U. PengaruhEkstrakEtanolDaunKerehau (*Callicarpa longifolia* Lamk.) Pada Kadar MalondialdehidHewan Yang DiinduksiAloksan. *Media Informasi* **2019,** *15* (1), 81–88.

Jawaid, T.; Kamal, M.; Maddheshiya, P. Hypoglycemic Effect of Methanolic Extract of *Callicarpa macrophylla* Fruits on STZ Induced Diabetic Rats. *Bull. Env. Pharmacol. Life Sci.* **2016,** *5*, 42–48.

Jia, A.; Yang, Y. F.; Kong, D. Y. Isolation and Structural Identification of C-Glycosylflavones from *Callicarpa kwangtungensis* Chun. and Their Preliminary Anti-Inflammatory Activities. *Chin. Pharm. J.* **2012,** *43*, 263–267.

Jones, W. P.; Kinghorn, A. D. Biologically Active Natural Products of the Genus *Callicarpa*, *Curr. Bioact. Compd.* **2008,** *4*, 15–32.

Jones, W. P.; Lobo-Echeverri, T.; Mi, Q.; Chai, H. B.; Soejarto, D. D.; Cordell, G. A.; Swanson, S. M.; Kinghorn, A. D. Cytotoxic Constituents from the Fruiting Branches of *Callicarpa americana* Collected in Southern Florida. *J. Nat. Prod.* **2007,** *70* (3), 372–377.

Junejo, J. A.; Rudrapal, M.; Nainwal, L. M.; Zaman, K. Anti-Diabetic Activity of Hydro-Alcoholic Stem Bark Extract of *Callicarpa. Biomed. Pharmacother*. **2017,** *95*, 84–94.

Kim, Y. S.; Shin, D. H. Volatile Constituents from the Leaves of *Callicarpa japonica* Thunb. and Their Antibacterial Activities. *J. Agric. Food Chem*. **2004,** *52*, 781–787.

Kobaisy, M.; Tellez, M. R.; Dayan, F. E.; Duke, S. O. Phytotoxicity and Volatile Constituents from Leaves of *Callicarpa japonica* Thunb. *Phytochemistry* **2002,** *61*, 37–40.

Koo, A. K.; Sung, S. H.; Park, J. H.;Kim, S. H.; Lee, K. Y.; Kim, Y. C. In Vitro Neuroprotective Activities of Phenylethanoid Glycosides from *Callicarpa dichotoma. Planta Med.* **2005,** *71*, 778–780.

Koo, K. A.; Kim, S. H.; Oh, T. H.; Kim, Y. C. Acteoside and Its Aglycones Protect Primary Cultures of Rat Cortical Cells from Glutamate-Induced Cytotoxicity. *Life Sci.* **2006,** *79*, 709–716.

Kumar P. K.; Lingadurai, S.; Nath, L. K.; Nanda, B. Evaluation of Wound Healing Activity of *Urtica parviflora* Roxb. and *Callicarpa arborea* Roxb. in Rats. *Pharmacologyonline* **2009,** *1*, 1095–1103.

Lalawmpuii, R.; Lalhriatpuii, T. C.; Lalzikpuii.; Lalengliani.; Ghosh, S. K. Qualitative Phytochemical Screening and Evaluation of the In Vitro Antioxidant Activity of *Callicarpa arborea* Roxb, an Ethnomedicinal Plant of Mizoram, Northeast India. *Asian J. Pharm. Clin. Res.* **2015,** *8*, 202–205.

Lalfakzuala, R.; Lalramnghinglova, H.; Kayang, H. Ethnobotanical Usages of Plants in Western Mizoram. *Indian J. Trad. Knowl.* **2007,** *6*, 486–493.

Lalthanpuii, P. B.; Lalchhandama, K. Beautyberry (*Callicarpa arborea*) as an Antiparasitic Agent Against *Raillietina echinobothrida*, an Intestinal Tapeworm. *Pharmacogn. J.* **2020,** *12*, 66–70.

Liang, J. J.; Qi, J. L.; Li, L. Flavonoids from *Callicarpa nudiflora* Leaves. *Chem. Nat. Compd.* **2011,** *47*, 110–111.

Liang, J. J.; Xu, K.; Li, L. F.; Wang, Z. Y.; Mo, F. K. Study of Total Flavonoids of *Callicarpa nudiflora* on Anti-Inflammatory and Hemostasis Effects. *Modern J. Integr. Trad. Chin. Western Med.* **2009,** *18*, 3161–3162.

Lin, C. Z.; Zhu, C. C.; Zhang, C. X.; Zhao, Z. X.; He, W. J.; Li, J. K.; Chai, L.; Deng, J. W.; Wei, Y. L. Chemical Constituents and Antioxidant Activity of the Essential Oils from the Leaves of *Callicarpa formosana* Rolfe. *J. Trop. Subtrop. Bot.* **2009,** *17*, 401–405.

Lin, C. Z.; Chai, L.; Cai, Z. X.; Zhao, Z. X.; Zhang, C. X.; Zhu, C. C. Component Analysis of Phenylethanoid Glycosides from *Callicarpa kochiana* Makino. *Curr. Clin. Pharmacol.* **2010,** *21*, 276–284.

Lin, C. Z.; Zhu, C. C.; Zhao, Z. X.; Li, X. H.; Xiong, T. Q.; Xia, Y. Y.; Ning, Y. Two New Abietanediterpenoids from the Caulis and Leaves of *Callicarpa kochiana. Fitoterapia* **2012,** *83*, 1–5.

Liu, C. H.; Cao, H.; Xing, J. B. HPLC Content Determination of Luteolin in *Callicarpa kwangtungensis* and Gongfukang Capsule. *Zhong Cao Yao* **2006,** *37*, 539–541.

Liu, Y. W.; Cheng, Y. B.; Liaw, C. C.; Chen, C. H.; Guh, J. H.; Hwang, T. L.; Tsai, J. S.; Wang, W. B.; Shen, Y. C. Bioactive Diterpenes from *Callicarpa longissima. J. Nat. Prod.* **2012,** *75* (4), 689–693.

Lu, C. H.; Shen, Y. M. Water-Soluble Constituents from *Callicarpa pedunculata. Chin. J. Nat. Med.* **2008,** *6*, 176–178.

Lu, Y.; Hua, Y. Abietanediterpenes from *Callicarpa pedunculata. Nat. Prod. Res. Dev.* **2011,** *23*, 66–68.

Luo, Y. H.; Zhou, Z. Q.;Ma, S. C.; Fu, H. Z. Three New Antioxidant Furofuranlignans from *Callicarpa nudiflora. Phytochem. Lett.* **2014,** *7*, 194–197.

Mei, W. L.; Han, Z.; Cui, H. B.; Zhao, Y. X.; Deng, Y. Y.; Dai, H. F. A New Cytotoxic Iridoid from *Callicarpa nudiflora. Nat. Prod. Res.* **2010,** *24* (10), 899–904.

Mi, H. M.; Zhang, C. H.; Su, Z. W.; Li, C. G. A SEM Observation on the Leaves of the Genus *Callicarpa* of 16 Chinese Species. *Acta. Pharm. Sin.* **1984,** *19*, 381–386.

Nghakliana, F.; Fanai, J.L; Tochhawng, L.; Balachandar, V; Zothansiama. Anticancer Activity of *Callicarpa arborea* Roxb. Extracts Against Type-II Human Lung Adenocarcinoma Cell Line, A549. *J. Environ. Biol.* **2020,** *41*, 901–907.

Novaryatiin, S.; Sari, A. A.; Mulyani, E. Antibacterial Activity of Ethanolic Extract of Sangkareho (*Callicarpa longifolia* Lam.) Against *Staphylococcus epidermidis. Borneo J. Pharm.* **2018,** *1* (2), 85–88.

Ono, M.; Chikuba, T.; Mishima, K.; Yamasaki, T.; Ikeda, T.; Yoshimitsu, H.; Nohara, T. A New Diterpenoid and a New Triterpenoid Glucosyl Ester from the Leaves of *Callicarpa japonica* Thunb. var. *luxurians* Rehd. *J. Nat. Med.* **2009,** *63*, 318–322.

Pakyari, M.; Farrokhi, A.; Maharlooei, M. K.; Ghahary, A. Critical Role of Transforming Growth Factor Beta in Different Phases of Wound Healing. *Adv. Wound Care* **2013,** *2* (5), 215–224

Pan, P.; Sun, Q. S. Chemical Constituents of *Callicarpa macrophylla* Vahl. *Shenyang Yaoke Daxue Xuebao* **2006,** *23*, 565–567.

Porras G.; Bacsa, J.; Tang, H.; Quave, C. L. Characterization and Structural Analysis of Genkwanin, a Natural Product from *Callicarpa americana*. *Crystals* **2019,** *9* (10), 491.
Qadrie, Z. L.; Balasubramanian, R.; Kavimani, S. Antioxidant Activity of Ethanolic Extracts of *Callicarpa linata* Leaf. *Pharmacologyonline* **2014,** *3* (5), 121–125.
Ren, F. Z.; He, B. K.; Luan, X. H.; Zhao, Y. M. Study on the Chemical Constituents of *Callicarpa bodinieri*. *Zhongguo Yaoxue Zazhi* **2004,** *39*, 17–19.
Ren, F. Z.; Niu, G. Y.; Luan, X. H.; Zhang, L.; Zhao, Y. M.; Liu, G. S. Study on the Analgesic Effect of *Callicarpa bodinieri* Levl. *Nat. Prod. Res.* **2003,** *15*, 155–156.
Ren, F. Z.; Qu, H. H.; Luan, X. H.; Zhao, Y. M. Studies on the Chemical Constituents of *Callicarpa bodinieri* Levl. *Nat. Prod. Res. Develop.* **2001,** *13*, 33–34.
Roy, S.; Zomuanpuia, J. H.; Laithangliani; Malsawmdawngkimi, R. L.; Hulianthanga; Lalhriatzuali, J.; Hauhnar, L.; Lalchhandama, L.; Lalhminghlui, K.; Lalthanpuii, P. B. Screening of *Callicarpa arborea* and *Hemigraphis alternata* for Antibacterial Activity. *Sci. Vis.* **2020,** *20*, 72–77.
Shankar, P. A.; Bhavana, S.; Manish, M. W.; Pandey, N. K.; Jadhav, A. D. *Callicarpa macrophylla*: Review of Its Phytochemistry, Pharmacology, Folklore Claims and Ayurvedic Studies. *Global J. Res. Med. Plants Indigen. Med.* **2014,** *3*, 91–100.
Shao, Y.; Hu, L. H.; Sim, K. Y.; Goh, S. H. Lignanoids and Diterpenoids from *Callicarpa furfuracea*. *Helv. Chim. Acta* **2006,** *89*, 64–72.
Sharma, G. N.; Harjinder, K.; Birendra, S.; Satish, C. A. Comparative In Vitro Antioxidant Activity of Ethyl Acetate and Ethanol Extracts of *Callicarpa macrophylla*. *Asian J. Pharm. Clin. Res.* **2021,** 31–35.
Sharma, H. K.; Gogoi, B.; Nainwal, L. M. A Review on Some Antioxidant Plant Species Growing in North East India. *Int. J. Pharm. Sci. Rev.* **2016,** *37*, 224–229.
Shihan, M. H.; Mahmud, Z. A.; Qais, N.; Riaz, M. Pharmacological Evaluation of Stem Bark of *Callicarpa arborea* Roxb. Dhaka. *Univ. J. Pharm. Sci.* **2015,** *14*, 111–116.
Shrilakshmi, G.; Kumar, N. H.; Jagannath, S. Preliminary Phytochemical Analysis, Antioxidant and Antibacterial Activities of *Callicarpa tomentosa* L.—An Ethnomedicinal Plant of Western Ghats, Karnataka, India. *Plant Cell Biotechnol. Mol. Biol.* **2011,** 157–166.
Siang, S. T. Use of Combined Traditional Chinese and Western Medicine in the Management of Burns. *Panminerva Med.* **1983,** *25*, 197–202.
Singh, A. K.; Agrawal, P. K. 16α, 17-Isopropylideno-3-oxophyllocladane, a Diterpenoid from *Callicarpa macrophylla*. *Phytochemistry* **1994,** *37*, 587–588.
Singh, A. K.; Chanotiya, C. S.; Yadav, A. Volatiles of *Callicarpa macrophylla*: Arich Source of Selinene Isomers. *Nat. Prod. Commun.* **2010,** *5*, 269–272.
Subramanian, S. S.; Nair, A. G. R.; Vedantham, T. N. C. Terpenoids and Flavones of *Callicarpa macrophylla* and *C. longifolia*. *Phytochemistry* **1974,** *13*, 306–307.
Suffness, M.; Abbott, B. J.; Statz, D.; Wonilowicz, E.; Spjut, R. The Utility of P388 Leukemia Compared to B16 Melanoma and Colon Carcinoma 38 for In Vivo Screening of Plant Extracts. *Phytother. Res.* **1988,** *2*, 89–97.
Taoubi, K.; Fauvel, M. T.; Gleye, J.; Fouraste, I. Etude des esters heterosidiques de l'acidecafeique chez des Verbenaceae. (Phenylpropanoid Glycosides from Verbenaceae). *Bull. Liaison Groupe Polyphenols*.**1992,** *16*, 174–177.
Tripathi, Y. C.; Nishat A.; Upadhyay, L. In Vitro Evaluation of Antifungal Activity of *Callicarpa macrophylla* Vahl Leaves. *OMCIJ* **2017,** *1* (4), 168–171.

Tu, Y.; Sum, L.; Guo, M.; Chen, W. The Medicinal Uses of *Callicarpa* L. in Traditional Chinese Medicine: An Ethnopharmacological, Phytochemical and Pharmacological Review. *J. Ethnopharmacol.* **2013,** *146*, 465–481.

Wang, Y. M.; Wang, F.; Xiao, H. Chemical Constituents of *Callicarpa arborea*Roxb. *Shizhen Guoyi Guoyao* **2011a,** *22*, 1341–1342.

Wang, Y. M.; Wang, F.; Xiao, H. Chemical Constituents of *Callicarpa formosana*. *Zhong Cao Yao* **2011b,** *42*, 1696–1698.

Wang, Z. N.; Han, Z.; Cui, H. B.; Dai, H. F. Chemical Constituents from *Callicarpa nudiflora*. *J. Trop. Subtrop. Bot.* **2007,** *15*, 359–362.

WenYu, M.; LePing, M.; Bo, Y.; Min, Z.; ShiXiu, F.; LiPing, T. Antidiabetic Activity of *Callicarpa nudiflora* Extract in Type 2 Diabetic Rats via Activation of the AMPK-ACC Pathway. *Asian Pac. J. Trop. Biomed.* **2019,** *9* (11), 456–466.

Wu, A. Z.; Lin, C. Z.; Zhai, Y. J.; Zhuo, J. L.; Zhu, C. C. Investigation of the Interaction Between Two Phenylethanoid Glycosides and Bovine Serum Albumin by Spectroscopic Methods. *J. Pharm. Anal.* **2012,** *3*, 61–65.

Xie, E. L.; Zhou, G. P.; Ji, T. F. A Novel Phenylpropanoid Glycoside from *Callicarpa kwangtungensis* Chun. *Chin. Chem. Lett.* **2009,** *20*, 827–829.

Xu, J.; Harrison, L. J.; Vittal, J. J.; Xu, Y. J.; Goh, S. H. Four New Clerodanediterpenoids from *Callicarpa pendandra. J. Nat. Prod.* **2000,** *63*, 1062–1065.

Yadav, V.; Jayalakshmi, S.; Singla, R. K.;Patra, A. Ex Vivo Screening of Stem Extracts of *Callicarpa macrophylla* Vahl for Antifungal Activity. *Indo Glob. J. Pharm.* **2012,** *2* (2), 103–107.

Yang, Y.; Li, Z. Y.; Shao, J. J.; Wang, G.; Wen, R.; Tian, J. Z.; Hou, L. *Callicarpa nudiflora* Hook. & Arn.: A Comprehensive Review of Its Phytochemistry and Pharmacology. *J. Ethnopharmacol.* **2021,** *264*, 113123.

Zhang, J.; Li, B. Q.; Feng, F.; Tang, Y. P.; Liu, W. Y. Chemical Constituents from *Callicarpa nudiflora* and Their Hemostatic Activity. *Zhongguo Zhong Yao ZaZhi* **2010,** *35*, 3297–3301.

Zhi-ying, Y.; Huan, Y.; Ying, Z.; Chao-zhan, L.; Guang-tian, P.; Ai-zhi, W. U.; Chen-chen, Z. Phenylpropanoid Compounds Isolated from *Callicarpa kwangtungensis* and Antibacterial Activity Research. *Nat. Prod. Res. Dev.* **2019,** *31* (11), 1928–1933.

Zhou, B. T.; Li, X. Z.; Xu, S. P. Chemical Composition from *Callicarpa cathayana*H. T. Chang. *Guangdong Yaoxueyuan Xuebao* **2005,** *21*, 695–696.

Zhou, B. T.; Li, X. Z.; Zhong, G. R.; Zheng, G. Y.; Chen, Z. Y. Elementary Pharmacodynamical Studies of Aerial Part of *Callicarpa kwungtungensis* Chun. *Chin. J. Mod. Med.* **2006,** *16*, 204–206.

Zhou, L. Y. Chemical Constituents in Aerial Parts of *Callicarpa pedunculata. Zhong Cao Yao* **2011,** *42*, 454–456.

Zhu, C. C.; Gao, L.; Zhao, Z. X.; Lin, C. Z. Triterpenes from *Callicarpa integerrima* Champ. *Acta Pharm. Sin. B.* **2012,** *47*, 77–83.

CHAPTER 27

PHYTOCHEMICAL CONSTITUENTS AND PHARMACOLOGY OF *OCIMUM GRATISSIMUM* L.

THADIYAN PARAMBIL IJINU[1,2], JENNY JACOB[3], SASIDHARAN LATHIKAKUMARI SREEJITH[4], SUDARSAN PARVATHY[5], and PALPU PUSHPANGADAN[1]

[1]*Amity Institute for Herbal and Biotech Products Development, Thiruvananthapuram, Kerala, India*

[2]*Naturæ Scientific, Kerala University Business Innovation and Incubation Centre, Karyavattom Campus, Thiruvananthapuram, Kerala, India*

[3]*School of Biosciences, Mar Athanasios College for Advanced Studies Tiruvalla, Pathanamthitta, Kerala, India*

[4]*Central Analytical Facility, Sree Chitra Tirunal Institute for Medical Sciences and Technology, Thiruvananthapuram, Kerala, India*

[5]*Department of Agricultural Biotechnology and Molecular Biology, Dr. Rajendra Prasad Central Agricultural University, Pusa, Bihar, India*

ABSTRACT

Ocimum gratissimum L. is an herbaceous perennial shrub that belongs to the family Lamiaceae. It is native to tropical Africa, India, and South East Asia. It is introduced and cultivated extensively across tropical and subtropical regions of the world. Traditionally, the leaves have been used to treat diarrhea, conjunctivitis, headaches, and as a general tonic. The essential oil

Bioactives and Pharmacology of Lamiaceae. T. Pullaiah, PhD (Ed.)

from the leaves is mixed with alcohol and used against infections; the oil is taken internally to treat bronchitis. The major essential oil components of *O. gratissimum* reported in various publications include eugenol, thymol, γ-terpinene, cis-ocimene, γ-muurolene p-cymene, 1,8-cineole, etc. Further, *O. gratissimum* is rich in polyphenols and flavonoids. *O. gratissimum* showed hepatoprotective, antioxidant, neuroprotective, antihypertensive, antiobesity, antidiabetic, anti-inflammatory, antinociceptive, wound healing, gastroprotective, anticancer, antidiarrheal, antiprotozoal, antifungal, and antibacterial properties.

27.1 INTRODUCTION

Ocimum gratissimum L. (Syn.: *Geniosporum discolor* Baker, *Ocimum arborescens* Bojer ex Benth., *Ocimum dalabaense* A.Chev., *Ocimum febrifugum* Lindl., *Ocimum guineense* Schumach. & Thonn., *Ocimum heptodon* P.Beauv., *Ocimum holosericeum* J.F.Gmel., *Ocimum petiolare* Lam., *Ocimum robustum* B.Heyne ex Hook.f., *Ocimum sericeum* Medik., *Ocimum suave* Willd., *Ocimum trichodon* Baker ex Gürke, *Ocimum urticifolium* Roth, *Ocimum viride* Willd., *Ocimum viridiflorum* Roth, *Ocimum zeylanicum* Medik.) belongs to the family Lamiaceae. The vernacular names of this plant include Ram tulsi, ban tulsi (Hindi), ajaka, ajeka, bilvaparni, doshakleshi (Sanskrit), Rama tulasi, kattei-tulluva, kattutrittavu, anathuasi, karpoorathulasi (Malayalam), elumichantulasi, peruntulasi, elumiccam tulaci (Tamil), ajagandhika, biliya kargoli, kaadu thulasi, lavanga thulasi, nimbe thulasi, rama-tulashi (Kannada), nimma-tulasi, rama-tulasi (Telugu), ajavala, ramatulasi, tanatulasu (Marathi), Sondabhogohulono (Oriya), Ram tulsi (Bengali), Ram tulsi (Manipuri), Ram tulsi (Assamese), faranjmishk, furanjmishk (Arabic), tukhm faranjmushk (Urdu), and Ram tulsi, wild basil, African basil, clove basil, shrubby basil, tree basil (English).

It is an herbaceous perennial shrub, up to 2 m high; stem glandular-scabrid. The leaves are elliptic to obovate, base attenuate serrate, acuminate at apex; nerves six pairs; thinly tomentose below; petiole up to 6 cm long. The racemes are terminal, panicled; bracts oblanceolate. The flowers are 4–6 at each node; pedicel 4 mm long; calyx 5 mm long, glabrous; upper lip ovate, obtuse, lobes of lower lip acuminate, glandular; corolla is white, tube 2–3 mm long, lobes obtuse; Filaments are glabrous or villous at base; Anthers are sagittate at base. The nutlets are 1.5 × 1 mm, pitted, brown

(Sasidharan, 2004, 2011). It is native to tropical Africa, India, and South East Asia. It is introduced and cultivated extensively across tropical and subtropical regions of the world. It has the potential to grow forming dense monospecific thickets that outcompetes the native species and reduce local habitat. The leaves have been used to treat diarrhea and conjunctivitis, and used as a general tonic. The essential oil from the leaves is mixed with alcohol and used against infections, and the oil is taken internally to treat bronchitis. The dried leaves are snuffed to alleviate headaches and fever (Iwu, 1993). In Kenya, *O. gratissimum* is used to treat blocked nose, snake bite (Kokwaro, 1993; Owuor et al., 2005), and as a mosquito repellent (Githinji and Kokwaro, 1994).

27.2 PHYTOCHEMICAL CONSTITUENTS

Chimnoi et al. (2018) found that the essential oil obtained from the fresh leaves of *O. gratissimum* contains mainly eugenol (55.6%), cis-ocimene (13.9%), γ-muurolene (11.6%), (Z, E)-α-farnesene (5.6%), α-trans-bergamotene (4.1%), and β-caryophyllene (2.7%). Mith et al. (2016) found that *O. gratissimum* essential oil contains thymol (28.3–37.7%) and γ-terpinene (12.5–19.3%) as the major constituents. Lima et al. (2018) found that the main compounds present in the *O. gratissimum* essential oil are thymol (33.4–47.9%), γ-terpinene (26.2–36.8%), and p-cymene (4.3–17.0%). The other components like germacrene, bisabolene, citral, geraniol, linalool, 1,8-cineole, β-pinene, cis-ocimene, trans-ocimene, camphor, methyl eugenol, caryophyllene, and germacrene-D are also present in the *O. gratissimum* essential oil (Sulistiarini et al., 1999; Keita et al., 2000; Vieira et al., 2001; Lemos et al., 2005; Matasyoh et al., 2007). Benitez et al. (2009) found that the essential oil components of *O. gratissimum* are eugenol (43%), 1,8-cineole (12%), and β-selinene (9%). Further polyphenols such as rosmarinic acid, salvigenin, and trans-ferulic acid; and flavonoids such as xanthomicrol, cirsimaritin, rutin, kaempferol 3-O-rutinoside, vicenin-2, luteolin 5-O-glucoside, luteolin 7-O-glucoside, apigenin 7-O-glucoside, vitexin, isovitexin, quercetin 3-O-glucoside, and isothymusin are also present in the *O. gratissimum* (Grayer et al., 2000; Costa et al., 2012; Ouyang et al., 2013, Venuprasad et al., 2014).

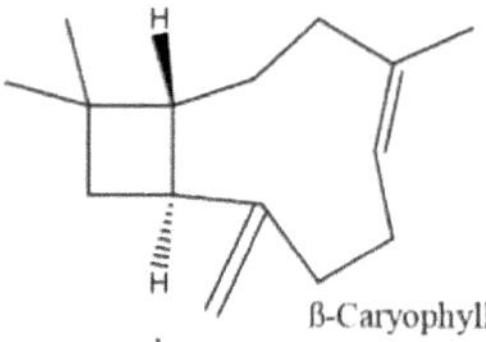

ß-Caryophyllene

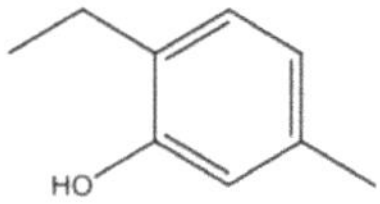

Thymol

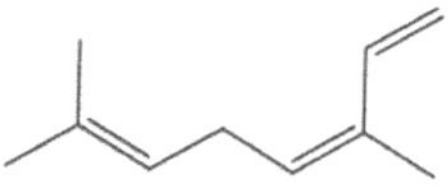
cis-Ocimene

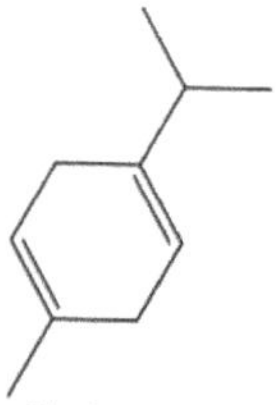
γ-Terpinene

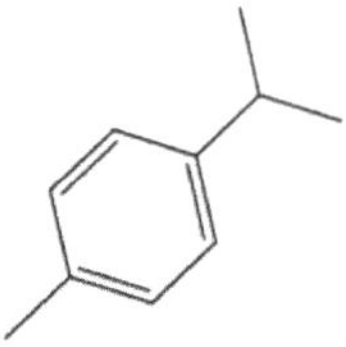
p-Cymene

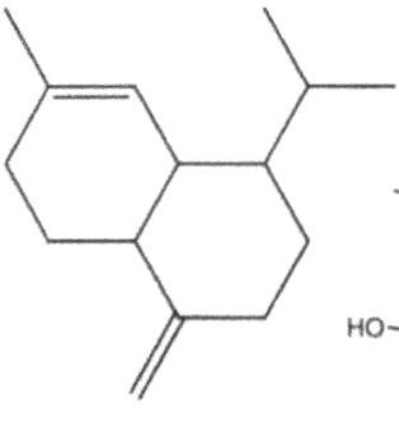
γ-Muurolene

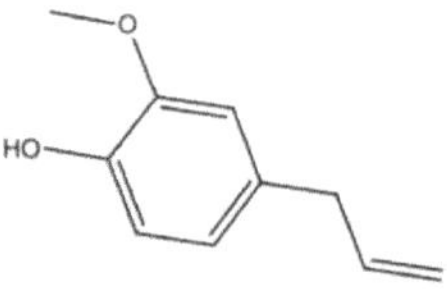

Eugenol

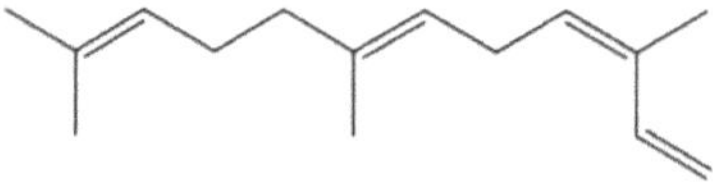
(Z,E)-α-Farnesene

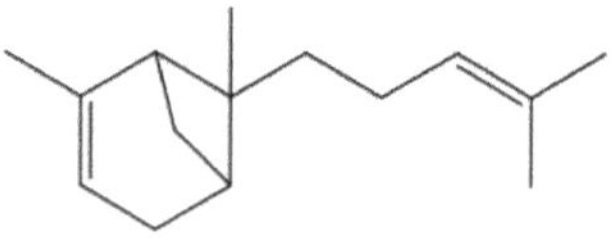
α-trans-Bergamotene

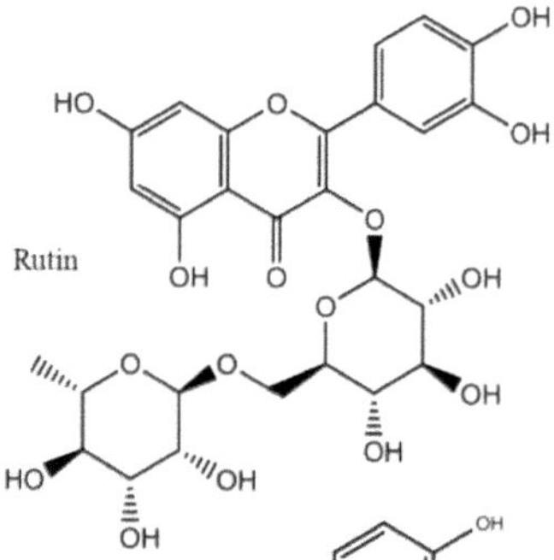

Rutin

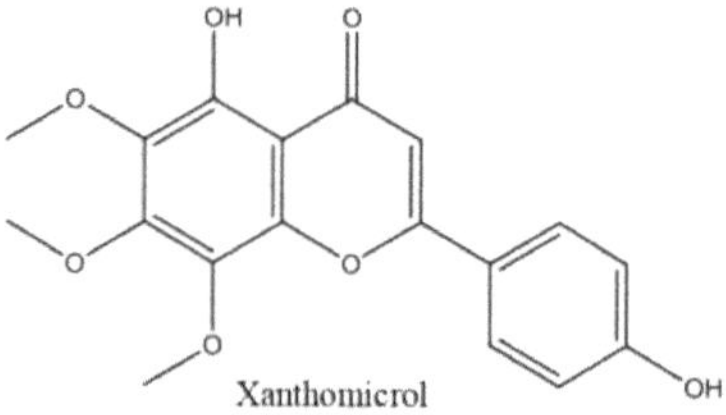

Xanthomicrol

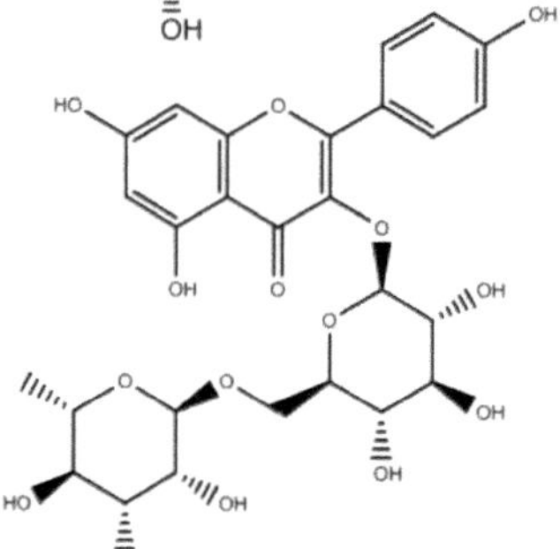

Kaempferol 3-O-rutinoside

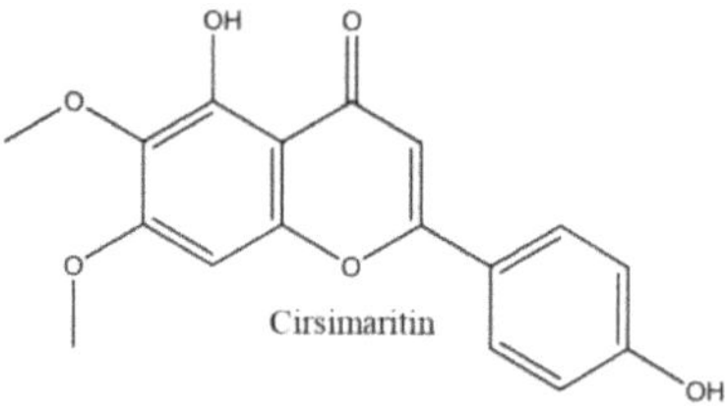

Cirsimaritin

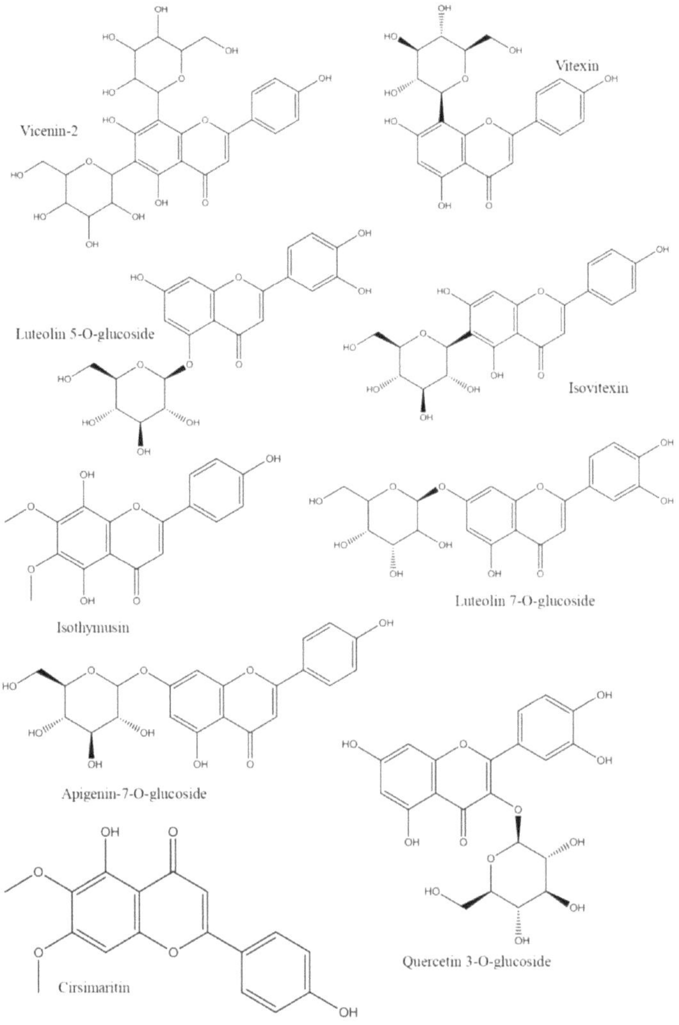

27.3 PHARMACOLOGICAL STUDIES

27.3.1 HEPATOPROTECTIVE ACTIVITY

O. gratissimum aqueous extract (0.2 mg/kg) showed hepatoprotective activity against carbon tetrachloride-induced toxicity in the male Wistar rats. The extract-treated group showed significant reduction in the expression of heat shock protein 70, induced nitric oxide synthase and urokinase-type plasminogen activator proteins along with a significant decrease in the matrix metalloproteinases-9/MMP-2 ratio (Chiu et al., 2012). Liver fibrosis induced by carbon tetrachloride in rats was protected from oxidative damage with the use of polyphenolic extract from *O. gratissimum*. The hepatic enzymes, alanine aminotransferase, aspartate aminotransferase, and catalase levels were improved, and malondialdehyde production was significantly reduced. The physiological changes in the liver were also maintained at normal levels including the weight and steatosis (Chen et al., 2015). Chiu et al. (2014) report on the preventive action of the *O. gratissimum* extracts against liver fibrosis in the male Wistar rats and primary cultures of hepatic stellate cells. The parameters studied were similar to the earlier study by Chen et al. (2015) showing a dose-dependent reduction in hepatic enzymes and other physiological changes. In the primary culture of the stellate cells, the expression of smooth muscle actin and collagen was significantly inhibited. Ighodaro and Ebuehi (2009) also reported the protective effect of *O. gratissimum* aqueous leaf extract.

27.3.2 ANTIOXIDANT ACTIVITY

The oral administration of 20 mg of *O. gratissimum* aqueous extract caused significant dose-dependent decline in the basal serum levels of total acid phosphatase and prostatic acid phosphatase, and the maximum inhibition was at 4 h (Aprioku and Obianime, 2008). Oyem et al. (2021) reported the protective activity of *O. gratissimum* against lead acetate-induced oxidative damage in Wistar rats by increasing antioxidant status. At the doses of 125 and 250 mg/kg, *O. gratissimum* extract was able to reverse reduction in superoxide dismutase and catalase activities to normal, increase glutathione concentrations, and normalize blood parameters reducing overall lead acetate-induced toxicity significantly. The leaf extract of *O. gratissimum* showed reduction in toxicity and anemia caused by phenyl hydrazine in Wistar rats. After 14 days of treatment, the levels of hepatic enzymes like aspartate aminotransferase, alanine aminotransferase, and alkaline phosphatase were significantly reduced as compared to the standard drug. Other marked effects were the increase in blood parameters like erythrocyte count and packed cell volume followed by an increase in hemoglobin concentration, thereby reducing anemia (Akara et al., 2021).

Mahapatra et al. (2009) initially reported on the antioxidant activity of the methanol extract of *O. gratissimum* in free radical-generated cell damage during nicotine toxicity. They have further reported that *O. gratissimum* methanol and aqueous extracts were less potent compared to eugenol (Mahapatra and Roy, 2014). Comparatively in cobalt chloride-induced cardiovascular and renal toxicity, *O. gratissimum* chloroform extract at a dose of 200 mg/kg was able to reverse the damages and provide antioxidant protection in Wistar rats (Akinrinde et al., 2016). The hydrogen peroxide-induced damage in the neural Schwann cells (RSC96) was also significantly minimized by the aqueous extract of *O. gratissimum* with downregulation of caspase-3 and Bax, followed by upregulation of Bcl-2 (Chao et al., 2017a, 2017b). The cell apoptosis was significantly inhibited by the aqueous extracts of *O. gratissimum*, after hydrogen peroxide induction, by the reduction of caspase activity and upregulation of apoptotic protease activating factor 1 and B-cell lymphoma 2 (Lee et al., 2011). Igbinosa et al. (2013) studied the in vitro antioxidant activity of different extracts of *O. gratissimum* through the 2,2-diphenyl-1-picrylhydrazyl and 2,2′-azino-bis(3-ethylbenzothiazoline-6-sulfonic acid) assays. The oral administration of *O. gratissimum* (500 and 1000 mg/kg) increased the red blood cells, packed cell volume, hemoglobin, platelet count, and neutrophils; and decreased the platelet indices like mean platelet volume, platelet-large cell ratio, and platelet distribution width (Ofem et al., 2012a, 2012b).

27.3.3 NEUROPROTECTIVE ACTIVITY

Eugenol and 1,8-cineole rich *O. gratissimum* essential oil showed good anticonvulsant and anxiolytic activities by protecting mice against electroshock-induced seizures (Freire et al., 2006). Galindo et al. (2010) studied the effect of eugenol and 1,8-cineole on anticonvulsant activity in animal models and concluded that the plant extract is better than individual compounds. A study conducted by Okoli et al. (2010) used the leaf extract of *O. gratissimum* for anticonvulsant and anxiolytic activities. The study revealed 50% protection from mortality in mice, while the acute toxicity testing revealed an LD_{50} value of methanol extract of *O. gratissimum* to be 5000 mg/kg. Ibironke and Modupe (2015) found the significant participation of the cholinergic system in *O. gratissimum* induced neurobehavioral alterations in mice. The pretreatment with *O. gratissimum* essential oil (150 or 300 mg/kg) significantly attenuated brain oxidative stress, damage, and neurological deficits (Bora et al., 2011).

27.3.4 ANTIHYPERTENSIVE AND ANTIOBESITY ACTIVITIES

Lahlou et al. (2004) reported the hypotensive activity of *O. gratissimum* essential oil, which is independent of autonomic nervous system but results in bradycardia. These cardiovascular effects were further studied in hypertension-induced rats, where it produced the same effects. The hypotensive effect of the essential oil of *O. gratissimum* was reported by Interaminense et al. (2007). An enhanced hypotensive effect was observed in rats and the mechanism of action was confirmed to be through smooth muscle relaxation of the blood vessels (Interaminense et al., 2005). The leaf extract of *O. gratissimum* inhibited the angiotensin-converting enzyme (IC_{50} 29.44 μg/mL) and pancreatic lipase (IC_{50} 20.69 μg/mL), resulting in control of obesity and hypertension (Irondi et al., 2016). *O. gratissimum* extract treatment in ovariectomized rats was conducted by Chao et al. (2017a, 2017b). The study indicates that the extract was able to reduce weight gain in the female rats with little or no effect on body fat composition.

27.3.5 ANTIDIABETIC ACTIVITY

The hypoglycemic potential of the methanolic extract of *O. gratissimum* (400 mg/kg) was demonstrated in alloxan-induced diabetic rats (Aguiyi et al., 2000). In streptozotocin-induced diabetic rats, the water extract of *O. gratissimum* leaf at a dose of 500 mg/kg showed 81.3% reduction in glucose after 24 h (Mohammed et al., 2007). A significant reduction in blood glucose concentrations was found in similar streptozotocin-induced diabetic rats after 28 days of treatment with aqueous leaf extract of *O. gratissimum* (Egesie et al., 2006). *O. gratissimum* aqueous leaf extract significantly reduced the blood glucose concentration in alloxan-induced diabetic rats, while other hematological parameters like red blood cell concentration and packed cell volume were increased (Shittu et al., 2016). Enyievi et al. (2020) reported the normalization of reduced blood cell parameters in streptozotocin-induced diabetic rats after a 28 day treatment with *O. gratissimum* leaf extract. The insulin sensitivity was reversed by 38.6% in streptozotocin-induced diabetic rats together with increased insulin availability (74.8%) (Okoduwa et al., 2017). Okon et al. (2015) observed the reduction in fecal glucose and protein concentrations in streptozotocin-induced diabetic rats. *O. gratissimum* (1500 mg/kg) treatment reduced the serum pancreatic amylase level in diabetic rats and limited the extent of structural damage (Akpan et al., 2014).

27.3.6 ANTI-INFLAMMATORY, ANTINOCICEPTIVE AND WOUND HEALING ACTIVITIES

Antinociceptive and wound healing potential of hydroethanolic extract of *O. gratissimum* leaves in rats was evaluated by Ajayi et al. (2014). The methanolic leaf extract of *O. gratissimum* showed 43.2% reduction of paw edema in rats induced by carageenan (Ajayi et al., 2017a). The extract significantly reduced the parameters like leukocyte count, tumor necrosis factor-α, xanthine oxidase, myeloperoxidase activity, and increased the glutathione level. Similar study of antinociception of the essential oil of *O. gratissimum* was conducted in male Swiss mice, reporting a 60% decrease in the pain response at a dose of 300 mg/kg (Rabelo et al., 2003). Orafidiya et al. (2003) found the wound healing potential of the essential oil of *O. gratissimum*, as evidenced by increased proliferation in albino rabbits. The wound healing potential of *O. gratissimum* leaf methanolic extract was studied by Osuagwu et al. (2004). The methanolic extract showed excellent wound healing in 10 days' time with significant wound contraction and lesser inflammatory cell infiltration. The antinociceptive activity of the aqueous extract of *O. gratissimum* leaves was studied by Tanko et al. (2008). The aqueous extract of *O. gratissimum* caused significant inhibition of inflammatory edema and pain. The LD_{50} value of the extract was found to be 1264.9 mg/kg body weight of mice.

The ethylacetate extract of *O. gratissimum* rich in phenolics reduced rat paw edema by 61% at 100 mg/kg and reduced oxidative stress by reducing tumor necrosis factor-α, myeloperoxidase activity, leucocyte count, and exudate volume (Ajayi et al., 2017b). A similar study of the flavonoid-rich extract of *O. gratissimum* also displayed similar results at 100 mg/kg in lipopolysaccharide-induced inflammation (Ajayi et al., 2019). Alabi et al. (2018) reported that the polyphenol extract of *O. gratissimum* regulates the production of tumor necrosis factor-α and other proinflammatory cytokines in dextran sodium sulfate-induced colitic rats. It also reduced the myeloperoxidase, cyclooxygenase-2, nitric oxide, and increased superoxide dismutase and catalase. The flavonoid-rich extract of *O. gratissimum* was also studied in lipopolysaccharide-induced RAW264.7 cells and mice peritonitis. The extract significantly reduced the concentration of inflammatory cells and proinflammatory cytokines in mice peritonitis (Ajayi et al., 2017c).

O. gratissimum aqueous extract at a dose of 10 μg/mL exhibited immunomodulation in macrophages by the decreasing the Th1 cytokine response and activating the Th2 response. The nitric oxide concentration was reduced together with intracellular nitric oxide synthase. Other functions of macrophages like phagocytosis, chemotaxis, and cell surface adherence were also found to be greatly enhanced (Mahapatra et al., 2011). The essential oil of *O. gratissimum* produced a vasodilator effect on aorta and mesenteric vessels of rats depending on the release of nitric oxide by the endothelium (Pires et al., 2012). Costa et al. (2012) reported the effect of *O. gratissimum* in allergic asthma and anaphylactic reactions. The extract reduces mucus production in the respiratory tract, eosinophilic cell concentrations, and its peroxidase activity in bronchial alveoli. The essential oil extracted from *O. gratissimum* and its active components were studied for its antinociceptive properties in adult male mice. Both eugenol and myrcene were able to reduce pain in the hot plate test and formalin test (Paula-Freire et al., 2013). Eugenol and myrcene were also effective against neuropathic pain from chronic injury of the sciatic nerve and seemed to possess antinociception, as evident in hot plate test and von Frey test (Paula-Freire et al., 2016). *O. gratissimum* extract inhibited cellular proliferation in RAW264.7 cells. The activity guided isolation leads to the identification of pomolic and tormentic acids (pentacyclic triterpenes). These compounds induced apoptosis as well as reduced the secretion of gamma interferons and cyclooxygenase (Dzoyem et al., 2021). Madeira et al. (2002) found that *O. gratissimum* essential oil showed relaxant effect on isolated ileum of the guinea pig. The studies in rabbits found that *O. gratissimum* shows analgesic activity and also protects from spasms (Aziba et al., 1999).

27.3.7 GASTROPROTECTIVE ACTIVITY

The treatment of *O. gratissimum* in rabbit gastrointestinal system significantly increased protection of the gastric mucosa with reduced ulceration (Nwangwu et al., 2013). *O. gratissimum* leaf extract reported gastroprotective and ulcer-inhibiting activities in albino Wistar rats (Ofem et al., 2012a, 2012b).

27.3.8 ANTICANCER ACTIVITY

The *O. gratissimum* aqueous extract of leaves was found to induce apoptosis in human hepatocellular carcinoma cells. The study proposes that the mechanism of apoptosis induction is through control of caspase-3, poly(ADP-ribose) polymerase, and phosphorylated extracellular signal-related kinase (Huang et al., 2020). The preliminary study on *O. gratissimum* leaf aqueous extract demonstrated its potential in preventive and therapeutic action against breast cancer (Nangia-Makker et al., 2007). The proposed action was based on inhibition of cancer cell proliferation, migration, and morphogenesis. Another study by Nangia-Makker et al. (2013) emphasized that the inhibition of chemotaxis and tumor growth is through the inhibition of matrix metalloproteases, MMP-9 and MMP-2. The effectiveness of *O. gratissimum* aqueous extract was studied in human adenocarcinoma A549 cells (Chen et al., 2011). The result showed that the reduction in cell viability is through the induction of apoptotic mechanisms like activated caspase enzymes, increased phosphorylation in kinases like p38 MAP kinase and c-Jun N-terminal kinase. *O. gratissimum* reduced the viability of osteosarcoma cells by the induction of apoptotic enzyme caspase-3. Other genes targeted include the stress-related proteins, HSPA1B and DDIT4, and proteins expressed in certain cancers like PPP1R15A and SQSTM1 (Lin et al., 2014). Ekunwe et al. (2010) reported the antiproliferative activity of fractions of the leaf extract of *O. gratissimum*. The fraction was further reported with antiproliferative activity in prostate cancer by the inhibition of survivin and androgen receptor (Ekunwe et al., 2013). Caffeic acid isolated from *O. gratissimum* at a concentration of 10 mM reduced cellular proliferation of the cervical cancer HeLa cells (Ye et al., 2010).

27.3.9 ANTIDIARRHEAL ACTIVITY

The aqueous leaf extract of *O. gratissimum* was able to prevent diarrhea in castor-oil treated rats and slowed the responses in castor oil-induced guinea pigs. The study measured the number of wet feces by the treated animals. The effect was mediated through the inhibition of intestinal motility and muscarinic receptors (Offiah and Chikwendu, 1999). Ezekwesili et al. (2004) found that the aqueous extract of *O. gratissimum* (25, 50, and 100 mg/kg) showed antidiarrheal activity against castor oil-induced diarrhea in Wistar albino rats. Iwalokun et al. (2003) studied the effect of leaf extract against diarrheal agent, *Shigella* strain isolates. The leaf extract was able to successfully reduce the virulence and invasivity of the multidrug-resistant strains through reduced expression of O-lipopolysaccharide content.

27.3.10 ANTIPROTOZOAL ACTIVITY

The essential oil of *O. gratissimum* showed inhibition in the development of *Herpetomonas samuelpessoai* (Holets et al., 2003). *H. samuelpessoai* was treated with 100 μg/mL of the essential oil and showed a deteriorating effect of succinate cytochrome c reductase catalyst. There is a critical increase in the quantity of cristae in the inner mitochondrial layer and the mitochondrial grids turn out to be less thick in few cells. Tchoumbougnang et al. (2005) used the essential oil of *O. gratissimum* against *Plasmodium berghei* infected mice in a 4 days treatment protocol, which showed effective reduction in the parasite count in the blood. The antimalarial activity was further investigated in the malarial vector, *Anopheles gambiae* with the *O. gratissimum* ethanolic extract whereby both larvae and pupae were exposed for 24 h. *O. gratissimum* extracts were found to be less effective compared to *O. basilicum*, but both were able to prevent the vector bites considerably. *Haemonchus contortus* and *Heligmosomoides polygyrus* larvae (15–16%) were paralyzed at 8 mg/mL concentration of *O. gratissimum* leaf extract and also produced 55.8% anthelmintic effect in *Ascaridia galli* infected chicks (Njoku and Asuzu, 1998). *H. contortus* eggs were subjected to *O. gratissimum* essential oil treatment and the extract was reported with anthelmintic activity at 0.5% concentration (Pessoa et al., 2002). Ueda-Nakamura et al. (2006) studied the effect of *O. gratissimum* essential oil on *Leishmania amazonensis*, indicating a minimum inhibitory concentration of 150 μg/mL at different stages of the life cycle of *Leishmania* spp.

27.3.11 ANTIFUNGAL ACTIVITY

O. gratissimum essential oil showed to control *Aspergillus flavus* growth and AFB1 synthesis, and the minimum inhibitory concentration value was found to be 800 mg/L (Tatsadjieu et al., 2008). Lima et al. (1993) and Nwosu and Okafor (1995) found that *O. gratissimum* showed inhibitory activity against *Trichophyton rubrum* and *T. mentagrophytes*. Dubey et al. (2000) found that thymol component of *O. gratissimum* showed good activity against *T. rubrum*, *T. mentagrophytes*, *Candida neoformans*, *C. albicans*, and *Malassezia pachydermatis*. Nakamura et al. (2004) demonstrated that the essential oil of *O. gratissimum* was effective against *C. albicans*, *C. krusei*, *C. parapsilosis*, and *C. tropicalis*. Silva et al. (2005) used the hexane fraction of *O. gratissimum* leaves and eugenol against *T. rubrum*, *T. mentagrophytes*, *Microsporum canis*, and *M. gypseum*. At 125 μg/mL concentration, the hexane fraction showed complete inhibition and the eugenol showed 80% of inhibition. Lemos et al. (2005) found that the chloroform fraction of *O. gratissimum* at concentration of 62.5 μg/mL inhibited 92% of *Cryptococcus* isolates; the hexane fraction inhibited eight (40%) isolates at a concentration of 250 μg/mL and the eugenol inhibited the growth of 16% of isolates of *Cryptococcus neoformans* at concentration of 0.9 μg/mL. The essential oil of *O. gratissimum* showed an inhibition of 8% of isolates at a concentration of 125 μg/mL. The essential oil obtained from *O. gratissimum* showed activity against *Botryosphaeria rhodina, Rhizoctonia* sp., and two strains of *Alternaria* sp. (Faria et al., 2006). The essential oil of *O. gratissimum* exhibited high sensitivity of minimum inhibitory concentration of 31.25 μg/mL against both *Fusarium oxysporum f.* sp. *lycopersici* and *Rhizoctonia solani* (Mohr et al., 2017).

27.3.12 ANTIBACTERIAL ACTIVITY

The essential oil from *O. gratissimum* showed good antibacterial effect against *Staphylococcus aureus* at a concentration of 0.75 mg/mL (Nakamura et al., 1999). The ethanolic extract of *O. gratissimum* leaves showed good antibacterial effect against *Escherichia coli* and *Listeria monocytogenes.* The minimum bactericidal concentration for *L. monocytogenes* was 2.15 mg/mL, and for *E. coli,* 2.50 mg/mL (Koche et al., 2012). *O. gratissimum* essential oil showed good activity against *Staphylococcus aureus, Streptococcus mutans, Enterococcus faecalis,* and *Staphylococcus epidermidis* with zone of inhibition 16, 14, 10, and 8 mm, respectively (Verma et al., 2011). The essential oil obtained from the leaves of *O. gratissimum* showed good antibacterial effect against both Gram-positive (*Staphylococcus aureus* and *Bacillus* spp.) and Gram-negative (*Escherichia coli, Pseudomonas aeruginosa, Salmonella typhi, Klebsiella pneumoniae,* and *Proteus mirabilis*) bacteria, and the minimum inhibitory concentration ranged from 93.7 to 150 mg/mL for Gram-positive bacteria and 107 to 750 mg/mL for Gram-negative bacteria (Matasyoh et al., 2007). A similar study was conducted with *O. gratissimum* collected from different population zones in Nigeria where the Meru oil was found to be more consistently effective than the Mombasa oil and had a higher concentration (92.48%) of monoterpene content (Matasyoh et al., 2008). The essential oil of *O. gratissimum* was also studied in the microbial isolates from dental patients, which revealed that it was effective against *K. pneumoniae* (0.5%), *P. aeruginosa* (10%), *Streptococcus viridans* (1.1%), *Staphylococcus albus* (1.1%), and *Proteus vulgaris* (0.53%) (Ahonkhai et al., 2009). Joshi (2013) reported that the eugenol-rich essential oil of *O. gratissimum* was highly effective against *Serratia marcescens*. The multidrug-resistant *S. aureus* and *E. coli* were inhibited by *O. gratissimum* essential oil extract at a minimum inhibitory concentration of 1000 μg/mL (Melo et al., 2019).

27.3.13 TOXICITY STUDIES

The aqueous leaves extract of *O. gratissimum* showed inhibitory activity on enzymes like phosphodiesterase-5, angiotensin I converting enzyme, acetylcholinesterase, and arginase, relevant to erectile dysfunction. The inhibition is higher in penile tissue than that of testicular tissue homogenate (Ojo et al., 2019). Male fertility was studied in Wistar rats with treatment of *O. gratissimum* methanol extract (25, 100, and 400 mg/kg) of leaves leading to increased levels of testosterone and sperm count (Njan et al., 2019). Joseph et al. (2019) reported that the treatment of methanol extract of *O. gratissimum* in rats did not cause any problems or toxicity with male fertility. The *O. gratissimum* essential oil stimulates the immune system and triggers the chronic inflammatory response on continuous exposure (Orafidiya et al., 2004). The effect of aqueous leaf extract of *O. gratissimum* on the renal function of Wistar rats with gentamicin-induced nephrotoxicity was investigated by Ogundipe et al. (2017). The extract showed significant protection from renal toxicity by the decrease in creatinine clearance. Investigations have led to the finding that *O. gratissimum* leaf extracts may aggravate infertility in alloxan-induced rats through altered cellular architecture and sperm count in testis (Shittu et al., 2019).

ACKNOWLEDGMENTS

The authors express their sincere thanks to Dr. Ashok K. Chauhan, Founder President, Ritnand Balved Education Foundation (RBEF) and Amity Group of Institutions, and Dr. Atul Chauhan, Chancellor, Amity University Uttar Pradesh (AUUP) for facilitating this work. Thadiyan Parambil Ijinu has received Young Scientist Fellowship from the Department of Science and Technology, Government of India (SP/YO/413/2018).

KEYWORDS

- **traditional medicinal use**
- **eugenol**
- **polyphenols**
- **flavonoids**
- **hepatoprotective**
- **antidiabetic**
- **antihypertensive**

REFERENCES

Aguiyi, J. C.; Obi, C. I.; Gang, S. S.; Igweh, A. C. Hypoglycaemic Activity of *Ocimum gratissimum* in rats. *Fitoterapia.* **2000,** *71*, 444–446.

Ahonkhai, I.; Ba, A.; Edogun, O.; Mu, U. Antimicrobial Activities of the Volatile Oils of *Ocimum basilicum* L. and *Ocimum gratissimum* L. (Lamiaceae) Against Some Aerobic Dental Isolates. *Pak. J. Pharm. Sci.* **2009,** *22*, 405–409.

Ajayi, A. M.; Ben-Azu, B.; Onasanwo, S. A.; Adeoluwa, O.; Eduviere, A.; Ademowo, O. G. Flavonoid-Rich Fraction of *Ocimum gratissimum* Attenuates Lipopolysaccharide-Induced Sickness Behavior, Inflammatory and Oxidative Stress in Mice. *Drug Res. (Stuttg).* **2019,** *69*, 151–158.

Ajayi, A. M.; Ologe, M. O.; Ben-Azu, B.; Okhale, S. E.; Adzu, B.; Ademowo, O. G. *Ocimum gratissimum* Linn. Leaf Extract Inhibits Free Radical Generation and Suppressed Inflammation in Carrageenan-Induced Inflammation Models in Rats. *J. Basic Clin. Physiol. Pharmacol.* **2017a,** *27*, 531–541.

Ajayi, A. M.; Umukoro, S.; Ben-Azu, B.; Adzu, B.; Ademowo, O. G. Toxicity and Protective Effect of Phenolic-Enriched Ethylacetate Fraction of *Ocimum gratissimum* (Linn.) Leaf Against Acute Inflammation and Oxidative Stress in Rats. *Drug Dev. Res.* **2017b,** *78*, 135–145.

Ajayi, A. M.; Martins, D. T. O.; Balogun, S. O.; Oliveira, R. G.; Ascêncio, S. D.; Soares, I. M.; Barbosa, R. D. S.; Ademowo, O. G. *Ocimum gratissimum* L. Leaf Flavonoid-Rich Fraction Suppress LPS-Induced Inflammatory Response in RAW 264.7 Macrophages and Peritonitis in Mice. *J. Ethnopharmacol.* **2017c,** *23*, 169–178.

Ajayi, A. M.; Tanayen, J. K.; Ezeonwumelu, J.; Dare, S.; Okwanachi, A.; Adzu, B.; Ademowo, O. G. Anti-inflammatory, Anti-Nociceptive and Total Polyphenolic Content of Hydroethanolic Extract of *Ocimum gratissimum* L. Leaves. *Afr. J. Med. Med. Sci.* **2014,** *43*, 215–224.

Akara, E. U.; Emmanuel, O.; Ude, V. C.; Uche-Ikonne, C.; Eke, G.; Ugbogu, E. A. *Ocimum gratissimum* Leaf Extract Ameliorates Phenylhydrazine-Induced Anaemia and Toxicity in Wistar Rats. *Drug Metab Pers Ther.* **2021.** Doi: 10.1515/dmdi-2020–0185

Akinrinde, A. S.; Oyagbemi, A. A.; Omobowale, T. O.; Asenuga, E. R.; Ajibade, T. O. Alterations in Blood Pressure, Antioxidant Status and Caspase 8 Expression in Cobalt

Chloride-Induced Cardio-Renal Dysfunction Are Reversed by *Ocimum gratissimum* and Gallic Acid in Wistar Rats. *J. Trace Elem. Med. Biol.* **2016,** *36*, 27–37.

Akpan, O. U.; Bassey, R. B.; Agba, B. S.; Edegha, I. A. Elevation of Serum Pancreatic Amylase and Distortion of Pancreatic Cyto-Architecture in Type 1 Diabetes Mellitus Rats Treated with *Ocimum gratissimum*. *Niger. Med. J.* **2014,** *55*, 34–38.

Alabi, Q. K.; Akomolafe, R. O.; Omole, J. G.; Adefisayo, M. A.; Ogundipe, O. L.; Aturamu, A.; Sanya, J. O. Polyphenol-Rich Extract of *Ocimum gratissimum* Leaves Ameliorates Colitis via Attenuating Colonic Mucosa Injury and Regulating Pro-Inflammatory Cytokines Production and Oxidative Stress. *Biomed. Pharmacother.* **2018,** *103*, 812–822.

Aprioku, J. S.; Obianime, A. W. Antioxidant Activity of the Aqueous Crude Extract of *Ocimum gratissimum* Linn. Leaf on Basal and Cadmium-Induced Serum Levels of Phosphatises in Male Guinea-Pigs. *J. Appl. Sci. Environ. Manage.* **2008,** *12*, 33–39.

Aziba, P. I.; Bass, D.; Elegbe, Y. Pharmacological Investigation of *Ocimum gratissimum* in Rodents. *Phytother. Res.* **1999,** *13*, 427–429.

Benitez, N. P.; León, E. M. M.; Stashenko, E. E. Eugenol and Methyl Eugenol Chemotypes of Essential Oil of Species *Ocimum gratissimum* L. and *Ocimum campechianum* Mill. from Colombia. *J. Chromatogr. Sci.* **2009,** *47*, 800–803.

Bora, K. S.; Shri, R.; Monga, J. Cerebroprotective Effect of *Ocimum gratissimum* Against Focal Ischemia and Reperfusion-Induced Cerebral Injury. *Pharm. Biol.* **2011,** *49*, 175–181.

Chao, P. Y.; Chiang, T. I.; Chang, I. C.; Tsai, F. L.; Lee, H. H.; Hsieh, K.; Chiu, Y. W.; Lai, T. J.; Liu, J. Y.; Hsu, L. S.; Shih, Y. C. Amelioration of Estrogen-Deficiency-Induced Obesity by *Ocimum gratissimum*. *Int. J. Med. Sci.* **2017a,** *14*, 896–901.

Chao, P. Y.; Lin, J. A.; Ye, J. C.; Hwang, J. M.; Ting, W. J.; Huang, C. Y.; Liu, J. Y. Attenuation of Oxidative Stress-Induced Cell Apoptosis in Schwann RSC96 Cells by *Ocimum gratissimum* aqueous extract. *Int. J. Med. Sci.* **2017b,** *14*, 764–771.

Chen, H. M.; Lee, M. J.; Kuo, C. Y.; Tsai, P. L.; Liu, J. Y.; Kao, S. H. *Ocimum gratissimum* Aqueous Extract Induces Apoptotic Signalling in Lung Adenocarcinoma Cell A549. *Evid. Based Complem. Altern. Med.* **2011,** 739093.

Chen, Y. H.; Chiu, Y. W.; Shyu, J. C.; Tsai, C. C.; Lee, H. H.; Hung, C. C.; Hwang, J. M.; Liu, J. Y.; Wang, W. H. Protective Effects of *Ocimum gratissimum* Polyphenol Extract on Carbon Tetrachloride-Induced Liver Fibrosis in Rats. *Chin. J. Physiol.* **2015,** *58*, 55–63.

Chimnoi, N.; Reuk-Ngam, N.; Chuysinuan, P.; Khlaychan, P.; Khunnawutmanotham, N.; Chokchaichamnankit, D.; Thamniyom, W.; Klayraung, S.; Mahidol, C.; Techasakul, S. Characterization of Essential Oil from *Ocimum gratissimum* Leaves: Antibacterial and Mode of Action Against Selected Gastroenteritis Pathogens. *Microb. Pathog.* **2018,** *118*, 290–300.

Chiu, C. C.; Huang, C. Y.; Chen, T. Y.; Kao, S. H.; Liu, J. Y.; Wang, Y. W.; Tzang, B. S.; Hsu, T. C. Beneficial Effects of *Ocimum gratissimum* Aqueous Extract on Rats with CCl_4 Induced Acute Liver Injury. *Evid. Based Complem. Altern. Med.* **2012,** 736752.

Chiu, Y. W.; Chao, P. Y.; Tsai, C. C.; Chiou, H. L.; Liu, Y. C.; Hung, C. C.; Shih, H. C.; Lai, T. J.; Liu, J. Y. *Ocimum gratissimum* Is Effective in Prevention Against Liver Fibrosis In Vivo and In Vitro. *Am. J. Chin. Med.* **2014,** *42*, 833–852.

Costa, R. S.; Carneiro, T. C.; Cerqueira-Lima, A. T.; Queiroz, N. V.; Alcântara-Neves, N. M.; Pontes-de-Carvalho, L. C.; Velozo, E. S.; Oliveira, E.J; Figueiredo, C. A. *Ocimum gratissimum* Linn. and Rosmarinic Acid, Attenuate Eosinophilic Airway Inflammation in an Experimental Model of Respiratory Allergy to *Blomia tropicalis*. *Int. Immunopharmacol.* **2012,** *13*, 126–134.

Dubey, N. K.; Tiwari, T. N.; Mandin, D.; Andriamboavonjy, H.; Chaumont, J. P. Antifungal Properties of *Ocimum gratissimum* Essential Oil (Ethyl Cinnamate Chemotype). *Fitoterapia*. **2000,** *71*, 567–569.

Dzoyem, J. P.; Nganteng, D. N. D.; Melong, R.; Wafo, P.; Ngadjui, B.; Allémann, E.; Delie, F. Bioguided Identification of Pentacyclic Triterpenoids as Anti-Inflammatory Bioactive Constituents of *Ocimum gratissimum* Extract. *J. Ethnopharmacol*. **2021,** *268*, 113637.

Egesie, U. G.; Adelaiye, A. B.; Ibu, J. O.; Egesie, O. J. Safety and Hypoglycaemic Properties of Aqueous Leaf Extract of *Ocimum gratissimum* in Streptozotocin Induced Diabetic Rats. *Niger. J. Physiol. Sci.* **2006,** *21*, 31–35.

Ekunwe, S. I.; Hall, S. M.; Luo, X.; Wang, H.; Begonia, G. B. Fractionated *Ocimum gratissimum* Leaf Extract Inhibit Prostate Cancer (PC3·AR) Cells Growth by Reducing Androgen Receptor and Survivin Levels. *J. Health Care Poor Underserved* **2013,** *24*, 61–69.

Ekunwe, S. I.; Thomas, M. S.; Luo, X.; Wang, H.; Chen, Y.; Zhang, X.; Begonia, G. B. Potential Cancer-Fighting *Ocimum gratissimum* (OG) Leaf Extracts: Increased Anti-Proliferation Activity of Partially Purified Fractions and Their Spectral Fingerprints. *Ethn. Dis*. **2010,** *20*, S1–12–16.

Enyievi, P. B.; Mgbeje, B. I. A.; Nja, G. M. E.; Edu, B. C.; Ejemot-Nwadiaro, R. I. Effect of *Ocimum gratissimum* Leaf-Extract on Hematological Indices and Lipid Profile of Streptozotocin-Induced Diabetic Wistar Rats. *Pak. J. Biol. Sci.* **2020,** *23*, 1523–1529.

Ezekwesili, C. N.; Obiora, K. A.; Ugwu, O. P. Evaluation of Anti-Diarrhoeal Property of Crude Aqueous Extract of *Ocimum gratissimum* L. (Labiatae) in Rats. *Biokemistri*. **2004,** *16*, 122–131.

Faria, T. D. J.; Ferreira, R. S.; Yassumoto, L.; Souza, J. R. P. D.; Ishikawa, N. K.; Barbosa, A. D. M. Antifungal Activity of Essential Oil Isolated from *Ocimum gratissimum* L. (Eugenol Chemotype) Against phytopathogenic Fungi. *Braz. Arch. Biol. Technol.* **2006,** *49*, 867–871.

Freire, C. M.; Marques, M. O.; Costa, M. Effects of Seasonal Variation on the Central Nervous System Activity of *Ocimum gratissimum* L. Essential Oil. *J. Ethnopharmacol.* **2006,** *105*, 161–166.

Galindo, L. A.; Pultrini, A. M.; Costa, M. Biological Effects of *Ocimum gratissimum* L. Are Due to Synergic Action Among Multiple Compounds Present in Essential Oil. *J. Nat. Med.* **2010,** *64*, 436–441.

Githinji, C. W.; Kokwaro, J. O., Ethnomedicinal Study of Major Species in the Family Labiatae from Kenya. *J. Ethnobiol.* **1994,** *39*, 197–203.

Grayer, R. J.; Kite, G. C.; Abou-Zaid, M.; Archer, L. J. The Application of Atmospheric Pressure Chemical Ionisation Liquid Chromatography-Mass Spectrometry in the Chemotaxonomic Study of Flavonoids: Characterisation of Flavonoids from *Ocimum gratissimum* var. *gratissimum*. *Phytochem. Anal.* **2000,** *11*, 257–267.

Holets, F. B.; Ueda-Nakamura, T.; Filho, B. P. D.; Cortez, D. A. G.; Morgado-Díaz, J. A.; Nakamura, C. V. Effect of Essential Oil of *Ocimum gratissimum* on the Trypanosomatid *Herpetomonas samuelpessoai*. *Acta Protozool.* **2003,** *42*, 269–276.

Huang, C. C.; Hwang, J. M.; Tsai, J. H.; Chen, J. H.; Lin, H.; Lin, G. J.; Yang, H. L.; Liu, J. Y.; Yang, C. Y.; Ye, J. C. Aqueous *Ocimum gratissimum* Extract Induces Cell Apoptosis in Human Hepatocellular Carcinoma Cells. *Int. J. Med. Sci.* **2020,** *17*, 338–346.

Ibironke, G. F.; Modupe, O. G. Non Cholinergic Dependent Mechanism of *Ocimum gratissimum* Induced Neurobehavioural Alterations in Mice. *Afr. J. Med. Med. Sci.* **2015,** *44*, 213–220.

Igbinosa, E. O.; Uzunuigbe, E. O.; Igbinosa, I. H.; Odjadjare, E. E.; Igiehon, N. O.; Emuedo, O. A. In Vitro Assessment of Antioxidant, Phytochemical and Nutritional Properties of Extracts from the leaves of *Ocimum gratissimum* (Linn). *Afr. J. Tradit. Complement. Altern. Med.* **2013,** *10*, 292–298.

Ighodaro, O. M.; Ebuehi, O. A. Aqueous Leaf Extract of *Ocimum gratissimum* Potentiates Activities of Plasma and Hepatic Antioxidant Enzymes in Rats. *Nig. Q. J. Hosp. Med.* **2009,** *19*, 106–109.

Interaminense, L. F.; Jucá, D. M.; Magalhães, P. J.; Leal-Cardoso, J. H.; Duarte, G. P.; Lahlou, S. Pharmacological Evidence of Calcium-Channel Blockade by Essential Oil of *Ocimum gratissimum* and Its Main Constituent, Eugenol, in Isolated Aortic Rings from DOCA-Salt Hypertensive Rats. *Fundam. Clin. Pharmacol.* **2007,** *21*, 497–506.

Interaminense, L. F.; Leal-Cardoso, J. H.; Magalhães, P. J., Duarte, G. P.; Lahlou, S. Enhanced Hypotensive Effects of the Essential Oil of *Ocimum gratissimum* Leaves and Its Main Constituent, Eugenol, in DOCA-Salt Hypertensive Conscious Rats. *Planta Med.* **2005,** *71*, 376–378.

Irondi, E. A.; Agboola, S. O.; Oboh, G.; Boligon, A. A. Inhibitory Effect of Leaves Extracts of *Ocimum basilicum* and *Ocimum gratissimum* on Two Key Enzymes Involved in Obesity and Hypertension In Vitro. *J. Intercult. Ethnopharmacol.* **2016,** *5*, 396–402.

Iwalokun, B. A.; Gbenle, G. O.; Adewole, T. A.; Smith, S. I.; Akinsinde, K. A.; Omonigbehin, E. O. Effects of *Ocimum gratissimum* L Essential Oil at Subinhibitory Concentrations on Virulent and Multidrug-Resistant *Shigella* Strains from Lagos, Nigeria. *APMIS J. Path. Microbiol. Immunol.* **2003,** *111*, 477–482.

Iwu, M. M. *Handbook of African Medicinal Plants*; CRC Press Inc.: Boca Raton, 1993.

Joseph, I. E.; Jaja, I. F.; Boyi, A. H.; Olugbenga, O. M. Comparative Effects of Methanol and Oil Extracts of *Ocimum gratissimum* on Testicular Morphology and Epididymal Sperm Reserve of Adult Male Albino Rats (Wistar Strain). *Toxicol. Rep.* **2019,** *17*, 1127–1134.

Joshi, R. K. Chemical Composition, In Vitro Antimicrobial and Antioxidant Activities of the Essential Oils of *Ocimum gratissimum, O. sanctum* and Their Major Constituents. *Indian J. Pharm. Sci.* **2013,** *75*, 457–462.

Keita, S. M.; Vincent, C.; Jean-Pierre, S.; Bélanger, A. Essential Oil Composition of *Ocimum basilicum* L., *O. gratissimum* L., and *O. suave* L. in the Republic of Guinea. *Flavour Fragr. J.* **2000,** *15*, 339–341.

Koche, L. D. K.; Kokate, P. S.; Suradkar, S. S.; Bhadange, D. G. Preliminary Phytochemistry and Antibacterial Activity of Ethanolic Extract of *Ocimum gratissimum. Biosci. Discov.* **2012,** *3*, 20–24.

Kokwaro, J. O. *Medicinal Plants of East Africa*, 2nd ed.; Kenya Literature Bureau: Nairobi, 1993.

Lahlou, S.; Interaminense, Lde F.; Leal-Cardoso, J. H.; Morais, S. M.; Duarte, G. P. Cardiovascular Effects of the Essential Oil of *Ocimum gratissimum* Leaves in Rats: Role of the Autonomic Nervous System. *Clin. Exp. Pharmacol. Physiol.* **2004,** *31*, 219–225.

Lee, M. J.; Chen, H. M.; Tzang, B. S.; Lin, C. W.; Wang, C. J.; Liu, J. Y.; Kao, S. H. *Ocimum gratissimum* Aqueous Extract Protects H9c2 Myocardiac Cells from H_2O_2 Induced Cell Apoptosis Through Akt Signalling. *Evid. Based Complem. Altern. Med.* **2011,** 578060.

Lemos, J. A.; Passos, X. S.; Fernandes, O. F.; Paula, J. R.; Ferri, P. H.; Souza, L. K.; Lemos, A. A.; Silva, M. R. Antifungal Activity from *Ocimum gratissimum* L. Towards *Cryptococcus neoformans. Mem. Inst. Oswaldo Cruz.* **2005,** *100*, 55–58.

Lima, A. S.; Milhomem, M. N.; Monteiro, O. S.; Arruda, A. C. P.; de Castro, J. A. M.; Fernandes, Y. M. L.; Maia, J. G. S.; Costa-Junior, L. M. Seasonal Analysis and Acaricidal Activity of the Thymol-Type Essential Oil of *Ocimum gratissimum* and Its Major Constituents Against *Rhipicephalus microplus* (Acari: Ixodidae). *Parasitol. Res.* **2018,** *117*, 59–65.

Lima, E. O.; Gompertz, O. F.; Giesbrecht, A. M.; Paulo, M. Q. In Vitro Antifungal Activity of Essential Oils Obtained from Officinal Plants Against Dermatophytes. *Mycoses* **1993,** *36*, 333–336.

Lin, C. C.; Chao, P. Y.; Shen, C. Y.; Shu, J. J.; Yen, S. K.; Huang, C. Y.; Liu, J. Y. Novel Target Genes Responsive to Apoptotic Activity by *Ocimum gratissimum* in Human Osteosarcoma Cells. *Am. J. Chin. Med.* **2014,** *42*, 743–767.

Madeira, S. V.; Matos, F. J.; Leal-Cardoso, J. H.; Criddle, D. N. Relaxant Effects of the Essential Oil of *Ocimum gratissimum* on Isolated Ileum of the Guinea Pig. *J. Ethnopharmacol.* **2002,** *81*, 1–4.

Mahapatra, S. K.; Chakraborty, S. P.; Das, S.; Roy, S. Methanol Extract of *Ocimum gratissimum* Protects Murine Peritoneal Macrophages from Nicotine Toxicity by Decreasing Free Radical Generation, Lipid and Protein Damage and Enhances Antioxidant Protection. *Oxid. Med. Cell Longev.* **2009,** *2*, 222–230.

Mahapatra, S. K.; Chakraborty, S. P.; Roy, S. Immunomodulatory Role of *Ocimum gratissimum* and Ascorbic Acid Against Nicotine-Induced Murine Peritoneal Macrophages In Vitro. *Oxid. Med. Cell Longev.* **2011,** *2011*, 734319.

Mahapatra, S. K.; Roy, S. Phytopharmacological Approach of Free Radical Scavenging and Anti-Oxidative Potential of Eugenol and *Ocimum gratissimum* Linn. *Asian Pac. J. Trop. Med.* **2014,** *S1*, S391–397.

Matasyoh, L. G.; Matasyoh, J. C.; Wachira, F. N.; Kinyua, M. G.; Muigai, A. W.; Mukiama, T. K. Antimicrobial Activity of Essential Oils of *Ocimum gratissimum* L. from Different Populations of Kenya. *Afr. J. Trad. Complem. Altern. Med.* **2008,** *5*, 187–193.

Matasyoh, L. G.; Matasyoh, J. C.; Wachira, F. N.; Kinyua, M. G.; Muigai, A. W. T.; Mukiama, T. K. Chemical Composition and Antimicrobial Activity of the Essential Oil of *Ocimum gratissimum* L. Growing in Eastern Kenya. *Afr. J. Biotechnol.* **2007,** *6*, 760–765.

Melo, R. S.; Albuquerque, A. Á. M.; Gomes, P. A. M.; Rocha, R. R.; Bastos, C. R. M.; Carneiro, M. M. N.; Ribeiro, L. P. H.; Gomes, G. A.; Soares, R. T. H.; Santos, H. S. D.; Ponte, I. L.; Costa, R. A.; Brito, G. S.; Catunda, J. F. E. A.; Carneiro, V. A. Chemical Composition and Antimicrobial Effectiveness of *Ocimum gratissimum* L. Essential Oil Against Multidrug-Resistant Isolates of *Staphylococcus aureus* and *Escherichia coli*. *Mols* **2019,** *24*, 3864.

Mith, H.; Yayi-Ladékan, E.; Kpoviessi, S. D. S.; Bokossa, I. Y.; Moudachirou, M.; Daube, G.; Clinquart, A. Chemical Composition and Antimicrobial Activity of Essential Oils of *Ocimum basilicum*, *Ocimum canum* and *Ocimum gratissimum* in Function of Harvesting Time. *J. Essent. Oil Bear. Plants*. **2016,** *19*, 1413–1425.

Mohammed, A.; Tanko, Y.; Okasha, M. A.; Magaji, R. A.; Yaro, A. H. Effects of Aqueous Leaves Extract of *Ocimum gratissimum* on Blood Glucose Levels of Streptozotocin Induced Diabetic Wistar Rats. *Afr. J. Biotechnol.* **2007,** *6*, 2087–2090.

Mohr, F. B.; Lermen, C.; Gazim, Z. C.; Gonçalves, J. E.; Alberton, O. Antifungal Activity, Yield, and Composition of *Ocimum gratissimum* essential oil. *Genet. Mol. Res.* **2017,** *16*, gmr16019542.

Nakamura, C. V.; Ishida, K.; Faccin, L. C.; Filho, B. P.; Cortez, D. A.; Rozental, S.; de Souza, W.; Ueda-Nakamura, T. In Vitro Activity of Essential Oil from *Ocimum gratissimum* L. Against Four *Candida* Species. *Res. Microbiol.* **2004,** *155*, 579–586.

Nakamura, C. V.; Nakamura, T. U.; Bando, E.; Melo, A. F. N.; Cortez, D. A. G.; Filho, B. P. D. Antibacterial Activity of *Ocimum gratissimum* Essential Oil. *Memórias do Instituto Oswaldo Cruz*. **1999,** *94*, 675–678.

Nangia-Makker, P.; Raz, T.; Tait, L.; Shekhar, M. P.; Li, H.; Balan, V.; Makker, H.; Fridman, R.; Maddipati, K.; Raz, A. *Ocimum gratissimum* Retards Breast Cancer Growth and Progression and Is a Natural Inhibitor of Matrix Metalloproteases. *Cancer Biol. Ther.* **2013,** *14*, 417–427.

Nangia-Makker, P.; Tait, L.; Shekhar, M. P.; Palomino, E.; Hogan, V.; Piechocki, M. P.; Funasaka, T.; Raz, A. Inhibition of Breast Tumor Growth and Angiogenesis by a Medicinal Herb: *Ocimum gratissimum*. *Int. J. Cancer*. **2007,** *121*, 884–894.

Njan, A. A.; Olaoye, S. O.; Afolabi, S. O.; Ejimkonye, B. C.; Soje, A.; Olorundare, O. E.; Iwalewa, E. O. Safety Effect of Fractions from Methanolic Leaf Extract of *Ocimum gratissimum* on Reproduction in Male Wistar Rats. *Toxicol. Rep.* **2019,** *1*, 496–504.

Njoku, C. J.; Asuzu, I. U. The Anthelmintic Effects of the Leaf Extract of *Ocimum gratissimum* (L.). *Phytomed.* **1998,** *5*, 485–488.

Nwangwu, S.; Ajeigbe, K.; Omotoso, D.; Josiah, S.; Olayanju, A.; Bolaji, A. Acid Buffering Effects of Aqueous Leaf Extract of *Ocimum gratissimum* L. in the Rabbit Stomach. *Turk. J. Gastroenterol.* **2013,** *24*, 204–210.

Nwosu, M. O.; Okafor, J. Preliminary Studies of the Antifungal Activities of Some Medicinal Plants Against *Basidiobolus* and Some other Pathogenic Fungi. *Mycoses* **1995,** *8*, 191–195.

Ofem, O.; Ani, E.; Eno, A. Effect of Aqueous Leaves Extract of *Ocimum gratissimum* on Hematological Parameters in Rats. *Int. J. Appl. Basic Med. Res.* **2012a,** *2*, 38–42.

Ofem, O. E.; Eno, A. E.; Antai, A. B. Gastric Acid Anti-Secretory, Anti-Ulcerogenic and Mucogenic Effects of Aqueous Leaves' Extract of *Ocimum gratissimum* in Rats. *Niger. J. Physiol. Sci.* **2012b,** *27*, 41–47.

Offiah, V. N.; Chikwendu, U. A. Antidiarrhoeal Effects of *Ocimum gratissimum* Leaf Extract in Experimental Animals. *J. Ethnopharmacol.* **1999,** *68*, 327–330.

Ogundipe, D. J.; Akomolafe, R. O.; Sanusi, A. A.; Imafidon, C. E.; Olukiran, O. S.; Oladele, A. A. *Ocimum gratissimum* Ameliorates Gentamicin-Induced Kidney Injury But Decreases Creatinine Clearance Following Sub-Chronic Administration in Rats. *J. Evid. Based. Complem. Altern. Med.* **2017,** *22*, 592–602.

Ojo, O. A.; Ojo, A. B.; Oyinloye, B. E.; Ajiboye, B. O.; Anifowose, O. O.; Akawa, A.; Olaiya, O. E.; Olasehinde, O. R.; Kappo, A. P. *Ocimum gratissimum* Linn. Leaves Reduce the Key Enzymes Activities Relevant to Erectile Dysfunction in Isolated Penile and Testicular Tissues of Rats. *BMC Complem. Altern. Med.* **2019,** *19*, 71.

Okoduwa, S. I. R.; Umar, I. A.; James, D. B.; Inuwa, H. M. Anti-Diabetic Potential of *Ocimum gratissimum* Leaf Fractions in Fortified Diet-Fed Streptozotocin Treated Rat Model of type-2 Diabetes. *Med. (Basel).* **2017,** *4*, 73.

Okoli, C. O.; Ezike, A. C.; Agwagah, O. C.; Akah, P. A. Anticonvulsant and Anxiolytic Evaluation of Leaf Extracts of *Ocimum gratissimum*, a Culinary Herb. *Pharmacog. Res.* **2010,** *2*, 36–40.

Okon, U. A.; Davies, K. G.; Olubobokun, T. H. Improvement in Nutrient Handling in STZ Induced Diabetic Rats Treated with *Ocimum gratissimum*. *Int. J. Appl. Basic Med. Res.* **2015,** *5*, 49–53.

Orafidiya, L. O.; Agbani, E. O.; Abereoje, O. A.; Awe, T.; Abudu, A.; Fakoya, F. A. An Investigation Into the Wound-Healing Properties of Essential Oil of *Ocimum gratissimum* linn. *J. Wound Care*. **2003,** *12*, 331–334.
Orafidiya, L. O.; Agbani, E. O.; Iwalewa, E. O.; Adelusola, K. A.; Oyedapo, O. O. Studies on the Acute and Sub-Chronic Toxicity of the Essential Oil of *Ocimum gratissimum* L. Leaf. *Phytomed.* **2004,** *11*, 71–76.
Osuagwu, F. C.; Oladejo, O. W.; Imosemi, I. O.; Adewoyin, B. A.; Aiku, A.; Ekpo, O. E.; Oluwadara, O. O.; Ozegbe, P. C.; Akang, E. E. Wound Healing Activities of Methanolic Extracts *Ocimum gratissimum* Leaf in Wistar Rats—A Preliminary Study. *Afr. J. Med. Med. Sci*. **2004,** *33*, 23–26.
Ouyang, X.; Wei, L.; Pan, Y.; Huang, S.; Wang, H.; Begonia, G. B.; Ekunwe, S. I. Antioxidant Properties and Chemical Constituents of Ethanolic Extract and Its Fractions of *Ocimum gratissimum*. *Med. Chem. Res.* **2013,** *22*, 1124–1130.
Owuor, B. O.; Mulemi, B. A.; Kokwaro, J. O., Indigenous Snake Bite Remedies of the Luo of Western Kenya. *J. Ethnobiol.* **2005,** *25*, 129–141.
Oyem, J. C.; Chris-Ozoko, L. E.; Enaohwo, M. T.; Otabor, F. O.; Okudayo, V. A.; Udi, O. A. Antioxidative Properties of *Ocimum gratissimum* Alters Lead Acetate Induced Oxidative Damage in Lymphoid Tissues and Hematological Parameters of Adult Wistar Rats. *Toxicol. Rep*. **2021,** *8*, 215–222.
Paula-Freire, L. I.; Andersen, M. L.; Molska, G. R.; Köhn, D. O.; Carlini, E. L. Evaluation of the Antinociceptive Activity of *Ocimum gratissimum* L. (Lamiaceae) Essential Oil and Its Isolated Active Principles in Mice. *Phytother. Res*. **2013,** *27*, 1220–1224.
Paula-Freire, L. I.; Molska, G. R.; Andersen, M. L.; Carlini, E. L. *Ocimum gratissimum* Essential Oil and Its Isolated Compounds (Eugenol and Myrcene) Reduce Neuropathic Pain in Mice. *Planta Med*. **2016,** *82*, 211–216.
Pessoa, L. M.; Morais, S. M.; Bevilaqua, C. M.; Luciano, J. H. Anthelmintic Activity of Essential Oil of *Ocimum gratissimum* Linn. and Eugenol Against *Haemonchus contortus*. *Vet. Parasitol*. **2002,** *109*, 59–63.
Pires, A. F.; Madeira, S. V.; Soares, P. M.; Montenegro, C. M.; Souza, E. P.; Resende, A. C.; Soares, M. R., Assreuy, A. M.; Criddle, D. N. The Role of Endothelium in the Vasorelaxant Effects of the Essential Oil of *Ocimum gratissimum* in Aorta and Mesenteric Vascular Bed of Rats. *Can. J. Physiol. Pharmacol.* **2012,** *90*, 1380–1385.
Rabelo, M.; Souza, E. P.; Soares, P. M.; Miranda, A. V.; Matos, F. J.; Criddle, D. N. Antinociceptive Properties of the Essential Oil of *Ocimum gratissimum* L. (Labiatae) in Mice. *Braz. J. Med. Biol. Res*. **2003,** *36*, 521–524.
Sasidharan, N. *Biodiversity Documentation for Kerala, Part 6: Flowering Plants*; Kerala Forest Research Institute: Kerala, India, 2004.
Sasidharan, N. *Flowering plants of Kerala*, Ver. 2.0 (DVD), Serial Number 698329520, Kerala Forest Research Institute, Thrissur 680653, Kerala, India, 2011.
Shittu, S. T.; Oyeyemi, W. A.; Lasisi, T. J.; Shittu, S. A.; Lawal, T. T.; Olujobi, S. T. Aqueous Leaf Extract of *Ocimum gratissimum* Improves Hematological Parameters in Alloxan-Induced Diabetic Rats via Its Antioxidant Properties. *Int. J. Appl. Basic Med. Res*. **2016,** *6*, 96–100.
Shittu, S. T.; Shittu, S. A.; Olatunji, A. A.; Oyeyemi, W. A. *Ocimum gratissimum* Leaf Extract May Precipitate Infertility in Male Diabetic Wistar Rats. *JBRA Assist. Reprod.* **2019,** *23*, 34–44.

Silva, M. R.; Oliveira, J. G. Jr.; Fernandes, O. F.; Passos, X. S.; Costa, C. R.; Souza, L. K.; Lemos, J. A.; Paula, J. R. Antifungal Activity of *Ocimum gratissimum* Towards Dermatophytes. *Mycoses.* **2005,** *48*, 172–175.

Sulistiarini, D.; Oyen, L. P. A.; Nguyen, X. D. *Ocimum gratissimum* L. In *Plant Resources of South-East Asia. Essential Oils Plants*, Vol. 19; Bogor: Prosea Foundation, 1999; pp 140–142.

Tanko, Y.; Magaji, G. M.; Yerima, M.; Magaji, R. A.; Mohammed, A. Anti-Nociceptive and Anti-Inflammatory Activities of Aqueous Leaves Extract of *Ocimum gratissimum* (Labiate) in Rodents. *Afr. J. Tradit. Complem. Altern. Med.* **2008,** *5*, 141–146.

Tatsadjieu, N. L.; Etoa, F. X.; Mbofung, C. M. F.; Ngassoum, M. B. Effect of *Plectranthus glandulosus* and *Ocimum gratissimum* Essential Oils on Growth of *Aspergillus flavus* and Aflatoxin B1 Production. *Tropicultura.* **2008,** *26*, 780–783.

Tchoumbougnang, F.; Zollo, P. H.; Dagne, E.; Mekonnen, Y. In Vivo Antimalarial Activity of Essential Oils from *Cymbopogon citratus* and *Ocimum gratissimum* on Mice Infected with *Plasmodium berghei. Planta Med.* **2005,** *71*, 20–23.

Ueda-Nakamura, T.; Mendonça-Filho, R. R.; Morgado-Díaz, J. A.; Korehisa, M. P.; Prado, D.; Filho, B.; Aparício, G. C. D.; Alviano, D. S.; Rosa, M. S.; Lopes, A. H.; Alviano, C. S.; Nakamura, C. V. Antileishmanial Activity of Eugenol-Rich Essential Oil from *Ocimum gratissimum. Parasitol. Int.* **2006,** *55*, 99–105.

Venuprasad, M. P.; Kumar Kandikattu, H.; Razack, S.; Khanum, F. Phytochemical Analysis of *Ocimum gratissimum* by LC-ESI–MS/MS and Its Antioxidant and Anxiolytic Effects. *S. Afr. J. Bot.* **2014,** *92*, 151–158.

Verma, R. S.; Bisht, P. S.; Padalia, R. C.; Saikia, D.; Chauhan, A. Chemical Composition and Antibacterial Activity of Essential Oil from Two *Ocimum* spp Grown in Sub-Tropical India During Spring-Summer Cropping Season. *Asian J. Tradit. Med.* **2011,** *6*, 211–217.

Vieira, R. F.; Grayer, R. J.; Paton, A.; Simon, J. E. Genetic Diversity of *Ocimum gratissimum* L. Based on Volatile Oil Constituents, Flavonoids and RAPD Markers. *Biochem. Syst. Ecol.* **2001,** *29*, 287–304.

Ye, J. C.; Hsiao, M. W.; Hsieh, C. H.; Wu, W. C.; Hung, Y. C.; Chang, W. C. Analysis of Caffeic Acid Extraction from *Ocimum gratissimum* Linn. by High Performance Liquid Chromatography and Its Effects on a Cervical Cancer Cell Line. *Taiwan. J. Obstet. Gynecol.* **2010,** *49*, 266–271.

CHAPTER 28

PHYTOCHEMICAL CONSTITUENTS AND PHARMACOLOGY OF *OCIMUM TENUIFLORUM* L.

THADIYAN PARAMBIL IJINU[1,2], JENNY JACOB[3], PARAMESWARAN SASIKUMAR[4], TREESA VARGHESE[3], RAGESH RAVEENDRAN NAIR[5], and PALPU PUSHPANGADAN[1]

[1]*Amity Institute for Herbal and Biotech Products Development, Thiruvananthapuram, Kerala, India*

[2]*Naturæ Scientific, Kerala University Business Innovation and Incubation Centre, Karyavattom Campus, Thiruvananthapuram, Kerala, India*

[3]*School of Biosciences, Mar Athanasios College for Advanced Studies Tiruvalla, Pathanamthitta, Kerala, India*

[4]*Drug Testing Laboratory, Department of Rasasastra and Bhaishajyakalpana, Government Ayurveda College, Thiruvananthapuram, Kerala, India*

[5]*Department of Botany, NSS College Nilamel, Kollam, Kerala, India*

ABSTRACT

Ocimum tenuiflorum L. or holy basil, is an erect, much-branched, strongly aromatic, and softly hairy annual herb that belongs to the Lamiaceae family. It is naturally found from India and Sri Lanka eastward through south-east Asia to Malaysia. In *Ayurveda*, tulsi is described as *Dashemani*, *Shwasaharni* (antiasthmatic), and *Kaphaghna* (antikaphic drugs). The leaves and flowers of tulsi are used for the treatment of heart and blood diseases,

Bioactives and Pharmacology of Lamiaceae. T. Pullaiah, PhD (Ed.)

leukoderma, strangury, asthma, bronchitis, lumbago, and purulent discharges of the ear. The terpenes and phenolic derivatives are rich in *O. tenuiflorum*. The major essential oil components reported in various studies include methyl eugenol, methyl chavicol, eugenol, β-caryophyllene, β-elemene, germacrene A, germacrene D, etc. Other major phytochemicals identified are kaempferol, rosmarinic acid, caffeic acid, 3,4-dimethoxycinnamic acid, luteolin, diosmetin, apigenin, and genistein. *O. tenuiflorum* possesses immunomodulatory, antidiabetic, hepatoprotective, cardioprotective, analgesic, anti-inflammatory, anticancer, neuroprotective, antifertility, wound healing, dental protective, and antimicrobial properties.

28.1 INTRODUCTION

Ocimum tenuiflorum L. (Syn.: *Geniosporum tenuiflorum* (L.) Merr., *Lumnitzera tenuiflora* (L.) Spreng., *Moschosma tenuiflorum* (L.) Heynh., *O. flexuosum* Blanco, *O. hirsutum* Benth., *O. inodorum* Burm.f., *O. monachorum* L., *O. sanctum* L., *O. scutellarioides* Willd. ex Benth., *O. subserratum* B.Heyne ex Hook.f., *O. tenuiflorum* f. *villicaulis* Domin, *O. tomentosum* Lam., *O. villosum* Roxb., *Plectranthus monachorum* (L.) Spreng.) belongs to Lamiaceae family. *O. tenuiflorum* is an erect, much-branched, strongly aromatic, and softly hairy annual herb generally up to 60 cm high. The leaves are elliptic and oblong. The flowers are small, purplish or crimson, and borne in close whorled racemes. The fruits are minute, sub-globose or broadly ellipsoid (Sasidharan, 2011). The vernacular names of this plant include surasa, bahumaniri, bhutaghn (Sanskrit), tulasi (Hindi), thulasi, krishnathulasi, nallathulasi, perinthulasi, nallatrittavu, shiva-tulasi (Malayalam), tulasi, thulasi, thiruthazhai, cinkattulaci, malkotait-tulaci, manchi tulasi, nalla tulasi, nallagaggera, tuvaicicceti, ulaikkuruntu (Tamil), tulasi, sri tulasi, daeva thulasi, kaala thulasi, kappu tulasi, kapputulachi (Kannada), thulasi, manchi thulasi, nalla thulasi, gaggera-chettu, krishna thulsi (Telugu), thulasa, thulasichajadha, chojharr, tulasa, tulasi, tulasichajadha, tulsi (Marathi), bye u rug pa dkar po, byeu rug pa nag po, byi rug pa (Tibetan), raihan (Urdu), and holy basil, sacred basil (English). It occurs naturally from India and Sri Lanka eastwards through south East Asia to Malaysia. It is found throughout the India.

It is a holy plant in Hinduism and this religious application is closely connected with the medicinal use. In Ayurveda, tulsi is described as Dashemani, Shwasaharni (antiasthmatic), and Kaphaghna (antikaphic drugs). The leaves and flowers of tulsi are used for the treatment of heart and blood diseases, leukoderma, strangury, asthma, bronchitis, lumbago, and purulent discharges of the ear (Puri, 2002; Cohen, 2014). The leaf juice also possesses diaphoretic, antiperiodic, adaptogenic, antipyretic, hepatoprotective, stimulant, and expectorant properties; it is used to treat infantile cough, cold, catarrh, bronchitis, diarrhea, and dysentery (Prakash and Gupta, 2000; Mondal et al., 2009). Seeds are used in genitourinary diseases. Roots are antimalarial. The essential oil of tulsi is used as an antimicrobial.

The leaf juice of *O. tenuiflorum* along with *Triphala* is recommended for glaucoma, cataract and chronic conjunctivitis. The fresh juice of leaf of *O. tenuiflorum* is given to patients to treat chronic fever, dysentery, hemorrhage, and dyspepsia (Chopra et al., 1956, 1993; Sen, 1993; Prakash and Gupta, 2000). The leaf decoction is a popular remedy for cold. The aqueous decoction of the whole plant lowers the blood glucose level and is said to control diabetes mellitus. The paste of tulsi leaves is found effective in the treatment of ringworm and other skin diseases. The aqueous decoction of leaves is given to patients suffering from gastric and hepatic disorders. Tulsi has been also recommended for use as antidote for dog bite, scorpion sting, and insect bite in traditional system of medicine (Kirtikar and Basu, 1965). The fresh leaves and flower tops have been used as a muscle relaxant. The seeds are mucilaginous and demulcent, and are given in disorders of the genitourinary system. A paste prepared from the leaves mixed with roots of *Calotropis procera* is taken to treat menorrhagia.

28.2 PHYTOCHEMICAL CONSTITUENTS

The terpenes and phenolic derivatives are the major constituents of *O. tenuiflorum*. Joshi and Hoti (2014) found that the essential oil of *O. tenuiflorum* is rich in phenyl derivative (83.8%), sesquiterpene hydrocarbons (11.1%), oxygenated monoterpenes (3.1%) along with minor monoterpene hydrocarbons (0.6%), and oxygenated sesquiterpene (0.3%). More than 25 compounds were identified according to their mass spectra and relative retention indices such as methyl eugenol (82.9%), β-caryophyllene (4.1%), borneol (2.4%), germacrene D (2.3%), δ-cadinene (1.1%), eugenol (0.9%), germacrene A (0.7%), linalool (0.5%), and α-elemene (0.5%) along with many other minor components. The studies conducted in different research groups around the world showed that there is drastic variation in the essential oil composition of *O. tenuiflorum*. The quantitative and qualitative differences in the major constituents of plant could be due to the drastic change in climate, season, and soil conditions. In most of the results, the major component present in the *O. tenuiflorum* essential oil was methyl eugenol. The varying amounts of methyl eugenol along with diverse chemotypes from different regions has been reported (Joshi, 2013; Gbolade and Lockwood, 2008; Kothari et al., 2005). Kothari et al. (2005) reported the methyl eugenol as the marker component (72.5, 75.3, 83.7, and 65.2%) from the essential oils of whole herb, leaf, stem, and inflorescence, respectively. Whereas, the second most dominant component was the *β*-caryophyllene with a respective concentration of 5.5, 6.4, 2.7, and 12.0%.

It is interesting to note that chemotypes containing other compounds have also been reported. The leaf oil of *O. tenuiflorum* collected from Australia contained methyl chavicol as the major compound (87.7%), followed by *β*-caryophyllene (5.3%) and camphor (4.3%) along with some minor terpenes (Brophy et al., 1993). Pino et al. (1998) found eugenol as the marker compound (34.3%) from the stalks and leaves of *O. tenuiflorum* collected from Cuba, followed by *β*-caryophyllene (23.1%) and *β*-elemene (18%). The oil obtained by steam and microwave distillation of *O. tenuiflorum* collected from Brazil contained eugenol (79.0–82.7%) as a major component, followed by *β*-caryophyllene (7.9–9.8%), *β*-elemene (5.0–7.6%), and germacrene A (4.7–1.7%) (Machado et al., 1999). Mousavi et al. (2018) reported the presence of polyphenolic active constituents from the crude methanol extract of the leaves of *O. tenuiflorum* using liquid chromatography mass spectrometry. The major constituents identified were kaempferol, rosmarinic acid, caffeic acid, 3,4-dimethoxycinnamic acid, luteolin, diosmetin, apigenin, and genistein.

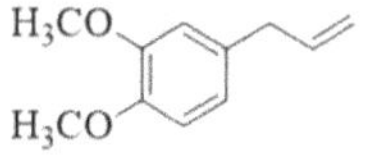

Methyl eugenol

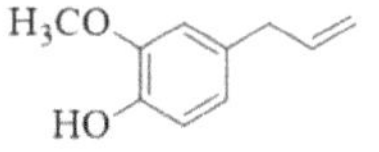

Eugenol

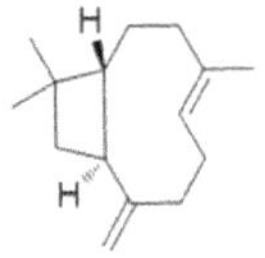

β-Caryophyllene

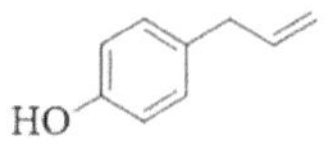

Chavicol

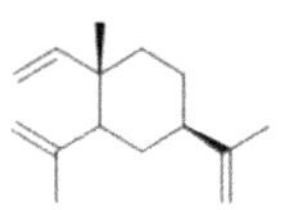
β-Elemene

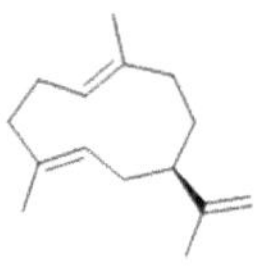
Germacrene A

Camphor

Borneol

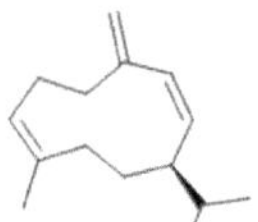
Germacrene D

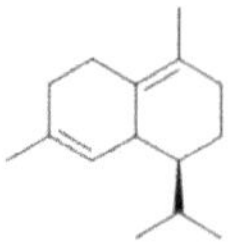
δ-Cadinene

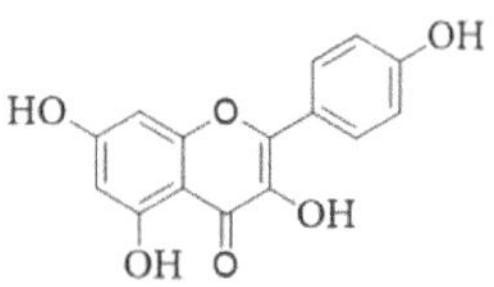

Kaempferol

28.3 PHARMACOLOGICAL PROPERTIES

Ocimum tenuiflorum showed wide range of biological properties and are discussed below.

28.3.1 IMMUNOMODULATORY ACTIVITY

The clinical evaluation of immunomodulatory effects of ethanolic leaf extract of *O. tenuiflorum* (300 mg) was carried out by Mondal et al. (2011) in 24 healthy volunteers. Upon administration of the extract, the Th1 and Th2 cytokines were significantly increased in the blood samples of *Ocimum* extract intervention group in contrast to the placebo group, after stimulation with lipopolysaccharide and phytohemagglutinin. Other immunological parameters like percentages of T-helper and NK-cells were also considerably elevated after 4 weeks. The immunomodulation of *O. tenuiflorum* was earlier reported by Mediratta et al. (2002) in the seed oil. The study evaluated that the seed oil was able to increase antibody production against sheep red blood cells and reduce mast cell degranulation in rats. The cell-mediated immunity was also modulated by the seed oil through consequent inhibition of leukocyte migration.

28.3.2 ANTIDIABETIC ACTIVITY

Dandekar et al. (2021) evaluated the antidiabetic effect of dehydrodieugenol B from the leaves of *O. tenuiflorum* and its ability to inhibit human pancreatic α-amylase. The compound inhibited human pancreatic α-amylase with an IC_{50} value of 29.6 μM for starch, and apparent Ki of 2.49 and 47.6 μM for starch and maltopentaose as substrates. Dehydrodieugenol B was thermodynamically favorable with a dissociation constant (Kd) of 1.97 μM and calculated association constant (Ka) of 0.507 μM. The antihyperglycemic effect of methanol extract of leaves of *O. tenuiflorum* and its fraction was studied by Mousavi et al. (2020) in diabetic rats. The highest dose of the extract (500 mg/kg) showed significant reduction in the blood glucose level. The ethyl acetate–butanol fraction showed a better antihyperglycemic activity. The effect of hydroalcoholic extract of *O. tenuiflorum* on streptozotocin- and nicotinamide-induced diabetes mellitus in rat was determined by Parasuraman et al. (2015). The results of the study showed that the extract at doses of 250 and 500 mg/kg showed significant antidiabetic and antihyperlipidemic activity, comparable to the standard glibenclamide.

Arenal et al. (2012) evaluated the antihyperglycemic property of aqueous extract of *O. tenuiflorum* in *Oreochromis niloticus* (tilapia fish) and the result showed dropped serum glucose level similar to control group and was dose-dependent (0, 40, 80, 200, and 400 mg/L of pond water). Suanarunsawat et al. (2016) showed the protective effect of oil extracted from *O. tenuiflorum* leaves by reducing the levels of blood glucose and serum lipid in streptozotocin-induced diabetic rats. The treatment with oil also significantly decreased the elevated levels of serum creatinine and blood urea nitrogen. Husain et al. (2015) found that oral administration of leaf extract of *O. sanctum* (500 mg/kg) for 15 days in streptozotocin-induced diabetic rats significantly decreased the blood glucose and lipid parameters of plasma lipoproteins, adipose, and liver. Patil et al. (2011) evaluated the antidiabetic activity of 10 fractions separated from the hydroalcoholic extract of *O. tenuiflorum* aerial part in alloxan-induced diabetic rats. The fifth fraction with rich amount of a tetracyclic triterpenoid namely 16-hydroxy-4,4,10,13-tetramethyl-17-(4-methyl-pentyl)-hexadeca hydro-cyclopenta[a]phenanthren-3-one showed potent activity.

Reddy et al. (2008) evaluated the delayed development of insulin resistance in rats on a fructose diet and prevention of all fructose-induced alterations including increased body weight, high retention of blood glucose concentrations, and increased insulin levels. Hannan et al. (2006) studied the stimulated release of insulin from beta cell line of rat pancreas, isolated islets of Langerhans from rat pancreas, and perfused rat pancreas by the treatment with *O. tenuiflorum* leaf extracts. Vats et al. (2004) found that a 30 days treatment with *O. tenuiflorum* ethanolic leaf extract (200 mg/kg) in strepto-zotocin-induced diabetic rats significantly reduced the blood glucose level and decreased the renal weight. The increased hexokinase, glucokinase, and phosphofructokinase enzymes activity in diabetic rats was partially reduced by the leaf extract. A similar study in streptozotocin-induced diabetic rats by Hussain et al. (2001) also showed the evidence of reduced blood glucose and lipid concentrations with increased antioxidant activity as measured by increase in superoxide dismutase and catalase, by administration of 200 mg/kg of *O. sanctum* aqueous leaf extract.

28.3.3 HEPATOPROTECTIVE ACTIVITY

Yuniarti et al. (2021) found that oral administration of *O. tenuiflorum* leaf extract for 25 days significantly protected the mice from lead acetate-induced toxicity in kidney and liver. The extract decreased the hydropic degeneration, tubular necrosis, and glomerular necrosis. Further, it also decreased the levels of malondialdehyde, serum glutamic-oxaloacetic transaminase, serum glutamic-pyruvic transaminase, and creatinine in a dose-dependent manner (140, 280, and 560 mg/kg). Lahon et al. (2011) evaluated the hepatoprotective activity of alcoholic extract of *O. sanctum* on paracetamol-induced toxicity in albino rats. The results indicated that the extract at a dose of 200 mg/kg body weight and at 100 mg/kg along with silymarin (50 mg/kg) groups has almost normal levels of liver enzymes, albumin/globulin ratio, and normalized liver histopathology comparable to standard and normal groups. Geetha and Vasudevan (2004) found that the *O. tenuiflorum* leaf extracts showed inhibition of lipid peroxidation in different experimental models.

28.3.4 CARDIOPROTECTIVE ACTIVITY

Kavitha et al. (2015) found that pretreatment with *O. tenuiflorum* leaf extract lowered the levels of cardiac markers like phospholipases, phospholipid content, and activities of 5-lipoxygenase, cycloxygenase-2, and levels of leukotriene B4, thromboxane B2 along with mRNA expressions of nuclear factor kappa-B, 5-lipoxygenase activating protein and receptor for leukotriene B4 in isoproterenol-induced myocardial infarction in rats. Sood et al. (2005) have reported the cardioprotective effect of *O. tenuiflorum* in isoproterenol-induced myocardial necrosis in rats. The study has demonstrated that a dosage of 50 mg/kg of the *O. tenuiflorum* for 30 days leads to an increase in antioxidant enzyme activity including superoxide dismutase and catalase without any toxicity. Sharma et al. (2001) found that *O. sanctum* at a dose of 50 mg/kg showed more protection in isoproterenol-induced myocardial infarction in rats.

Hydroalcoholic extract of *O. tenuiflorum* administered at a dose of 50 mg/kg exhibited a reduction in antioxidant enzyme levels like superoxide dismutase and lactate dehydrogenase, and further substantiated by histopathological studies. Lipid peroxidation was also significantly reduced by *O. tenuiflorum* extract. Suanarunsawat et al. (2010) reported the effect of *O. tenuiflorum* essential oil in rats, given a high cholesterol diet for 7 weeks. The administration of *O. tenuiflorum* oil to the rats for 3 weeks resulted in a decreased serum lipid profile, lactate dehydrogenase, and creatine kinase levels. The essential oil was also found to induce an antioxidant effect by decreasing thiobarbituric acid reactive substances, glutathione peroxidase, and superoxide dismutase in cardiac tissue. A similar study of *O. tenuiflorum* seed oil (0.8 g/kg) conducted in experimental alloxan-induced diabetic rabbits showed a drastic reduction in the lipid parameters resulting in reduced lipid peroxidation levels (Gupta et al., 2006).

28.3.5 ANALGESIC ACTIVITY

Kaur et al. (2015) found that administration of *O. tenuiflorum* extract (100 and 200 mg/kg) and its saponin fraction for 14 days considerably attenuated chronic constriction injury-induced neuropathic pain, oxidative stress, and calcium levels in rats. Kaur et al. (2010) found that administration of *O. tenuiflorum* extract and its saponin-rich fraction at doses 100 and 200 mg/kg, drastically decreased the neuropathic pain, oxidative stress, and calcium levels in vincristine-induced (50 mg/kg) peripheral neuropathic pain in rats. The alcoholic leaf extract of *O. tenuiflorum* (50, 100, and 200 mg/kg) showed potent analgesic effect, both in glacial acetic acid-induced writhing and tail flick tests. The leaf extract significantly reduced number of writhes and increased the tail withdrawal latency (Khanna and Bhatia, 2003).

28.3.6 ANTI-INFLAMMATORY ACTIVITY

Choudhury et al. (2014) evaluated the anti-inflammatory effect of alcoholic and water extracts (1 mg/mL) of *O. tenuiflorum* and its constituent eugenol (60 μg/mL) in human monocytic cells (THP-1). Eugenol showed marked inhibition of LPS-induced tumor necrosis factor-α secretion and gene expression of cytokines and chemokines like interleukin-6, tumor necrosis factor-α, macrophage inflammatory protein-1α, monocyte chemoattractant protein-1, and translocation of nuclear factor-κB-p65 to the nuclei. Manaharan et al. (2014) evaluated the antimetastatic and anti-inflammatory activities of *O. tenuiflorum* essential oil and inferred that the essential oil at a concentration of 250 μg/mL and above showed a significant decrease in the number of migrated cancer cells and reduced MMP-9 activity in lipopolysaccharide-induced inflammatory cells. The anti-inflammatory activity of essential oil of *O. tenuiflorum* was compared with ascorbic acid and verapamil in cocaine exposed animal models (Bhattacharyya et al., 2009). The oil decreased the nitric oxide production to normal, decreased malondialdehyde, and increased superoxide dismutase concentration.

Mediratta et al. (2008) found that the seed oil at a dose of 1 mg/kg reduced lindane-induced oxidative stress, producing an increase in sheep red blood cell antibody titer and inhibited lindane-induced damage. Charoo et al. (2008) found that *O. tenuiflorum* increased the bioavailability (3.8 times more than control) of the nonsteroidal anti-inflammatory drug, flurbiprofen through its bioenhanced transdermal patches compared to oral administration. The oil of *O. tenuiflorum* was also able to increase the vascular permeability for leukocyte migration to the carrageenan-induced inflamed site in rats and inhibit the generalized inflammation in peritoneal inflammation in mice (Singh and Majumdar, 1999a). Kelm et al. (2000) reported the antioxidant and cyclooxygenase-2 inhibitory activity of the phenolic extracts of *O. tenuiflorum* leaves and stem, particularly eugenol. The aqueous leaf extract of *O. tenuiflorum* (150 μg/mL) ameliorated the damage induced in epithelial cells of the lens due to hydrogen peroxide exposure (Halder et al., 2009). The cellular integrity was maintained with the formation of pyknotic nuclei and chromatin clumping by the leaf extract. Singh et al. (2001) studied the effect of fixed oil of *O. sanctum* on blood parameters like increased clotting time resulting from the inhibition of platelet aggregation.

28.3.7 ANTICANCER ACTIVITY

Luke et al. (2021) evaluated the effect of aqueous and dry leaves extract of *O. tenuiflorum* on oral squamous cell carcinoma using Ca9-22 cell line. The minimum cell inhibitory concentration value was 5 mg/L, whereas the highest permissive concentration was 30 mg/L for the plant extract in aqueous state, and 5 and 35 mg/L for dry extract. The IC_{50} value obtained for the aqueous extract was 14.79 mg/L for 3-[4,5-dimethylthiazol-2-yl]-2,5 diphenyl tetrazolium bromide (MTT) and neutral red uptake (NRU) assay, whereas for dry extract it was 20.89 mg/L for MTT and 23.44 mg/L for NRU assay. Wihadmadyatami et al. (2020) performed the antiangiogenic activity of *O. tenuiflorum* ethanolic extract on A549 lung adenocarcinoma cells. The results indicated that the ethanolic extract significantly inhibits tube formation of A549 cells; suppresses the expression of integrin alpha-v beta-3, matrix metalloproteinase-2 (MMP-2), and MMP-9; and disrupts angiogenesis cell migration and proliferation. Utispan et al. (2020) found that the *O. tenuiflorum* leaf extract significantly inhibited the invasion, and MMP-2 and MMP-9 activities of squamous carcinoma cell lines, HN4 and HN12.

Harsha et al. (2020) evaluated the cytotoxic effect of *O. tenuiflorum* dry and aqueous extracts on leukoerythroid progenitor leukemic cell lines (K562).

The results showed that both extracts had potent cytotoxicity on K562 cell lines. Wihadmadyatami et al. (2019) found that the ethanolic extract of *O. tenuiflorum* showed apoptotic activity in human alveolar pulmonary adenocarcinoma A549 cells via mitochondrial intrinsic pathway. Flegkas et al. (2019) found that the (-)-rabdosiin isolated from *Ocimum* is highly toxic to MCF-7, SKBR3, and HCT-116 cell lines, and low toxic to normal peripheral blood mononuclear cells. Manaharan et al. (2016) evaluated anticancer and apoptosis mechanisms of *O. tenuiflorum* essential oil in MCF-7 cells. The result showed that the oil inhibited proliferation of MCF-7 cells with an IC_{50} of 170 μg/mL and increased the apoptotic activity by 84%. Shivpuje et al. (2015) found that the *O. tenuiflorum* leaf extract showed dose-dependent cytotoxic activity in KB mouth cell lines.

Magesh et al. (2009) found that *O. sanctum* leaf ethanolic extract induced apoptosis in lung carcinoma cells (A549), as evidenced by presence of apoptotic bodies, upregulation of preapoptotic Bax protein, and inhibition of protein kinase B (Akt)–extracellular-regulated kinase (ERK) pathway. Rastogi et al. (2007) found that the alcoholic leaf extract of *O. tenuiflorum* effectively reduced the number of tumors with increased immunogenic cell infiltration to the tumors and increased proinflammatory cytokines like interleukin-1 beta and tumor necrosis factor alpha. Other effects include a decrease in gamma-glutamyl transferase and glutathione S-transferase, followed by decrease in phase I enzymes and increase in phase II enzymes. The effect of *O. tenuiflorum* alcoholic leaf extract against induced gastric tumorigenesis in rats was studied by Manikandan et al. (2007b). The mechanism of action of the leaf extract includes decrease in proliferating cell nuclear antigen, glutathione S-transferase, and creatine kinase; and increase in Bax, caspase 3, and cytochrome. Karthikeyan et al. (1999) found that *O. tenuiflorum* leaf extract increased the survival rate of hamsters with buccal pouch cancer. Prakash and Gupta (2000) evaluated the effect of the seed oil of *O. tenuiflorum* in fibrosarcoma tumor of Swiss albino mice. The extract significantly increased survival rate, changes in liver enzymes, and lipid peroxidation.

28.3.8 NEUROPROTECTIVE ACTIVITY

Sarangi et al. (2020) investigated the effect of *O. tenuiflorum* on chronic seizure model and its interaction with levetiracetam in pentylenetetrazole kindling model of epilepsy. The hydroalcoholic extract of *Ocimum* (1000 mg/kg) together with levetiracetam (300 mg/kg) showed least seizure score and highest seizure protection of 85.72% during kindling with pentylenetetrazole challenge. The extract-treated groups showed better memory retention potential, low oxidative stress, and least neuronal degeneration as shown in histopathology of the brain. *Ocimum* has very minimal pharmacokinetic interaction with levetiracetam and acts as an adjuvant. Mataram et al. (2021) evaluated the effect of *O. tenuiflorum* ethanolic extract on the neurons of the cornu ammonis 1 (CA1), CA3, and dentate gyrus (DG) hippocampal areas using in vivo models of Alzheimer's disease in rats. The results showed a significant increase in the number of neurons and neuropeptide Y expression, and promoted neuronal density in drug-treated groups. In the in vitro model of neurodegeneration, the ethanolic extract inhibited apoptosis and nuclear fragmentation in TMT-induced HEK-293 cells.

Raditya et al. (2020) demonstrated the potential of *O. tenuiflorum* extract as a neuroprotective agent and its effect on expression of serotonergic nerve in the hippocampal area in Alzheimer's-induced rats. The extract-treated rats showed increased neuronal density and serotonin immunoreactivity. Pasangulapati et al. (2020) investigated the neuroprotective activity of ethanolic extract of *O. tenuiflorum* leaves (100, 200, and 400 mg/kg) against hyperhomocysteinemia-induced vascular dementia in Wistar rats. The results showed that the treatment with extract significantly improved the spatial learning, working memory performance, increased nitric oxide bioavailability, and significantly altered all serum and brain biochemical parameters in a dose-dependent manner, when compared to standard donepezil (0.1 mg/kg). Hening et al. (2018) analyzed the neuroprotective effect of *O. tenuiflorum* ethanolic extract on human embryonic kidney 293 (HEK-293) cells. Pretreatment with extract significantly increased the viability of trimethyltin (10 μM) induced apoptotic cells, maintained cell viability of the normal HEK-293 cells, and markedly stabilized ChAT gene expression compared to the standard camptothecin (500 μM). Kusindarta et al. (2018) found that the *O. tenuiflorum* ethanolic extract at a dose of 100 mg/kg for 45 days induced the cognitive ability, and observed a significant increase in density of granular and pyramidal cells in dentate gyrus, CA1 and CA3.

Siddique et al. (2014) found that dietary supplementation of leaf extract of *O. tenuiflorum* showed a dose-dependent significant delay in the loss of climbing ability and reduction in oxidative stress in the brain of transgenic Drosophila model of Parkinson's disease. Venuprasad et al. (2013) found that hydroalcoholic extract of *O. sanctum* exhibited strong antioxidant activity against 2,2-diphenyl-1-picrylhydrazyl radical (IC_{50} 395 μg/mL), metal chelating (IC_{50} 241 μg/mL), hydroxyl radical (IC_{50} 188.6 μg/mL), and

2,2-azinobis-(3-ethylbenzothiazoline-6-sulfonate) radical (IC_{50} 97.6 μg/mL). Pretreatment with extract, 2 h before hydrogen peroxide-induced oxidative challenge in SH-SY5Y human neuronal cells, reduced the levels of lactate dehydrogenase leakage, lipid peroxidation, DNA damage, reactive oxygen species generation, and depolarization of the mitochondrial membrane. Ahmad et al. (2012) studied the neuroprotective effect of *O. tenuiflorum* to reduce brain injury after middle cerebral artery occlusion in male Wistar rats. The administration of extract at 200 mg/kg once daily for 15 days before brain injury showed marked reduction in infarct size, reduced neurological deficits, and suppressed neuronal loss in rats. Giridharan et al. (2011) found that the aqueous (300 and 500 mg/kg) and alcoholic (300 and 500 mg/kg) extracts of *O. tenuiflorum* leaves significantly decreased the anticholinesterase activity in different parts of the brain and increased the immune status of the animals.

Muthuraman et al. (2008) found that the methanolic extract of *O. sanctum* reduced thiobarbituric acid reactive species and total calcium content, and increased the glutathione levels. In fact, all the neuropathy-induced effects were inhibited particularly the nociception and motor incoordination. Yanpallewar et al. (2004) reported that *O. tenuiflorum* methanolic extract of leaves induced neuroprotective effect in long-term hypoperfusion and cerebral reperfusion. Pretreatment of *O. tenuiflorum* led to the reduction of lipid peroxidation and superoxide dismutase levels preventing reperfusion-induced injury. Similar effects were also seen in long-term hypoperfusion with a reversal of memory deficits and altered behavior, thus preventing hypoperfusion-induced changes in function and structure. Joshi and Parle (2006) investigated the efficacy of the aqueous extract of *O. sanctum* as a brain performance enhancing drug for use in cognitive disorders like Alzheimer's or dementia. The neuroprotective effect of *O. tenuiflorum* ethanolic extract in noise-induced stress culminates in normalization of neurotransmitters like norepinephrine, epinephrine, dopamine, and serotonin (Ravindran et al., 2005).

Saxena et al. (2012) conducted a randomized, double-blind, placebo-controlled study to evaluate the efficacy of OciBest, an extract of *O. tenuiflorum* in controlling general stress. They found 1.6 times or 39% overall improvement in stress symptoms in OciBest group with respect to placebo and no adverse events observed during the study. Jothie et al. (2016) investigated the antistress potential of *O. tenuiflorum* in chronic variable stress rats model and observed that the oral administration of extract significantly increased body weight gain, decreased the immobility time, inhibited cortisol release, exhibited significant CRHR1 receptor antagonist activity, and had specific inhibitory activity towards 11β-HSD1 and COMT inducing antistress activity. Bhattacharyya et al. (2008) reported that *O. tenuiflorum* was able to reduce anxiety disorders and improve the quality of life in 35 human subjects at a dosage of 500 mg twice daily. Jyoti et al. (2007) administered *O. tenuiflorum* fresh leaves to rabbits leading to increased enzymatic and nonenzymatic antioxidants. Ocimumoside A, ocimumoside B, and 4-allyl-1-O-β-d-glucopyronosyl-2-hydroxybenzene isolated from *O. tenuiflorum* leaves exhibited varied actions with a generalized antistress activity in Sprague–Dawley rats (Gupta et al., 2007). Samson et al. (2007) studied the reduction in noise exposure-induced oxidative stress by *O. tenuiflorum* ethanolic leaf extract as evidenced by increased antioxidant enzymes and reduced oxidation products. Sembulingam et al. (2005) found that *O. sanctum* leaf extract reduced the oxidative stress by reducing acetylcholine and acetylcholine esterase in noise exposure-induced animals. Sood et al. (2006) found that the leaf extract of *O. tenuiflorum* can reduce the chronic restraint stress with increased superoxide dismutase.

28.3.9 ANTIFERTILITY ACTIVITY

Poli and Challa (2019) evaluated the antifertility efficacy of *O. tenuiflorum* leaf extract (500 mg/kg) and eugenol (0.4 mL/day) in female albino rats. The results showed that the administration of eugenol prolonged the total duration of estrous cycle, and elevated serum estradiol and progesterone levels. The leaf extract had no significant changes in estrous cycle, but significantly enhanced the serum progesterone levels leading to reduced frequency of ovulation and resulted in the impairment of fertility. Sethi et al. (2010) evaluated the effect of fresh leaves of *O. tenuiflorum* on male reproductive function in albino rabbits. After 30 days of study a significant decrease in the sperm count, increased testosterone levels, lowered follicle-stimulating hormone, and luteinizing hormone levels was observed in the drug-treated animals indicating contraceptive potential of *Ocimum*. Ahmed et al. (2002) reported the reversible effect of *O. tenuiflorum* leaf extract on rats after a 48-days treatment reducing the sperm count and motility. The effect was reversed to normal in 2 weeks after discontinuance of *O. tenuiflorum* extract from the diet. Further study by Ahmed et al. (2009) indicated that the antifertility effect of *O. tenuiflorum* was due to the high disruption of the general architecture of the cauda epididymis in testis.

28.3.10 WOUND HEALING ACTIVITY

Goel et al. (2010) studied the wound healing potential of cold aqueous extract of *O. sanctum* on excision wound repair model in Wistar rats. It was observed that topical application of 10% of extract along with oral administration of 250 mg/kg of extract daily for 20 consecutive days improved the rate of epithelization wound contraction. Shetty et al. (2008) evaluated the wound healing potential of alcoholic and aqueous extracts of *O. tenuiflorum* in Wistar rats. The wound contraction was completed in 14 and 16 days in aqueous and alcoholic extracts, respectively as compared to the 21 days in untreated control group. Other parameters like hexuronic acid, hexosamine, and hydroxyproline as part of the extracellular matrix and collagen synthesis at wounds were found to be significantly increased as compared to control group. The lipid peroxidation was also reduced by the extracts. Wound healing activity in both excision and incision models evaluated with the aqueous extract of *O. tenuiflorum* showed a better wound contraction percentage (Shetty et al., 2006). The study also used the dead space wound model exhibiting increased tissue granulation, and increased activity of superoxide dismutase and catalase. Udupa et al. (2006) compared the wound healing potential of *O. tenuiflorum* ethanolic leaf extract to dexamethasone standard using animal models. Increased rates of wound breaking strengths, faster epithelialization, significant granulation tissue formation, and hydroxyproline content together indicate the potential of *O. tenuiflorum* in wound healing.

Ethanol-induced peptic ulcer was controlled by the hydroalcoholic extract of *O. tenuiflorum* leaves as evidenced by reduced oxidative damage causing malondialdehyde and increased superoxide dismutase levels (Kath and Gupta, 2006). The study also covered guinea pigs induced with duodenal ulcers resulting in similar results compared to controls. Goel et al. (2005) reported the effect of *O. sanctum* on induced gastric ulcers. Despite being inactive against aspirin-induced ulcers, the extract was better against acetic acid-induced ulcers at 100 mg/kg concentrations after a maximum of 10 days treatment. Dharmani et al. (2004) reported the ulcer healing properties of *O. tenuiflorum* in Sprague–Dawley rats. At a dose of 100 mg/kg, the extract was more effective in cold restraint-induced ulcer (65.07%) and least effective in alcohol-induced ulcer (53.87%). Acidity and peptic activity were significantly reduced together with mucus production providing evidence to its use in ulcer treatment. Singh and Majumdar (1999b) demonstrated the antiulcer and anti-inflammatory activities of the fixed oil of *O. tenuiflorum* in animal models.

28.3.11 DENTAL PROTECTION

Nadar et al. (2020) evaluated the antiplaque and antigingivitis effect of *O. tenuiflorum* leaf extract dentifrice among 14–15-year-old school children and found significant reduction in the plaque and gingival score was observed in 4% *O. tenuiflorum* leaf extract dentifrice group on the 21st day. Eswar et al. (2016) evaluated the effect of *O. tenuiflorum* leaf extract on a periodontal microorganism in human dental plaque with 0.2% chlorhexidine as the positive control and a zone of inhibition of 22 mm was obtained, comparable to the standard with 25 mm zone.

28.3.12 ANTIMICROBIAL ACTIVITY

Lahiri et al. (2021) evaluated the inhibition potential of eugenol and linalool isolated from the leaves of *O. tenuiflorum* on biofilm formation of *Pseudomonas aeruginosa*. These compounds reduced protein and carbohydrate content of extracellular polymeric substance, and inhibited the synthesis of quorum sensing proteins and virulence factors such as pyocyanin and rhamnolipids inhibiting the biofilm. Yamani et al. (2016) examined the antimicrobial potential of the essential oil of *O. tenuiflorum* collected from Australia. The oil at concentrations of 4.5 and 2.25% inhibited the growth of *Staphylococcus aureus* and *Escherichia coli*, but partly inhibited *Pseudomonas aeruginosa*. Camphor, eucalyptol, and eugenol were identified in the leaves and flower spikes of *O. tenuiflorum*. Jayanti et al. (2018) evaluated the antimicrobial activity of *O. sanctum* ethanolic extract on *Aggregatibacter actinomycetemcomitans* and *Porphyromonas gingivalis* using 0.2% chlorhexidine as a standard. The results indicated that 8% concentration of the *Ocimum* extract showed maximum zone of inhibition against *A. actinomycetemcomitans* and *P. gingivalis* (40.10 and 33.79 mm), followed by the 6, 4, and 2% concentrations, which is higher than that of positive control (39.80 and 32.28 mm). Mallikarjun et al. (2016) investigated the antimicrobial activity of *O. tenuiflorum* leaf extract on periodontal pathogens with doxycycline as standard, and the result for 5% and 10% concentrations of extracts demonstrated similar inhibition zones to doxycycline. Kochikar et al. (2015) evaluated the antimicrobial efficacy of *O. tenuiflorum* (Syn. *O. sanctum*). The extract at 10% concentration was most effective against *S. mutans* and *S. sanguis*.

Khan et al. (2014) evaluated the antivirulence activity of essential oil of *O. sanctum* on 22 strains of *C. albicans* isolated from both HIV-positive and

HIV-negative patients. The sub-minimum inhibitory concentrations of the oil inhibit morphological transition in *C. albicans* and decrease the secretion of hydrolytic enzymes involved in the early stage of infection by downregulating the associated genes. Balakumar et al. (2011) evaluated the antifungal activity of *O. tenuiflorum* leaf extract. The water fraction of the leaf extract at a dose of 200 μg/mL had maximum activity against dermatophytic fungi. Agarwal et al. (2010) evaluated the antimicrobial activity of 15 concentrations of *O. sanctum* leaves extract against *Streptococcus mutans*, and revealed that the most potent dose was 4% concentration with maximum zone of inhibition (22 mm). Kumar et al. (2010) reported the efficiency of eugenol rich *O. tenuiflorum* essential oils on the fungus particularly *Aspergillus flavus* at a minimum inhibitory concentration of 0.3 μg/mL. The essential oil exhibits broad spectrum fungicidal activity against 12 different fungi. Shokeen et al. (2008) studied the effect of eugenol against *Neisseria gonorrheae*, showed minimum inhibitory concentration of 85 mg/L. The LD_{50} of eugenol was also determined as 2 g/kg of body weight of rats. The ethyl acetate fraction of *O. tenuiflorum* was also found to exhibit antileishmanial activity (Suzuki et al., 2009).

28.4 TOXICITY STUDIES

Gautam et al. (2014) evaluated acute and subacute toxicities of orally administered 50% ethanolic leaves extract of *O. tenuiflorum* in rats. The results showed that in acute toxicity, the extract at doses of 200, 600, and 2000 mg/kg did not produce any adverse symptoms or death. In subacute toxicity study (28 days), the extract at the doses of 200, 400, and 800 mg/kg/day not showed any changes in biochemical, hematological, and histopathological parameters. Sharmila et al. (2009) studied the effect of *O. tenuiflorum* leaf extract on arsenic-induced toxicity in Wistar albino rats. At a dose of 200 mg/kg daily for 4 months, the leaf extract was able to reverse the inhibition of aminolevulinic acid dehydratase, decreased the reactive oxygen species generation and increased glutathione levels after arsenic exposure. Cyproterone acetate-induced toxicity in human lymphocyte cells was evaluated by Siddique et al. (2007). *O. tenuiflorum* infusion was found to exhibit a dose-dependent decrease in genotoxicity evidenced by reduced chromosomal aberrations, mitotic index, replication index, etc.

O. sanctum ethanolic extract showed protective effect on genotoxicity induced by 7,12-dimethylbenzanthracene indicated the protective effect on xenobiotic metabolizing enzymes in rats (Manikandan et al., 2007a). The study reports decrease in the formation of micronuclei and phase I detoxification enzymes while increasing the phase II enzymes and the overall antioxidant status of the cells at a dose of 300 mg/kg of *O. tenuiflorum* leaf extract. Dutta et al (2007) reported the antigenotoxic effect of leaf extract of *O. tenuiflorum* against chromosomal aberrations and formation of micronucleus. Bhartiya et al. (2006) found the protective effect of *O. tenuiflorum* aqueous extract on high concentration of iodine exposed mice. The study indicated a reduced lipid peroxidation, protected levels of glutathione in kidney and salivary glands, and no effect in stomach. A similar study was conducted in Swiss albino mice exposed to mercury causing renal damage (Sharma et al., 2005). Pretreatment of mice with *O. tenuiflorum* leaf extract for 30 days before mercuric induction, leads to a reduction in pathophysiology of the renal tissues followed by decreased lipid peroxidation and increased lactate dehydrogenase activity. Sharma et al. (2002) reported the protective effect of *O. tenuiflorum* leaf extract on mercury-induced toxicity by enhancing the production of reduced glutathione content.

ACKNOWLEDGMENTS

The authors express their sincere thanks to Dr. Ashok K. Chauhan, Founder President, Ritnand Balved Education Foundation (RBEF) and Amity Group of Institutions, and Dr. Atul Chauhan, Chancellor, Amity University Uttar Pradesh (AUUP) for facilitating this work. Thadiyan Parambil Ijinu has received Young Scientist Fellowship from the Department of Science and Technology, Government of India (SP/YO/413/2018).

KEYWORDS

- **traditional medicinal use**
- **methyl eugenol**
- **methyl chavicol**
- **rosmarinic acid**
- **neuroprotective activity**
- **anticancer activity**

REFERENCES

Agarwal, P.; Nagesh, L.; Murlikrishnan, V. Evaluation of the Antimicrobial Activity of Various Concentrations of Tulsi (*Ocimum sanctum*) Extract Against *Streptococcus mutans*: An *In Vitro* Study. *Indian J. Dent. Res.* 2010, *21*, 357–359.

Ahmad, A.; Khan, M. M.; Raza, S. S.; Javed, H.; Ashafaq, M.; Islam, F.; Safhi, M. M.; Islam, F. *Ocimum sanctum* Attenuates Oxidative Damage and Neurological Deficits Following Focal Cerebral Ischemia/Reperfusion Injury in Rats. *Neurol. Sci.* 2012, *33*, 1239–1247.

Ahmed, M.; Ahamed, R. N.; Aladakatti, R. H. Effect of Benzene Extract of *Ocimum sanctum* Leaves on Cauda Epididymis of Albino Rats. *J. Basic Clin. Physiol. Pharmacol.* 2009, *20*, 29–41.

Ahmed, M.; Ahamed, R. N.; Aladakatti, R. H.; Ghosesawar, M. G. Reversible anti-fertility effect of benzene extract of *Ocimum sanctum* leaves on sperm parameters and fructose content in rats. *J. Basic Clin. Physiol. Pharmacol.* 2002, *13*, 51–59.

Arenal, A.; Martín, L.; Castillo, N. M.; de la Torre, D.; Torres, U.; Gonzâlez, R. Aqueous extract of *Ocimum tenuiflorum* decreases levels of blood glucose in induced hyperglycemic tilapia (*Oreochromis niloticus*). *Asian Pac. J. Trop. Med.* 2012, *5*, 634–637.

Balakumar, S.; Rajan, S.; Thirunalasundari, T.; Jeeva, S. Antifungal activity of *Ocimum sanctum* Linn. (Lamiaceae) on clinically isolated dermatophytic fungi. *Asian Pac. J. Trop. Med.* 2011, *4*, 654–657.

Bhartiya, U. S.; Raut, Y. S.; Joseph, L. J. Protective Effect of *Ocimum sanctum* L After High-dose 131iodine Exposure in Mice: An *In Vivo* Study. *Indian J. Exp. Biol.* 2006, *44*, 647–652.

Bhattacharya, S. K.; Rathi, N.; Mahajan, P.; Tripathi, A. K.; Paudel, K. R.; Rauniar, G. P.; Das, B. P. Effect of *Ocimum sanctum*, Ascorbic Acid, and Verapamil on Macrophage Function and Oxidative Stress in Mice Exposed to Cocaine. *Indian J. Pharmacol.* 2009, *41*, 134–139.

Bhattacharyya, D.; Sur, T. K.; Jana, U.; Debnath, P. K. Controlled Programmed Trial of *Ocimum sanctum* Leaf on Generalized Anxiety Disorders. *Nepal Med. Coll. J.* 2008, *10*, 176–179.

Brophy, J. J.; Goldsack, R. J.; Clarkson, J. R. The Essential Oil of *Ocimum tenuiflorum* L. (Lamiaceae) Growing in Northern Australia. *J. Essent. Oil Res.* 1993, *5*, 459–461.

Charoo, N. A.; Shamsher, A. A.; Kohli, K.; Pillai, K.; Rahman, Z. Improvement in Bioavailability of Transdermally Applied Flurbiprofen Using Tulsi (*Ocimum sanctum*) and Turpentine Oil. *Colloids Surf. B. Biointerfaces* 2008, *65*, 300–307.

Chopra, R. N.; Chopra, I. C.; Handa, K. L.; Kapoor, L. D. *Indigenous Drugs of India*; Calcutta: U.N. Dhur Pvt. Ltd., 1993.

Chopra, R. N.; Nayar, S. I.; Chopra, I. C. *Glossary of Indian Medicinal Plants*; New Delhi: Council of Scientific and Industrial Research, 1956.

Choudhury, S. S.; Bashyam, L.; Manthapuram, N.; Bitla, P.; Kollipara, P.; Tetali, S. D. *Ocimum sanctum* Leaf Extracts Attenuate Human Monocytic (THP-1) Cell Activation. *J. Ethnopharmacol.* 2014, *154*, 148–155.

Cohen, M. M. Tulsi - *Ocimum sanctum*: A Herb for All Reasons. *J. Ayurveda Integr. Med.* 2014, *5*, 251–259.

Dandekar, P. D.; Kotmale, A. S.; Chavan, S. R.; Kadlag, P. P.; Sawant, S. V.; Dhavale, D. D.; RaviKumar, A. Insights into the Inhibition Mechanism of Human Pancreatic α-amylase, A Type 2 Diabetes Target, by Dehydrodieugenol B Isolated from *Ocimum tenuiflorum*. *ACS Omega* 2021, *6*, 1780–1786.

Dharmani, P.; Kuchibhotla, V. K.; Maurya, R.; Srivastava, S.; Sharma, S.; Palit, G. Evaluation of Anti-ulcerogenic and Ulcer-healing Properties of *Ocimum sanctum* Linn. *J. Ethnopharmacol.* 2004, *93*, 197–206.

Dutta, D.; Devi, S. S.; Krishnamurthi, K.; Kumar, K.; Vyas, P.; Muthal, P. L.; Naoghare, P.; Chakrabarti, T. Modulatory Effect of Distillate of *Ocimum sanctum* Leaf Extract (Tulsi) on Human Lymphocytes Against Genotoxicants. *Biomed. Environ. Sci.* 2007, *20*, 226–234.

Eswar, P.; Devaraj, C. G.; Agarwal, P. Anti-microbial Activity of Tulsi [*Ocimum sanctum* (Linn.)] Extract on a Periodontal Pathogen in Human Dental Plaque: An *In Vitro* Study. *J. Clin. Diagn. Res.* 2016, *10*, ZC53–56.

Flegkas, A.; Milosević Ifantis, T.; Barda, C.; Samara, P.; Tsitsilonis, O.; Skaltsa, H. Antiproliferative Activity of (-)-rabdosiin Isolated from *Ocimum sanctum* L. *Med. (Basel)*. 2019, *6*, 37.

Gautam, M. K.; Goel, R. K. Toxicological Study of *Ocimum sanctum* Linn Leaves: Hematological, Biochemical, and Histopathological Studies. *J. Toxicol.* 2014, *135654*, 1–9.

Gbolade, A. A.; Lockwood, G. B. Toxicity of *Ocimum sanctum* L. Essential Oil to *Aedes aegypti* Larvae and Its Chemical Composition. *J. Essent. Oil-Bear. Plants* 2008, *11*, 148–153.

Geetha, R. K.; Vasudevan, D. M. Inhibition of Lipid Peroxidation by Botanical Extracts of *Ocimum sanctum*: *In Vivo* and *In Vitro* Studies. *Life Sci.* 2004, *76*, 21–28.

Giridharan, V. V.; Thandavarayan, R. A.; Mani, V.; Ashok D. T.; Watanabe, K.; Konishi, T. *Ocimum sanctum* Linn. leaf extracts inhibit acetylcholinesterase and improve cognition in rats with experimentally induced dementia. *J. Med. Food* 2011, *14*, 912–919.

Goel, A.; Kumar, S.; Singh, D. K.; Bhatia, A. K. Wound Healing Potential of *Ocimum sanctum* Linn. with Induction of Tumor Necrosis Factor-alpha. *Indian J. Exp. Biol.* 2010, *48*, 402–406.

Goel, R. K.; Sairam, K.; Dorababu, M.; Prabha, T.; Rao, C. V. Effect of Standardized Extract of *Ocimum sanctum* Linn. on Gastric Mucosal Offensive and Defensive Factors. *Indian J. Exp. Biol.* 2005, *43*, 715–721.

Gupta, P.; Yadav, D. K.; Siripurapu, K. B.; Palit, G.; Maurya, R. Constituents of *Ocimum sanctum* with Antistress Activity. *J. Nat. Prod.* 2007, *70*, 1410–1416.

Gupta, S.; Mediratta, P. K.; Singh, S.; Sharma, K. K.; Shukla, R. Antidiabetic, Antihypercholesterolaemic and Antioxidant Effect of *Ocimum sanctum* (Linn) Seed Oil. *Indian J. Exp. Biol.* 2006, *44*, 300–304.

Halder, N.; Joshi, S.; Nag, T. C.; Tandon, R.; Gupta, S. K. *Ocimum sanctum* Extracts Attenuate Hydrogen Peroxide Induced Cytotoxic Ultrastructural Changes in Human Lens Epithelial Cells. *Phytother. Res.* 2009, *23*, 1734–1737.

Hannan, J. M.; Marenah, L.; Ali, L.; Rokeya, B.; Flatt, P. R.; Abdel-Wahab, Y. H. *Ocimum sanctum* Leaf Extracts Stimulate Insulin Secretion from Perfused Pancreas, Isolated Islets and Clonal Pancreatic Beta-cells. *J. Endocrinol.* 2006, *189*, 127–136.

Harsha, M.; Mohan Kumar, K. P.; Kagathur, S.; Amberkar, V. S. Effect of *Ocimum sanctum* Extract on Leukemic cell Lines: A Preliminary *In-vitro* Study. *J. Oral Maxillofac. Pathol.* 2020, *24*, 93–98.

Hening, P.; Mataram Auriva, M. B.; Wijayanti, N.; Kusindarta, D. L.; Wihadmadyatami, H. The Neuroprotective Effect of *Ocimum sanctum* Linn. Ethanolic Extract on Human Embryonic Kidney-293 Cells as *In Vitro* Model of Neurodegenerative Disease. *Vet. World* 2018, *11*, 1237–1243.

Husain, I.; Chander, R.; Saxena, J. K.; Mahdi, A. A.; Mahdi, F. Antidyslipidemic Effect of *Ocimum sanctum* Leaf Extract in Streptozotocin Induced Diabetic Rats. *Indian J. Clin. Biochem.* 2015, *30*, 72–77.

Hussain, E. H.; Jamil, K.; Rao, M. Hypoglycaemic, Hypolipidemic and Antioxidant Properties of Tulsi (*Ocimum sanctum* Linn.) on Streptozotocin Induced Diabetes in Rats. *Indian J. Clin. Biochem.* 2001, *16*, 190–194.

Jayanti, I.; Jalaluddin, M.; Avijeeta, A.; Ramanna P. K.; Rai, P. M.; Nair, R. A. *In Vitro* Antimicrobial Activity of *Ocimum sanctum* (Tulsi) Extract on *Aggregatibacter actinomycetemcomitans* and *Porphyromonas gingivalis. J. Contemp. Dent. Pract.* 2018, *19*, 415–419.

Joshi, H.; Parle, M. Evaluation of Nootropic Potential of *Ocimum sanctum* Linn. in Mice. *Indian J. Exp. Biol.* 2006, *44*, 133–136.

Joshi, R. K. Chemical Composition, *In Vitro* Antimicrobial and Antioxidant Activities of the Essential Oils of *Ocimum gratissimum, O. sanctum* and Their Major Constituents. *Indian J. Pharm. Sci.* 2013, *75*, 457–462.

Joshi, R. K.; Hoti, S. L. Chemical Composition of the Essential Oil of *Ocimum tenuiflorum* L. (Krishna Tulsi) from North West Karnataka, India. *Plant Sci. Today* 2014, *1*, 99–102.

Jothie, R. E.; Illuri R.; Bethapudi, B.; Anandhakumar, S.; Bhaskar, A.; Chinampudur, V. C.; Mundkinajeddu, D.; Agarwal, A. Anti-stress Activity of *Ocimum sanctum*: Possible Effects on Hypothalamic-pituitary-adrenal Axis. *Phytother. Res.* 2016, *30*, 805–814.

Jyoti, S.; Satendra, S.; Sushma, S.; Anjana, T.; Shashi, S. Antistressor Activity of *Ocimum sanctum* (Tulsi) Against Experimentally Induced Oxidative Stress in Rabbits. *Methods Find. Exp. Clin. Pharmacol.* 2007, *29*, 411–416.

Karthikeyan, K.; Ravichandran, P.; Govindasamy, S. Chemopreventive effect of *Ocimum sanctum* on DMBA-induced Hamster Buccal Pouch Carcinogenesis. *Oral Oncol.* 1999, *35*, 112–119.

Kath, R. K.; Gupta, R. K. Antioxidant Activity of Hydroalcoholic Leaf Extract of *Ocimum sanctum* in Animal Models of Peptic Ulcer. *Indian J. Physiol. Pharmacol.* 2006, *50*, 391–396.

Kaur, G.; Bali, A.; Singh, N.; Jaggi, A. S. Ameliorative Potential of *Ocimum sanctum* in Chronic Constriction Injury-induced Neuropathic Pain in Rats. *An. Acad. Bras. Cienc.* 2015, *87*, 417–429.

Kaur, G.; Jaggi, A. S.; Singh, N. Exploring the Potential Effect of *Ocimum sanctum* in Vincristine-induced Neuropathic Pain in Rats. *J. Brachial Plex. Peripher. Nerve Inj.* 2010, *5*, 3.

Kavitha, S.; John, F.; Indira, M. Amelioration of Inflammation by Phenolic Rich Methanolic Extract of *Ocimum sanctum* Linn. Leaves in Isoproterenol Induced Myocardial Infarction. *Indian J. Exp. Biol.* 2015, *53*, 632–640.

Kelm, M. A.; Nair, M. G.; Strasburg, G. M.; DeWitt, D. L. Antioxidant and Cyclooxygenase Inhibitory Phenolic Compounds from *Ocimum sanctum* Linn. *Phytomedicine* 2000, *7*, 7–13.

Khan, A.; Ahmad, A.; Xess, I.; Khan, L. A.; Manzoor, N. *Ocimum sanctum* Essential Oil Inhibits Virulence Attributes in *Candida albicans. Phytomedicine* 2014, *21*, 448–452.

Khanna, N.; Bhatia, J. Antinociceptive Action of *Ocimum sanctum* (Tulsi) in Mice: Possible Mechanisms Involved. *J. Ethnopharmacol.* 2003, *88*, 293–296.

Kirtikar, K. R.; Basu, B. D. *Ocimum sanctum.* In *Indian Medicinal Plants*; Allahabad: LB Basu, 1965.

Kochikar, P. R.; Bhat, S. S.; Salman, A.; Chandra, J. Use of an Extract of Indian Sacred Plant *Ocimum sanctum* as an Anticariogenic Agent: An *In Vitro* Study. *Int. J. Clin. Pediatr. Dent.* 2015, *8*, 99–101.

Kothari, S. K.; Bhattacharya, A. K.; Ramesh, S.; Garg, S. N.; Khanuja, S. P. S. Volatile Constituents in Oil from Different Plant Parts of Methyl Eugenol-rich *Ocimum tenuiflorum* L. f. (Syn. *O. sanctum* L.) Grown in South India. *J. Essent. Oil Res.* 2005, *17*, 656–658.

Kumar, A.; Shukla, R.; Singh, P.; Dubey, N. K. Chemical Composition, Antifungal and Antiaflatoxigenic Activities of *Ocimum sanctum* L. Essential Oil and Its Safety Assessment as Plant Based Antimicrobial. *Food Chem. Toxicol.* 2010, *48*, 539–543.

Kusindarta, D. L.; Wihadmadyatami, H.; Jadi, A. R.; Karnati, S.; Lochnit, G.; Hening, P.; Haryanto, A.; Auriva, M. B.; Purwaningrum, M. Ethanolic Extract *Ocimum sanctum* Enhances Cognitive Ability from Young Adulthood to Middle Aged Mediated by Increasing Choline Acetyl Transferase Activity in Rat Model. *Res. Vet. Sci.* 2018, *118*, 431–438.

Lahiri, D.; Nag, M.; Dutta, B.; Dey, S.; Mukherjee, D.; Joshi, S. J.; Ray, R. R. Antibiofilm and Anti-quorum Sensing Activities of Eugenol and Linalool from *Ocimum tenuiflorum* Against *Pseudomonas aeruginosa* Biofilm. *J. Appl. Microbiol.* 2021, 34077580.

Lahon, K.; Das, S. Hepatoprotective Activity of *Ocimum sanctum* Alcoholic Leaf Extract Against Paracetamol-induced Liver Damage in Albino Rats. *Pharmacognosy Res.* 2011, *3*, 13–18.

Luke, A. M.; Patnaik, R.; Kuriadom, S. T.; Jaber, M.; Mathew, S. An *In Vitro* Study of *Ocimum sanctum* as a Chemotherapeutic Agent on Oral Cancer Cell-line. *Saudi J. Biol. Sci.* 2021, *28*, 887–890.

Machado, M. I. L.; Silva, M. G. V.; Matos, F. J. A.; Craveiro, A. A.; Alencar, J. W. Volatile Constituents from Leaves and Inflorescence oil of *Ocimum tenuiflorum* L. f. (Syn. *O. sanctum* L.) Grown in Northeastern Brazil. *J. Essent. Oil Res.* 1999, *11*, 324–326.

Magesh, V.; Lee, J. C.; Ahn, K. S.; Lee, H. J.; Lee, H. J.; Lee, E. O.; Shim, B. S.; Jung, H. J.; Kim, J. S.; Kim, D. K.; Choi, S. H.; Ahn, K. S.; Kim, S. H. *Ocimum sanctum* Induces Apoptosis in A549 Lung Cancer Cells and Suppresses the In Vivo Growth of Lewis Lung Carcinoma Cells. *Phytother. Res.* 2009, *23*, 1385–1391.

Mallikarjun, S.; Rao, A.; Rajesh, G.; Shenoy, R.; Pai, M. Antimicrobial Efficacy of Tulsi Leaf (*Ocimum sanctum*) Extract on Periodontal Pathogens: An *In Vitro* Study. *J. Indian Soc. Periodontol.* 2016, *20*, 145–150.

Manaharan, T.; Thirugnanasampandan, R.; Jayakumar, R.; Kanthimathi, M. S.; Ramya, G.; Ramnath, M. G. Purified Essential Oil from *Ocimum sanctum* Linn. Triggers the Apoptotic Mechanism in Human Breast Cancer Cells. *Pharmacogn. Mag.* 2016 12, 327–331.

Manaharan, T.; Thirugnanasampandan, R.; Jayakumar, R.; Ramya, G.; Ramnath, G.; Kanthimathi, M. S. Antimetastatic and Anti-inflammatory Potentials of Essential Oil from Edible *Ocimum sanctum* Leaves. *Sci. World J.* 2014, 239508.

Manikandan, P.; Murugan, R. S.; Abbas, H.; Abraham, S. K.; Nagini, S. *Ocimum sanctum* Linn. (Holy Basil) Ethanolic Leaf Extract Protects Against 7,12-dimethylbenz(a)anthracene-induced Genotoxicity, Oxidative Stress, and Imbalance in Xenobiotic-metabolizing Enzymes. *J. Med. Food.* 2007a, 10, 495–502.

Manikandan, P.; Vidjaya, P. L.; Prathiba, D.; Nagini, S. Proliferation, Angiogenesis and Apoptosis-associated Proteins are Molecular Targets for Chemoprevention of MNNG-induced Gastric Carcinogenesis by Ethanolic *Ocimum sanctum* Leaf Extract. *Singapore Med. J.* 2007b, 48, 645–651.

Mataram, M. B. A.; Hening, P.; Harjanti, F. N.; Karnati, S.; Wasityastuti, W.; Nugrahaningsih, D. A. A.; Kusindarta, D. L.; Wihadmadyatami, H. The Neuroprotective Effect of Ethanolic Extract *Ocimum sanctum* Linn. in the Regulation of Neuronal Density in Hippocampus Areas as a Central Autobiography Memory on the Rat Model of Alzheimer's Disease. *J. Chem. Neuroanat.* 2021, 101885.

Mediratta, P. K.; Sharma, K. K.; Singh, S. Evaluation of Immunomodulatory Potential of *Ocimum sanctum* Seed Oil and Its Possible Mechanism of Action. *J. Ethnopharmacol.* 2002, *80*, 15–20.

Mediratta, P. K.; Tanwar, K.; Reeta, K. H.; Mathur, R.; Benerjee, B. D.; Singh, S.; Sharma, K. K. Attenuation of the Effect of Lindane on Immune Responses and Oxidative Stress by *Ocimum sanctum* Seed Oil (OSSO) in Rats. *Indian J. Physiol. Pharmacol.* 2008, *52*, 171–177.

Mondal, S.; Mirdha, B. R.; Mahapatra, S. C. The Science Behind Sacredness of Tulsi (*Ocimum sanctum* Linn.). *Indian J. Physiol. Pharmacol.* 2009, *53*, 291–306.

Mondal, S.; Varma, S.; Bamola, V. D.; Naik, S. N.; Mirdha, B. R.; Padhi, M. M.; Mehta, N.; Mahapatra, S. C. Double-blinded Randomized Controlled Trial for Immunomodulatory Effects of Tulsi (*Ocimum sanctum* Linn.) Leaf Extract on Healthy Volunteers. *J. Ethnopharmacol.* 2011, *136*, 452–456.

Mousavi, L.; Salleh, R. M.; Murugaiyah, V. Antidiabetic and *In Vitro* Enzyme Inhibition Studies of Methanol Extract of *Ocimum tenuiflorum* Linn Leaves and Its Fractions. *Trop. Life Sci. Res.* 2020, *31*, 141–158.

Mousavi, L.; Salleh, R. M.; Murugaiyah, V. Phytochemical and Bioactive Compounds Identification of *Ocimum tenuiflorum* Leaves of Methanol Extract and Its Fraction with an Antidiabetic Potential. *Int. J. Food Prop.* 2018, *21*, 2390–2399.

Muthuraman, A.; Diwan, V.; Jaggi, A. S.; Singh, N.; Singh, D. Ameliorative Effects of *Ocimum sanctum* in Sciatic Nerve Transection-induced Neuropathy in Rats. *J. Ethnopharmacol.* 2008, *120*, 56–62.

Nadar, B. G.; Usha, G. V.; Lakshminarayan, N. Comparative Evaluation of Efficacy of 4% Tulsi Extract (*Ocimum sanctum*), Fluoridated and Placebo Dentifrices Against Gingivitis and Plaque Among 14–15 Years School Children in Davangere City, India - A Triple Blinded Randomized Clinical Trial. *Contemp. Clin. Dent.* 2020, *11*, 67–75.

Parasuraman, S.; Balamurugan, S.; Christapher, P. V.; Petchi, R. R.; Yeng, W. Y.; Sujithra, J.; Vijaya, C. Evaluation of Antidiabetic and Antihyperlipidemic Effects of Hydroalcoholic Extract of Leaves of *Ocimum tenuiflorum* (Lamiaceae) and Prediction of Biological Activity of Its Phytoconstituents. *Pharmacogn. Res.* 2015, *7*, 156–165.

Pasangulapati, J. P.; Ravula, A. R.; Kanala, D. R.; Boyina, S.; Gangarapu, K.; Boyina, H. K. *Ocimum sanctum* Linn: A Potential Adjunct Therapy for Hyperhomocysteinemia-induced Vascular Dementia. *Adv. Exp. Med. Biol.* 2020, *1195*, 213–225.

Patil, R.; Patil, R.; Ahirwar, B.; Ahirwar, D. Isolation and Characterization of Anti-diabetic Component (Bioactivity-guided Fractionation) from *Ocimum sanctum* L. (Lamiaceae) Aerial Part. *Asian Pac. J. Trop. Med.* 2011, *4*, 278–282.

Pino, J. A.; Rosado, A.; Rodriguez, M.; Garcia, D. Composition of the Essential Oil of *Ocimum tenuiflorum* L. Grown in Cuba. *J. Essent. Oil Res.* 1998, *10*, 437–438.

Poli, V.; Challa, C. A comparative Study of Eugenol and *Ocimum sanctum* Linn. Leaf Extract on the Antifertility Effect in Female Albino Rats. *J. Chin. Med. Assoc.* 2019, *82*, 231–234.

Prakash, J.; Gupta, S. K. Chemopreventive Activity of *Ocimum sanctum* Seed Oil. *J. Ethnopharmacol.* 2000, *72*, 29–34.

Puri, H. S. *Rasayana Ayurvedic Herbs for Longevity and Rejuvenation*; CRC Press, 2002; pp 272–280.

Raditya, M. N.; Bagus, A. M. M.; Kustiati, U.; Wihadmadyatami, H.; Kusindarta, D. L. Data of the Expression of Serotonin in Alzheimer's Disease (AD) Rat Model Under Treatment of Ethanolic Extract *Ocimum sanctum* Linn. *Data Brief* 2020, 105654.

Rastogi, S.; Shukla, Y.; Paul, B. N.; Chowdhuri, D. K.; Khanna, S. K.; Das, M. Protective Effect of *Ocimum sanctum* on 3-methylcholanthrene, 7,12-dimethylbenz(a)anthracene and Aflatoxin B1 Induced Skin Tumorigenesis in Mice. *Toxicol. Appl. Pharmacol.* 2007, *224*, 228–240.

Ravindran, R.; Rathinasamy, S. D.; Samson, J.; Senthilvelan, M. Noise-stress-induced Brain Neurotransmitter Changes and the Effect of *Ocimum sanctum* (Linn) Treatment in Albino Rats. *J. Pharmacol. Sci.* 2005, *98*, 354–360.

Reddy, S. S.; Karuna, R.; Baskar, R.; Saralakumari, D. Prevention of Insulin Resistance by Ingesting Aqueous Extract of *Ocimum sanctum* to Fructose-fed Rats. *Horm. Metab. Res.* 2008, *40*, 44–49.

Samson, J.; Sheeladevi, R.; Ravindran, R. Oxidative Stress in Brain and Antioxidant Activity of *Ocimum sanctum* in Noise Exposure. *NeuroToxicology* 2007, *28*, 679–685.

Sarangi, S. C.; Pattnaik, S. S.; Katyal, J.; Kaleekal, T.; Dinda, A. K. An Interaction Study of *Ocimum sanctum* L. and Levetiracetam in Pentylenetetrazole Kindling Model of Epilepsy. *J. Ethnopharmacol.* 2020, 112389.

Sasidharan, N. *Flowering Plants of Kerala,* Ver. 2.0 (DVD), Serial Number 698329520, Kerala Forest Research Institute, Thrissur 680653, Kerala, India, 2011.

Saxena, R. C.; Singh, R.; Kumar, P.; Negi, M. P.; Saxena, V. S.; Geetharani, P.; Allan, J. J.; Venkateshwarlu, K. Efficacy of an Extract of *Ocimum tenuiflorum* (OciBest) in the Management of General Stress: A Double-blind, Placebo-Controlled Study. *Evid. Based Complement. Alternat. Med.* 2012, 894509.

Sembulingam, K.; Sembulingam, P.; Namasivayam, A. Effect of *Ocimum sanctum* Linn on the Changes in Central Cholinergic System Induced by Acute Noise Stress. *J. Ethnopharmacol.* 2005, *96*, 477–482.

Sen, P. Therapeutic Potentials of Tulsi: from Experience to Facts. *Drugs News and Views* 1993, *1*, 15–21.

Sethi, J.; Yadav, M.; Sood, S.; Dahiya, K.; Singh, V. Effect of tulsi (*Ocimum sanctum* Linn.) on Sperm Count and Reproductive Hormones in Male Albino Rabbits. *Int. J. Ayurveda Res.* 2010, *1*, 208–210.

Sharma, M.; Kishore, K.; Gupta, S. K.; Joshi, S.; Arya, D. S. Cardioprotective Potential of *Ocimum sanctum* in Isoproterenol Induced Myocardial Infarction in Rats. *Mol. Cell. Biochem.* 2001, *225*, 75–83.

Sharma, M. K.; Kumar, M.; Kumar, A. *Ocimum sanctum* Aqueous Leaf Extract Provides Protection Against Mercury Induced Toxicity in Swiss Albino Mice. *Indian J. Exp. Biol.* 2002, *40*, 1079–1082.

Sharma, M. K.; Madhu, K.; Ashok, K. Protection Against Mercury-induced Renal Damage in Swiss Albino Mice by *Ocimum sanctum*. *Environ. Toxicol. Pharmacol.* 2005, *19*, 161–167.

Sharmila, B. G.; Kumar, G.; Murugesan, A. G. Effects of Leaves Extract of *Ocimum sanctum* L. on Arsenic-induced Toxicity in Wistar Albino Rats. *Food Chem. Toxicol.* 2009, *47*, 490–495.

Shetty, S.; Udupa, S.; Udupa, L. Evaluation of Antioxidant and Wound Healing Effects of Alcoholic and Aqueous Extract of *Ocimum sanctum* Linn. in Rats. *Evid. Based Complement. Alternat. Med.* 2008, *5*, 95–101.

Shetty, S.; Udupa, S.; Udupa, L.; Somayaji, N. Wound Healing Activity of *Ocimum sanctum* Linn. with Supportive Role of Antioxidant Enzymes. *Indian J. Physiol. Pharmacol.* 2006, *50*, 163–168.

Shivpuje, P.; Ammanangi, R.; Bhat, K.; Katti, S. Effect of *Ocimum sanctum* on Oral Cancer Cell Line: An *In Vitro* Study. *J. Contemp. Dent. Pract.* 2015, *16*, 709–714.

Shokeen, P.; Bala, M.; Singh, M.; Tandon, V. *In vitro* Activity of Eugenol, an Active Component from *Ocimum sanctum*, Against Multiresistant and Susceptible Strains of *Neisseria gonorrhoeae*. *Int. J. Antimicrob. Agents.* 2008, *32*, 174–179.

Siddique, Y. H.; Ara, G.; Beg, T.; Afzal, M. Anti-genotoxic Effect of *Ocimum sanctum* L. Extract Against Cyproterone Acetate Induced Genotoxic Damage in Cultured Mammalian Cells. *Acta Biol. Hung.* 2007, *58*, 397–409.

Siddique, Y. H.; Faisal, M.; Naz, F.; Jyoti, S.; Role of *Ocimum sanctum* Leaf Extract on Dietary Supplementation in the Transgenic Drosophila Model of Parkinson's Disease. *Chin. J. Nat. Med.* 2014, *12*, 777–781.

Singh, S.; Majumdar, D. K. Effect of *Ocimum sanctum* Fixed Oil on Vascular Permeability and Leucocytes Migration. *Indian J. Exp. Biol.* 1999a, 37, 1136–1138.

Singh, S.; Majumdar, D. K. Evaluation of the Gastric Antiulcer Activity of Fixed Oil of *Ocimum sanctum* (Holy Basil). *J. Ethnopharmacol.* 1999b, 65, 13–19.

Singh, S.; Rehan, H. M.; Majumdar, D. K. Effect of *Ocimum sanctum* Fixed Oil on Blood Pressure, Blood Clotting Time and Pentobarbitone-induced Sleeping Time. *J. Ethnopharmacol.* 2001, *78*, 139–143.

Sood, S.; Narang, D.; Dinda, A. K.; Maulik, S. K. Chronic Oral Administration of *Ocimum sanctum* Linn. Augments Cardiac Endogenous Antioxidants and Prevents Isoproterenol-induced Myocardial Necrosis in Rats. *J. Pharm. Pharmacol.* 2005, *57*, 127–133.

Sood, S.; Narang, D.; Thomas, M. K.; Gupta, Y. K.; Maulik, S. K. Effect of *Ocimum sanctum* Linn. on Cardiac Changes in Rats Subjected to Chronic Restraint Stress. *J. Ethnopharmacol.* 2006, *108*, 423–427.

Suanarunsawat, T.; Anantasomboon, G.; Piewbang, C. Anti-diabetic and Anti-oxidative Activity of Fixed Oil Extracted from *Ocimum sanctum* L. Leaves in Diabetic Rats. *Exp. Ther. Med.* 2016, *11*, 832–840.

Suanarunsawat, T.; Devakul, N.; Ayutthaya, W.; Songsak, T.; Thirawarapan, S.; Poungshompoo, S. Antioxidant Activity and Lipid-lowering Effect of Essential oils Extracted from *Ocimum sanctum* L. Leaves in Rats Fed with a High Cholesterol Diet. *J. Clin. Biochem. Nutr.* 2010, *46*, 52–59.

Suzuki, A.; Shirota, O.; Mori, K.; Sekita, S.; Fuchino, H.; Takano, A.; Kuroyanagi, M. Leishmanicidal Active Constituents from Nepalese Medicinal Plant Tulsi (*Ocimum sanctum* L.). *Chem. Pharm. Bull.* 2009, *57*, 245–251.

Udupa, S. L.; Shetty, S.; Udupa, A. L.; Somayaji, S. N. Effect of *Ocimum sanctum* Linn. on Normal and Dexamethasone Suppressed Wound Healing. *Indian J. Exp. Biol.* 2006, *44*, 49–54.

Utispan, K.; Niyomtham, N.; Yingyongnarongkul, B. E.; Koontongkaew, S. Ethanolic extract of *Ocimum sanctum* Leaves Reduced Invasion And Matrix Metalloproteinase activity of Head and Neck Cancer Cell Lines. *Asian Pac. J. Cancer Prev.* 2020, *21*, 363–370.

Vats, V.; Yadav, S. P.; Grover, J. K. Ethanolic Extract of *Ocimum sanctum* Leaves Partially Attenuates Streptozotocin-induced Alterations in Glycogen Content and Carbohydrate Metabolism in Rats. *J. Ethnopharmacol.* 2004, *90*, 155–160.

Venuprasad, M. P.; Hemanth, K. K.; Khanum, F. Neuroprotective Effects of Hydroalcoholic Extract of *Ocimum sanctum* Against H_2O_2 Induced Neuronal Cell Damage in SH-SY5Y Cells via Its Antioxidative Defence Mechanism. *Neurochem. Res.* 2013, *38*, 2190–2200.

Wihadmadyatami, H.; Hening, P.; Kustiati, U.; Kusindarta, D. L.; Triyono, T.; Supriatno, S. *Ocimum sanctum* Linn. Ethanolic Extract Inhibits Angiogenesis in Human Lung Adenocarcinoma (a549) Cells. *Vet. World* 2020, *13*, 2028–2032.

Wihadmadyatami, H.; Karnati, S.; Hening, P.; Tjahjono, Y.; Maharjanti, F.; Kusindarta, D. L.; Triyono, T.; Ethanolic Extract *Ocimum sanctum* Linn. Induces an Apoptosis in Human Lung Adenocarcinoma (A549) Cells. *Heliyon* 2019, *14*, 02772.

Yamani, H. A.; Pang, E. C.; Mantri, N.; Deighton, M. A. Antimicrobial Activity of Tulsi (*Ocimum tenuiflorum*) Essential Oil and Their Major Constituents Against Three Species of Bacteria. *Front Microbiol.* 2016, *7*, 681.

Yanpallewar, S. U.; Rai, S.; Kumar, M.; Acharya, S. B. Evaluation of Antioxidant and Neuroprotective Effect of *Ocimum sanctum* on Transient Cerebral Ischemia and Long-term Cerebral Hypoperfusion. *Pharmacol. Biochem. Behav.* 2004, *79*, 155–164.

Yuniarti, W. M.; Krismaharani, N.; Ciptaningsih, P.; Celia, K.; Veteriananta, K. D.; Ma'ruf, A.; Lukiswanto, B. S. The Protective Effect of *Ocimum sanctum* Leaf Extract Against Lead Acetate-induced Nephrotoxicity and Hepatotoxicity in Mice (*Mus musculus*). *Vet World.* 2021, *14*, 250–258.

CHAPTER 29

ETHNOBOTANY, PHYTOCHEMISTRY, AND PHARMACOLOGICAL OVERVIEW OF ROSEMARY (*ROSMARINUS OFFICINALIS* L.)

SAVALIRAM G. GHANE[1], PRADEEP BHAT[2], HARSHA V. HEGDE[2], RAHUL ZANAN[3], and SANTOSHKUMAR JAYAGOUDAR[4]

[1]*Department of Botany, Shivaji University, Vidyanagar, Kolhapur, Maharashtra, India*

[2]*ICMR- National Institute of Traditional Medicine, Nehru Nagar, Belagavi, Karnataka, India*

[3]*Department of Botany, Elphinstone College, Dr. Homi Bhabha State University, Mumbai, Maharashtra, India*

[4]*Department of Botany, G. S. S. College & Rani Channamma University P. G. Centre, Belagavi, Karnataka, India*

ABSTRACT

Rosmarinus officinalis L., commonly called as Rosemary, is a perennial woody shrub belongs to the family Lamiaceae. The present review highlights traditional uses, bioactive compounds and their structures as well as pharmacological activities. The essential oil and aerial part of the plant is traditionally used to treat various ailments such as hypertension, dysmenorrhea, relieving respiratory and digestive disorders, kidney stones, diabetes and inflammatory diseases. The essential oil of the plant is also having several industrial applications in making shampoos, bathing essence, hair lotions, and cologne waters. The major chemical constituents of the essential

Bioactives and Pharmacology of Lamiaceae. T. Pullaiah, PhD (Ed.)

oil include camphor, α-pinene, 1,8-cineole, myrcene, rosmarinic acid, rosmadial, rosmaridiphenol, carnosic acid, and carnosol etc. The extracts (ethanol, hydro-ethanol, aqueous, and methanol) and essential oil from different parts of the plant is reported to have various biological activities such as wound healing, anthelmintic, anticoccidial, hepatoprotective, antioxidant, antimicrobial, anti-inflammatory, and antiproliferative, etc. Apart from these, the detailed clinical research approaches are also needed to establish its safety and efficacy against the diseases which it has been claimed.

29.1 INTRODUCTION

Rosmarinus officinalis L. belongs to the family Lamiaceae and is generally known as Rosemary. *R. angustifolius* Mill., *R. flexuosus* Jord. & Fourr., *R. latifolius* Mill., *R. laxiflorus* Noë, *R. palaui* (O.Bolòs & Molin.) Rivas Mart. & M. J. Costa, *R. prostratus* Mazziari, *R. rigidus* Jord. & Fourr., *R. serotinus* Loscos, *R. tenuifolius* Jord. & Fourr., *Salvia rosmarinus* Schleid., and *S. fasciculata* Fernald are some of the synonyms of this species (www.theplantlist.org). It is a perennial woody shrub, grows up to 2 m tall. The bark is irregularly fissured, exfoliating, dark gray in color; young branches are compactly covered with white stellate-tomentose hairs. The leaves are evergreen, needle-like (1–2.5 cm × 1–2 mm), fragrant, tufted, almost sessile or short petiolated, white stellate-tomentose at adaxial surface and somewhat shiny abaxially. The nutlets are 2.5 × 1.5 mm in size, become mucilaginous when wetted (www.flowersofindia.net; www.worldfloraonline.org). The plant is native to Mediterranean region and in many islands such as Baleari, Corsica, Elba Europe, Sicily, Sardinia; also occurs in Africa, Europe and some Asian countries. Currently, it is distributed throughout the world as an ornamental and aromatic plant (Akrout et al., 2010; Rašković et al., 2014).

In the traditional system of Mediterranean region, both fresh and dried leaves are used as spices and flavoring agents (Akrout et al., 2010). The aerial parts have a long history of traditional use as an antirheumatic, stimulant, mild analgesic, hepatoprotective, stimulator of blood circulation, nervous system and antispasmodic in renal colic. It is also used in the treatment of hypertension, dysmenorrhea, relieving respiratory and digestive disorders, kidney stones, diabetes, and inflammatory diseases. It is also used to treat headache, physical and mental fatigue, skin diseases, wounds, and infections (Abu-Al-Basal, 2010; Rašković et al., 2014; Ielciu et al., 2021). The plant contains high amount of essential oil (0.5–2.5% v/w), which has pulmonary antiseptic, antirheumatic, choloretic, stomachic, and antidiarrhoic properties. The oil is also used in cosmetic industry to produce shampoos, bathing essence, hair lotions, and cologne waters (Akrout et al., 2010).

29.2 BIOACTIVES

R. officinalis is an aromatic plant with several biological activities (Neves et al., 2018). Dellacassa et al. (1999) analyzed the essential oils of rosemary cultivars grown in different areas of Rio Grande do Sul State of South Brazil and Uruguay. The GC and GC–MS analysis showed that the oil obtained from Uruguay cultivar was reported with higher contents of α-pinene (37.8–46.2%), followed by 1,8-cineole (13.4–13.8%). Similarly, the essential oil composition of rosemary plants cultivated in Brazil was noted with higher concentrations of α-pinene (32.2%) and 1,8-cineole (14.7%). However, compounds such as α-pinene, myrcene, and 1,8-cineole were the major compounds found in the wild plants in Brazil with 12.4, 22.7, and 15.3% contents. Porte et al. (2000) isolated the essential oil from fresh leaves through hydrodistillation method, wherein camphor (26.0%), 1,8-cineole (22.1%), myrcene (12.4%), and α-pinene (11.5%) were the major components detected. Akrout et al. (2010) obtained the essential oils from the leaves of cultivated plants in different areas of Djerba (Southern region of Tunisian island) and their chemical compositions were explored using the GC and GC–MS. They noted three chemotypes: 1,8-cineole, 1,8-cineole/camphor/α-pinene/camphene, and 1,8-cineole/camphor/α-pinene/verbenone/borneol which were not detected in Tunisian genotypes. Borras-Linares et al. (2014) found the essential oil components of fresh rosemary leaves collected from different biogeographical regions of Serbia. The study revealed higher concentrations of rosmarinic acid, rosmadial, rosmaridiphenol, carnosic acid, and carnosol. Binzet et al. (2020) studied the chemical composition of rosemary oil obtained from Mersin, Turkey. Among 45 compounds detected, oxygenated monoterpenes were the main components (64.78%) with eucalyptol (33.15%) and camphor (10.31%).

Mena et al. (2016) studied the phytochemical profiling of a proprietary rosemary extract through ultrahigh performance liquid chromatography–electrospray ionization mass spectrometry (UHPLC-ESI-MS). The quantification of 57 compounds was carried out of which 14 were detected for the first time. The extract contained 24 flavonoids (flavones, flavanones, and flavonols), 24 diterpenoids (rosmanol derivatives, carnosic acid, and carnosol), 3 lignans (derivatives of medioresinol), 5 phenolic acids, and a triterpenoid compound (betulinic acid). Amongst all, carnosic acid derivatives were the major compounds with 77.1% of total phenolics (Fig. 29.1). The study was also supported by Ali et al. (2019).

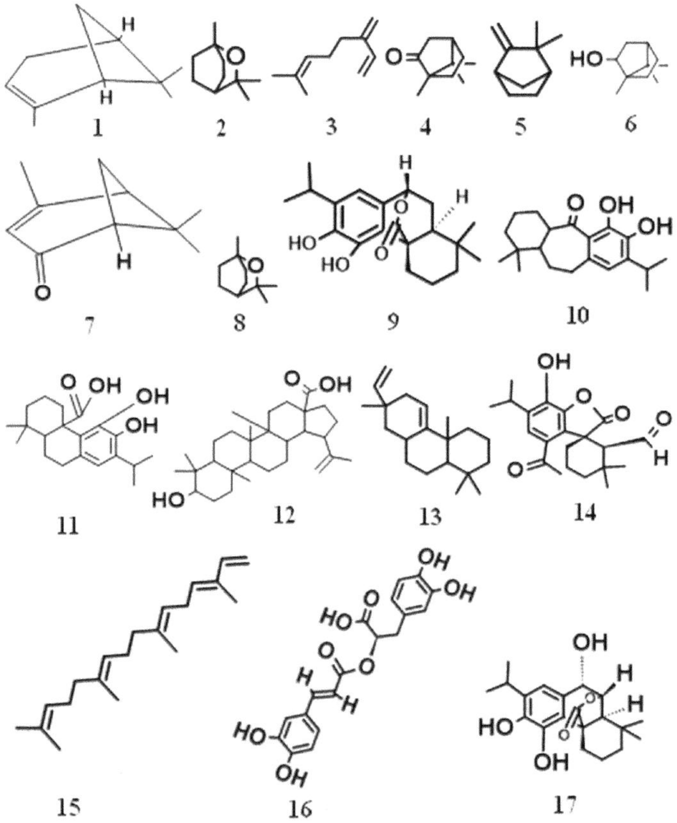

FIGURE 29.1 Structures of α-pinene [1], 1,8-cineole [2], myrcene [3], camphor [4], camphene [5], borneol [6], verbenone [7], eucalyptol [8], carnosol [9], rosmaridiphenol [10], carnosic acid [11], betulinic acid [12], isopimara-9 (11), 15-diene [13], rosmadial [14], α-springene [15], rosmarinic acid [16], and rosmanol [17].

29.3 PHARMACOLOGY

29.3.1 WOUND HEALING ACTIVITY

Abu-Al-Basal (2010) assessed the wound healing property of water extract and essential oil of aerial parts of the plant on alloxan-induced diabetic

BALB/c mice. Two full-thickness circular excision wounds were created on the dorsal surface of mice and treated with aqueous extract (10% intraperitoneal route for 3 days) and essential oil (25 μL/wound twice a day for 3 days). A higher percentage of wound contraction was found in the essential oil-treated animal group (92.91 ± 1.06%), followed by aqueous extract (88.83 ± 1.24%) compared to the negative diabetic control group (64.25 ± 1.41%) on the 15th post-wounding day.

29.3.2 ANTHELMINTIC ACTIVITY

Zoral et al. (2017) examined in vitro and in vivo anthelmintic activity of hydroethanolic (75%) and aqueous extracts against monogenean fish parasite (*Dactylogyrus minutus*) through oral and bath treatment methods. The parasitic survival time of aqueous extract was dose-dependent (100 and 200 g/L) with the mean survival time of 61.8 ± 5.6 and 7.8 ± 1.4 min, respectively. Whereas, the hydroethanolic extract at 1 and 32 g/L showed the mean survival time of 166.7 ± 48.2 and 5.4 ± 1.01 min. Aouadi et al. (2021) evaluated anthelmintic property of the leaf essential oil against *Haemonchus contortus*, a parasitic gastrointestinal nematode of sheep. The percentage inhibition of egg hatch assay was found to be 73.76% at 16 mg/mL concentration (IC_{50} 11.41 mg/mL). Similarly, complete inhibition (100%) of test organism was found in the adult worm's motility assay and it could be due the presence of three monoterpenes, namely, 1,8-cineole, α-pinene, and camphor.

29.3.3 ANTICOCCIDIAL EFFECTS

Aouadi et al. (2021) evaluated anticoccidial activity of the leaf essential oil against *Eimeria* oocysts of sheep using different concentrations of the essential oil (0.125, 0.250, 0.5, and 1 mg/mL). The study revealed promising anticoccidial activity with IC_{50} value of 1.82 μg/mL.

29.3.4 HEPATOPROTECTIVE ACTIVITY

The hepatoprotective effect of tincture prepared from young shoots of *R. officinalis* (50 and 500 mg/kg b.w.) was investigated through liver toxicity induced by carbon tetrachloride (1 mL/kg b.w.) in Swiss mice model. It was found that the oral administration of tincture at 50 mg/kg dose exhibited better protective effect with significant increase in aspartate aminotransaminase (ASAT) enzyme. Meanwhile, it also played a major role in the restoration of antioxidant enzymes and in alleviation of lipid peroxidation (Ielciu et al., 2021).

29.3.5 ANTIOXIDANT ACTIVITY

Bourhia et al. (2019) assessed the antioxidant activity of ethanol extract of rosemary plants collected from different regions of Morocco, namely, Marrakesh, Beni Mellal, Taounate, and El Jadida through DPPH radical scavenging assay. The IC_{50} values against DPPH radicals were found to be 0.176, 0.236, 0.258, and 0.302 mg/mL, respectively. Raskovic et al. (2014) also found appreciable DPPH free radical scavenging activity of the essential oil obtained from aerial parts (IC_{50} 77.6 μL/mL) as compared to α-tocopherol (IC_{50} 25.3 μg/mL). Hussain et al. (2010) have also reported the potential DPPH radical scavenging activity of the essential oil (59.1% inhibition; IC_{50} 20.9 μg/mL).

29.3.6 ANTIMICROBIAL ACTIVITY

Hussain et al. (2010) carried out the antibacterial activity of leaf essential oil through disc diffusion and resazurin microtiter plate methods. A higher antibacterial activity was found against *Bacillus cereus* (24.2 ± 1.1 mm inhibition zone; MIC 0.26 ± 0.01 mg/mL), followed by *B. subtilis* (23.0 ± 0.9 mm; 0.20 ± 0.01 mg/mL). Jafari-Sales and Hossein-Nezhad (2020) performed the antimicrobial activity of methanol extract using agar well diffusion method. A higher antimicrobial activity was found against *P. aeruginosa* (19.8 ± 1.64 mm zone of inhibition; MIC: 6.25 mg/mL), followed by *E. coli* (15.6 ± 0.54 mm; MIC: 25 mg/ml) and *S. aureus* (14.4 ± 1.14 mm; MIC: 25 mg/mL).

Bomfim et al. (2019) found significant antifungal potentiality of the essential oil obtained from dried leaves against *Aspergillus flavus* with minimum inhibitory and fungicidal concentration of 500 μg/mL. They noted reduction in the mycelial growth by 15.3%; they also observed reduction in the size of conidiophores and thickness of hyphae at 250 μg/mL concentration.

29.3.7 ANTI-INFLAMMATORY ACTIVITY

Takaki et al. (2008) evaluated anti-inflammatory effect of the leaf essential oil through carrageenan-induced pleurisy and paw edema tests in Swiss mice model. The oral supplementation of the essential oil (250, 500, and 750 mg/kg) significantly downregulated the number of acetic acid-induced writhes with 28, 34, and 28% inhibitions, respectively. However, the essential oil at same concentrations failed to extend the latency period compared to control group (Meperidine, 50 mg/kg, i.p.) in hot plate assay.

29.3.8 ANTIPROLIFERATIVE EFFECT

Bourhia et al. (2019) tested the antiproliferative effect of the ethanol extracts of *R. officinalis* plants grown in different regions of Morocco against human prostate cancer cell line (LNCaP). The extract (0–250 µg/mL) significantly reduced the cell viability of LNCaP with IC_{50} values of 15.04, 14.95, 14.93, and 14.15 µg/mL from Marrakesh, Beni Mellal, Taounate, and El Jadida localities, respectively. Hussain et al. (2010) evaluated antiproliferative activity of the leaf essential oil against MCF-7 and LNCaP human cancer cells and found prominent activity against both the cancer cell lines (IC_{50}: 190.1 and 180.9 µg/mL, respectively).

KEY WORDS

- **Lamiaceae**
- **rosemary**
- **essential oil**
- **traditional uses**
- **pharmacology**
- **myrcene**
- **rosmarinic acid**

REFERENCES

Abu-Al-Basal, M. A. Healing Potential of *Rosmarinus officinalis* L. on Full-thickness Excision Cutaneous Wounds in Alloxan-induced-diabetic BALB/c Mice. *J. Ethnopharmacol.* **2010,** *131*, 443–450.

Akrout, A.; Hajlaoui, H.; Mighri, H.; Najjaa, H.; Jani, H. E.; Zaidi, S.; Neffati, M. Chemical and Biological Characteristics of Essential Oil of *Rosmarinus officinalis* Cultivated in Djerba. *J. Essent. Oil Bear. Plants* **2010,** *13* (4), 398–411.

Ali, A.; Chua, B. L.; Chow, Y. H. An Insight into the Extraction and Fractionation Technologies of the Essential Oils and Bioactive Compounds in *Rosmarinus officinalis* L.: Past, Present and Future. *Trend. Anal. Chem.* **2019,** *118*, 338–351.

Aouadi, M.; Sebai, E.; Saratsis, A.; Kantzoura, V.; Saratsi, K.; Msaada, K.; Sotiraki, S.; Akkari, H. Essential Oil of *Rosmarinus officinalis* Induces *In Vitro* Anthelmintic and Anticoccidial Effects Against *Haemonchus contortus* and *Eimeria* spp. in Small Ruminants. *Vet. Med-Czech.* **2021,** *66* (4), 146–155.

Binzet, G.; Binzet, R.; Arslan, H. The Essential Oil Compositions of *Rosmarinus officinalis* L. Leaves Growing in Mersin, Turkey. *Eur. J. Chem.* **2020,** *11* (4), 370–376.

Bomfim, N. D. S.; Kohiyama, C. Y.; Nakasugi, L. P.; Nerilo, S. B.; Mossini, S. A. G.; Romoli, J. C. Z.; Mikcha, J. M. G.; Filho B. A. D. A.; Machinski Jr, M. Antifungal and Antiaflatoxigenic Activity of Rosemary Essential Oil (*Rosmarinus officinalis* L.) Against *Aspergillus flavus*. *Food Addit. Contam.* **2019.** Part A. DOI: 10.1080/19440049.2019.1678771.

Borras-Linares, I.; Stojanovic, Z.; Quirantes-Pine, R.; Arraez-Roman, D.; Svarc-Gajic, J.; Fernandez-Gutierrez A.; Segura-Carretero A. *Rosmarinus officinalis* Leaves as a Natural Source of Bioactive Compounds. *Int. J. Mol. Sci.* **2014,** *15*, 20585–20606.

Bourhia, M.; Laasri, F. E.; Aourik, H.; Boukhris, A.; Ullah, R.; Bari, A.; Ali, S. S.; Mzibri, M. E.; Benbacer, L.; Gmouh, S. Antioxidant and Antiproliferative Activities of Bioactive Compounds Contained in *Rosmarinus officinalis* Used in the Mediterranean Diet. *Evid. Based. Complement. Altern. Med.* **2019,** 1–7.

Dellacassa, E.; Lorenzo, D.; Moyna, P.; Frizzo, C. D.; Serafini, L. A.; Dugo, P. *Rosmarinus officinalis* L. (Labiatae) Essential Oils from the South of Brazil and Uruguay. *J. Essent. Oil Res.* **1999,** *11*, 1, 27–30.

Hussain, A. I.; Anwar, F.; Chatha, S. A. S.; Jabbar, A.; Mahboob, S.; Nigam, P. S. *Rosmarinus officinalis* Essential Oil: Antiproliferative, Antioxidant and Antibacterial Activities. *Braz. J. Microbiol.* **2010,** *41*, 1070–1078.

Ielciu, I.; Sevastre, B.; Olah, N. K.; Turdean, A.; Chise, E.; Marica, R.; Oniga, I.; Uifalean, A.; Sevastre-Berghian, A. C.; Niculae, M.; Benedec, D.; Hanganu, D. Evaluation of Hepatoprotective Activity and Oxidative Stress Reduction of *Rosmarinus officinalis* L. Shoots Tincture in Rats with Experimentally Induced Hepatotoxicity. *Molecules* **2021,** *26*, 1737, 1–15.

Jafari-Sales, A.; Hossein-Nezhad, P. Antimicrobial Effects of *Rosmarinus officinalis* Methanolic Extract on *Staphylococcus aureus*, *Bacillus cereus*, *Escherichia coli* and *Pseudomonas aeruginosa* in Laboratory Conditions. *J. Med. Chem. Sci.* **2020,** *3*, 103–108.

Mena, P.; Cirlini, M.; Tassotti, M.; Herrlinger, K. A.; Dall'Asta, C.; Rio, D. D. Phytochemical Profiling of Flavonoids, Phenolic Acids, Terpenoids, and Volatile Fraction of a Rosemary (*Rosmarinus officinalis* L.) Extract. *Molecules* **2016,** *21*, 1576.

Neves, J. A.; Neves J. A.; Oliveira, R. D. C. M. Pharmacological and Biotechnological Advances with *Rosmarinus officinalis* L. *Expert. Opin. Ther. Pat.* **2018,** *28* (5), 399–413.

Porte, A.; Godoy, R. L. D. O.; Lopes, D.; Koketsu, M.; Goncalves, S. L.; Torquilho, H. S. Essential Oil of *Rosmarinus officinalis* L. (Rosemary) from Rio de Janeiro, Brazil. *J. Essent. Oil Res.* **2000,** *12* (5), 577–580.

Raskovic, A.; Milanovic, I.; Pavlovic, N.; Cebovic, T.; Vukmirovic, S.; Mikov, M. Antioxidant Activity of Rosemary (*Rosmarinus officinalis* L.) Essential Oil and Its Hepatoprotective Potential. *BMC Complement. Altern. Med.* **2014,** *14* (225), 1–9.

Takaki, I.; Bersani-Amado, L. E.; Vendruscolo, A.; Sartoretto, S. M.; Diniz, S. P.; Bersani-Amado, C. A.; Cuman, R. K. N. Anti-inflammatory and Antinociceptive Effects of *Rosmarinus officinalis* L. Essential Oil in Experimental Animal Models. *J. Med. Food.* **2008,** *11* (4), 741–746.

Zoral, M. A.; Futami, K.; Endo, M.; Maita, M.; Katagiri T. Anthelmintic Activity of *Rosmarinus officinalis* Against *Dactylogyrus minutus* (Monogenea) Infections in *Cyprinus carpio*. *Vet. Parasitol.* **2017,** *247*, 1–6.

INDEX

N

O

R

S